Lecture Notes in Computer Science 15448

The series Lecture Notes in Computer Science (LNCS), including its subseries Lecture Notes in Artificial Intelligence (LNAI) and Lecture Notes in Bioinformatics (LNBI), has established itself as a medium for the publication of new developments in computer science and information technology research, teaching, and education.

LNCS enjoys close cooperation with the computer science R & D community, the series counts many renowned academics among its volume editors and paper authors, and collaborates with prestigious societies. Its mission is to serve this international community by providing an invaluable service, mainly focused on the publication of conference and workshop proceedings and postproceedings. LNCS commenced publication in 1973.

Oscar Camara · Esther Puyol-Antón ·
Maxime Sermesant · Avan Suinesiaputra ·
Jichao Zhao · Chengyan Wang · Qian Tao ·
Alistair Young
Editors

Statistical Atlases and Computational Models of the Heart. Workshop, CMRxRecon and MBAS Challenge Papers

15th International Workshop, STACOM 2024
Held in Conjunction with MICCAI 2024
Marrakesh, Morocco, October 10, 2024
Revised Selected Papers

Springer

Editors
Oscar Camara
Universitat Pompeu Fabra
Barcelona, Spain

Esther Puyol-Antón
King's College London
London, UK

Maxime Sermesant
Inria
Paris, France

Avan Suinesiaputra
King's College London
London, UK

Jichao Zhao
University of Auckland
Auckland, New Zealand

Chengyan Wang
Fudan University
Shanghai, China

Qian Tao
Delft University of Technology
Delft, Zuid-Holland, The Netherlands

Alistair Young
King's College London
London, UK

ISSN 0302-9743 ISSN 1611-3349 (electronic)
Lecture Notes in Computer Science
ISBN 978-3-031-87755-1 ISBN 978-3-031-87756-8 (eBook)
https://doi.org/10.1007/978-3-031-87756-8

This Springer imprint is published by the registered company Springer Nature Switzerland AG
The registered company address is: Gewerbestrasse 11, 6330 Cham, Switzerland

Preface

Cardiac image analysis has remained an active research field in medical imaging. Numerous studies have focused on developing statistical atlases and computational models of the heart to enhance our understanding of cardiac structure and function. These models provide valuable insights into the complex dynamics and behaviour of the heart, allowing for improved diagnosis and treatment of various cardiac conditions. Additionally, the use of artificial intelligence and machine learning algorithms in cardiac image analysis has further revolutionized the field.

However, clinical translations of these methods are still constrained by the lack of complete and rigorous technical and clinical validations, as well as benchmarking with common data. The Statistical Atlases and Computational Modelling of the Heart (STACOM) workshop aims to achieve these goals by providing a platform for researchers to exchange ideas, present their advancements, and collaborate on the development of robust and validated computational models of the heart.

The 15th edition of STACOM (https://stacom.github.io/stacom2024/), was held in conjunction with the 27th MICCAI conference in Marrakesh, Morocco on 10 October 2024. It followed the fourteen successful previous editions (https://stacom.github.io/): STACOM 2010 (Beijing, China), STACOM 2011 (Toronto, Canada), STACOM 2012 (Nice, France), STACOM 2013 (Nagoya, Japan), STACOM 2014 (Boston, USA), STACOM 2015 (Munich, Germany), STACOM 2016 (Athens, Greece), STACOM 2017 (Quebec City, Canada), STACOM 2018 (Granada, Spain), STACOM 2019 (Shenzhen, China), STACOM 2020 (Lima, Peru), STACOM 2021 (Strasbourg, France), STACOM 2022 (Singapore), and STACOM 2023 (Vancouver, Canada). Throughout these fifteen years, the STACOM workshop has provided a forum to discuss the latest developments in various areas of computational cardiac imaging, cardiac modelling, application of artificial intelligence and machine learning to cardiac image analysis, electromechanical modelling of the heart, novel methods in preclinical/clinical imaging for tissue characterization and image reconstruction, as well as statistical cardiac atlases.

The STACOM 2024 workshop attracted 32 paper submissions, of which 26 were accepted for presentation during the workshop. Topics ranged from cardiac segmentation, modelling, motion estimation, statistical shape analysis, and quality control. Deep learning methods were still the predominant approach to performing automated cardiac image analysis. Left atrial image analysis and modelling gained more attention in this workshop with atrial fibrillation being the most common area of interest.

The workshop awarded the best oral to Bruno Viti from University of Graz, Austria, for his paper entitled Gaussian Process Emulators for Few-Shot Segmentation in Cardiac MRI.

The best poster presenter was awarded to Patryk Rygiel from University of Twente, Netherlands, for his paper entitled Global Control for Local SO(3)-Equivariant Scale-Invariant Vessel Segmentation.

This year STACOM 2024 also hosted two challenges. Firstly, a challenge on Multi-class Bi-Atrial Segmentation

MBAS (https://codalab.lisn.upsaclay.fr/competitions/18516), including both left and right atria as well as their walls, focused on multi-class machine learning from LGE-MRIs to improve ablation outcomes for patients with persistent or long-standing persistent atrial fibrillation. This built on our highly successful left atrium segmentation challenge held at STACOM 2018. The new challenge objective was to benchmark methods for bi-atrial segmentation and biomarker identification using 200 multi-center 3D late gadolinium enhanced MRIs with each scan meticulously labelled by three experts. The MBAS challenge attracted 14 paper submissions of which 12 are presented in this volume. The second challenge was the Cardiac MRI Reconstruction Challenge (CMRxRecon 2024, https://cmrxrecon.github.io/2024), which aimed to provide a benchmark that encourages the broader research community to contribute to the critical work of accelerated CMR imaging. This initiative seeks to establish universal approaches that facilitate diverse applications and enhance performance in real-world deployment across various environments. The challenge released multi-contrast CMR imaging data collected from a total of 330 healthy volunteers. The dataset includes k-space data encompassing cine, T1/T2 mapping, tagging, phase-contrast, and dark-blood imaging. There were two tasks in this challenge: multi-contrast CMR reconstruction and random sampling CMR reconstruction. The CMRxRecon challenge attracted 18 paper submissions, of which 10 are presented in this volume.

Finally, the STACOM workshop organizers would like to express our gratitude to external reviewers from King's College London (UK), University of Auckland (New Zealand), Inria (France), University of Oxford (UK), Fudan University (China), Imperial College London (UK), Technische Universiteit Delft (the Netherlands), and Universitat Pompeu Fabra (Spain). Special thanks to Jieyun Bai, Edward Ferdian, Xinqi Li, Jun Lyu, Cheng Ouyang, Chen Qin Michael Tanzer, Fanwen Wang, Shuo Wang, Vicky Wang, Zi Wang, Hao Xu, Chenyu Yue, Yi Zhang, Debbie Zhao and Yidong Zhao, who spent their time to meticulously review papers in these proceedings. Ultimately, we would like to thank all authors who participated in this workshop.

November 2024

Alistair Young
Maxime Sermesant
Oscar Camara
Esther Puyol-Antón
Avan Suinesiaputra
Qian Tao
Chengyan Wang
Jichao Zhao

Organization

Program Committee Chairs

Camara, Oscar	Universitat Pompeu Fabra, Spain
Puyol-Anton, Esther	King's College London, UK
Sermesant, Maxime	Inria, Sophia Antipolis, France
Suinesiaputra, Avan	King's College London, UK
Tao, Qian	Delft University of Technology, The Netherlands
Wang, Chengyan	Fudan University, China
Young, Alistair	King's College London, UK
Zhao, Jichao	University of Auckland, New Zealand

Program Committee Members

Camara, Oscar	Universitat Pompeu Fabra, Spain
Ferdian, Edward	University of Auckland, New Zealand
Li, Xinqi	Max Delbrück Center, Germany
Lyu, Jun	The Hong Kong Polytechnic University, Hong Kong
Ouyang, Cheng	University of Oxford, UK
Puyol-Antón, Esther	King's College London, UK
Puyol-Antón, Esther	King's College London, UK
Qin, Chen	Imperial College London, UK
Sermesant, Maxime	Inria, Sophia Antipolis, France
Suinesiaputra, Avan	King's College London, UK
Tanzer, Michael	Imperial College London, UK
Tao, Qian	Delft University of Technology, The Netherlands
Wang, Vicky	Stanford University, USA
Wang, Zi	Imperial College London, UK
Wang, Fanwen	Imperial College London, UK
Wang, Shuo	Fudan University, China
Wang, Chengyan	Fudan University, China
Xu, Hao	King's College London, UK
Young, Alistair	King's College London, UK
Yue, Chenyu	Fudan University, China

Zhang, Yi	Delft University of Technology, The Netherlands
Zhao, Yidong	Delft University of Technology, The Netherlands
Zhao, Jichao	University of Auckland, New Zealand
Zhao, Debbie	University of Auckland, New Zealand

Reviewers

Bai, Jieyun	Jinan University, China
Camara, Oscar	Universitat Pompeu Fabra, Spain
Ferdian, Edward	University of Auckland, New Zealand
Li, Xinqi	Max Delbrück Center, Germany
Lyu, Jun	The Hong Kong Polytechnic University, Hong Kong
Ouyang, Cheng	University of Oxford, UK
Puyol-Anton, Esther	King's College London, UK
Qin, Chen	Imperial College London, UK
Sermesant, Maxime	Inria, Sophia Antipolis, France
Suinesiaputra, Avan	King's College London, UK
Tanzer, Michael	Imperial College London, UK
Tao, Qian	Delft University of Technology, The Netherlands
Wang, Vicky	Stanford University, USA
Wang, Zi	Imperial College London, UK
Wang, Fanwen	Imperial College London, UK
Wang, Shuo	Fudan University, China
Wang, Chengyan	Fudan University, China
Xu, Hao	King's College London, UK
Young, Alistair	King's College London, UK
Yue, Chenyu	Fudan University, China
Zhang, Yi	Delft University of Technology, The Netherlands
Zhao, Yidong	Delft University of Technology, The Netherlands
Zhao, Jichao	University of Auckland, New Zealand
Zhao, Debbie	University of Auckland, New Zealand

Contents

Single-Source Domain Generalization for Coronary Vessels Segmentation in X-Ray Angiography 1
Mohammad Atwany, Mojtaba Lashgari, Robin P. Choudhury, Vicente Grau, and Abhirup Banerjee

Constraint-Based Model in Multimodal Learning to Improve Ventricular Arrhythmia Prediction 12
Evariste Njomgue Fotso, Buntheng Ly, Hubert Cochet, and Maxime Sermesant

Automated Estimation of Cardiac Stroke Volumes from Computed Tomography 22
Ana-Teodora Radutoiu, Yasmin W. E. Youssef, Paula López Diez, Kristine A. Sørensen, Mathias B. Lindeskog, Michael H. C. Pham, Jonas J. Pedersen, Andreas Fuchs, Klaus F. Kofoed, and Rasmus R. Paulsen

Peridevice Leaks Following Left Atrial Appendage Occlusion - Analysis with Morphology Descriptive Centerlines and Explainable Graph Attention Network 33
Paula López Diez, Jan Margeta, Javier Gómez-Herrero, Davorka Lulic, Yannick Willemen, Klaus F. Kofoed, Ole De Backer, and Rasmus R. Paulsen

Improved 3D Whole Heart Geometry from Sparse CMR Slices 43
Yiyang Xu, Hao Xu, Matthew Sinclair, Esther Puyol-Antón, Steven A. Niederer, Amedeo Chiribiri, Steven E. Williams, Michelle C. Williams, and Alistair A. Young

CavityBASNet: Cavity-Focused Biatrial Automatic Segmentation on LGE-MRI with Augmented Input Channel and Left-Right Myocardium Splitting 53
Susana Merino-Caviedes, Emmanuel Chibuikem Nnadozie, Elisa Moya-Sáez, Marcos Martín-Fernández, and Carlos Alberola-López

Exploring Chemotherapy-Induced Cardiotoxicity Combining A 3D Computational Model and Preclinical Cardiac Imaging Data 64
Javier Villar-Valero, Jesus Jairo Rodríguez Padilla, Buntheng Ly, Juan F. Gomez, Mihaela Pop, Beatriz Trenor, and Maxime Sermesant

Quantitative Comparison of Blood Flow Patterns from in Silico Simulations and 4D Flow Data Before and After Left Atrial Occlusion 75
Paula Casademunt, Xabier Morales, Pedro Cepas-Guillén, Ada Doltra, Jordi Mill, Marta Saiz-Vivó, Eduardo Flores-Umanzor, Laura Sanchis, Xavier Freixa, and Oscar Camara

Panoramic Anatomical Context in 3D Intracardiac Echocardiography (ICE) with 3D Registration and Geometry-Based Image Fusion 87
Julia Hickler, Lydia Neary-Zajiczek, Raja Sekhar Bandaru, Marcos Balle Sánchez, Christoph Hennersperger, and Stefan Wörz

Physics-Informed Neural Networks Can Accurately Model Cardiac Electrophysiology in 3D Geometries and Fibrillatory Conditions 98
Ching-En Chiu, Aditi Roy, Sarah Cechnicka, Ashvin Gupta, Arieh Levy Pinto, Christoforos Galazis, Kim Christensen, Danilo Mandic, and Marta Varela

Beyond the Standards: Fully-Automated Aortic Annulus Segmentation on Contrast-Free Magnetic Resonance Imaging Using a Computational Aorta Unwrapping Method 110
Enrique Almar-Munoz, Mathias Pamminger, Christian Kremser, Markus Haltmeier, and Agnes Mayr

Coronary Artery Calcium Scoring from Non-contrast Cardiac CT Using Deep Learning with External Validation 122
Hyunho Mo, Daniel Bos, Maryam Kavousi, Maarten J. G. Leening, and Esther E. Bron

Effective Approach Based on Student-Teacher Self-supervised Deep Learning for Multi-class Bi-atrial Segmentation Challenge 132
Moona Mazher, Abdul Qayyum, and Steven A. Niederer

Sampling-Pattern-Agnostic MRI Reconstruction Through Adaptive Consistency Enforcement with Diffusion Model 140
Anurag Malyala, Zhenlin Zhang, Chengyan Wang, and Chen Qin

HyperCMR: Enhanced Multi-contrast CMR Reconstruction with Eagle Loss 152
Ruru Xu, Caner Özer, and Ilkay Oksuz

A Multi-contrast Cardiac MRI Reconstruction Method Using an Advanced Unrolled Network Architecture 164
Yajing Zhang, Yanxin Huang, Zhixin Xu, and Jin Qi

Implicit Neural Representations for Registration of Left Ventricle Myocardium During a Cardiac Cycle ... 172
Mathias Micheelsen Lowes, Jonas Jalili Pedersen, Bjørn S. Hansen, Klaus Fuglsang Kofoed, Maxime Sermesant, and Rasmus R. Paulsen

Deep Multi-contrast Cardiac MRI Reconstruction via vSHARP with Auxiliary Refinement Network ... 183
George Yiasemis, Nikita Moriakov, Jan-Jakob Sonke, and Jonas Teuwen

Multi-model Ensemble Approach for Accurate Bi-Atrial Segmentation in LGE-MRI of Atrial Fibrillation Patients ... 193
Lucas Beveridge and Le Zhang

Two-Stage nnU-Net for Automatic Multi-class Bi-atrial Segmentation from LGE-MRIs ... 200
Y. On, C. Galazis, C. Chiu, and M. Varela

An Ensemble of 3D Residual Encoder UNet Models for Solving Multi-class Bi-atrial Segmentation Challenge ... 209
Alexander Zolotarev, Kiane Johnson, Abbas Khan, Gregory Slabaugh, and Caroline Roney

Evaluating Convolution, Attention, and Mamba Based U-Net Models for Multi-class Bi-Atrial Segmentation from LGE-MRI ... 214
Claas Thesing, Alfonso Bueno-Orovio, and Abhirup Banerjee

On the Foundation Model for Cardiac MRI Reconstruction ... 226
Chi Zhang, Michael Loecher, Cagan Alkan, Mahmut Yurt, Shreyas S. Vasanawala, and Daniel B. Ennis

Multi-loss 3D Segmentation for Enhanced Bi-atrial Segmentation ... 236
Enrique Almar-Munoz, Markus Tiefenthaler, Nohemi Sofia Leon Contreras, Adria Aguilar-Minguez, Christian Kremser, Markus Haltmeier, and Agnes Mayr

Classification of Mitral Regurgitation from Cardiac Cine MRI Using Clinically-Interpretable Morphological Features ... 245
Y. On, K. Vimalesvaran, S. Zaman, M. Shun-Shin, J. Howard, N. Linton, G. Cole, A. A. Bharath, and M. Varela

Gaussian Process Emulators for Few-Shot Segmentation in Cardiac MRI ... 257
Bruno Viti, Franz Thaler, Kathrin Lisa Kapper, Martin Urschler, Martin Holler, and Elias Karabelas

Global Control for Local SO(3)-Equivariant Scale-Invariant Vessel Segmentation 269
Patryk Rygiel, Dieuwertje Alblas, Christoph Brune, Kak Khee Yeung, and Jelmer M. Wolterink

A Self-distillation Bi-atrial Segmentation Network for Cardiac MRI 280
Waqas Anwaar, Manh The Van, Jiongtong Hu, Zhurong Chen, Wufeng Xue, and Dong Ni

Enhanced Deep Unrolled Models Applied to the CMRxRecon2024 Challenge 289
Bingyu Xin, Meng Ye, Leon Axel, and Dimitris N. Metaxas

Reducing the Number of Leads for ECG Imaging with Graph Neural Networks and Meaningful Latent Space 301
Giacomo Verardo, Daniel F. Perez-Ramirez, Samuel Bruchfeld, Magnus Boman, Marco Chiesa, Sabine Koch, Gerald Q. Maguire Jr., and Dejan Kostic

Rotor Core Projection Ablation (RCPA): Novel Computational Approach to Catheter Ablation Therapy for Atrial Fibrillation 313
George Obada, Shaheim Ogbomo-Harmitt, Maria Deprez, Robin Van Den Abeelen, Nele Vandersickel, and Oleg Aslanidi

Automated Pipeline for Regional Epicardial Adipose Tissue Distribution Analysis in the Left Atrium 323
Èric Lluch, Eduardo Castañeda, Maximilian E. R. Weiss, Anamaria Vizitiu, Athira Jacob, Jitin Jami, Chloé Audigier, Felix Meister, Viorel Mihalef, and Tiziano Passerini

Low-Rank Conjugate Gradient-Net for Accelerated Cardiac MR Imaging 334
Jaykumar H. Patel, Brenden T. Kadota, Calder D. Sheagren, Mark Chiew, and Graham A. Wright

SegBAW-Net: Segmentation of Bi-Atria and Wall Network Offering Valuable Insights into Challenge Data 345
Arthur L. Lefebvre, Ishan Vatsaraj, Carolyna A. P. Yamamoto, Kensuke Sakata, Brock Tice, Natalia A. Trayanova, and Eugene G. Kholmovski

ResNet-Based Convolutional Framework for Segmenting Left Atrial Scars and Cavities 357
Malitha Gunawardhana, Fangqiang Xu, Yun Gu, and Jichao Zhao

EAT-Mamba: Epicardial Adipose Tissue Segmentation from Multi-modal Dixon MRI ... 368
Fan Feng, Abdallah Hasaballa, Fangqiang Xu, Jiayuan Yang, Yun Gu, Grace Wen, Carl-Johan Carlhäll, and Jichao Zhao

Neural Fields for Continuous Periodic Motion Estimation in 4D Cardiovascular Imaging ... 378
Simone Garzia, Patryk Rygiel, Sven Dummer, Filippo Cademartiri, Simona Celi, and Jelmer M. Wolterink

Exploring CNN and Transformer Architectures for Multi-class Bi-Atrial Segmentation from Late Gadolinium-Enhanced MRI ... 390
Yichi Zhang, Yuan Cheng, and Yuan Qi

EigenBoundaries for Temporally Regularized Segmentation of Echocardiographic Images ... 398
Georgios Tziritas

Dynamic Cardiac MRI Reconstruction via Separate Optimization of K-Space and Hybrid-Domian Spatial-Temporal Feature Fusion ... 411
Xinrui Guo, Liwen Li, and Fumin Guo

Can Interpretable Learning of Risk Explain Ventricular Arrhythmia Mechanism? ... 421
Buntheng Ly, Sonny Finsterbach, Marta Nuñez-Garcia, Hubert Cochet, and Maxime Sermesant

3D Left Ventricular Reconstruction from 2D Echocardiograms for Reliable Volume Estimation ... 433
Moïse Z. Jansen, Debbie Zhao, Edward Ferdian, Stephen A. Creamer, Joshua R. Dillon, Mathilde A. Verlyck, Gina M. Quill, Nicola C. Edwards, Malcolm E. Legget, Robert N. Doughty, Alistair A. Young, and Martyn P. Nash

Comparing Left Atrial Spontaneous Echo Contrast Intensity with Gaussian Process Emulator Predictions ... 443
Paolo Melidoro, Malak Sabry, Abdel Rahman Amr Sultan, Ahmed Qureshi, Gregory Y. H. Lip, Natalie Montarello, Nishant Lahoti, Ronak Rajani, Magdalena Klis, Steven E. Williams, Oleg Aslanidi, and Adelaide De Vecchi

UPCMR: A Universal Prompt-Guided Model for Random Sampling Cardiac MRI Reconstruction ... 453
Donghang Lyu, Chinmay Rao, Marius Staring, Matthias J. P. van Osch, Mariya Doneva, Hildo J. Lamb, and Nicola Pezzotti

An All-in-One Approach for Accelerated Cardiac MRI Reconstruction 464
Kian Anvari Hamedani, Narges Razizadeh, Shahabedin Nabavi, and Mohsen Ebrahimi Moghaddam

Improving the Scan-Rescan Precision of AI-Based CMR Biomarker Estimation 476
Dewmini Hasara Wickremasinghe, Yiyang Xu, Esther Puyol-Antón, Paul Aljabar, Reza Razavi, and Andrew P. King

Author Index 487

Single-Source Domain Generalization for Coronary Vessels Segmentation in X-Ray Angiography

Mohammad Atwany[1], Mojtaba Lashgari[1], Robin P. Choudhury[2], Vicente Grau[1], and Abhirup Banerjee[1,2(✉)]

[1] Institute of Biomedical Engineering, Department of Engineering Science, University of Oxford, Oxford OX3 7DQ, UK
abhirup.banerjee@eng.ox.ac.uk

[2] Division of Cardiovascular Medicine, Radcliffe Department of Medicine, University of Oxford, Oxford OX3 9DU, UK

Abstract. Deployment of medical image segmentation models remains a challenge largely because of changes in the distribution of image domains for differences in acquisition protocols, devices, and patient demographics. These factors limit model generalization on new unseen datasets, termed out-of-distribution (OOD) datasets. To address this, we introduce "DomainFlow", a novel single-source domain generalization (DG) method for coronary vessels segmentation in x-ray angiography. In our method, we hypothesize that training a model to predict connectivity masks instead of conventional binary masks preserves model performance against domain shift, by capturing domain-invariant spatial relationships and structural consistency between vessels, reducing the impact of domain-specific noise. To further encourage DG through model robustness, our model learns a Gaussian posterior at the latent space, which is further refined through a supervised prior, therefore, exposing the model to a broader range of data variations during training. Our extensive evaluation on three x-ray angiography datasets shows that DomainFlow outperforms the state-of-the-art methods with 0.85–3.85% and 2.86–5.09% Dice score improvements in OOD and in-domain datasets, respectively, thus, demonstrating its potential for real-time clinical applications.

Keywords: X-ray Angiography · Coronary Vessels Segmentation · Domain Generalization · Connectivity · Regularization

1 Introduction

Cardiovascular diseases remain a major cause of death in the world, with coronary vessel stenosis accounting for 19.1 million deaths [14,15] as of 2020. Invasive coronary angiography (ICA) is the gold standard for assessing coronary stenosis, and can greatly benefit from real-time coronary vessels segmentation for faster and easier interpretation [12]. Several machine learning techniques have previously been explored, including cross-anatomy learning [20], mean student-teacher

O. Camara et al. (Eds.): STACOM 2024, LNCS 15448, pp. 1–11, 2025.
https://doi.org/10.1007/978-3-031-87756-8_1

learning [7], graph neural networks [8], etc. These models are trained and tested on the same domain where feature distributions are similar. However, they fail during deployment on new datasets due to distribution shift between the training datasets and new datasets, termed out-of-distribution (OOD) datasets. X-ray quantum noise, electronic noise, scattering effects, and motion artifacts can cause a significant domain shift in coronary x-ray angiography images, since different hospitals and regions have different equipment quality, imaging protocols, and patient demography. Additionally, administration of varying contrast levels at different hospitals and different protocols can lead to substantially varying level of contrast in ICA images. Domain generalization (DG), in this regard, aims to address domain shifts induced by the aforementioned sources. Several DG methods for medical segmentation have been proposed previously, such as DG-based data augmentations [21,24], contrastive learning [9], etc. But to the best of our knowledge, no previous method has addressed the problem of coronary vessels segmentation in x-ray angiography images.

In this work, we address domain shift in coronary vessels segmentation from x-ray angiography by introducing "DomainFlow", a novel single-source DG technique. We consider two regularization perspectives. First, for model-level regularization, we utilize localized inter-pixel relationships through connectivity prediction [18,19] instead of a direct binary mask prediction. This regularizes model training by focusing on directional probability prediction between pixels instead of the usual pixel overlap prediction only. Second, for feature-level regularization, we introduce Minimal Connectivity Dimension (MCD) priors at the bottleneck. This regularizes features at the bottleneck by supervised training on a directional prior for enhanced domain-invariant feature learning during upsampling. After training on a large-scale ICA dataset, we carry out in-domain testing on a test set of the same dataset and out-of-domain testing on two different ICA datasets. Our extensive analysis demonstrates that the proposed DomainFlow yields significantly improved and robust segmentation performance on both the OOD and in-domain test sets, with Dice score improvements of 0.85–3.85% and 2.86–5.09%, respectively.

2 Methodology

In this section, we provide an overview of the datasets used in this study, the proposed method pipeline, description of the losses used and finally, implementation details of our model. An overview of the pipeline is summarized in Fig. 1.

2.1 Source and Target Datasets

Using a single-source training strategy, we train on the SJTU dataset containing 323 images (train: 173, validation: 82, test: 68) from the Renji Hospital of Shanghai Jiao Tong University [6]. The first target dataset is a private dataset that consists of 56 ICA images obtained from the John Radcliffe Hospital, Oxford,

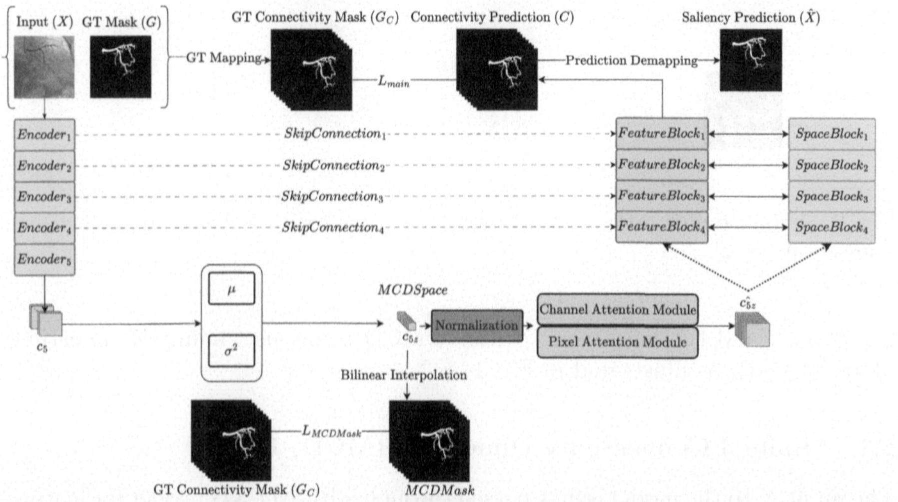

Fig. 1. Overview of the proposed method pipeline. The input image X is first processed by the encoder for feature extraction. Then, MCD priors are applied at the bottleneck, and the supervised MCDMask serves as a directional prior for the decoder. Finally, the connectivity prediction C is converted from an 8-channel to a 1-channel saliency prediction mask $\hat{X}$.

including 26 images of the right coronary artery and 30 images of the left anterior descending artery with fine-grained vessel annotations. As the second target dataset, we include the UMAE T1-León dataset with 130 ICA images obtained from 25 different areas in Mexico [3].

2.2 Binary to Connectivity Mask Ground Truth Conversion

Learning a connectivity mask [11] instead of directly learning a binary segmentation mask as a model-wide task allows the model to learn the direction between different pixels and how they are connected to each other. This can be achieved through an 8-way connectedness approach, where a pixel at (x, y) considers the neighboring pixels at coordinates $[x \pm 1, y \pm 1]$. As a result, each neighboring pixel or entry in the connectivity vector represents a direction between two neighboring pixels, allowing for a more accurate representation of how every pixel is connected to its neighbours. In this way, the model is encouraged to learn domain-invariant consistent structural features that represent the vessel anatomy rather than domain-specific noise or artifacts in the background, leading to robust and improved performance on OOD datasets. The generation of a connectivity mask is illustrated in Fig. 2.

By converting the single-channel original ground truth mask G into an eight-channel ground truth connectivity mask G_C, we use it to supervise training at both the prior and final prediction level. Throughout this paper, we refer to the prior predicted connectivity mask as $MCDMask$, final prediction connectivity

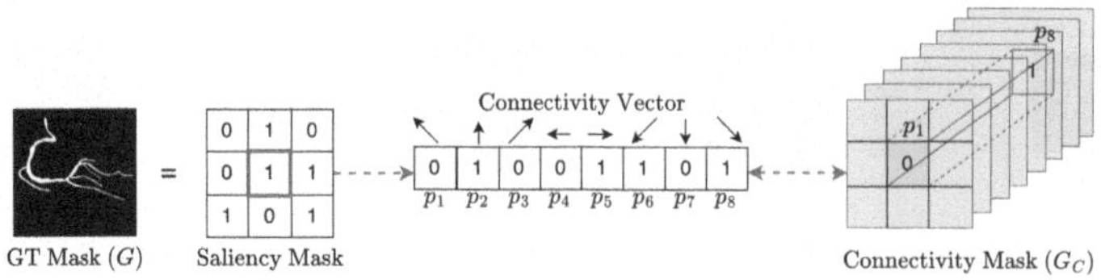

Fig. 2. Simplified illustration of mapping a binary mask with 1-channel to an 8-channel connectivity mask.

mask as C, and the base model without MCD priors as Channel Connectivity Model (CCM), as illustrated in Fig. 1.

2.3 Minimal Connectivity Dimension (MCD) Priors

The input X to the model is first passed through a ResNet-34 encoder for features extraction, where each encoder block is denoted as $Encoder_i, i = [1, \cdots, 5]$. This outputs a low-dimensional feature space at the bottleneck of the last encoder layer c_5. In order to enhance the feature representations of this bottleneck, we compute the posterior distribution of these features by using two fully connected layers, one for the mean and one for the variance. This "MCDSpace", defined as a Gaussian distribution, allows for the capturing of variability in the data, exposing the model to more variations during training and therefore improving its domain generalization ability. We empirically chose a spatial dimension of $(3, 3)$ spanning 8 channels for the MCDSpace.

Since the latent space is low-dimensional, it primarily captures the high-level context of the data, which is typically domain-invariant. This is because low-level features that often contain domain-specific information, such as texture and local patterns, are processed in the initial layers of the encoder. In contrast, the latent space captures more abstract, high-level representations that are less sensitive to domain-specific variations. In order to enhance the representation of these domain-invariant features in the latent space, the second prior MCDMask is derived by upsampling the learned MCDSpace with bilinear interpolation to the size of ground truth connectivity map G_C. This MCDMask is then supervised against G_C, refining low-dimensional high-level context representation in the latent space to further enhance DG performance.

2.4 Directional Prior and Disentanglement

To enhance upsampling and segmentation in the expansion path, MCD-Mask refines coarse representations of features at the bottleneck from a low-dimensional MCDSpace; therefore, MCDMask can act as directional prior for the decoder. This is achieved by passing MCDMask through a global average pooling layer followed by a 1×1 convolutional layer. This reduction condenses the context information into a compact form, ensuring that only the most relevant features are retained. Subsequently, this vector is passed through a sigmoid layer to obtain a vector of probabilities. These probabilities are multiplied with the original latent space output c_{5z} to modulate the latent features, enhancing their relevance for the final segmentation task.

In order to further process the modulated output, we incorporate a Dual Attention Network [5] including a Channel Attention Module (CAM) and Pixel Attention Module (PAM). The CAM module can highlight important channels dynamically by learning channel dependencies, enabling the model to focus more on informative features and suppress less useful ones. Concurrently, the PAM module generates a position attention map that highlights important spatial features, which can be combined with the original feature map to produce more informative representations of how different pixels are related to each other.

The attention network output $\hat{c_5}_z$ with dense directional representation of features is then duplicated to initialize the directional flow $SpaceBlock_i$ and feature flow $FeatureBlock_i$ in the decoder. A normalized direction attention map θ_i is created using a sigmoid channel-wise dot product in four consecutive sequences. The subsequent $FeatureBlock_{i+1}$ contains features multiplied by the previous θ_i, allowing the directional information to propagate through the network. Skip-connections are used to preserve important low-level features in each feature flow block from the corresponding encoder stage, ensuring that essential information is retained throughout the network. Finally, the highest pixel values across the 8 channels are selected, and a sigmoid activation followed by thresholding at 0.5 is applied to determine each pixel's saliency. This converts the decoder output C from a connectivity map into a binary saliency map $\hat{X}$. A summary of the overall process can be found in Fig. 1.

2.5 Losses

The first prior loss $L_{MCDSpace}$ is defined as Kullback-Leiber (KL) divergence loss of the form:

$$L_{MCDSpace} = -\frac{1}{2}\sum_{i=1}^{N}\left(1 + \log(\sigma_i^2) - \mu_i^2 - \sigma_i^2\right), \tag{1}$$

where μ_i and σ_i^2 are mean and variance of the i-th dimension of MCD latent space vector c_{5z} with dimensionality N. We use the same hybrid loss for both the second prior loss $L_{MCDMask}$ and the main model prediction loss L_{main}, where the hybrid loss [19] contains an edge-decoupled loss $L_{decouple}$ and a connectivity consistency loss L_{con_const}.

$$L_{decouple} = \begin{cases} L_{bce}(1 - \min\{\tilde{C}_i(x,y)\}_{i=1}^{8}, G(x,y)), & \text{if } (x,y) \in P_{edge} \\ L_{bce}(Mean\{\tilde{C}_i(x,y)\}_{i=1}^{8}, G(x,y)), & \text{if } (x,y) \notin P_{edge} \end{cases} \tag{2}$$

where L_{bce} is a binary cross-entropy (BCE) loss, P_{edge} the set of edge pixels, and $G(x,y) \in \{0,1\}$ is the binary ground truth mask. $\tilde{C}$ represents a regularized predicted connectivity map, where for every pixel, the values of two opposite neighbors are multiplied, and the result is assigned to both pixels in the pair.

This is done to represent the bi-directional probability between two opposing pixels equally, which was shown empirically to improve segmentation performance in [19]. L_{con_const} is based on the weighted sum of BCE losses applied to the predicted connectivity map C and regularized connectivity map $\tilde{C}$, as follows:

$$L_{con\_const} = L_{bce}(C, G_C) + \omega L_{bce}(\tilde{C}, G_C), \tag{3}$$

where ω is a regularization hyperparameter. The total loss L_{total} is defined as

$$L_{total} = \lambda_1 L_{MCDSpace} + \lambda_2 L_{MCDMask} + L_{main}, \tag{4}$$

where λ_1 and λ_2 are the regularization control parameters.

2.6 Implementation Details

For training, we use a batch size of 4, a poly scheduler and an initial learning rate of 0.0038. We use the SJTU dataset with the same split provided by the dataset originally. Dice score is used as the primary metric to evaluate the segmentation performance against state-of-the-art methods. The images all have a size of (512, 512, 3) except UMAE T1-León, which has been zero-padded to (512, 512, 3). Since ICA images are in grayscale, we collapse the images in the channel direction to (512, 512, 1) to remove redundancy. Data augmentations include hue (−30, 30), saturation (−5, 5), value shift (−15, 15), horizontal and vertical flipping, random shift (−0.05, 0.05), scaling (−0.1, 0.1), aspect ratio (−0.1, 0.1), and rotation (−30, 30). Finally, the values of ω, λ_1, and λ_2 are determined empirically as 0.2, 0.9, and 0.3, respectively.

3 Results and Discussion

Incorporation of the CCM and MCD priors into the proposed DomainFlow significantly improves coronary vessel segmentation on in-domain test set. This indicates the model focuses on learning domain-invariant features that mitigate overfitting to noises in training data. Using DomainFlow, the standard deviation (SD) of the Dice score is reduced by a significant margin from 3.66–6.28% to 0.33–0.81% when compared against state-of-the-art models for the in-domain testing, showcasing the model's focus on regularization and learning domain-invariant features. This also translates visually to minimization of oversegmentation of edge pixels and undersegmentation of foreground pixels and clearly enhanced connectivity in vessel segmentation with significantly less discontinuities, as evidenced by the qualitative results in Fig. 3.

The efficacy of DomainFlow is also evident in the quantitative results presented in Table 1. In the in-domain SJTU test set, DomainFlow achieves Dice scores of 83.48% with a significantly lower SD of 0.81%, which illustrates its reliability and balance between sensitivity and precision while also delineating smaller vessels that other state-of-the-art methods miss. A relative improvement of 2.86–5.09% in Dice score can be observed with DomainFlow when compared

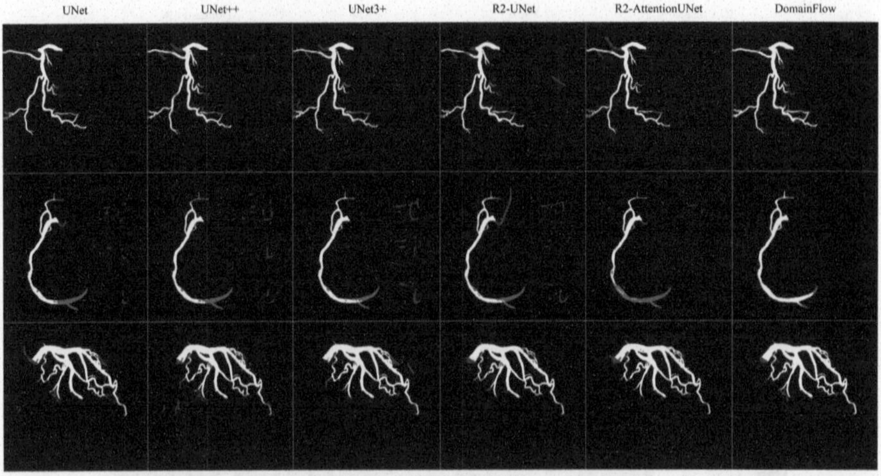

Fig. 3. Qualitative results comparison on the in-domain test set of SJTU for UNet, UNet++, UNet3+, R2-UNet, R2-AttentionUNet, and DomainFlow. Adopting connectivity prediction as a model task in DomainFlow has improved vessel connectivity and delineation of smaller vessels, as visible through increased true positive (TP, in white) and reduced false positive (FP, in blue) and false negative (FN, in red). (Color figure online)

Table 1. Comparative performance evaluation in terms of Dice score using SJTU dataset as source dataset. Values represent Mean ± Standard Deviation (SD).

Model	In-domain (SJTU)	Private	UMAE T1-León
State-of-the-art			
UNet [13]	79.62 ± 4.69	–	–
UNet++ [25]	80.92 ± 3.66	–	–
UNet3+ [10]	81.16 ± 4.58	–	–
R2-UNet [1]	80.51 ± 4.71	–	–
R2-AttentionUNet [26]	79.43 ± 6.28	–	–
Domain Generalization			
MixStyle [23]	82.62 ± 0.18	76.40 ± 0.94	64.86 ± 4.02
EFDMix [22]	82.39 ± 0.67	76.59 ± 1.50	66.28 ± 3.41
TriD [4]	83.04 ± 0.40	76.61 ± 1.25	66.77 ± 4.30
Ablation			
CCM	83.23 ± 0.33	77.67 ± 2.66	65.96 ± 6.30
DomainFlow	83.48 ± 0.81	78.18 ± 1.98	67.34 ± 4.88

to the state-of-the-art models. In addition, the SD is reduced from 3.66–6.28% to 0.81%, translating to consistency improvement by 78–87%.

For OOD performance comparison of DomainFlow against other DG methods, in the second half of Table 1 we report the state-of-the-art DG methods MixStyle, EFDMix, and TriD for single-source DG scenario. DomainFlow outperforms these methods on both the private and UMAE T1-León datasets, with Dice scores of 78.18% and 67.34% respectively. This translates to a relative Dice

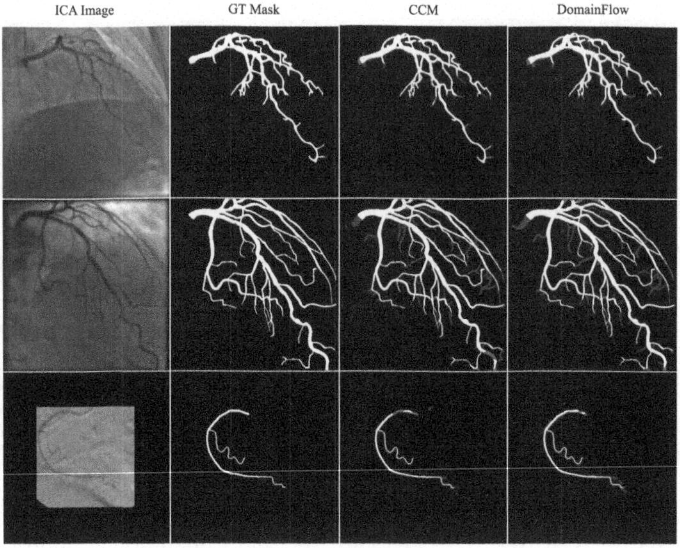

Fig. 4. Qualitative performance of the CCM alone and CCM with MCD priors (DomainFlow) on in-domain, private, and UMAE T1-León (top to bottom). By exposing the model to more variations during training using MCD priors and utilizing channel connectivity, DomainFlow demonstrates improved vessel connectivity and delineation of smaller vessels, achieving improved results on OOD datasets. True positives are shown in white, false positives in blue, and false negatives in red. (Color figure online)

score improvement of 2.05–2.32% on the private dataset and 0.85–3.85% on the UMAE T1-León dataset. This improvement in OOD performance is also evidenced by the qualitative results in Fig. 4. Additionally, in terms of in-domain test improvement compared to other DG methods, our method achieves a relative improvement of 0.53–1.32%. Overall, this improvement of the in-domain and OOD performance can be attributed to the observation that while DG methods like MixStyle, EFDMix, and TriD augment the feature space to expose the model to more variations during training, they may not explicitly encourage the learning of domain-invariant features. Our method circumvents this by encouraging the learning of 8-channel connectivity while also exposing the model to different data variations through MCD priors, leading to improved in-domain and OOD performance.

In terms of ablation, incorporating MCD priors into CCM shows a relative Dice score improvement of 0.66–2.09% for OOD and 0.4% for in-domain test set. This demonstrates that the inclusion of MCD priors helps the model to learn a more comprehensive latent space representation by capturing data variability. Consequently, this exposure to more data variations during training leads to improved generalization performance on OOD datasets and consistency.

4 Conclusion

In this paper, we have presented DomainFlow, a novel DG method for coronary vessels segmentation from x-ray angiography that utilizes connectivity-based regularization with MCD priors. In comparison to the state-of-the-art methods, our approach not only improves DG performance on OOD datasets, but also outperforms other models on the in-domain test sets. In real-time clinical applications where imaging conditions are unpredictable, a generalizable model like DomainFlow can serve as a reliable tool for accurate delineation of coronary vasculatures to support clinical decision making. This generalizability also potentiates to be integrated into an automated clinical pipeline for non-invasive measurements of the severity of the coronary arterial disease through patient-specific 3D coronary vessel reconstruction [2,16,17] as well as *in silico* studies [12].

Acknowledgments. The authors acknowledge the use of the facilities and services of the Institute of Biomedical Engineering (IBME), Department of Engineering Science, University of Oxford, in carrying out this work. AB is supported by the Royal Society University Research Fellowship (Grant No. URF\R1\221314). MA is supported by a Research Studentship supported by the Royal Society Enhanced Research Expenses Grant, awarded to AB. The work of ML, RC, and AB is partially supported by the British Heart Foundation (BHF) Project under Grant PG/20/21/35082.

References

1. Ambrosini, P., Ruijters, D., Niessen, W.J., Moelker, A., van Walsum, T.: Fully automatic and real-time catheter segmentation in x-ray fluoroscopy. In: Medical Image Computing and Computer-Assisted Intervention- MICCAI 2017: 20th International Conference, Quebec City, QC, Canada, 11–13 September 2017, Part II, pp. 577–585. Springer (2017)
2. Banerjee, A., Galassi, F., Zacur, E., De Maria, G.L., Choudhury, R.P., Grau, V.: Point-cloud method for automated 3D coronary tree reconstruction from multiple non-simultaneous angiographic projections. IEEE Trans. Med. Imaging **39**(4), 1278–1290 (2019)
3. Cervantes-Sanchez, F., Cruz-Aceves, I., Hernandez-Aguirre, A., Hernandez-Gonzalez, M.A., Solorio-Meza, S.E.: Automatic segmentation of coronary arteries in x-ray angiograms using multiscale analysis and artificial neural networks. Appl. Sci. **9**(24), 5507 (2019)
4. Chen, Z., Pan, Y., Ye, Y., Cui, H., Xia, Y.: Treasure in distribution: a domain randomization based multi-source domain generalization for 2D medical image segmentation. In: Medical Image Computing and Computer Assisted Intervention – MICCAI 2023, pp. 89–99. Springer, Cham (2023)
5. Fu, J., et al.: Dual attention network for scene segmentation. In: Proceedings of the IEEE/CVF Conference on Computer Vision and Pattern Recognition, pp. 3146–3154 (2019)
6. Hao, D., et al.: Sequential vessel segmentation via deep channel attention network. Neural Netw. **128**, 172–187 (2020)

7. He, H., Banerjee, A., Beetz, M., Choudhury, R.P., Grau, V.: Semi-supervised coronary vessels segmentation from invasive coronary angiography with connectivity-preserving loss function. In: 2022 IEEE 19th International Symposium on Biomedical Imaging (ISBI), pp. 1–5. IEEE (2022)
8. He, H., Banerjee, A., Choudhury, R.P., Grau, V.: Automated coronary vessels segmentation in x-ray angiography using graph attention network. In: Statistical Atlases and Computational Models of the Heart. Regular and CMRxRecon Challenge Papers, pp. 209–219. Springer, Cham (2024)
9. Hu, S., Liao, Z., Xia, Y.: Devil is in channels: contrastive single domain generalization for medical image segmentation. In: Greenspan, H., et al. (eds.) Medical Image Computing and Computer Assisted Intervention - MICCAI 2023, pp. 14–23. Springer, Cham (2023)
10. Huang, H., et al.: UNet 3+: a full-scale connected UNet for medical image segmentation. In: ICASSP 2020 - 2020 IEEE International Conference on Acoustics, Speech and Signal Processing (ICASSP), pp. 1055–1059 (2020)
11. Kampffmeyer, M., Dong, N., Liang, X., Zhang, Y., Xing, E.P.: ConnNet: a long-range relation-aware pixel-connectivity network for salient segmentation. IEEE Trans. Image Process. **28**(5), 2518–2529 (2018)
12. Lashgari, M., Choudhury, R.P., Banerjee, A.: Patient-specific In Silico 3D coronary model in cardiac catheterisation laboratories. Front. Cardiovas. Med. **11**, 1398290 (2024)
13. Ronneberger, O., Fischer, P., Brox, T.: U-net: convolutional networks for biomedical image segmentation. In: Medical Image Computing and Computer-assisted Intervention–MICCAI 2015: 18th international conference, Munich, Germany, 5–9 October 2015, Part III, pp. 234–241. Springer (2015)
14. Tsao, C.W., et al.: Heart disease and stroke statistics-2022 update: a report from the American heart association. Circulation **145**(8), e153–e639 (2022)
15. Vaduganathan, M., Mensah, G.A., Turco, J.V., Fuster, V., Roth, G.A.: The global burden of cardiovascular diseases and risk. J. Am. Coll. Cardiol. **80**(25), 2361–2371 (2022)
16. Wang, Y., Banerjee, A., Choudhury, R.P., Grau, V.: Deep learning-based 3D coronary tree reconstruction from two 2D non-simultaneous x-ray angiography projections. arXiv preprint arXiv:2407.14616 (2024)
17. Wang, Y., Banerjee, A., Grau, V.: NeCA: 3D coronary artery tree reconstruction from two 2D projections by neural implicit representation. arXiv preprint arXiv:2409.04596 (2024)
18. Yang, Z., Farsiu, S.: Directional connectivity-based segmentation of medical images. In: Proceedings of the IEEE/CVF Conference on Computer Vision and Pattern Recognition, pp. 11525–11535 (2023)
19. Yang, Z., Soltanian-Zadeh, S., Farsiu, S.: BiconNet: an edge-preserved connectivity-based approach for salient object detection. Pattern Recogn. **121**, 108231 (2022)
20. Zhang, J., Gu, R., Wang, G., Xie, H., Gu, L.: SS-CADA: a semi-supervised cross-anatomy domain adaptation for coronary artery segmentation. In: 2021 IEEE 18th International Symposium on Biomedical Imaging (ISBI), pp. 1227–1231 (2021)
21. Zhang, L., et al.: Generalizing deep learning for medical image segmentation to unseen domains via deep stacked transformation. IEEE Trans. Med. Imaging **39**(7), 2531–2540 (2020)
22. Zhang, Y., Li, M., Li, R., Jia, K., Zhang, L.: Exact feature distribution matching for arbitrary style transfer and domain generalization. In: Proceedings of the IEEE/CVF Conference on Computer Vision and Pattern Recognition, pp. 8035–8045 (2022)

23. Zhou, K., Yang, Y., Qiao, Y., Xiang, T.: Domain generalization with mixstyle. In: International Conference on Learning Representations (2021)
24. Zhou, Z., Qi, L., Yang, X., Ni, D., Shi, Y.: Generalizable cross-modality medical image segmentation via style augmentation and dual normalization. In: Proceedings of the IEEE/CVF Conference on Computer Vision and Pattern Recognition, pp. 20856–20865 (2022)
25. Zhou, Z., Rahman Siddiquee, M.M., Tajbakhsh, N., Liang, J.: UNet++: a nested U-net architecture for medical image segmentation. In: Deep Learning in Medical Image Analysis and Multimodal Learning for Clinical Decision Support: 4th International Workshop, DLMIA 2018, and 8th International Workshop, ML-CDS 2018, Held in Conjunction with MICCAI 2018, Granada, Spain, 20 September 2018, pp. 3–11. Springer (2018)
26. Zuo, Q., Chen, S., Wang, Z.: R2au-net: attention recurrent residual convolutional neural network for multimodal medical image segmentation. Secur. Commun. Netw. **2021**(1), 6625688 (2021)

Constraint-Based Model in Multimodal Learning to Improve Ventricular Arrhythmia Prediction

Evariste Njomgue Fotso[1(✉)], Buntheng Ly[2], Hubert Cochet[2], and Maxime Sermesant[1,2]

[1] Inria, Université Côte d'Azur, Epione team, Sophia Antipolis, France
{evariste.njomgue-fotso,maxime.sermesant}@inria.fr
[2] IHU Liryc, Université de Bordeaux, Bordeaux, France

Abstract. Cardiac disease evaluation depends on multiple diagnostic modalities: electrocardiogram (ECG) to diagnose abnormal heart rhythms, and imaging modalities such as Magnetic Resonance Imaging (MRI), Computed Tomography (CT) and echocardiography to detect signs of structural abnormalities. Each of these modalities brings complementary information for a better diagnosis of cardiac dysfunction. However, training a machine learning (ML) model with data from multiple modalities is a challenging task, as it increases the dimension space, while keeping constant the number of samples. In fact, as the dimension of the input space increases, the volume of data required for accurate generalisation grows exponentially. In this work, we address this issue, for the application of Ventricular Arrhythmia (VA) prediction, based on the combined clinical and CT imaging features, where we constrained the learning process on medical images (CT) based on the prior knowledge acquired from clinical data. The VA classifier is fed with features extracted from a 3D myocardium thickness map (TM) of the left ventricle. The TM is generated by our pipeline from the imaging input and a Graph Convolutional Network is used as the feature extractor of the 3D TM. We introduce a novel Sequential Fusion method and evaluate its performance against traditional Early Fusion techniques and single-modality models. The cross-validation results show that the Sequential Fusion model achieved the highest average scores of $80.7\% \pm 4.4$ Sensitivity and $73.1\% \pm 6.0$ F1 score, outperforming the Early Fusion model at $65.0\% \pm 8.9$ Sensitivity and $63.1\% \pm 6.3$ F1 score. Both fusion models achieved better scores than the single-modality models, where the average Sensitivity and F1 score are $62.8\% \pm 10.1; 52.1\% \pm 6.5$ for the clinical data modality and $62.9\% \pm 6.3; 60.7\% \pm 5.3$ for the medical images modality.

Keywords: Multimodal deep learning · Arrhythmia · Imaging

O. Camara et al. (Eds.): STACOM 2024, LNCS 15448, pp. 12–21, 2025.
https://doi.org/10.1007/978-3-031-87756-8_2

1 Introduction

Sudden Cardiac Death (SCD), when the heart stops beating suddenly, is recognised as a high-priority public health topic. SCD is the most common cause of death worldwide, accounting for 4.25 million deaths every year [1]. Among causes of SCD, ischaemic heart disease (also called coronary heart disease) is the most common cause [1]. Ischaemic heart disease is a significant risk factor for the development of Ventricular Arrhythmia, such as Ventricular Tachycardia (VT) and Ventricular Fibrillation (VF). Despite intensive research on the topic, over the past 25 years, the absolute number of Cardiovascular disease (CVD) cases has increased in Europe and in the EU, with increases in the number of new CVD cases found in most countries [2]. The most significant advance in the prevention of SCD has been the development of the implantable cardioverter-defibrillator (ICD) [3]. On the one hand people are still dying of SCD and on the other hand, the current prescription guidelines of ICDs which is left ventricular ejection fraction (LVEF) <30–35% [7] capture only a mere 20% all SCD [8].

To enhance early risk identification of SCD, Dakun Lai et al. have proposed a machine learning model on measurable arrhythmic markers derived from ECG signals [11]. O'Mahony et al., have presented an individualised risk estimates for SCD in hypertrophied cardiomyopathy patients, considering various clinical parameters [10]. Global Electric Heterogeneity (GEH) parameters, quantifying the abnormal electrophysiological substrate, have also been identified as independent factors associated with SCD risk, improving risk prediction when combined with other clinical characteristics [12]. For paediatric patients, the HCM Risk-Kids model estimates the risk of SCD and highlights the importance of considering both clinical and genetic factors in risk assessment [13]. In order to improve the current clinical criteria for ICD candidacy, a combination (fusion) of clinical characteristics with other markers may significantly improve risk stratification [9], but feature fusion is still a challenging research topic in machine learning [5]. In the literature, feature concatenation is the primary technique used for data fusion in multimodal classification [15]. This concatenation is typically performed either at the beginning of the learning process (Early Fusion) or at the end of the learning process (Late Fusion).

In this study, we propose a novel fusion technique for multimodal VA classification (positive VA+ or negative VA−). The technique can also handle the class imbalance issue which is particularly prevalent in medical diagnosis, where the occurrence of positive cases (VA+) is significantly rarer than negative cases. Our Sequential Fusion technique progressively constrains the learning process on higher dimensional data (medical images modality), based on knowledge acquired on lower dimensional data (clinical data). The knowledge acquired on clinical data, called prior modality, is used in the loss function of the learning process on medical images modality. We also study the contribution of each modality in our fusion technique.

2 Method

2.1 Sequential Fusion Model

Inspired by constraints-based models, we propose the Sequential Fusion model. A constraints-based model refers to a type of modelling approach that incorporates constraints to guide the learning process or restrict the possible solutions. These constraints can be mathematical expressions, logical statements, or any other form of limitation that helps narrow down the solution space and ensure that the model produces results that adhere to certain predefined criteria.

To setup a Sequential Fusion model with two modalities (Fig. 1), a first classifier is trained only with the first modality (the prior modality). The classification result on the prior modality is used to compute the prior modality constraint as defined in the next chapter. This constraint is then added into the learning process with the second modality to get the final output.

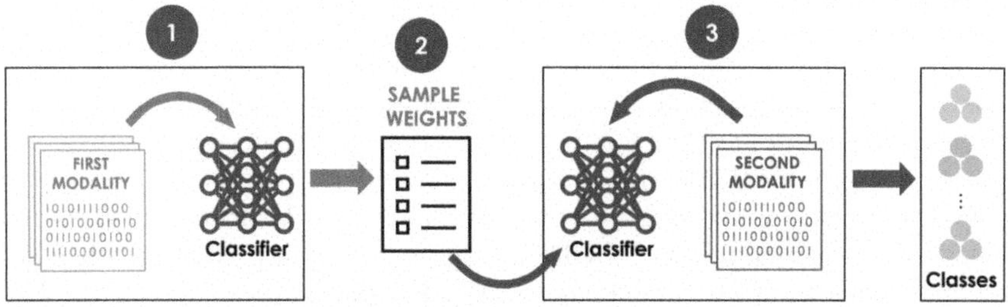

Fig. 1. Sequential Fusion Model: general setting.

2.2 Prior Modality Constraint Formulation

The first model is trained using only clinical data which is the prior modality: it provides prior knowledge for the final classifier on imaging data from the classification errors on the clinical data. We account for this prior knowledge with the following weighting strategy.

Let us consider two sets s_1 and s_2, where s_1 is the set of well-classified samples and s_2 the set of misclassified samples using the prior modality. The prior modality weights ($prior\_weight_i$) is computed using the Eq. 1, where $prior\_weight_i$ is the weight for each set s_i, n is the total number of samples, k is the number of sets (two in our setting) and n_{s_i} is the number of samples in each set s_i.

$$prior\_weight_i = \frac{n}{k * n_{s_i}} \tag{1}$$

The prior modality weights ($prior\_weight_i$) are the mathematical formulation of the prior-knowledge, where more importance is given to misclassified samples by the prior modality so that the final classifier is more focused on these samples.

Moreover, in our setting, the data set is unbalanced. To tackle the unbalanced class issue, we use the same samples weighting technique, to give higher weight to minority class (VA−) and lower weight to the majority class (VA+).

To control both of these effects, we define two hyper-parameters $\alpha, \beta \in [0,1]^2$ and two weighting strategies: same weighting, defined in Eq. 2, and stratified weighting, defined in Eq. 3.

$$w_l = \left\{ \alpha * class\_weight_j + \beta * prior\_weight_i \right. \tag{2}$$

$$w_l = \begin{cases} \alpha * class\_weight_j & \text{for} s_1 (i = 1) \\ \alpha * class\_weight_j + \beta * prior\_weight_i & \text{for} s_2 (i = 2) \end{cases} \tag{3}$$

In Eqs. 2 and 3, $class\_weight_j$ corresponds to weights which correct the imbalance effect. The weight of each class j ($class\_weight_j$) is computed by the same formulation in Eq. 1, where n is the total number of samples, k is the number of classes (two in our setting) and n_{c_j} is the number of samples in each class c_j. In Eq. 1, n_{s_i} is replaced by n_{c_j} for the computation of $class\_weight_j$.

In the same weighting strategy (Eq. 2), equal importance is given to both the effects of unbalanced class and misclassified samples with the prior modality, whereas in stratified weighting strategy (Eq. 3) more importance is given to positive (VA+) class and misclassified samples by the prior modality.

The loss function (binary cross-entropy) of the learning process on the medical images modality is then defined in Eq. 4, where y_l is the true label of the l-th sample (1 for the positive class VA+ and 0 for the negative class VA− and p_l is the predicted probability of the l-th sample belonging to the positive class.

$$Loss_{BCE} = w_l * y_l log(p_l) + w_l * (1 - y_l) log(1 - p_l) \tag{4}$$

The Early Fusion model is trained with both weighting strategies and only the best one in terms of our performance metrics is kept (F1 score and Sensitivity). The novelty of our approach lies in these weights ($prior\_weight_i$), which is computed based on the knowledge acquired from the prior modality. This formulation can be generalised in multi-class classification with any additional constraints required in the learning process.

3 Materials

3.1 Data Set and Features Extraction

We used a retrospective data set of myocardial infarction patients collected at a hospital between 2010 and 2020. Patients with history of surgical procedure on the LV were excluded from the cohort. In this work, patient's cardiac CT image, with the scan date more than 1 month after the infarction, and clinical descriptions were collected. We classified as VA+ the patients, who have experienced sustained VT, VF and abort cardiac arrest. Finally, we obtained a study data set of 600 patients, with 165 and 435 of VA+ and VA− patients, respectively. Table 1 gives details on available clinical data modality.

Table 1. Clinical Data Modality (N = 600)

Feature Name	Statistics: [min, max] mean or proportion
LVEF %	[10, 78] 45.07 ± 13.36
Age (years)	[27, 99] 72.69 ± 12.09
Smoking	270 (45%)
Dyslipidaemia	436 (73%)
Diabete	159 (27%)
Hypertension	394 (66%)
Sex	100 Females/500 Males
VA	Positive (VA+) 165 (28%)

Description: LVEF % - Left Ventricular Ejection Fraction, Age - Patient age at the scan date

To extract the thickness map features from CT images, we used the automatic pipeline proposed by a cardiac imaging group in 2022. [4].

3.2 Experimental Setup

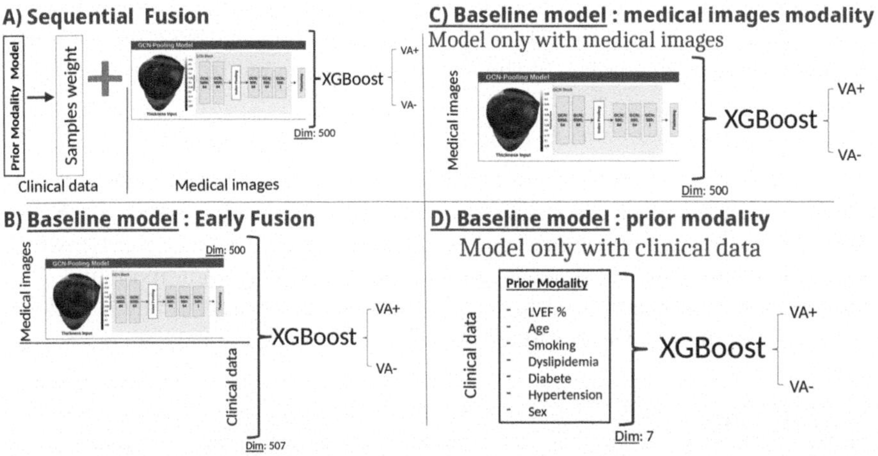

Fig. 2. Different model configurations compared.

The Early Fusion model (see Fig. 2) is the concatenation of extracted thickness features with patients' clinical data. The GCN-Pooling Model is used to extract thickness features. The data tensor is flattened and then given as input to a XGBoost classifier. The Early Fusion is the baseline model with both modalities.

The Sequential Fusion and Early Fusion models are compared with two others models used as a baseline with single modality. The first single modality baseline model is the prior model train with XGBoost classifier and only with clinical data (prior modality). The second single modality baseline model is the GCN-Pooling Model [4] where the "FCN block" is replaced by an XGBoost classifier; this model is trained only with the medical images modality. We use the XGBoost model implemented in the xgboost python package [14] as the base classifier. Same hyper-parameters are used for the four models with the number of estimators set to 300 and the maximum depth is set to 6. The default values, as proposed by the [14] package are kept for the other hyper-parameters.

To isolate the impact of multimodal learning, we employed the XGBoost classifier across our model settings, acknowledging that the architecture of deep learning models (such as "FCN block") can vary depending on the input data. This approach ensures that any observed differences in performance are attributable to the multimodal learning techniques rather than variations in classifier architecture. We chose to utilise the TM rather than raw CT images for the VA classifier, as myocardial scar characterisation-derived from the TM-is a well-recognised substrate for VA [16,17].

3.3 Controlled Cross-Validation Setup

In the realm of machine learning, the quality and structure of the training data significantly impact the performance and generalisation of a machine learning model. In our context, in addition to keeping the same ratio VA− over VA+ in each split, we must also check that each split (training, validation and testing set) has the same data distribution as the whole data set (used as reference). We use the T-test (uni-variate hypothesis testing on each patients' clinical data) to only keep triple splits (training, validation and testing set) which have the same distribution as our reference. The p-value significance threshold for the T-tests was set to 0.05.

To compare the four models in our setting we have selected Sensitivity and F1 Score for performance metrics. To measure the variability of our performance metrics, 10 cross-validations are performed, then we compute the mean and standard deviation of each metric.

4 Results

As shown in Table. 2, the Sequential Fusion model outperforms the Early Fusion model in term of both metrics F1 Score (+10%) and Sensitivity (+15%). On the F1 Score, it performs 21% better than the prior modality model and 12% better than the Medical images modality model. Looking at the Sensitivity metric, the Sequential Fusion performs 18% better than the prior modality model and 18% better than the Medical images modality model.

Table 2. Model performance results. Best performance metrics in **bold**.

Model	Accuracy	F1 Score	Sensitivity	Specificity
(A) Sequential Fusion	**0.833:** ±0.04	**0.731:** ±0.06	**0.807:** ±0.04	**0.844:** ±0.06
(B) Early Fusion	0.790: ±0.03	0.631: ±0.06	0.650: ±0.08	**0.844:** ±0.04
(C) Medical Images Modality	0.773: ±0.04	0.607: ±0.05	0.629: ±.06	0.828: ±0.05
(D) Prior modality	0.682: ±0.04	0.521: ±0.06	0.628: ±0.10	0.703: ±0.05

5 Discussion

The objective of modalities fusion is to get the best from each modality according to the purpose of interest. When we look at the details, with Fig. 3-A, which shows samples well-classified by each modality (statistics on tests set), we can see that the Sequential Fusion is able to capture almost all knowledge from medical images modality. However, some samples well-classified by the prior modality are still misclassified by the Sequential Fusion. This point highlights the fact

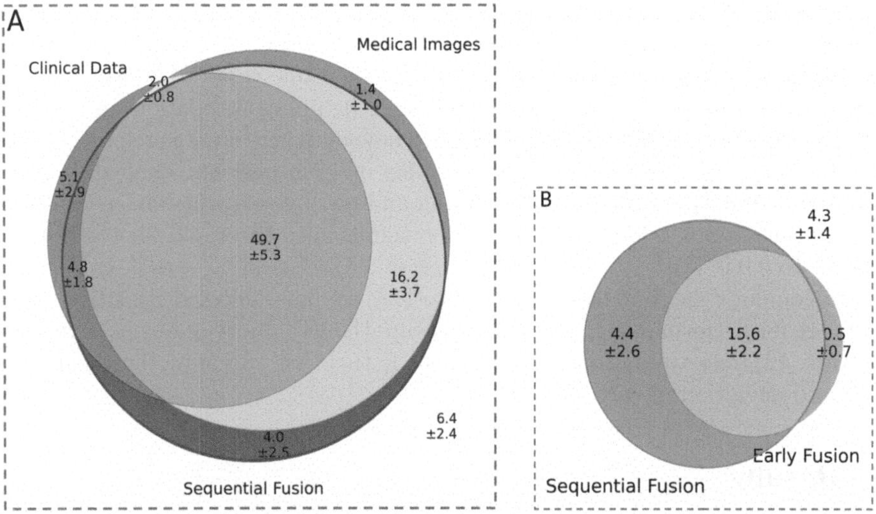

Fig. 3. **A:** Venn diagram comparing well-classified samples by each modalities and Sequential Fusion model. Values are means and standard deviations on tests set. The clinical data classifier is red, the medical images classifier is green, and the Sequential Fusion classifier is blue. Overlapping regions are samples well-classified by corresponding overlapping classifiers. Overlapping regions colour are the mix of corresponding colours red, green or blue, except the intersection of all models which is in gray. Out of circles is the statistic of misclassified samples by all models. **B:** Venn diagram comparing well-classified positive samples (VA+) by Early Fusion and Sequential Fusion models. Sequential Fusion classifier is in light red, Early Fusion in light green and the intersection in gray. (Color figure online)

that there is still a room of progress to fuse modalities. When comparing the classification results of Early Fusion and Sequential Fusion on positive samples (VA+), we observe an interesting point: Sequential Fusion captures nearly all the correctly classified VA+ samples identified by Early Fusion, as shown in Fig. 3-B. In addition, we can also observed that some samples are misclassified by all models. Those samples may required another modality to improve the Sequential Fusion model.

Figure 4 gives an insight into the contribution of the medical images modality to the classification of VA+ samples, by comparing the decision tree of prior modality and Sequential Fusion. In fact, we can notice that the medical images modality has improved the classification of positive samples (VA+) with Left Ventricular (LV) Ejection Fraction greater than 46.5%. This point is in agreement with the fact that cardiac imaging techniques have allowed improved SCD risk stratification, especially in the group of samples with an LVEF > 35% [6].

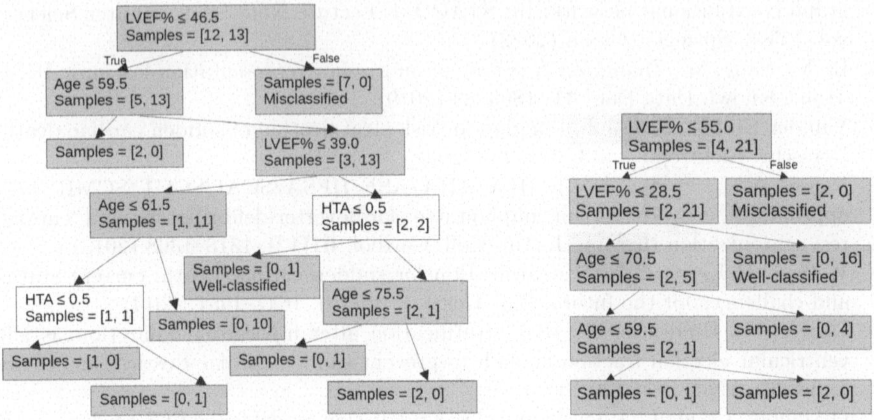

(a) **Prior Modality** (b) **Sequential Fusion**

Fig. 4. Classification decision tree, of positive samples (VA+), on test set.

6 Conclusion

We have presented a novel technique to fuse two modalities by sequential learning on each modality. The main novelty of this technique is the weights computation based on knowledge acquired from the prior modality. Our Sequential Fusion outperforms the Early Fusion with the ability to get most information from both modalities. When we look at misclassified samples by the Sequential Fusion and misclassified samples by all modalities, there is potential for growth, on the one hand, for fusion techniques, and for the inclusion of additional modalities on the other hand. For future work on features fusion techniques, we believe that constraining progressively higher dimensional data based on lower dimensional data can improve the robustness of the classification model.

Acknowledgments. This work has been supported by the French government through France 2030, the National Research Agency (ANR) Investments in the Future with 3IA Côte d'Azur (ANR-19-P3IA-000) and LIRYC (ANR-10-IAHU-04).

References

1. Srinivasan, N.T., Schilling, R.J.: Sudden cardiac death and arrhythmias. Arrhythm. Electrophysiol. **7**(2), 111–117 (2018)
2. Wilkins, E., et al.: European cardiovascular disease statistics. Eu. Heart Netw. (2017)
3. Nanthakumar, K., Epstein, A.E., Kay, G.N., Plumb, V.J., Lee, D.S.: Prophylactic implantable cardioverter-defibrillator therapy in patients with left ventricular systolic dysfunction: a pooled analysis of 10 primary prevention trials. J. Am. Coll. Cardiol. **44**(11), 2166–2172 (2004)
4. Ly, B., et al.: Interpretable prediction of post-infarct ventricular arrhythmia using graph convolutional network. In: STACOM. Lecture Notes in Computer Science, vol. 13593. Springer, Cham (2022)
5. Li, Y., Yang, M., Zhang, Z.: A survey of multi-view representation learning. IEEE Trans. Knowl. Data Eng. **31**, 1863–83 (2019)
6. Van der Bijl, P., Bax, J.J.: Imaging for risk stratification of sudden cardiac death. Herzschr. Elektrophys. **33**, 261–267 (2022)
7. Russo, A.M., et al.: ACCF/HRS/AHA/ASE/HFSA/SCAI/SCCT/SCMR 2013 appropriate use criteria for implantable cardioverter-defibrillators and cardiac resynchronization therapy. J. Am. Coll. Cardiol. **61**(12), 1318–1368 (2013)
8. Wellens, H.J., et al.: Risk stratification for sudden cardiac death: current status and challenges for the future. Eur. Heart J. **35**(25), 1642–1651 (2014)
9. Dagres, N., Hindricks, G.: Risk stratification after myocardial infarction: is left ventricular ejection fraction enough to prevent sudden cardiac death? Eur. Heart J. **34**(26), 1964–1971 (2013)
10. O'Mahony, C., et al.: A novel clinical risk prediction model for sudden cardiac death in hypertrophic cardiomyopathy (HCM Risk-SCD). Eur. Heart J. **2013** (2014)
11. Lai, D., Zhang, Y., Zhang, X., Su, Y., Bin Heyat, M.B.: An automated strategy for early risk identification of sudden cardiac death by using machine learning approach on measurable arrhythmic risk markers. IEEE Access **7**, 94701–94716 (2019)
12. Waks, J.W., et al.: Global electric heterogeneity risk score for prediction of sudden cardiac death in the general population: the atherosclerosis risk in communities (aric) and cardiovascular health (CHS) studies. Circulation **133**(23), 2222–2234 (2016)
13. Norrish, G., et al.: Development of a novel risk prediction model for sudden cardiac death in childhood hypertrophic cardiomyopathy (HCM Risk-Kids). JAMA cardiology **4**(9), 918–927 (2019)
14. Chen, T., Guestrin, C.: XGBoost: a scalable tree boosting system. In: Proceedings of the 22nd ACM SIGKDD International Conference on Knowledge Discovery and Data Mining (KDD 2016), pp. 785–794 (2016)
15. Sleeman, W.C., Kapoor, R., Ghosh, P.: Multimodal classification: current landscape, taxonomy and future directions. ACM Comput. Surv. **55**(7), 1–31 (2022)

16. Noordman, A.B.P., Maass, A.H., Groenveld, H., Mulder, B.A., Rienstra, M., Blaauw, Y.: Myocardial scar characterization and future ventricular arrhythmia in patients with ischemic cardiomyopathy and an implantable cardioverter-defibrillator. Front. Cardiovas. Med. **8** (2021)
17. Pandozi, C., et al.: The scar: the wind in the perfect storm-insights into the mysterious living tissue originating ventricular arrhythmias. J. Interv. Card. Electrophysiol. **66**(1), 27–38 (2023)

Automated Estimation of Cardiac Stroke Volumes from Computed Tomography

Ana-Teodora Radutoiu[1(✉)], Yasmin W. E. Youssef[2], Paula López Diez[1], Kristine A. Sørensen[1,3], Mathias B. Lindeskog[2], Michael H. C. Pham[2], Jonas J. Pedersen[2], Andreas Fuchs[2], Klaus F. Kofoed[2], and Rasmus R. Paulsen[1]

[1] Visual Computing, DTU Compute, Technical University of Denmark, Lyngby, Denmark
s194324@dtu.dk

[2] Department of Cardiology, Righospitalet, University of Copenhagen, Copenhagen, Denmark

[3] Novo Nordisk A/S, Copenhagen, Denmark

Abstract. Accurate assessment of left and right ventricular stroke volume is crucial for diagnosing and managing cardiac diseases such as heart failure, mitral regurgitation, and pulmonary embolism. The ventricular stroke volumes can be assessed from accurate ventricle segmentations in temporal 4D computed tomography (CT) images. Due to limited contrast and complicated sub-structures, the challenge of segmenting precise heart chamber volumes from CT images makes the estimation of both left and right ventricles a complicated task. Manual segmentation is time-consuming, necessitating the development of high-performing automatic segmentation methods for clinical use and large-scale studies. Existing models offer robust whole-heart segmentations but often include non-blood pool structures, leading to discrepancies in stroke volume calculations. This study proposes a data-driven method to correct left and right ventricle segmentations by accurately delineating the pure blood volume, leveraging Hounsfield Unit distributions. Our approach validates segmentation accuracy beyond traditional manual annotations by focusing on achieving equilibrium between left and right ventricular stroke volumes in healthy individuals. This method demonstrates the potential to improve cardiac function assessments from CT, facilitating more reliable and efficient clinical evaluations.

Keywords: Stroke volume · Heart segmentation · RANSAC · HUV distribution · CT

1 Introduction and Background

The human heart is responsible for circulating blood to the entire human body. The heart can functionally be divided into a left and right side. On the left side, the left ventricle (LV) receives blood from the left atrium and pumps it out into the systemic circulation. On the right side, the right ventricle (RV) receives

O. Camara et al. (Eds.): STACOM 2024, LNCS 15448, pp. 22–32, 2025.
https://doi.org/10.1007/978-3-031-87756-8_3

blood from the right atrium and pumps it into the pulmonary system. The heart cycle is divided into a systole (active phase) and a diastole (relaxing phase). The amount of blood ejected from each ventricle in the active phase is the difference between end-diastole (ED) volume and end-systole (ES) volume, termed the stroke volume (SV) ($SV = V_{ED} - V_{ES}$). Whereas the ejection fraction (EF) reflects the relative volume difference ($EF = (V_{ED} - V_{ES})/V_{ED}$). In healthy individuals, the left and right ventricular stroke volumes (LVSV and RVSV, respectively) should almost always be the same [7]. Discrepancies between these volumes could indicate diseases such as mitral regurgitation, where up to 50% of the blood in the LV may backflow through the diseased mitral valve leading to a decrease in the effective forward stroke volume [1]. Measuring EF is important for diagnostic and prognostic purposes. Many cardiac diseases such as coronary artery disease, hypertension, valvular heart disease, or cardiomyopathies may lead to heart failure. Distinguishing between heart failure with preserved and reduced EF (HFpEF and HFrEF) is important for understanding the pathophysiology and providing the optimal treatment [22]. In general, most research has been concerned around the left ventricle. The right ventricle is less well described, but literature suggests that RV EF provides additional prognostic information for evaluating the severity and prognosis of pulmonary hypertension [9] and for diagnosing Arrytmogen Right Ventricular Cardiomyopathy (ARVC) [21]. Reliable and non-invasive techniques for measuring the LVSV and RVSV are thus of great interest for many clinical purposes.

Cardiac magnetic resonance (cMR) is currently considered the gold standard for assessing EF, and the majority of past studies assess EF on either cMR or ultrasound (i.e., transthoracic echocardiography (TTE)). Cardiac Computed Tomography (CCT) is, however, becoming increasingly available and advances in CT imaging technology have enabled the acquisition of temporal 4D CT, where high-resolution images can be acquired over time to assess the cardiac motion during a heartbeat. Such image series is referred to as a cardiac functional analysis (CFA) and can be used to assess the LV and RV volumes and function. CFA series have been shown to provide comparable results to those of cMR [8,18,26,28] and TTE [15], making it a valuable alternative for patients who cannot undergo cMR or in patients who already have undergone CT examination for other purposes such as coronary artery assessment.

Manual annotation of the LV on CFA series is a time-consuming process, with annotation times being reported as high as 63 min for a full manual segmentation [3] and 16 min for semi-automatic methods [5]. High-performing automatic segmentation methods are thus required to use CT-derived functional parameters in daily clinical practice and for larger population studies. Automatic whole heart segmentation has been an active research field for many years and many solutions have been proposed in i.e. the Multi-Modality Whole Heart Segmentation (MM-WHS) challenge [29]. Recently, TotalSegmentator [27] was released, which is an out-of-the-box pre-trained CT segmentation model promising accurate segmentation of 104 anatomic structures including a high-resolution cardiac segmentation model. Both TotalSegmentator and the training data in MM-WHS consider the LV as the combination of the inner blood pool, trabeculated mass, and papillary muscles (Fig. 1(a)). Including the trabeculae and papillary muscles in LV has been shown to increase the LV ED volume with 11.9% [14] and affected the calculation of EF to a degree where it masked a significant difference in EF between patients with LV non-compaction (LVNC) and a healthy control group [16]. Being able to accurately determine the pure blood volume is therefore needed to correctly estimate chamber volumes and function. Robustly detecting the trabeculated mass has the additional benefit of being able to further investigate the relation between excessive trabeculation and cardiac diseases [25]. Being able to jointly analyze the right and left side heart function is further challenged by the fact that cardiac CT images usually are acquired when the contrast-filled blood reaches the aorta, which ensures a good contrast between cardiac tissue and the blood in the LA, LV, and coronary arteries. While images with contrast-filled right-sided chambers are possible through triggering the acquisition in the pulmonary arteries, these images are less common, and methods for accurate estimation of both LV and RV dynamics from images acquired at the arterial phase are thus of interest.

Previous work on segmenting the LV trabeculation and estimating the pure blood volumes are conducted with MRI, but CCT have been shown to provide equivalent results in terms of manually quantifying the LV trabeculated layer [20]. Bartoli et al. [2] presented a deep learning-based automatic method using a DenseNet to segment the LV blood cavity, LV myocardium, and LV trabeculation from MRI in ED. The method, however, requires a large amount of training data and achieves underwhelming results in segmenting the trabeculation. Other methods on MRI make use of adaptive thresholding methods [17] and combinations of region-growing techniques [4]. For CT imaging, we do not have large amounts of annotated data of the trabeculation and instead rely on data-driven methods based solely on the CT images.

In CT images, the tissue attenuation inside each voxel is measured in Hounsfield Units (HU), which is a calibrated measure. Ideally, the contrast-enhanced blood would have the same HU distribution across all individuals in the population and should be easily separated from the darker voxels representing the cardiac tissue. In practice, differences in body composition, contrast agent flow, scan protocol, and CT reconstruction result in significant HU variations over a population. To estimate the true organ-specific blood HU distribution, data-driven methods are necessary, which is one of the aims of this study. A comprehensive study of HU value variations in the aorta is presented in [11], where it is also argued that data-driven adaptive modeling of the HU distribution is needed. Figure 1(b) shows the distribution over HU in the two ventricles in the segmentation obtained from TotalSegmentator. The bimodal HU distribution in the LV suggests that the segmentation includes voxels belonging to both the pure blood volume and the cardiac tissue.

We make a simplified assumption that the Hounsfield unit value (HUV) distribution of the contrast-enhanced blood is approximately Gaussian distributed. By assuming that the HUV distribution of the trabeculae and muscles also are

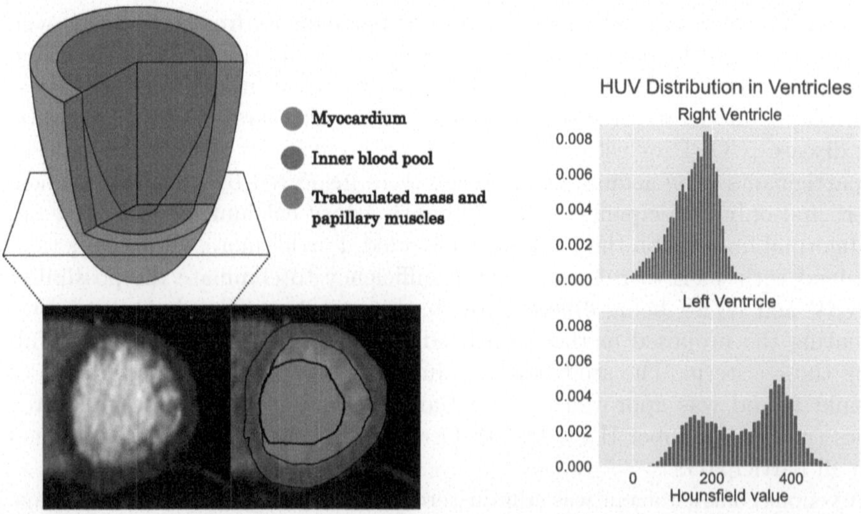

(a) (b)

Fig. 1. **(a)** Sketch of the studied anatomy on the top and an illustrative example from one of the images in our study of the ventricular short-axis view, inspired by [25]. **(b)** Hounsfield unit histogram for both the right and the left ventricle using TotalSegmentator's segmentation of one of the CT images in our dataset.

Gaussian distributed, one could model the total set of HUV as a mixture of Gaussians (GMM) [10]. We have performed extensive experiments with GMMs using the Aikekes and Bayesian Information Criteria [10] for optimal model selection. However, it was not possible to achieve consistent and stable results using GMM. Particularly, in cases with very little trabeculation, the blood pool HUV would be modeled using several Gaussians resulting in an erroneous selection of the *true blood pool voxels*. The proposed method instead aims to estimate the distribution of HUV in the pure blood volume and exclude out-of-distribution samples from LV and RV segmentation obtained with TotalSegmentator [27]. We propose a random sampling consensus (RANSAC) based inlier/outlier detection framework to robustly estimate the parameters of the blood volume distribution. We demonstrate that the proposed method achieves a balanced LVSV and RVSV in a healthy test population, and argue that this higher-level objective can be used to validate segmentation accuracy by capturing anatomical plausibility in a manner that standard comparison to manual annotations cannot achieve.

2 Data

The data used to evaluate the proposed method is based on a subset of participants from the Copenhagen General Population Study (CT substudy); an extensive prospective cohort study including individuals from the general population. To demonstrate the efficacy of our correction method to obtain correctly balanced SV estimates, we chose a subset of participants from this study, who are confirmed healthy participants [23]. In this subgroup, the following exclusion criteria were used: 1) BMI > 35, 2) diabetes, 3) hypertension, 4) hypercholesterolemia, 5) moderate or heavy smoking (>20 pack years), 6) bronchospastic lung disease (COPD or asthma) as well as 7) pregnant or actively breastfeeding participants. The acquired CT images were inspected by a trained medical expert and only participants with a coronary artery calcium score of zero and no abnormal findings on the CCT were included. Furthermore, all patients were examined with TTE to rule out valve insufficiency to eliminate the possibility of LVSV and RVSV being different due to regurgitation. The dataset used for evaluating the proposed method consisted of 43 CFA series of individuals fulfilling these criteria. The study was conducted according to the Declaration of Helsinki II and was approved by The Danish Committee on Health Research Ethics (protocol number H-19043153). Oral and written consent were obtained from all participants.

A venous contrast agent was administered to all participants before the image acquisition and the CFA series was acquired on a Canon Aquilion ONE scanner. Each CFA series consists of 20 consecutive images acquired in incremental steps of 5% during a single R-R cycle. The imaging was automatically triggered with a region of interest in the descending aorta, ensuring proper contrast filling in the left side of the heart. The volumes we use in this study were reconstructed by the CT scanner software to a resolution of $0.49 \times 0.49 \times 2$ mm.

3 Methods

Automated Segmentations. We have used a pre-trained model from the TotalSegmentator tool [27] as a baseline. Specifically, the `heartchambers_highres` (Release 2.1.0) pre-trained model is used to segment seven structures in and around the heart. The segmentations used in this study are the LV and the RV. TotalSegmentator is based on the nn-Unet [12] framework and the pre-trained model has been trained on a non-public heart dataset with unknown acquisition parameters. We have visually inspected the segmentation from TotalSegmentator and found that they are consistent with the ones from trained cardiac readers.

Refinement Using RANSAC HU Selections. We aim to automatically determine the parameters of the underlying Gaussian distribution of the HUV in the ventricular blood pool, $\mathcal{N}(\mu, \sigma^2)$. The original segmentation by TotalSegmentator includes trabeculae, papillary muscles, and, for the RV, its myocardium. Their HUV are smaller than the inner blood pool. To refine a segmentation, a threshold-based method is applied by discarding voxels below, $\mu - 3\sigma$.

This study uses random sampling consensus (RANSAC) to estimate the parameters, μ and σ. RANSAC was originally proposed by Fischler and Bolles [6] to fit a model to data containing many outliers. In our case, the trabeculae, muscle, and myocardium voxels are outliers when considering the blood pool, why we believe RANSAC is a well-suited approach for this challenge. We have tailored the algorithm to our task, and aim to obtain a subset of voxels only belonging to the blood pool to estimate μ and σ. The pseudocode is provided in Algorithm 1. The 'maybe inliers' (Algo. 1) denotes a small subset of voxels that may be the inner blood pool, found by collecting voxels in the ventricle from a few small ellipsoids. An ellipsoid has a random center but a fixed size. We set the following criteria for a good model; a $\bar{\mu}$ at a higher HUV, but not exceeding the peak of the original HUV distribution, and a low $\bar{\sigma}$.

The SV is found from volume estimations. At each phase in the R-R cycle, the ventricle volumes are estimated from segmentation masks (Fig. 2, top left). The ventricular end-diastole is defined as the 0% phase, while the ventricular end-systole is estimated as the lowest ventricular volume in the R-R cycle. Usually, the ES will be around the 30% phase.

Algorithm 1. Tailored Random Sampling Consensus

Data ct_image, anatomy_mask (either LV or RV segmentation mask)
Parameters $N = 1000$, $k = 10$, $d = 100$, $n_{\mathcal{E}} = 3$, $R_{\mathcal{E}} = (3, 3, 2)$, $\text{tol}_\mu = 10$, $\text{tol}_\sigma = 5$
Preprocessing $\mathcal{S}_{\text{all}}$ = all voxels in the anatomy, c_{peak} = HUV distribution peak
Initialize $\bar{\mu}_{\text{best}}, \bar{\sigma}_{\text{best}} = -\infty, \infty$, iter $= 0$.
while iter $< k$ **do**
  Define $n_{\mathcal{E}}$ ellipsoids with random center and radius $R_{\mathcal{E}}$ inside the anatomy
  Maybe inliers, $\mathcal{S}_{\text{maybe}}$ = voxels belonging to the ellipsoids
  $\bar{\mu}_{\text{maybe}}, \bar{\sigma}_{\text{maybe}} = \text{mean}(\mathcal{S}_{\text{maybe}}), \text{std}(\mathcal{S}_{\text{maybe}})$
  Choose N random values from $\mathcal{S}_{\text{all}}$, denote the set $\mathcal{S}_N \subset \mathcal{S}_{\text{all}}$
  Confirmed inliers, $\mathcal{S}_{\text{conf}} = \{v \in \mathcal{S}_N \mid v \in \bar{\mu}_{\text{maybe}} \pm 3\bar{\sigma}_{\text{maybe}}\}$
  if $|\mathcal{S}_{\text{conf}}| \geq d$ **then**
    $\bar{\mu}_{\text{better}}, \bar{\sigma}_{\text{better}} = \text{mean}(\mathcal{S}_{\text{conf}}), \text{std}(\mathcal{S}_{\text{conf}})$
    if $(\bar{\mu}_{\text{best}} - \text{tol}_\mu < \bar{\mu}_{\text{better}} < c_{\text{peak}} + \text{tol}_\mu) \wedge (\bar{\sigma}_{\text{better}} < \bar{\sigma}_{\text{best}} + \text{tol}_\sigma)$ **then**
      $\bar{\mu}_{\text{best}}, \bar{\sigma}_{\text{best}} = \bar{\mu}_{\text{better}}, \bar{\sigma}_{\text{better}}$
    end if
  end if
  iter = iter + 1
end while
Return $(\bar{\mu}_{\text{best}}, \bar{\sigma}_{\text{best}})$

4 Results and Discussion

A visualization of our refined segmentations, along with the original one by TotalSegmentator, throughout the R-R cycle is seen in Fig. 2. The main difference between the two segmentations is within the LV, where our refined segmentation can exclude the trabeculation and papillary muscles. Conversely, only a small decrease in the RV volume is seen throughout the R-R cycle. The smaller corrections are likely due to the trabeculation and papillary muscles being almost invisible on the right side. From visual inspection, our segmentation approach discards volume from the RV wall, thus indicating that it removed at least some of the myocardium.

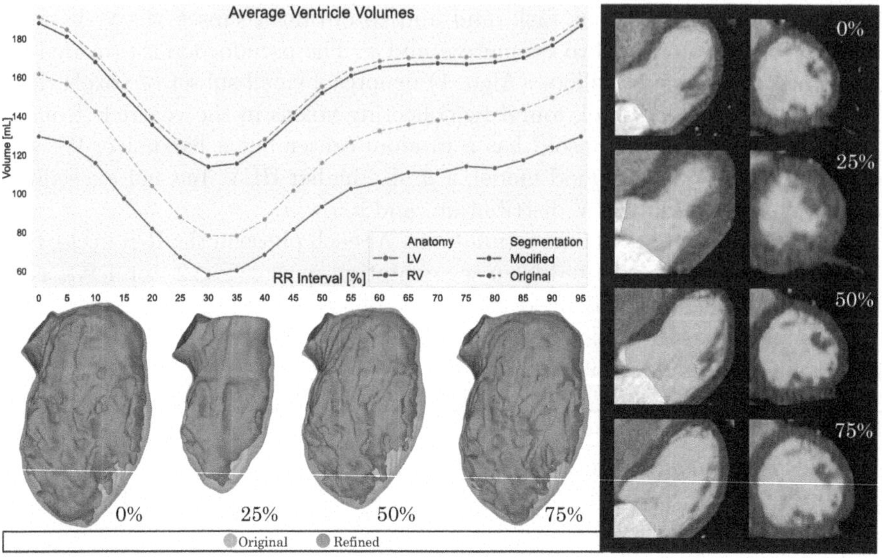

Fig. 2. Top left: Average ventricle volumes in the R-R cycle (LV in red and RV in blue) over our 43 samples before and after applying tailored random sampling consensus to fine-tune the segmentations. **Bottom Left:** Volumetric rendering of one sample in 0%, 25%, 50%, and 75% of the R-R cycle with the previous and fine-tuned blood pool segmentation. **Right:** Two views of the CT cross sections of the same sample with the overlying segmentations. (Color figure online)

Figure 3 shows a single short-axis slice image from six randomly selected individuals, which illustrates how our proposed correction method handles the papillary muscles and trabeculation. It is evident that the method correctly excludes regions with trabeculation and the papillary muscles, but the partial volume effect makes it difficult to judge whether too much blood volume has been removed. It is also noted that the amount of contrast in the RV varies across the individuals and that the proposed method only locates a small amount of trabeculation and papillary muscle. Evaluating the segmentation accuracy is challenged by the fact that detailed manual annotation of the trabeculation and papillary muscles is time-consuming and difficult, and no reliable ground truth is thus available. Expert manual annotation of the images in this dataset did not remove the trabeculation and papillary muscles and revealed a similar stroke volume discrepancy as was observed by the original segmentation algorithm, suggesting that the annotations are not suitable for evaluating the precision of the correction method. For the same reason, the volumes obtained from our method are underestimated compared to manual annotation. Even for an experienced medical reader, it is difficult to accurately delineate the trabeculation since the trabeculae can be smaller than the imaging resolution and the partial volume effect thus limits the ability to separate the blood from the surrounding muscle tissue.

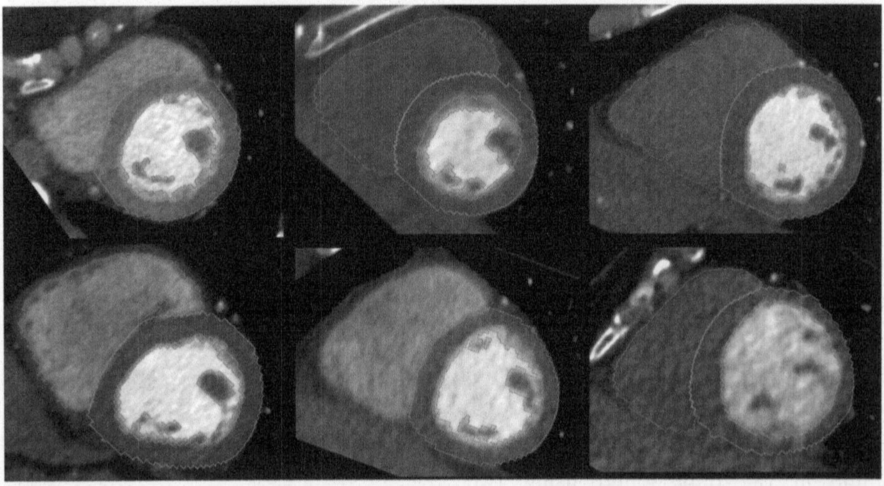

Fig. 3. Short axis slice from six randomly selected individuals illustrating the delineation of the right ventricle (red), LV myocardium (green) and inner LV blood volume (brown). (Color figure online)

Figure 4 compares the cardiac stroke volumes when using the original segmentation by TotalSegmentator and our refined version to estimate the volumes. An imbalance between the LVSV and RVSV is observed when considering the results with the original segmentations. To validate this imbalance, we perform a paired student-t test to investigate whether the means are different. The estimated mean LVSV and RVSV are 87.4 mL and 74.5 mL respectively, resulting in a p-value < 0.001 (Table 1), hence significantly different. In our population data, the individuals have undergone TTE which showed no leaks in the heart valves. Leakage in the heart valves is a factor for imbalanced SV between the LV and RV [1]; thus, we expect balanced SVs in our data. As the SV means differ significantly, we suspect the original segmentations need fine-tuning. Using our approach to refine the segmentations, we obtain a better balance between the left and right side of the heart (Fig. 4). Specifically, our data shows an estimated mean LVSV and RVSV of 79.2 mL and 77.3 mL, respectively. A paired student-t test further validates that these means are not significantly different (Table 1). The now balanced SVs are possibly mainly due to the refinements made in the LV segmentations, as only a small increase in the RV stroke volume is seen in Table 1. However, this is inevitable as only a few modifications are made to the RV segmentation on average (Fig. 2, top left). The myocardium is larger in the end-systole phase, why we suspect the RV volume might be overestimated in this phase and consequently, the stroke volume will be underestimated. Nevertheless, refining the RV segmentation is difficult due to the low contrast.

Considering clinically reported SVs, it is seen that the SV is larger for men than women and declines slightly with age [24]. Using cMR, the LVSV has been reported to be 91 mL and 73 mL on average for men and women, respectively,

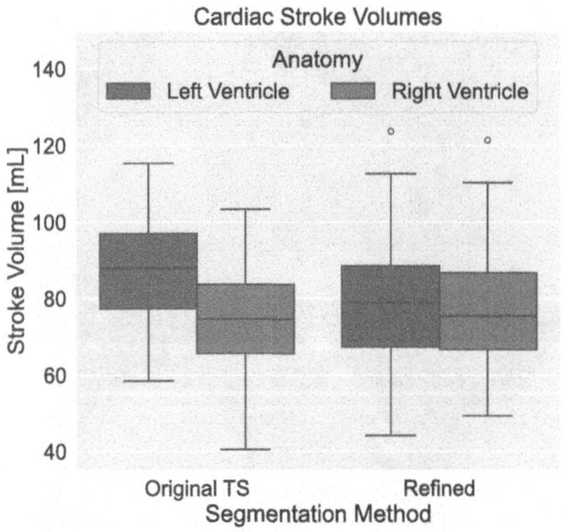

Fig. 4. Box plot of the ventricular stroke volume in the data found using two different segmentations; TotalSegmentator (original TS) and our (refined).

Table 1. Mean and standard deviation (SD) of LVSV and RVSV, where SV is found from two different segmentations; TotalSegmentator (original TS) and our (refined). A paired student t-test is included to test for equal means.

		mean ± SD	*p*-value
Original TS	LV	87.4 ± 12.8	< 0.001
	RV	74.4 ± 13.3	
Refined	LV	79.2 ± 16.1	0.428
	RV	77.3 ± 15.5	

in healthy individuals [13], while the RVSV is typically 88 mL and 75 mL for men and women, respectively [19]. Comparing the clinically reported SVs with our results obtained in Table 1, we observe that they are within the expected intervals for healthy individuals across gender and age groups.

The aim of this study is to highlight that stroke volume balance is a valuable higher-level objective to validate the anatomical plausibility of automatic segmentation algorithms, and we, therefore, tested the correction method in healthy participants where this objective is valid. Utilizing the method in clinical settings with diseased individuals requires additional testing. However, since the correction is based on estimation of the individual HUV-distribution, we believe it will generalize to the primary patient group, consisting of individuals with symptoms but no known heart disease. Special considerations should however be taken in special cases, where i.e. metal artifacts challenge the imaging.

5 Conclusion

We have demonstrated and evaluated a fully automated method for dual chamber cardiac stroke volume estimation. The proposed method achieves the necessary stroke volume balance in a healthy population. We argue that it is crucial and difficult to segment the true blood volume excluding trabecula and muscle to achieve this balance. While it is relatively easy to validate the result in the left ventricle, it is more challenging in the right ventricle, warranting further investigation in the future. We believe that the proposed method holds great potential for future diagnosis and risk assessment of complex cardiovascular diseases involving valve dysfunctions. Other clinically relevant measures as ejection fractions can also benefit from having a robust true blood pool estimate.

References

1. Apostolidou, E., Maslow, A.D., Poppas, A.: Primary mitral valve regurgitation: update and review. Glob. Cardiol. Sci. Pract. **2017**(1) (2017)
2. Bartoli, A., et al.: Deep learning-based automated segmentation of left ventricular trabeculations and myocardium on cardiac MR images: a feasibility study. Radiol.: Artif. Intell. **3**(1) (2021). https://doi.org/10.1148/ryai.2020200021
3. Boehm, T., et al.: Time-effectiveness, observer-dependence, and accuracy of measurements of left ventricular ejection fraction using 4-channel MDCT. Rofo Fortschritte Auf Dem Gebiet Der Rontgenstrahlen Und Der Bildgebenden Verfahren **176**(4), 529–537 (2004). https://doi.org/10.1055/s-2004-813012
4. Codella, N.C., Weinsaft, J.W., Cham, M.D., Janik, M., Prince, M.R., Wang, Y.: Left ventricle: automated segmentation by using myocardial effusion threshold reduction and intravoxel computation at MR imaging. Radiology **248**(3), 1004–1012 (2008). https://doi.org/10.1148/radiol.2482072016
5. Dewey, M., Müller, M., Teige, F., Hamm, B.: Evaluation of a semiautomatic software tool for left ventricular function analysis with 16-slice computed tomography. Eur. Radiol. **16**(1), 25–31 (2006). https://doi.org/10.1007/s00330-005-2817-7
6. Fischler, M.A., Bolles, R.C.: Random sample consensus: a paradigm for model fitting with applications to image analysis and automated cartography. Commun. ACM **24**(6), 381–395 (1981). https://doi.org/10.1145/358669.358692
7. Franklin, D.L., Vancitters, R.L., Rushmer, R.: Balance between right and left ventricular output. Circul. Res. **10**(1) (1962). https://doi.org/10.1161/01.RES.10.1.17
8. Fu, H., et al.: CT compared to MRI for functional evaluation of the right ventricle: a systematic review and meta-analysis. Eur. Radiol. **29**(12), 6816–6828 (2019). https://doi.org/10.1007/s00330-019-06228-2
9. Ghio, S., et al.: Prognostic relevance of the echocardiographic assessment of right ventricular function in patients with idiopathic pulmonary arterial hypertension. Int. J. Cardiol. **140**(3), 272–278 (2010). https://doi.org/10.1016/j.ijcard.2008.11.051
10. Hastie, T., Friedman, J.H., Tibshirani, R.: The Elements of Statistical Learning : Data Mining, Inference, and Prediction. Springer (2001)
11. Holcombe, S.A., Horbal, S.R., Ross, B.E., Brown, E., Derstine, B.A., Wang, S.C.: Variation in aorta attenuation in contrast-enhanced CT and its implications for calcification thresholds. PLoS One **17**(11) (2022)
12. Isensee, F., Jaeger, P.F., Kohl, S.A., Petersen, J., Maier-Hein, K.H.: nnU-net: a self-configuring method for deep learning-based biomedical image segmentation. Nat. Methods **18**(2), 203–211 (2021)
13. Kawel-Boehm, N., et al.: Reference ranges ("normal values") for cardiovascular magnetic resonance (CMR) in adults and children: 2020 update. J. Cardiovasc. Magn. Reson. **22**(1), 87 (2020). https://doi.org/10.1186/s12968-020-00683-3
14. Kim, M., You, S., Ha, T., Kim, T.H., Kang, D.K.: Effect of papillary muscle and trabeculae on left ventricular function analysis via computed tomography: a cross-sectional study. Medicine **102**(46) (2023). https://doi.org/10.1097/MD.0000000000036106
15. Ko, S.M., Kim, Y.J., Park, J.H., Choi, N.M.: Assessment of left ventricular ejection fraction and regional wall motion with 64-slice multidetector CT: a comparison with two-dimensional transthoracic echocardiography. Br. J. Radiol. **83**(985), 28–34 (2010). https://doi.org/10.1259/bjr/38829806

16. Laissy, J.P., Driss, A.B.: Left ventricular function assessment including or excluding trabeculations in left ventricular non-compaction, a preliminary case-control cardiac magnetic resonance study. Quant. Imaging Med. Surg. **14**(3), 2475–2484 (2024). https://doi.org/10.21037/qims-23-1274
17. Lu, Y., Radau, P., Connelly, K., Dick, A., Wright, G.A.: Segmentation of left ventricle in cardiac cine MRI: an automatic image-driven method. In: LNCS, vol. 5528, pp. 339–347 (2009). https://doi.org/10.1007/978-3-642-01932-6_37
18. Maffei, E., et al.: Left and right ventricle assessment with cardiac CT: validation study vs. cardiac MR. Eur. Radiol. **22**(5), 1041–1049 (2012). https://doi.org/10.1007/s00330-011-2345-6
19. Maffessanti, F., et al.: Age-, body size-, and sex-specific reference values for right ventricular volumes and ejection fraction by three-dimensional echocardiography. Circul.: Cardiovas. Imaging **6**(5), 700–710 (2013). https://doi.org/10.1161/CIRCIMAGING.113.000706
20. Manohar, A., et al.: Quantitative metrics of the LV trabeculated layer by cardiac CT and cardiac MRI in patients with suspected noncompaction cardiomyopathy. Eur. Radiol. 1–12 (2023). https://doi.org/10.1007/s00330-023-10526-1
21. Marcus, F.I., et al.: Diagnosis of arrhythmogenic right ventricular cardiomyopathy/dysplasia: proposed modification of the task force criteria. Circulation **121**(13), 1533–1541 (2010). https://doi.org/10.1161/CIRCULATIONAHA.108.840827
22. McDonagh, T.A., et al.: 2021 ESC guidelines for the diagnosis and treatment of acute and chronic heart failure. Eur. Heart J. **42**(36), 3599–3726 (2021). https://doi.org/10.1093/eurheartj/ehab368
23. Møller, M.B., Linde, J.J., Fuchs, A., Køber, L.V., Nordestgaard, B.G., Kofoed, K.F.: Normal values of myocardial blood flow measured with dynamic myocardial computed tomography perfusion. Eur. Heart J. - Cardiovas. Imaging **25**(7), 986–995 (2024). https://doi.org/10.1093/ehjci/jeae050
24. Raisi-Estabragh, Z., et al.: Cardiovascular magnetic resonance reference ranges from the healthy hearts consortium. JACC: Cardiovas. Imaging (2024). https://doi.org/10.1016/j.jcmg.2024.01.009
25. Sigvardsen, P.E., et al.: Left ventricular trabeculation and major adverse cardiovascular events: the Copenhagen general population study. Eur. Heart J. Cardiovasc. Imaging **22**(1), 67–74 (2021). https://doi.org/10.1093/ehjci/jeaa110
26. Takx, R.A., et al.: Quantification of left and right ventricular function and myocardial mass: comparison of low-radiation dose 2nd generation dual-source CT and cardiac MRI. Eur. J. Radiol. **81**(4) (2012). https://doi.org/10.1016/j.ejrad.2011.07.001
27. Wasserthal, J., et al.: Totalsegmentator: robust segmentation of 104 anatomic structures in CT images. Radiol.: Artif. Intell. **5**(5) (2023)
28. Yamasaki, Y., et al.: Quantitative assessment of right ventricular function and pulmonary regurgitation in surgically repaired tetralogy of Fallot using 256-slice CT: comparison with 3-Tesla MRI. Eur. Radiol. **24**(12), 3289–3299 (2014). https://doi.org/10.1007/s00330-014-3344-1
29. Zhuang, X., et al.: Evaluation of algorithms for multi-modality whole heart segmentation: an open-access grand challenge. Med. Image Anal. **58** (2019). https://doi.org/10.1016/j.media.2019.101537

Peridevice Leaks Following Left Atrial Appendage Occlusion - Analysis with Morphology Descriptive Centerlines and Explainable Graph Attention Network

Paula López Diez[1(✉)], Jan Margeta[2], Javier Gómez-Herrero[5], Davorka Lulic[3], Yannick Willemen[3], Klaus F. Kofoed[3,4], Ole De Backer[3], and Rasmus R. Paulsen[1]

[1] DTU Compute, Technical University of Denmark, Lyngby, Denmark
polly_146@hotmail.com

[2] Research & Development, KardioMe, Nová Dubnica, Slovakia

[3] Department of Cardiology, Copenhagen University Hospital - Rigshospitalet, Copenhagen, Denmark

[4] Department of Clinical Medicine, Faculty of Health and Medical Sciences, University of Copenhagen, Copenhagen, Denmark

[5] Servicio de Cardiología, Hospital Clínico Universitario de Valladolid, Valladolid, Spain

Abstract. Peridevice leaks (PDLs) following left atrial appendage occlusion can negate the protective effect of the procedure and are a challenge in clinical cardiology. In this study, we investigate how the morphology of the anatomy relevant to the procedure interacts with the risk of PDLs using pre-operative CT images. We construct 3D models of the anatomy involved in the transcatheter procedure from a dataset of 125 patients who underwent left atrial appendage occlusion, integrating an anatomical centerline that simulates the catheter trajectory during device placement and that is complemented with morphological features. Given the complex and variable nature of this anatomy, we utilize Graph Attention Networks to analyze these models. This architecture enables us to encode morphological features into a graph structure, capturing the intricate spatial relationships and dependencies. We predict the likelihood of potential PDLs and identify key morphological descriptors contributing to these predictions through attention scores. This method provides a standardized representation of the anatomy involved in the procedure, offering insights into the anatomical factors influencing PDL risk and aiding in pre-operative planning.

Keywords: LAAO · GAT · centerline · leak · explainability

O. Camara et al. (Eds.): STACOM 2024, LNCS 15448, pp. 33–42, 2025.
https://doi.org/10.1007/978-3-031-87756-8_4

1 Introduction

The left atrial appendage (LAA) is a small pouch in the left atrium of the heart, often implicated in thrombus formation, particularly in patients with atrial fibrillation (AF), in which around 90% of intracardiac thrombi are found in the LAA [20]. To mitigate stroke risk, left atrial appendage occlusion (LAAO) is performed to seal off the LAA and prevent thromboembolic events in AF patients. This transcatheter intervention involves placing a device that occludes the LAA from the LA. However, peridevice leaks (PDLs) can occur, where gaps between the occluder and the LAA wall allow blood to bypass the device.

PDLs are significant as they can negate the protective effect of LAAO. Recent studies have documented a relationship between PDLs and subsequent thromboembolic events [1]. These leaks arise from incomplete endothelialization, suboptimal device placement, or anatomical variations in the LAA. It has also been shown that CT is the most sensitive method to evaluate leaks because it provides a detailed perspective of the shape, size, and mechanism of the PDL [7]. Managing PDLs involves monitoring, pharmacological therapy, and sometimes secondary interventions. Predicting PDL occurrence and understanding their morphology remain challenging due to the complex geometries and dynamic nature of the LAA [3]. In this study, we want to explore to which degree the morphology of the anatomy is related to the potential risk of leakage. It is known that some anatomies are more difficult for the operator. We aim to find a standardized representation of the complex anatomy that captures the challenges involved in the procedure.

Given the high complexity and inter-patient variability of the LAA anatomy, standardization presents a significant challenge from an image analysis standpoint. However, many different methods have been developed in the past to optimize the LAAO planning. Some focus on the risk of thromboembolic events based on the morphology of the appendage itself [5], sometimes analyzing specific anatomical features of the LAA anatomy or the combination with hemodynamic properties [13,15]. Some research also focuses on helping clinicians choose the optimal device for a specific anatomy using both the morphology and computational fluid dynamics on the LA anatomy [2,9]. PDLs are a complex and multifactorial problem that has not yet been investigated in depth. It has been shown that possible predictors of PDLs are the length of the procedure or the number of attempts [8]. Even though these can be used to select patients for a closer follow-up, it is not applicable for pre-procedural planning. We focus on the analysis of how the anatomical centerline that mimics the expected catheter trajectory can help us detect morphologies with a higher risk of PDL after undergoing LAAO.

Recent advancements in machine learning, particularly graph attention networks (GATs), offer promising approaches for processing graph-structured data, capturing the intricate relationships and spatial dependencies [10]. Additionally, incorporating explainability features into these networks allows for a better understanding of the morphological descriptors contributing to a decision of the trained classifier.

In this paper, we present a novel method for characterizing the anatomy involved in the LAAO by extracting the morphological centerline that simulates the trajectory of the catheter for device placement. We build a graph based on this centerline and encode other relevant features of the morphology. Furthermore, we train a GAT model to predict the PDL likelihood given a pre-operative anatomy and use the attention scores to explain which parts of the graph had a higher influence on the predicted outcome.

2 Data

The dataset used for this study consists of 125 patients who have undergone LAAO. For each patient, pre-operative and post-operative CT images have been collected. The pre-operative CT images were used to extract the graph characterizing the anatomy. The post-operative CT images were analyzed by two expert annotators to determine whether a leak was present in the LAA by examining whether the contrast agent reached the distal area of the LAA. Furthermore, the different types of leaks have been classified according to their nature. The population distribution and examples of each type of leak are shown in Fig. 1.

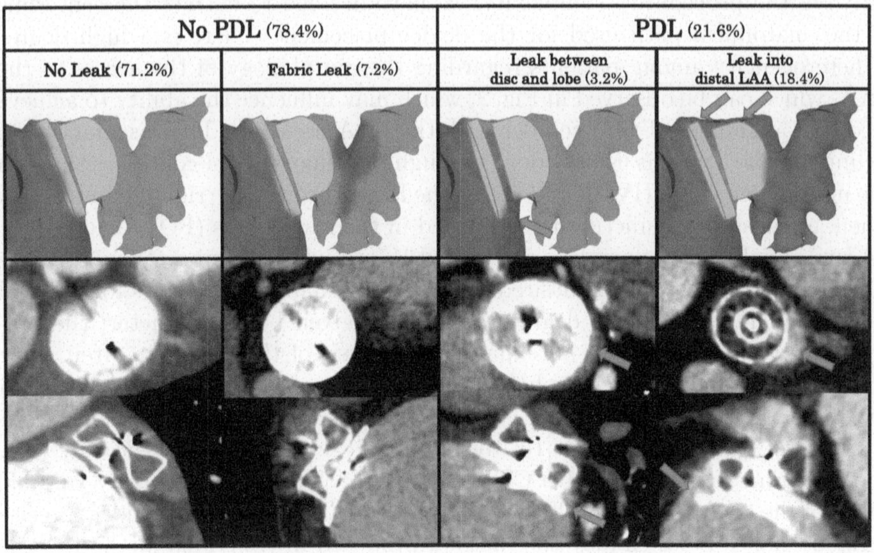

Fig. 1. Four classes defined based on the mechanisms of leaks after LAAO with a disc-and-lobe occluder. Prevalence of each class over the 125 subjects used for this study. The top row shows an illustration for conceptual understanding in 3D, underneath 2 views of an example CT image from our study. Blue arrows indicate the region in which the gap associated to the PDL can be observed. (Color figure online)

All patients were treated by the same medical team, utilizing three different types of devices. Two of the devices are disc-and-lobe occluders: Amplatzer

Amulet^{TM}(with an incidence of 32.8%) and Omega^{TM}(with an incidence of 36.8%); and the third type is a plug-in type: Watchman^{TM}(with an incidence of 30.4%). Leaks were classified into PDL and non-PDLs. A PDL is indicated by a visible gap between the device and the anatomy, even if the gap is only in the disc part and the lobe is sealed. Non-PDL cases include those with no patency or with patency but no visible gap; the latter are typically fabric leaks or micro leaks expected to seal over time [14]. Usually, in these cases the CT has been obtained before the process of endothelialization is completed and therefore is not considered a PDL in our dataset. It should be noted that, given the nature of the occluders, only the disc-and-lobe devices can have a leak between the disc and the lobe whereas fabric leaks or micro leaks are possible for all the devices. An example of each of these categories both in an anatomical illustration and a CT image can be observed in Fig. 1.

3 Methods

3.1 Graph Generation

To construct a graph that characterizes the anatomy in a way that is representative of the procedure's complexity, we have decided to extract the centerline of the anatomical path used for the device placement. There is a high degree of heterogeneity among patients regarding the morphology of the access to the LAA, which can be observed in Fig. 2, which may influence the ability to achieve a coaxial alignment of the occluder with the LAA neck [11]. The vascular access to implant the device is usually done through the femoral vein, continues through the inferior vena cava (IVC), and enters the heart through the right atrium (RA). Then, a transseptal puncture is performed in the fossa ovalis (FO) to access the LA and, finally, the appendage is reached [17]. Given the importance of reaching the appendage with the desired angle and coaxiality that allows for optimal placement and the limited degrees of freedom given by the catheter; the preoperative planning involves an exhaustive analysis of the anatomy, often using CT images to analyze its morphology and the potential risks associated with a specific patient.

We use TotalSegmentator [19] to obtain the segmentation of the LA, IVC, and RA from the pre-operative CT image. However, due to the highly variable shape of the LAA, which TotalSegmentator locates but does not provide a precise segmentation for, we employ the method proposed in [16]. This method learns the neural unsigned distance field and generates much more precise and detailed segmentations of the LAA.

Once all the relevant anatomies are extracted from the CT, we obtain the joint mesh by iso-surface extraction and proceed to generate a standardized opening that resembles the transseptal puncture. To do so, we locate the closest point between the RA and the LA, which is the thinnest point of the FO, and generate an opening of 5mm radius. After this, we extract the centerline with VMTK [4]. We use the geodesically furthest point in the IVC with respect to the RA as the initial point and the geodesically furthest point in the LAA with

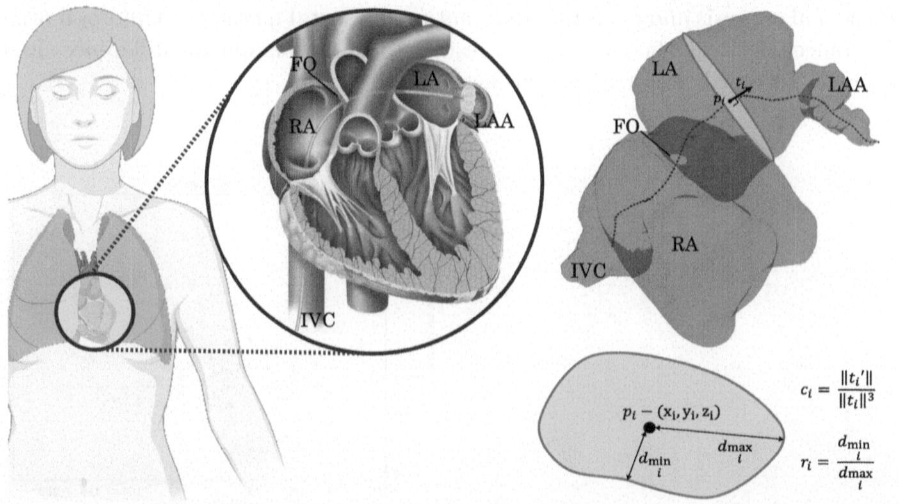

Fig. 2. Left: Illustration of the heart including an exemplified trajectory of the catheter to release the device in the neck of the LAA during a transcatheter LAAO procedure. **Right:** Anatomy and computed centerline extracted from the CT image including a sketch of the cross-sectional plane used to obtain $d_{\min_i}$ and $d_{\min_i}$, together with the equations used to calculate c_i and r_i.

respect to the LA as the endpoint. After extracting the centerline, we trim it so it starts 1 cm inside the IVC. Additionally, we fit a third-degree B-spline to this centerline and sample 100 points, which serve as the nodes of our graph. In Fig. 2, we can observe an example of the final centerline obtained with this method. To further characterize the anatomy and provide descriptive features to the nodes, we compute the tangent t_i for each point p_i. Using the tangent, we can obtain the estimated curvature in each point c_i using the equation shown in Fig. 2. Finally, we also extract the cross-sectional plane for each point p_i defined by the tangent t_i. From this plane, we obtain the shortest distance between the surface and the center point $d_{\min_i}$ and the largest one $d_{\max_i}$, as illustrated in Fig. 2. The last descriptive parameter associated with a point p_i is the ratio between the shortest and the largest distance in this cross-sectional plane, which we define as r_i and for which the computational details are shown in Fig. 2.

Figure 3 displays the 125 different graphs from our population, demonstrating the variability within our dataset. To explore this variability and see if there was any clear correlation between the extracted graphs and the presence of PDLs we project them into the 2D PCA space using the 100×7 features matrix of each graph, defined by the number of points and the corresponding 7 features associated with each of them. These features include the spatial coordinates x_i, y_i and z_i as well as derived features $d_{\min_i}$, $d_{\max_i}$, c_i, and r_i. We can see in Fig. 3 that both distributions are similar with no clusters being formed. After a careful statistical analysis and exploring different fundamental methods to find potential differences between the two groups, results ruled out significant differences and we moved towards more advanced methods.

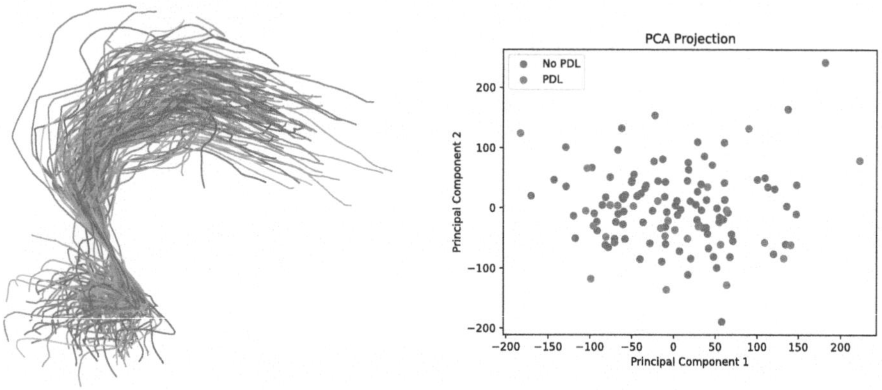

Fig. 3. Left: Display of the 125 computed centerlines. **Right:** PCA distribution of the 125 subjects (color-coded PDL/no PDL) using 7 features for each of the 100 points.

3.2 Graph Attention Network

For each subject, we can construct a structured graph using the method described in the previous section with 100 nodes and 7 features. All graphs are bidirectional and centered with respect to the FO.

GAT [18] is a type of graph neural network model that encodes edges into an attention mechanism and uses eigenvectors and eigenvalues of each node as positional embeddings for local structures. Due to the relatively simple structure of our graphs, we designed a streamlined architecture for graph-level predictions (Fig. 4). This architecture consists of 2 GAT convolutions followed by global mean pooling and a fully connected layer. We utilize multi-head attention with 6 heads that are concatenated.

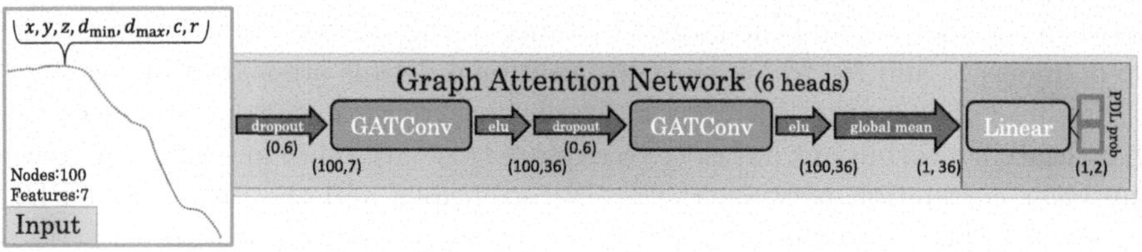

Fig. 4. Sketch of the GAT architecture used to classify the anatomical descriptive graphs of the centerline. From left to right, we present the input, the layers of the GAT architecture, and finally, the fully connected layers that map the output.

One advantage of using a multi-head GAT is its capability to generate explanations based on attention coefficients at the edge level. After processing a graph, we aggregate attention scores across layers and heads by selecting the maximum, generating a new graph where each edge contains its corresponding final attention score.

3.3 Implementation

We split our dataset into 70% for training and 30% for testing, maintaining the distribution of the different classes shown in Fig. 1. To address the class imbalance, we use a weighted version of the negative log-likelihood loss as our loss function.

The code used for generating graphs from the segmentation images, as well as for training, testing, and explaining the GAT model, is available at https://github.com/paulalopez10/GAT-graph_classification, where you can find all the implementation details. We trained the model on an NVIDIA GeForce RTX 4090 with a training time of 6 h. The code is built on PyTorch [12] and Pytorch Geometric [6].

4 Results and Discussion

We evaluate the performance of the GAT model in the graph-level binary classification task of predicting PDL or No PDL. The results, presented as a confusion matrix, are displayed in Table 1.

Table 1. Confusion matrix of the evaluation in the test set including the percentage of the corresponding true label.

		Predicted Label		Total
		No PDL	PDL	
True Label	No PDL	29(100%)	0(0%)	29
	PDL	5(62.5%)	3 (37.5%)	8
	Total	34	3	37

Given the multifactorial nature of the problem we are trying to tackle, we consider the results extremely positive. We observe that the model has been able to identify underlying patterns in the morphology described by our constructed graphs, which might indicate a higher chance of presenting a PDL after LAAO. In other words, the model has found some indication of the complexity of the procedure given a specific anatomy.

To analyze what the model is basing its predictions on, we choose the samples with the predicted highest probability of PDL and No PDL, respectively. Both cases are correctly classified according to our post-operative analysis. In

Fig. 5, we show the attention scores in the corresponding graphs together with an overlay of the given anatomy. We can observe that the model is focusing in the region between the IVC, RA, and FO, where the highest change in attention scores is present. It is shown that this area can exhibit a higher level of curvature compared with other regions of the centerline. This abrupt change is sometimes present if the angle of the IVC and the FO with respect to the LAA deviates significantly from 180°. Given the nature of the procedure, it appears that the high curvature in that area could be linked with more difficult alignment, making the achievement of the desired coaxiality between the device and the LAA more challenging.

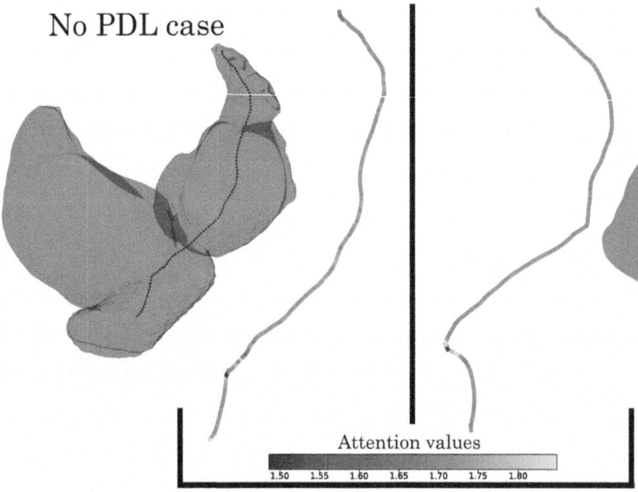

Fig. 5. Display of the attention scores associated with the edges of each graph generated during inference in two samples from the test set with our multi-headed GAT. On the left the case of the test set with the highest probability of No PDL (correctly classified), on the right the case with the highest probability of PDL (correctly classified).

5 Conclusion

We have presented a new method for characterizing the anatomy involved in the transcatheter LAAO procedure. We demonstrate that our standardized and automatic method, which characterizes the anatomy by generating a morphology-aware graph based on the centerline that mimics the catheter trajectory, successfully captures key elements to determine the complexity of the procedure.

We show how a graph attention network can be trained on these structures to predict PDLs, using the attention scores to highlight the regions of our graph that have a greater impact on the algorithm's decision. We observe that the curvature before the FO is the area of highest interest to our model for mapping the probability of a potential PDL. This allows us to observe underlying trends that would otherwise be very difficult to analyze given the high complexity of the studied anatomy and the significant inter-personal morphological variability. This paves the way toward characterizing the difficulty on a patient-specific basis and predicting potential procedure outcomes.

References

1. Afzal, M.R., et al.: Temporal changes and clinical implications of delayed peridevice leak following left atrial appendage closure. JACC: Clin. Electrophysiol. **8**(1), 15–25 (2022). https://doi.org/10.1016/j.jacep.2021.06.018
2. Aguado, A.M., et al.: In silico optimization of left atrial appendage occluder implantation using interactive and modeling tools. Front. Physiol. **10** (2019). https://doi.org/10.3389/fphys.2019.00237
3. Alkhouli, M., et al.: Peridevice leak after left atrial appendage occlusion: incidence, mechanisms, clinical impact, and management (2023). https://doi.org/10.1016/j.jcin.2022.12.006
4. Antiga, L., Piccinelli, M., Botti, L., Ene-Iordache, B., Remuzzi, A., Steinman, D.A.: An image-based modeling framework for patient-specific computational hemodynamics (2008). https://doi.org/10.1007/s11517-008-0420-1
5. Ferez, X.M., et al.: Deep learning framework for real-time estimation of in-silico thrombotic risk indices in the left atrial appendage. Front. Physiol. **12** (2021). https://doi.org/10.3389/fphys.2021.694945
6. Fey, M., Lenssen, J.E.: Fast graph representation learning with pytorch geometric (2019)
7. Korsholm, K., et al.: Peridevice leak following amplatzer left atrial appendage occlusion: cardiac computed tomography classification and clinical outcomes. JACC: Cardiovasc. Intervent. **14**(1), 83–93 (2021). https://doi.org/10.1016/j.jcin.2020.10.034
8. Lakkireddy, D., et al.: Mechanisms, predictors, and evolution of severe peri-device leaks with two different left atrial appendage occluders. EP Eur. **25**(9), euad237 (2023). https://doi.org/10.1093/europace/euad237
9. Mill, J., et al.: Domain expert evaluation of advanced visual computing solutions for the planning of left atrial appendage occluder interventions (2022). https://doi.org/10.1101/2022.04.11.22273553
10. Nerrise, F., Zhao, Q., Poston, K.L., Pohl, K.M., Adeli, E.: An explainable geometric-weighted graph attention network for identifying functional networks associated with gait impairment. In: International Conference on Medical Image Computing and Computer-Assisted Intervention, pp. 723–733. Springer (2023)
11. Pastormerlo, L.E., De Caterina, A.R., Esposito, A., Korsholm, K., Berti, S.: State-of-the-art of transcatheter left atrial appendage occlusion. J. Clin. Med. **13**(4) (2024). https://doi.org/10.3390/jcm13040939
12. Paszke, A., et al.: Pytorch: an imperative style, high-performance deep learning library (2019)
13. Saiz-Vivó, M., et al.: Unsupervised machine learning exploration of morphological and haemodynamic indices to predict thrombus formation in the left atrial appendage. In: Camara, O., et al. (eds.) STACOM 2022. LNCS, vol. 13593, pp. 200–210. Springer, Cham (2022). https://doi.org/10.1007/978-3-031-23443-9_19

14. Saw, J., et al.: Cardiac CT angiography for device surveillance after endovascular left atrial appendage closure. Eur. Heart J. - Cardiovasc. Imaging **16**(11), 1198–1206 (2015). https://doi.org/10.1093/ehjci/jev067
15. Smine, Z., et al.: Global sensitivity analysis of thrombus formation in the left atrial appendage of atrial fibrillation patients. In: Camara, O., et al. (eds.) STACOM 2023. LNCS, vol. 14507, pp. 55–65. Springer, Cham (2024). https://doi.org/10.1007/978-3-031-52448-6_6
16. Sørensen, K., Camara, O., Backer, O.d., Kofoed, K.F., Paulsen, R.R.: NUDF: neural unsigned distance fields for high resolution 3D medical image segmentation. In: 2022 IEEE 19th International Symposium on Biomedical Imaging (ISBI), pp. 1–5 (2022). https://doi.org/10.1109/ISBI52829.2022.9761610
17. Tzikas, A., et al.: Left atrial appendage occlusion with the amplatzer amulet device: an expert consensus step-by-step approach. EuroIntervention **11**(13), 1512–1521 (2016). https://doi.org/10.4244/EIJV11I13A292
18. Veličković, P., Cucurull, G., Casanova, A., Romero, A., Liò, P., Bengio, Y.: Graph attention networks (2018)
19. Wasserthal, J., et al.: TotalSegmentator: robust segmentation of 104 anatomic structures in CT images. Radiology: Artif. Intell. **5**(5) (2023)
20. Wunderlich, N.C., Beigel, R., Swaans, M.J., Ho, S.Y., Siegel, R.J.: Percutaneous interventions for left atrial appendage exclusion: options, assessment, and imaging using 2D and 3D echocardiography. JACC: Cardiovasc. Imaging **8**(4), 472–488 (2015). https://doi.org/10.1016/j.jcmg.2015.02.002

Improved 3D Whole Heart Geometry from Sparse CMR Slices

Yiyang Xu[1(✉)], Hao Xu[1,4], Matthew Sinclair[2,5], Esther Puyol-Antón[1,5], Steven A. Niederer[2], Amedeo Chiribiri[1], Steven E. Williams[1,3], Michelle C. Williams[3], and Alistair A. Young[1]

[1] School of Biomedical Engineering and Imaging Sciences, King's College London, London, UK
yiyang.1.xu@kcl.ac.uk
[2] Imperial College London, London, UK
[3] University/BHF Centre for Cardiovascular Science, University of Edinburgh, Edinburgh, UK
[4] College of Mathematical Medicine, Zhejiang Normal University, Jinhua, Zhejiang, China
[5] HeartFlow, Inc., Mountain View, USA

Abstract. Cardiac magnetic resonance (CMR) imaging and computed tomography (CT) are two common non-invasive imaging methods for assessing patients with cardiovascular disease. CMR typically acquires multiple sparse 2D slices, with unavoidable respiratory motion artefacts between slices, whereas CT acquires isotropic dense data but uses ionising radiation. In this study, we explore the combination of Slice Shifting Algorithm (SSA), Spatial Transformer Network (STN), and Label Transformer Network (LTN) to: 1) correct respiratory motion between segmented slices, and 2) transform sparse segmentation data into dense segmentation. All combinations were validated using synthetic motion-corrupted CMR slice segmentation generated from CT in 1699 cases, where the dense CT serves as the ground truth. In 199 testing cases, SSA-LTN achieved the best results for Dice score and Hausdorff distance (94.0% and 4.7 mm respectively, average over 5 labels) but gave topological errors in 8 cases. STN was effective as a plug-in tool for correcting all topological errors with minimal impact on overall performance (93.5% and 5.0 mm respectively). SSA also proves to be a valuable plug-in tool, enhancing performance over both STN-based and LTN-based models. The code for these different combinations is available at https://github.com/XESchong/STACOM2024.

Keywords: Motion Correction · Geometry Reconstruction · Cardiac Magnetic Resonance · Image Registration

1 Introduction

Accurate cardiac 3D geometry is required for evaluation of cardiovascular disease, the leading cause of death worldwide [1,19]. Reconstruction of accurate cardiac 3D geometry enables more precise and accurate evaluation of disease than

O. Camara et al. (Eds.): STACOM 2024, LNCS 15448, pp. 43–52, 2025.
https://doi.org/10.1007/978-3-031-87756-8_5

current clinical metrics [9,11]. Two common non-invasive imaging techniques for determining treatment are coronary computed tomography (CT) angiography [14] and cardiac magnetic resonance (CMR) imaging [13]. CT captures dense geometry information at high isotropic resolution (0.5 mm) but requires ionizing radiation. CMR is non-ionizing but is typically performed in sparse 2D slices, often in short axis (10 mm apart) and long axis orientations, requiring multiple breath-holds and introducing motion artefacts between slices [10] due to differences in patient breath-hold position. Two promising methods for 3D geometry reconstruction are label transformer networks (LTN) for shape reconstruction from sparse views [20] and spatial transformer networks (STN) for atlas registration [16]. However, an extensive evaluation of the performance of these methods, separately and combined, has not been performed. Here, we extensively evaluate the combination of slice shifting algorithm (SSA) for motion correction, LTN and STN modules for reconstruction of 3D geometry from CMR slice data. For the STN module, we evaluate two methods, one which registers an atlas to dense segmentation data (DSTN) and the other registers the atlas to sparse segmentation data (SSTN). We generate simulated CMR sparse segmentation data from a large CT cohort [14] so that all methods can be compared using the same ground truth.

Related Work. Methods to correct motion artefact in CMR data include Chandler et al. [3], a slice-to-volume registration approach employing rigid transformations only. Another approach proposed by Sinclair et al. [15] utilizes a segmentation-based method, where segmentations are iteratively registered between slices to determine the in-plane shift. Additionally, there is a growing interest in leveraging artificial intelligence (AI) models to either explicitly or implicitly eliminate motion artefacts across various organs [5,17]. Geometry reconstruction involves transforming a sparse label map into a dense label map [20]. Wang et al. [18] proposed an iterative latent optimisation network that simultaneously addresses super resolution and motion correction tasks for cardiac segmentation results, which demonstrates good performance over two CMR datasets. Besides that, Chen et al. [4] presented MR-Net, a mesh construction method relying on sparse information from point clouds. Furthermore, Kong et al. [7] incorporated a template mesh into a deep learning-based model for whole heart mesh reconstruction, leveraging prior knowledge of the surface topology.

Contributions. Our main contributions can be summarized as follows:

- We examine six combinations of SSA, LTN and STN modules to provide accurate and topologically correct reconstructions.
- We show that STN can serve as a plug-in module for addressing topological errors with minimal impact on overall performance.
- We provide an efficient implementation of SSA using *PyTorch* and show enhanced performance when combined with both SSTN-based and LTN-based models.

- We show that LTN-DSTN demonstrated increased robustness in scenarios where 3-chamber long axis information is missing, a common occurrence in real CMR data acquisition.

2 Methodology

(See Fig. 1).

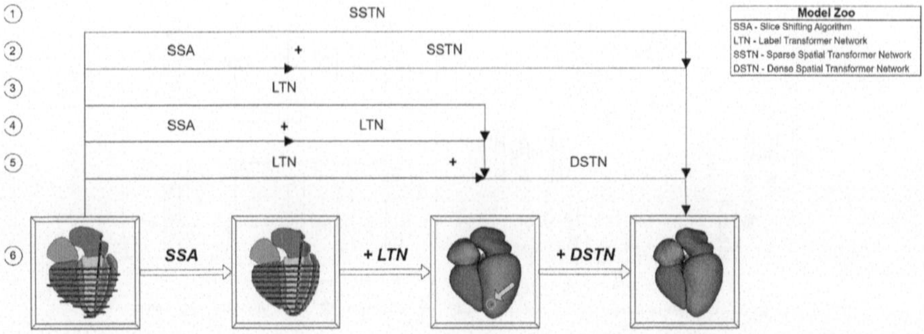

Fig. 1. Overview of experiments: six combinations of SSA, LTN and STN (sparse and dense versions) were examined. The yellow circle denotes a topological error from LTN, corrected by STN (zoom in for better view). (Color figure online)

Slice Shifting Algorithm (SSA). For the motion correction task, we adapted an iterative registration algorithm [15]. During each iteration, the label map undergoes adjustments aimed at optimizing its alignment with the overlapping regions of other slices. As in [15], adjustments are conducted solely with in-plane 2D translation. We also provide code for an efficient implementation of this method in *PyTorch*.

Spatial Transformer Network (STN). Derived from the Atlas-ISTN network [16], our approach involves the application of the spatial transformer network (STN) and transformation computation module as the core model for addressing the geometry reconstruction task in CMR data. The primary network architecture is similar to the registration component of the Atlas-ISTN framework [16]. However, the input label map now could be a sparse data volume obtained from the stack of short axis (SAX) and long axis (LAX) views, which is denoted as $\hat{y}_i$. Another input for STN is a dense atlas label map y^a, which can be derived from various sources: 1) mean of training data 2) sample data from training dataset 3) publicly available cardiac atlas. Mathematically, it can be represented as:

$$\{v_i, T_i\} = \boldsymbol{D}_{\theta_d}(\hat{y}_i, y^a) \tag{1}$$

where θ_d are the parameters for the STN. After passing through the transformation computation module $\boldsymbol{C}$, the affine transformation T_i and the stationary velocity field (SVF) v_i undergo additional processing to generate the final deformation fields denoted as Φ_i and Φ_i^{-1}. Specifically, Φ_i is the forward transformation from patient space to atlas space and Φ_i^{-1} is the inverse transformation from atlas space to the patient space. Finally, the forward pass of the STN framework with the input sparse volume $\hat{y}_i$ and atlas label map y^a is given as (Fig. 2):

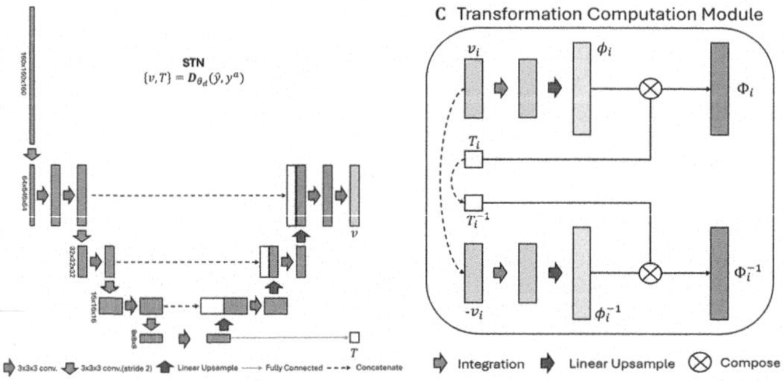

Fig. 2. Modified STN architecture (left) and Transformation Computation Module (right) [16]

$$\{\Phi_i, \Phi_i^{-1}\} = \boldsymbol{C}(\boldsymbol{D}_{\theta_d}(\hat{y}_i, y^a)) \tag{2}$$

where i is the index for the sample in the patient space. Given the dense ground truth label map volume y_i^{dense} for the CMR data, the STN continues to utilize the previous atlas-to-segmentation loss L_{a2s} and segmentation-to-atlas loss L_{s2a}, denoted as L_{a2s}^{dense} and L_{s2a}^{dense}in the context of CMR data.

$$L_{a2s}^{dense} = \sum_i \sum_{j=2}^{c} ||y_{i,j}^{dense} - y_j^a \circ \Phi_i^{-1}||^2 \tag{3}$$

$$L_{s2a}^{dense} = \sum_i \sum_{j=2}^{c} ||y_j^a - y_{i,j}^{dense} \circ \Phi_i||^2 \tag{4}$$

where j is the index for each channel of CMR data and c is the total number of channels. Additionally, a regularisation term is introduced to control the smoothness of the predicted non-rigid displacement fields u_i, i.e.,

$$L_{reg} = \sum_i ||\Delta u_i||^2 \tag{5}$$

$$\phi = I + u \tag{6}$$

where ϕ in the transformation computation module is represented as a combination of two components, i.e., identity transformation I and non-rigid displacement field u. Hence, the total loss is defined as:

$$L = L_{a2s} + L_{s2a} + \lambda L_{reg} \tag{7}$$

where λ is the weight for the regularisation term. A larger value of the hyperparameter λ imposes a greater penalty on the non-rigid displacement field u, which tends to encourage the learning of a more rigid transformation within the STN framework. Based on the input for the training of STN, we denoted SSTN for the sparse volume input and DSTN for the dense volume input respectively.

Label Transformer Network (LTN). Drawing inspiration from the LC-U-Net [20], we developed a 3D U-Net to perform the geometry reconstruction task, referred to as LTN in the subsequent sections. LTN is intended to learn the mapping between the sparse volume extracted from CMR data and their corresponding dense ground truth. During training, the mean square error (MSE) loss is employed to compute the total training loss, which compares the predictions y' from LTN with the dense ground truth y. Mathematically, it is computed as:

$$L = \sum_i \sum_j ||y'_{i,j} - y_{i,j}||^2 \tag{8}$$

where i is the index for CMR data and j is the index for each label of CMR data.

3 Experiments

Synthetic CMR Dataset. Based on the segmentation of SCOT-HEART CT data [14,21], a cardiac coordinate system is established by identifying specific anatomical landmarks, such as the mitral valve, tricuspid valve and apex [20]. After this, we define the location of the planes corresponding to the short and long axis for the synthetic CMR data. The CT data, originally in the patient coordinate space, is then resampled into the cardiac coordinate space. Using the location information of each plane, we generate multiple synthetic CMR slices from CT data. These synthetic slices include random shifts for slice positions to replicate CMR slice planning errors. Furthermore, we simulate respiratory motion artefacts by implementing a uniform random distribution for in-plane shifts. The main benefit of using synthetic CMR data is that we have paired motion-corrupted sparse data and dense ground truth volume for supervised-style training. Finally, the split of training, validation, and testing datasets in the synthetic CMR dataset is 1400, 100 and 199 cases, respectively.

Implementation Details. For SSA, each iteration begins by processing the long axis (LAX) slice of CMR data, with a maximum of 5 iterations in total. Usually, convergence is reached after approximately 3 iterations with our synthetic

CMR data. SSA is implemented in *PyTorch*, achieving a processing speed five times faster than the original implementation [15]. For the STN, the input size is [160, 160, 160], comprising one background channel and five foreground labels: left ventricle myocardium (LVM), left ventricle blood pool (LV), left atrium (LA), right ventricle blood pool (RV) and right atrium (RA). To check the topological errors, the mesh for each channel is generated by using the Marching Cubes algorithm [8]. The learning rate and lambda values are set to 0.001 and 2000, respectively. The initial atlas map is selected from an example case in the training dataset. Unlike Atlas-ISTN, the update procedure for the atlas label map is disabled during training. For the upcoming discussion, the model zoo is primarily categorized into two main groups: SSTN-based and LTN-based. Specifically, this includes SSA-SSTN and SSTN under the first group, and LTN, LTN-DSTN, SSA-LTN and SSA-LTN-DSTN under LTN-based models. All trainable models are trained for 50 epochs using the *Nvidia RTX A500* GPU.

Evaluation Metrics. We utilize the Dice score [2] and Hausdorff distance (HD) [6] to assess the similarity between the predicted label map and the ground truth. To quantify predictions in terms of topological accuracy, we employ the Euler characteristic to assess whether the predicted structure adheres to the topological constraints of heart. Specifically, the predicted mesh of each channel should satisfy the following formula to be considered topologically plausible [12]:

$$V - E + F = 2 \tag{9}$$

where V, E, and F denote the number of vertices, edges, and faces of the predicted mesh, respectively.

Results. In Table 1, we present the performance for the geometry reconstruction task and motion correction task across various models. A comparison between SSTN and SSA-SSTN reveals that SSA-SSTN consistently achieves a higher Dice score and lower Hausdorff distance across all foreground channels. This underscores the efficacy of separately addressing motion correction and geometry reconstruction tasks, as opposed to concurrently addressing both tasks, where SSA is mainly for the motion correction task and SSTN is primarily for the geometry reconstruction task. We can also observe the advantage of addressing two tasks separately by examining the LTN-based model. This also indicates that SSA can serve as a versatile tool for enhancing the capabilities of any models. Furthermore, by integrating a pre-trained registration-based model (DSTN) after the dense volume prediction from LTN, we observed a slight performance decrease compared to using LTN alone. Although statistically significant, we believe that a drop of <1% Dice and <1 mm Hausdorff distance will have minimal impact on resulting 3D shape statistics. However, DSTN proved beneficial in reducing cases of topological errors which has a big impact on computational shape analyses. Initially, 8 cases in the entire testing dataset exhibited topological errors based on the Euler characteristic for predictions from LTN. After

Table 1. Performance of various models across different foreground channels of synthetic CMR dataset: each entry is in *mean* ± *std* format, where the best model is highlighted in bold. † denotes $p < 0.001$ for the difference of the Dice score between SSA-SSTN and SSTN. ‡ denotes $p < 0.001$ for the difference of the Dice score between current model and SSA-LTN.

Models	Label					
	Metric	LVM	LV	RV	RA	LA
① SSTN	Dice (%) ↑	85.58 ± 3.60	93.18 ± 1.66	91.31 ± 2.68	87.10 ± 4.43	86.91 ± 4.18
	HD (*mm*) ↓	5.94 ± 1.79	5.31 ± 1.26	7.72 ± 2.90	8.36 ± 2.91	7.84 ± 2.14
② SSA-SSTN†	Dice (%) ↑	89.64 ± 3.98	94.87 ± 1.80	94.02 ± 2.63	88.97 ± 3.90	90.14 ± 3.32
	HD (*mm*) ↓	5.08 ± 1.89	4.65 ± 1.45	6.45 ± 2.90	8.13 ± 3.15	6.96 ± 2.25
③ LTN‡	Dice (%) ↑	90.10 ± 3.33	95.35 ± 1.39	94.85 ± 1.98	91.10 ± 2.87	92.09 ± 2.26
	HD (*mm*) ↓	4.59 ± 4.08	3.97 ± 1.11	4.84 ± 5.30	7.74 ± 8.58	5.71 ± 1.66
④ **SSA-LTN**	Dice (%) ↑	92.00 ± 4.0	96.25 ± 1.70	96.09 ± 2.19	92.13 ± 2.52	93.39 ± 2.02
	HD (*mm*) ↓	3.78 ± 2.21	3.61 ± 3.05	4.34 ± 7.05	6.16 ± 2.27	5.53 ± 2.00
⑤ LTN-DSTN‡	Dice (%) ↑	89.60 ± 3.36	95.06 ± 1.38	94.62 ± 1.92	91.00 ± 2.81	91.75 ± 2.39
	HD (*mm*) ↓	4.59 ± 1.56	4.25 ± 1.02	5.33 ± 2.23	6.55 ± 2.55	5.91 ± 2.16
⑥ SSA-LTN-DSTN‡	Dice (%) ↑	91.29 ± 3.94	95.87 ± 1.68	95.65 ± 2.16	91.75 ± 2.58	92.92 ± 2.07
	HD (*mm*) ↓	4.26 ± 1.71	3.85 ± 1.13	4.79 ± 2.06	6.36 ± 2.57	5.65 ± 1.97

applying DSTN, all these topological error cases were successfully rectified. Comparing the best models from SSTN-based and LTN-based approaches, SSA-LTN achieved the highest performance based on the Dice score and Hausdorff distance across all channels. Notably, the LTN-based model demonstrated superior performance in the atrium regions (LA and RA).

Based on the visual results shown in Fig. 3, the first row demonstrates that SSA-STN's predictions exhibit finer details compared to those produced by SSA-LTN. However, SSA-LTN's predictions closely resemble the ground truth shape of the testing data, presenting a smoother surface. Importantly, the SSA component effectively reduces motion artefacts in synthetic CMR data. In the second row of Fig. 3, a topological error is evident in SSA-LTN's prediction, specifically in the LA part. In contrast, SSA-LTN-DSTN, which incorporates atlas information during inference, successfully corrects the topological error and ensures topological plausibility.

Furthermore, to assess the models' generalisation ability, we modified the testing dataset by reducing the number of slices. Initially, we omitted one long axis (LAX) slice, specifically the 3-chamber view. Subsequently, we removed another LAX slice, the 2-chamber view, leaving only a single 4-chamber LAX slice in the testing data. We retained the 4-chamber view as it provides crucial information about the right atrium. We denoted the testing data obtained by the above procedures as SAX&2/3/4, SAX&2/4, and SAX&4, respectively. Evaluation from Table 2 indicates a consistent decline in the performance of SSTN and LTN-DSTN as the amount of information in the testing data decreases. However, when comparing the performance of each model between the first row and second row of Table 2, the decline in the performance of LTN-DSTN was less pronounced compared to SSTN across most channels. This trend was not

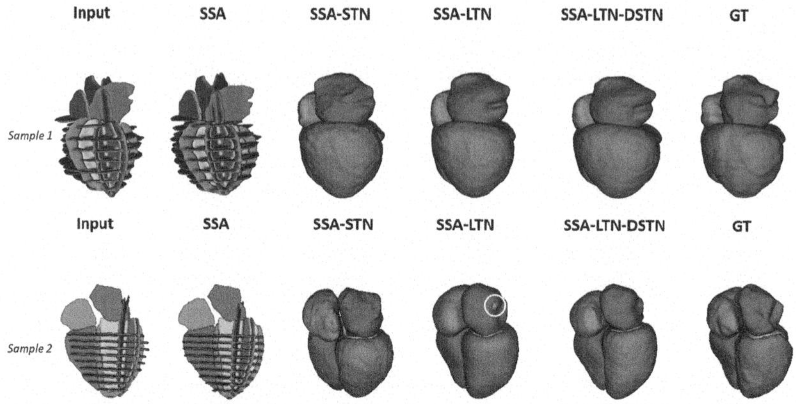

Fig. 3. Visualisation of the performance for SSA-STN, SSA-LTN and SSA-LTN-DSTN, where GT is ground truth dense volume from CT data.

Table 2. Robustness assessment between SSTN and LTN-DSTN, where Δ is the difference between the current row and the first row for a specific channel and bolded values signify a less substantial decrease.

Labels		LVM		LV		RV		RA		LA	
Model	Testing dataset	Dice	Δ	Dice	Δ	Dice	Δ	Dice	Δ	Dice	Δ
SSTN	SAX&2/3/4	85.58		93.18		91.31		87.10		86.91	
	SAX&2/4	83.87	1.71	92.11	1.07	90.80	0.51	85.42	1.68	83.77	**3.14**
	SAX & 4	80.87	**4.71**	90.59	2.59	90.10	**1.21**	83.84	3.26	76.67	**10.24**
LTN-DSTN	SAX&2/3/4	89.60		95.06		94.62		91.00		91.75	
	SAX&2/4	88.78	**0.82**	94.73	**0.33**	94.11	0.51	89.76	**1.24**	88.50	3.25
	SAX & 4	83.94	5.66	92.76	**2.30**	93.06	1.56	88.25	**2.75**	79.35	12.40

observed when considering the testing data containing only the sole 4-chamber LAX slice, where LTN-DSTN showed a notable decrease in performance for the LA channel.

4 Conclusion and Discussion

We present an extensive evaluation of the combination of SSA, LTN and STN modules for the first time. Rather than simultaneously addressing motion correction and geometry reconstruction tasks, as in SSTN or LTN separately, our findings indicate that a sequential modular approach yields superior performance. SSA-LTN provides the best Dice score, but also results in topological errors. An important advantage of STN is its ability to ensure the topological plausibility of the predicted heart shape, achieved through the incorporation of an atlas label map during both training and testing phases. Moreover, it can be seamlessly integrated as a plug-in tool for any models. As observed in the LTN-based model, it effectively corrected topological errors with minimal impact on overall performance. Furthermore, LTN-DSTN demonstrates increased robustness in scenarios where 3-chamber LAX information is missing, a common occurrence in real CMR data acquisition. Additionally, SSA serves as a valuable plug-in tool, which is validated for enhanced performance over both the SSTN-based and LTN-based models. Limitations of our work include reliance on segmentations, which poses challenges if the segmentation process yields inaccurate results. Moreover, as our motion correction task only accounts for in-plane shifting, it may alter the position of the heart within the 3D patient coordinate space after passing through SSA. In future endeavours, we plan to explore the integration of both rigid and non-rigid shifting techniques to effectively handle motion correction tasks. Also, we are using an example case as the atlas for STN training and testing. In the future, we could explore and apply different sources of atlas. Additionally, we aim to apply the combined modules to real-world datasets for further validation and practical implementation.

Acknowledgments. YX and AAY acknowledge funding from the EPSRC Centre for Doctoral Training in Smart Medical Imaging (EP/S022104/1) and the Wellcome/EPSRC Centre for Medical Engineering (WT203148/Z/16/Z). YX acknowledges stipend funding contribution from Heartflow Inc., CA, USA. MS and EP are employees from Heartflow Inc., CA, USA. SCOT-HEART was funded by The Chief Scientist Office of the Scottish Government Health and Social Care Directorates (CZH/4/588), with supplementary awards from Edinburgh and Lothian's Health Foundation Trust and the Heart Diseases Research Fund. MCW was supported by the British Heart Foundation FS/ICRF/20/26002 and CH/09/002. SEW is supported by the British Heart Foundation (FS/20/26/34952).

References

1. Barton, F., Saibh, A.: Death registration summary statistics, England and Wales (2022)
2. Bertels, J., et al.: Optimizing the dice score and jaccard index for medical image segmentation: theory and practice. In: Shen, D., et al. (eds.) MICCAI 2019. LNCS, vol. 11765, pp. 92–100. Springer, Cham (2019). https://doi.org/10.1007/978-3-030-32245-8_11
3. Chandler, A., et al.: Correction of misaligned slices in multi-slice MR cardiac examinations by using slice-to-volume registration. In: 3rd IEEE International Symposium on Biomedical Imaging: Nano to Macro, pp. 474–477 (2006). https://doi.org/10.1109/ISBI.2006.1624956
4. Chen, X., et al.: Shape registration with learned deformations for 3D shape reconstruction from sparse and incomplete point clouds. Med. Image Anal. **74**, 102228 (2021). https://doi.org/10.1016/j.media.2021.102228
5. Duffy, B.A., et al.: Retrospective motion artifact correction of structural MRI images using deep learning improves the quality of cortical surface reconstructions. Neuroimage **230**, 117756 (2021). https://doi.org/10.1016/j.neuroimage.2021.117756

6. Huttenlocher, D., Klanderman, G., Rucklidge, W.: Comparing images using the Hausdorff distance. IEEE Trans. Pattern Anal. Mach. Intell. **15**(9), 850–863 (1993). https://doi.org/10.1109/34.232073
7. Kong, F., Wilson, N., Shadden, S.: A deep-learning approach for direct whole-heart mesh reconstruction. Med. Image Anal. **74**, 102222 (2021). https://doi.org/10.1016/j.media.2021.102222
8. Lorensen, W.E., Cline, H.E.: Marching cubes: a high resolution 3D surface construction algorithm. SIGGRAPH Comput. Graph. **21**(4), 163–169 (1987). https://doi.org/10.1145/37402.37422
9. Mauger, C.A., et al.: Multi-Ethnic study of atherosclerosis: Relationship between left ventricular shape at cardiac MRI and 10-year outcomes. Radiology **306**(2), e220122 (2023)
10. McLeish, K., Hill, D., Atkinson, D., Blackall, J.M., Razavi, R.: A study of the motion and deformation of the heart due to respiration. IEEE Trans. Med. Imaging **21**(9), 1142–1150 (2002)
11. Muffoletto, M., et al.: Evaluation of deep learning estimation of whole heart anatomy from automated cardiovascular magnetic resonance short- and long-axis analyses in UK Biobank. Eur. Heart J. - Cardiovasc. Imaging jeae123 (2024). https://doi.org/10.1093/ehjci/jeae123
12. Richeson, D.: Euler's Gem: The Polyhedron Formula and the Birth of Topology. Princeton University Press (2008)
13. Schulz-Menger, J., et al.: Standardized image interpretation and post-processing in cardiovascular magnetic resonance - 2020 update: society for Cardiovascular Magnetic Resonance (SCMR): Board of Trustees Task Force on Standardized Post-Processing. J. Cardiovasc. Magn. Reson. **22**(1), 19 (2020)
14. SCOT-HEART Investigators: Coronary CT angiography and 5-year risk of myocardial infarction. New England J. Med. **379**(10), 924–933 (2018). https://doi.org/10.1056/NEJMoa1805971
15. Sinclair, M., Bai, W., Puyol-Antón, E., Oktay, O., Rueckert, D., King, A.P.: Fully automated segmentation-based respiratory motion correction of multiplanar cardiac magnetic resonance images for large-scale datasets. In: Descoteaux, M., Maier-Hein, L., Franz, A., Jannin, P., Collins, D.L., Duchesne, S. (eds.) Medical Image Computing and Computer-Assisted Intervention (MICCAI 2017). Springer, Cham (2017)
16. Sinclair, M., et al.: Atlas-ISTN: joint segmentation, registration and atlas construction with image-and-spatial transformer networks (2020)
17. Tamada, D.: Review: noise and artifact reduction for MRI using deep learning (2020). https://doi.org/10.48550/arXiv.2002.12889
18. Wang, S., et al.: Joint motion correction and super resolution for cardiac segmentation via latent optimisation (2021)
19. World Health Organization: The top 10 causes of death (2020). https://www.who.int/news-room/fact-sheets/detail/the-top-10-causes-of-death
20. Xu, H., Muffoletto, M., Niederer, S.A., Williams, S.E., Williams, M.C., Young, A.A.: Whole heart 3D shape reconstruction from sparse views: leveraging cardiac computed tomography for cardiovascular magnetic resonance. In: Bernard, O., Clarysse, P., Duchateau, N., Ohayon, J., Viallon, M. (eds.) Functional Imaging and Modeling of the Heart, pp. 255–264. Springer, Cham (2023)
21. Xu, H., Niederer, S.A., Williams, S.E., Newby, D.E., Williams, M.C., Young, A.A.: Whole heart anatomical refinement from CCTA using extrapolation and parcellation (2021)

CavityBASNet: Cavity-Focused Biatrial Automatic Segmentation on LGE-MRI with Augmented Input Channel and Left-Right Myocardium Splitting

Susana Merino-Caviedes(✉), Emmanuel Chibuikem Nnadozie, Elisa Moya-Sáez, Marcos Martín-Fernández, and Carlos Alberola-López

Laboratorio de Procesado de Imagen, ETSI Telecomunicación, Universidad de Valladolid, Valladolid, Spain
susana.merino@uva.es

Abstract. Late Gadolinium-Enhanced (LGE) Magnetic Resonance imaging (MRI) modality is of increasing relevance in clinical studies with atrial fibrillation patients due to its ability to detect scarring and fibrosis. However, segmentation of the atrial wall and cavities is a challenging task with respect to other heart chambers, especially on LGE-MRI images. The nnU-Net framework remains one of the top performing alternatives for medical image segmentation, whereas YOLOv8 provides fast identification of objects. In the context of the MBAS 2024 challenge, we employ a cascade of both models to segment the left and right atrial cavities and the biatrial myocardium. The spatial location of the biatrial region is identified using a YOLOv8 model and then the segmentation on the ROI is computed using a nnU-Net with residual connections. To provide robustness against magnetic field inhomogeneities, we add the logarithm of the LGE-MRI image as an extra input channel to the network, as a compromise between applying a time-consuming N4 bias correction algorithm and not correcting the images. Also, since there is a risk of the left and right myocardium merging in regions where they are very close together, the gold standard myocardium is split at training into left and right myocardium using an automatic method based on the Laplace equation (without any manual intervention). Both myocardial labels are converted into a single region at the output of the second network.

Keywords: MBAS24 challenge · nnU-Net · YOLOv8 · Biatrial segmentation

1 Introduction

Atrial fibrillation (AF) is the most frequently encountered cardiac arrhythmia in clinical practice [2]. The clinical implications of AF includes an increased risk of heart failure, stroke, and mortality [10]. In terms of treatment options, catheter

O. Camara et al. (Eds.): STACOM 2024, LNCS 15448, pp. 53–63, 2025.
https://doi.org/10.1007/978-3-031-87756-8_6

ablation of AF has shown more favorable outcomes compared to other methods [3]. In addition, several studies reported that atrial fibrosis is one of the main factors determining arrhythmia recurrence after catheter ablation [1,9,10]. Therefore, evaluating atrial fibrosis in AF patients is of increasing relevance for planning ablation strategies. In this context, late gadolinium-enhanced (LGE) magnetic resonance imaging (MRI) has recently been demonstrated to be a highly effective image modality for assessing the extent of atrial fibrosis and scarring [3].

Manual segmentations of the left and right atria as well as their walls require skilled manpower and are susceptible to subjectivity; thus, automating this process is of utmost importance. It remains a challenging task, however, due to the significant variations in atrial shape, intensity range, and the generally poor image quality [8].

YOLOv8 is a state-of-the-art single stage object deep learning model for detection, classification, and segmentation tasks [7]. At the time of the writing, it is one the latest of the YOLO series detection models, which in contrast to the use of sliding windows by traditional two-stage detectors, predict bounding boxes from the input image in one evaluation [12,14]. YOLOv8 has five versions at the time of writing - YOLOv8n, YOLOv8s, YOLOv8m, YOLOv8l and YOLOv8x. The depth and width of the networks increase from YOLOv8n to YOLOv8x. Also, increase in network size, complexity and accuracy follows the increase in width and depth, while the inference speed changes in a reverse fashion. To get a good tradeoff between speed, accuracy, and model size, the YOLOv8s was used in this work.

The nnU-Net framework was proposed for semantic segmentation of medical images and has achieved top performance on a wide variety of medical images [5]. It uses the U-Net architecture [13] at its core, but is able to automatically configure its hyperparameters after extracting a fingerprint of the training dataset, which are called plans. nnU-Net is able to provide plans for 2D and 3D U-Net architectures and, more recently, for Residual Encoder U-Net architectures [6]. However, its inference time is long compared to other models.

This paper proposes CavityBASNet, a cascade of a YOLOv8 model and an nnU-Net model with augmented input channels and left-right myocardium splitting. With the YOLO model, the biatrial region is detected and a bounding box to crop the image is provided to the second stage of the cascade, an nnU-Net model. The motivation for the augmented input channels is to provide robustness against non-uniform intensity inhomogeneity of MRI images without needing to resort to time-consuming bias correction algorithms such as [15]. The left-right myocardium splitting intends to help the model to not merge left and right atrial myocardium in locations where they are very close together.

The rest of the article is structured as follows. Section 2 describes the MBAS24 challenge and datasets, as well as the methodology proposed to segment the images. Section 3 contains the experimental results performed with the models and in Sect. 4 the conclusions of this work are found.

2 Materials and Methods

2.1 MBAS24 Challenge

For the MBAS dataset, data were collected across multiple centers using a uniform protocol. All training and validation images are either $640 \times 640 \times 44$ or $576 \times 576 \times 44$ resolution in scanner XYZ coordinates, with voxel spacings of $0.625 \times 0.625 \times 2.5$ mm^3. The overall dataset consists of 200 cases split into three subsets for training, validation and test. The datasets provided to the participants consisted in a training set including 70 3D LGE-MRI images and their associated expert segmentations. Each segmentation was provided as a label map with labels for the myocardium of both atria (MYO), the right atrial cavity (RA) and the left atrial cavity (LA). In addition to the training set, participants were also provided with a validation set of 30 3D LGE-MRI images whose segmentations were not provided to the participants. An additional test set of 100 cases was reserved by the organizers to elaborate the final ranking of the challenge and neither images nor segmentations were seen by the participants. Participants needed to submit a fully automatic method that fulfilled the following requirements: the method had to be trained exclusively with the training dataset and no additional manual annotations, training data or pre-trained models were allowed. Also, inference time for the test dataset should not be more than 10 min: that is, an average of 6 s per case.

2.2 CavityBASNet Framework

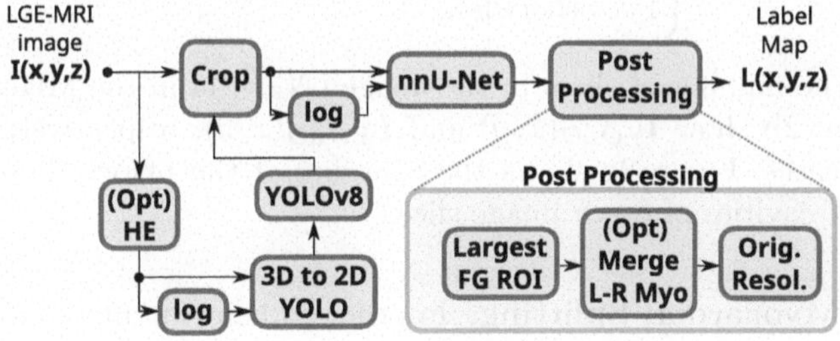

Fig. 1. General pipeline of the proposed CavityBASNet.

Figure 1 shows the pipeline of the proposed model. The first step is cropping every slice of the LGE-MRI image with a bounding box provided by the YOLOv8 model for the XY coordinates, which is padded in both in-plane axes and made square. Then, a channel augmentation is applied to the cropped image as explained in Sect. 2.3 and fed to an nnU-Net model, which outputs a segmentation which is postprocessed so that only the largest foreground region of interest (ROI) is kept and the segmentation is brought back to the LGE-MRI image resolution.

2.3 Data Preprocessing

Channel Augmentation. Let $I(\mathbf{x})$ be the 3D LGE-MRI image brightness and $L_{gt}(\mathbf{x})$ its associated expert segmentation. Histogram equalization may be applied to $I(\mathbf{x})$ using 1024 bins; hereafter, we refer to the image as $I_{heq}(\mathbf{x})$ whether histogram equalization was applied or not. $I_{heq}(\mathbf{x})$ is normalized so that $I_{heq}(\cdot) \in [0, 1]$. Apart from $I_{heq}(\cdot)$, an additional input channel $I_{aug}(\cdot)$ is generated from $I(\cdot)$ such that:

$$I_{aug}(\mathbf{x}) = \log(\varepsilon + I_{heq}(\mathbf{x})) \tag{1}$$

where $\varepsilon > 0$ is a small positive number ($\varepsilon = 2 \cdot 10^{-4}$ in our experiments). $I_{aug}(\mathbf{x})$ is also normalized so that $I_{aug}(\mathbf{x}) \in [0, 1]$.

Given that YOLOv8 models only accept 2D (with 1 or 3 channels) images or videos, and that the resolution along the Z axis was four times lower than the X and Y axes, the bounding box was restricted to the XY plane and is employed to crop every slice of the same image. Because of this, a 2D RGB image is generated from $I_{heq}(\cdot)$ and $I_{aug}(\cdot)$ by averaging them along the Z axis and using them as the red and green channels, with the blue channel set to zero. Label maps are similarly projected along the Z axis, so that $L_{YOLO}(x, y)$ is one if any of the voxels along the Z axis belongs to any of the foreground labels and zero otherwise:

$$\mathbf{I}_{YOLO}(x, y) = \left(\frac{1}{N_z} \sum_z I_{heq}(x, y, z), \frac{1}{N_z} \sum_z \log\left(I_{heq}(x, y, z)\right), 0 \right) \tag{2}$$

$$L_{YOLO}(x, y) = \begin{cases} 0 & \text{if } \sum_z L(x, y, z) = 0 \\ 1 & \text{otherwise.} \end{cases} \tag{3}$$

where N_z is the number of slices in the image ($N_z = 44$ in the MBAS dataset). Figures 2a and 2b show $\mathbf{I}_{YOLO}(x, y)$ and $L_{YOLO}(x, y)$, respectively, for one of the training cases. Figure 2c shows the contours of the myocardium and right and left atrial cavities of every image slice.

Left-Right Myocardial Splitting. In order to help the model to not merge left and right atrial myocardium in locations where they are very close together, we hypothesized that splitting the biatrial myocardium into left and right myocardium could help the model avoid merging the myocardium in those locations. For the training of the nnU-Net only, the expert segmentation of the biatrial myocardium is split into left atrial myocardium and right atrial myocardium using a fully automatic method proposed by Wang et al. [16], with some modifications. The outer biatrial surface Γ_{outer} and the right (Γ_{RA}) and left (Γ_{LA}) atrial cavity surfaces are computed using isosurfaces on the label map. These surfaces were rasterized on an upsampled isotropic grid of spacing 0.625 mm and an upsampled label map was built with them, taking care that slices on the original label map remained unchanged. Then, the following Laplace problem was solved on the upsampled grid:

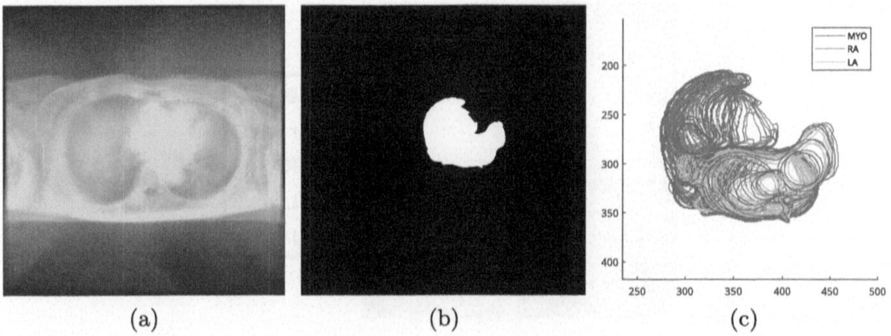

Fig. 2. Example of (a) a YOLOv8 input $\mathbf{I}_{YOLO}(x, y)$ and (b) its associated mask $L_{YOLO}(x, y)$. (c) Contours of the ROIs (MYO, RA, LA) on all slices of the volume.

$$\nabla^2 s(x, y, z) = 0, \ \textbf{s.t.} \ s(\Gamma_{outer}) = 0, s(\Gamma_{RA}) = 100, s(\Gamma_{LA}) = -100 \quad (4)$$

using the method proposed by Merino-Caviedes et al. [11] without extrapolating values on *Ghost* nodes. Then, voxels where $L(x, y, z) = 1$ and $s(x, y, z) \leq 0$ are assigned a new label $L(x, y, z) = 4$. The label map is returned to its original resolution and a connectivity analysis is carried out to suppress outliers. This connectivity analysis is applied to both the right atrium ($L(x, y, z) \in \{1, 2\}$) and the left atrium ($L(x, y, z) \in \{3, 4\}$). For each atrium, the connected components are evaluated using an in-plane 4-connected neighborhood. If the number of voxels of any connected component is below a certain threshold, the neighbors of that connected component are identified. If more neighbors belong to the opposite atrium (background voxels are excluded), then that connected component is assigned to the myocardium of the opposite atrium. Figure 3 contains an example of a slice where the myocardium of both atria was joined together.

The conversion of the labels with split myocardium to the label maps in the MBAS format is straightforward. It suffices to change the value of $L(x, y, z)$ on voxels where $L(x, y, z) = 4$ to $L(x, y, z) = 1$.

2.4 YOLOv8

The YOLOv8 architecture consists of three main blocks: the backbone, responsible for feature extraction from the image; the Neck, which aggregates features; and the Head, responsible for predictions. In more detail, the Backbone consists of a number of convolutional layers that perform the feature extraction from the images. The backbone also has a number of Cross stage modules, C2f, which enhance the learning ability of the network by combining high-level features with contextual information. At the end of the backbone is a Spatial Pyramid Pooling Fast (SPPF) layer which processes features at different scales. The feature aggregating capability of the YOLOv8 Neck is implemented using a series of

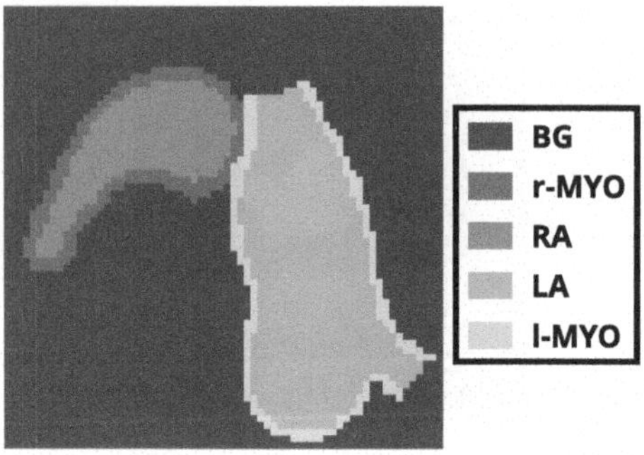

Fig. 3. Example of left-right myocardial splitting on an expert segmentation slice.

C2f, concatenation, upsampling, and convolutional layers. The resulting high-dimensional features are fed to the detection module at the YOLOv8 Head. The detection module consists of a set of convolutional and linear layers which are used to map these high-dimensional features to the output bounding boxes and object classes. The Non-maximum Suppression (NMS) is used to handle overlapping bounding boxes and to select the most relevant detections. YOLOv8 uses the Sigmoid Linear Unit (SiLU) also known as the Swish activation function, to improve gradient flow and feature expressiveness [4].

To detect the regions of interest on the dataset, the 70 training cases were initially split into 60:20:20% for training, validation, and testing respectively. After the initial training was completed with satisfactory results, the training data was modified to further increase the model generalizability. Thus, for the final training and validation, the data was split into 86:14% respectively. The validation was used for early stopping to avoid overfitting.

The built-in auto hyperparameter tuning feature of YOLOv8 was used to determine the best parameters for training. The determined learning rate was 0.002, with momentum of 0.9. The optimizer was the AdamW with weight decay varying from 0.0005 to 0. YOLOv8 uses an aggregation of the Class loss, Box (IOU) loss, and Distributional Focal Loss (DFL).

Both the initial and final training yielded mean average precision (mAP) of 0.995. The inference time, encompassing preprocessing, detection, and postprocessing was 9.4 ms per image.

2.5 nnU-Net

The nnU-Net framework methodology is organized in three steps. Firstly, a dataset needs to be built with the desired input channels and corresponding labels. Then, the nnU-Net computes a fingerprint of the dataset and generates plans according to that fingerprint. At the moment of writing, nnU-Net can provide plans for 2D and 3D U-Net architectures with and without residual connections. nnU-Net also randomly splits the training set using 80% for training and 20% for validation. The last step is training the model using a trainer which, by default, uses as loss function the combination of the cross entropy loss function and the dice loss function (both equally weighted). This trainer uses stochastic gradient descent (SGD) with initial learning rate of 0.01 and 1000 epochs. Although nnU-Net does not implement early stopping, the best model parameters (according to the validation metrics) are stored along with the last model parameters.

3 Experiments

Segmentations predicted by the model were evaluated using the Dice Similarity Coefficient (DSC) and the 95% Hausdorff distance (HD95) metrics on each of the three original expert ROIs. As in the challenge, models were ranked first according to each of the 6 metrics individually, then these 6 rankings were averaged and the models were ranked again according to this average.

Table 1 contains the different methodological variants that were tested experimentally for the nnU-Net model. Each variant represents a different preprocessing of the 70 training cases for training the nnU-Net. Unless specified otherwise, all trainings employed the same split of the training dataset, where 80% was used for training (56 cases) and 20% for validation (14 cases). The cropping for training the nnU-Net was done using for each image a bounding box based on its expert segmentation.

Table 1. Description of the different variants tested for nnU-Net. Ch. Aug.: Channel Augmentation. L-R Myo. Split: Left-Right Myocardium Split. HE: Histogram Equalization.

Variant	Ch. Aug.	L-R Myo. Split	Cropping (pad)	HE
D014	No	No	No	No
D015	Yes	No	No	No
D016	Yes	Yes	No	No
D019	Yes	No	Yes (50)	No
D020	Yes	Yes	Yes (50)	No
D021	Yes	Yes	Yes (50)	Yes

Table 2 shows the metrics, averaged on the 14 validation cases, and overall ranking for the variants without cropping, that is, the D014 (baseline), D015 (channel augmented only) and D016 (channel augmented, label split) datasets. For all of them, a U-Net with no residual connections was selected and models were trained with a SGD optimizer (initial learning rate of 0.01) and 1000 epochs, using as loss the sum of the cross entropy loss and the Dice loss. For the baseline variant (D014), both a 2D and a 3D U-Net were tested. It may be observed that the channel augmented, label split variant was ranked the first and achieved the highest Dice coefficient for the myocardium and left atrium, as well as the lowest

Table 2. Metrics, inference time and ranking of the nnU-Net without cropping the input. nnU-Net models were trained with the same trainer.

Variant	Plan	Dice			HD95 (mm)			Time (s)	Ranking
		MYO	RA	LA	MYO	RA	LA		
D014	nnUNetPlans 2D	0.706	0.915	0.924	3.334	4.556	3.814	8.016	4
D014	nnUNetPlans 3D	0.720	0.923	0.928	3.049	3.875	3.572	15.095	3
D015	nnUNetPlans 3D	0.721	0.922	0.930	3.169	4.172	3.191	15.233	2
D016	nnUNetPlans 3D	0.722	0.922	0.930	2.997	3.894	3.462	15.454	1

HD95 for the myocardium. Also, the variant with channel augmentation and no splitting achieved better results than the baseline variant. Finally, the 2D U-Net model achieved worse results than the 3D U-Net model in the baseline variant in every metric and was ranked the last. The average inference time per image for the 3D U-Net without YOLOv8 was between 8.016 s and 15.454 s. Therefore, these models by themselves (without YOLOv8) exceeded the inference time limit set in the challenge (6 s). It is worth noting that there is a high difference between the inference time of the 2D and the 3D U-Net models, but that the inference time of the channel augmented, label split variant (D016) was only slightly higher than its baseline variant counterpart (D014).

Table 3. Metrics and ranking obtained for the D015 variant with and without residual connections and training length of 250 or 1000 epochs.

Epochs	Plan (VRAM)	Dice			HD95 (mm)			Ranking
		MYO	RA	LA	MYO	RA	LA	
1000	nnUNetPlans	0.721	0.922	0.930	3.169	4.172	3.191	3
1000	ResEncM (16 GB)	0.721	0.922	0.931	2.920	3.991	3.169	1
250	nnUNetPlans	0.717	0.920	0.925	3.143	4.090	3.837	4
250	ResEncM (16 GB)	0.722	0.920	0.929	2.945	3.996	3.322	2

Table 3 compares the results obtained with the 14 validation cases of the D015 variant (augmented channels) using an nnU-Net with and without residual connections and training length of 250 or 1000 epochs. For both 250 and 1000 epochs, the best results were achieved by the U-Net with residual connections.

Table 4 shows the metrics and partial ranking of the variants trained on cropped images (D019, D020 and D021) using the nnU-Net plans with residual connections. In this case, the best alternative was D020, with the best results for the Dice coefficient of the myocardium and the left atrium and the lowest HD95 for the left atrium. D019, with no label splitting, was ranked last. Therefore, the variant D020 was chosen for CavityBASNet (i.e., YOLOc8s and nnU-net cascade). CavityBASNet, on the 14 validation cases, attained a Dice of 0.715

Table 4. Metrics and ranking yielded by the cropped datasets D019, D020 and D021, trained with 250 epochs and using U-Nets with residual connections.

Variant	Plan (VRAM)	Dice			HD95 (mm)			Ranking
		MYO	RA	LA	MYO	RA	LA	
D019	ResEncM (16 GB)	0.712	0.923	0.928	2.937	3.818	3.444	3
D020	ResEncM (16 GB)	0.715	0.921	0.930	2.893	3.800	2.984	1
D021	ResEncL (24 GB)	0.713	0.925	0.928	2.864	3.663	3.303	2

for MYO, 0.921 for RA and 0.930 for LA. The HD95 was 2.882 mm for MYO, 3.833 mm for RA and 3.016 mm for LA. Inference time was 2.7 s per image using a tile step size of 0.5, which met the challenge time limit for inference.

The metrics and ranking of the proposed methodology with a bounding box computed using the YOLOv8 model followed by the nnU-Net with augmented inputs and split labels, trained using all 70 training cases, 1000 epochs and ResEncL (24 GB) plans were given by the MBAS24 challenge organizers according to the evaluation on the validation set (cases 71–100). The methodology was ranked the 12th of 23 participants with the metrics shown in table 5. Also, CavityBASNet was ranked 4th for the right atrial HD95.

Table 5. Metrics and ranking of the proposed CavityBASNet model among the rest of the MBAS solutions, provided by the challenge organizers according to the validation dataset predictions.

ROI	Dice	Rank Dice	HD95 (mm)	Rank HD95
MYO	62.61	14	5.21	9
RA	86.92	11	6.99	4
LA	89.33	16	5.94	15

4 Conclusions

This paper describes a fully automatic method to segment LGE-MRI images of the biatrial myocardium and left and right atrial cavities. It is composed of two networks in cascade, a YOLOv8 model to provide a bounding box of the biatrial ROI and an nnU-Net with residual encoding connections to provide the predicted label map. This cascade augments the input channels with the logarithm of the LGE-MRI image as a trade-off between the time-consuming correction of bias inhomogeneities and only using the LGE-MRI as input. Also, the expert segmentation of the biatrial myocardium was automatically split into left myocardium and right myocardium for training the nnU-Net. The experimental results support that these modifications improve the segmentation quality.

Finally, the proposed channel augmentation and myocardial label splitting are not specific to the nnU-Net and may be applied to other models.

Acknowledgments. This work was supported in part by the spanish Agencia Estatal de Investigación, under Grants PID2020-115339RB-I00 TED2021-130090B-I00 and CPP2021-008880, and by Fundación La Caixa (HR22-00533) and ESAOTE Ltd.

Disclosure of Interests. The authors have no competing interests to declare that are relevant to the content of this article.

References

1. Gal, P., et al.: P1304: architecture of fibrosis predicts atrial fibrillation recurrences after ablation. Europace **7**, iii176 (2015)
2. Gal, P., Marrouche, N.F.: Magnetic resonance imaging of atrial fibrosis: redefining atrial fibrillation to a syndrome. Eur. Heart J. **38**(1), 14–19 (2017)
3. Higuchi, K., Akkaya, M., Akoum, N., Marrouche, N.F.: Cardiac MRI assessment of atrial fibrosis in atrial fibrillation: implications for diagnosis and therapy. Heart **100**(7), 590–596 (2014)
4. Hussain, M.: YOLOv5, YOLOv8 and YOLOv10: the go-to detectors for real-time vision (2024). https://arxiv.org/abs/2407.02988. Accessed 13 Aug 2024
5. Isensee, F., Jaeger, P.F., Kohl, S., Petersen, J., Maier-Hein, K.H.: nnU-Net: a self-configuring method for deep learning-based biomedical image segmentation. Nat. Methods **18**, 203–211 (2021)
6. Isensee, F., et al.: nnU-Net revisited: a call for rigorous validation in 3D medical image segmentation (2024). https://arxiv.org/abs/2404.09556
7. Jocher, G., Chaurasia, A., Qiu, J.: Ultralytics YOLO v8 (2023). https://github.com/ultralytics/ultralytics. Accessed 13 Aug 2024
8. Li, L., Zimmer, V.A., Schnabel, J.A., Zhuang, X.: Medical image analysis on left atrial LGE MRI for atrial fibrillation studies: a review. Med. Image Anal. **77**, 102360 (2022)
9. Marrouche, N.F., et al.: Association of atrial tissue fibrosis identified by delayed enhancement MRI and atrial fibrillation catheter ablation: the DECAAF study. JAMA **311**(5), 498–506 (2014)
10. McGann, C., et al.: Atrial fibrillation ablation outcome is predicted by left atrial remodeling on MRI. Circ. Arrhythmia Electrophysiol. **7**(1), 23–30 (2014)
11. Merino-Caviedes, S., et al.: Computing thickness of irregularly-shaped thin walls using a locally semi-implicit scheme with extrapolation to solve the Laplace equation: application to the right ventricle. Comput. Biol. Med. **169**, 107855 (2024)
12. Reis, D., Kupec, J., Hong, J., Daoudi, A.: Real-time flying object detection with YOLOv8 (2023). http://arxiv.org/abs/2305.09972. Accessed 13 Aug 2024
13. Ronneberger, O., Fischer, P., Brox, T.: U-Net: convolutional networks for biomedical image segmentation. In: Navab, N., Hornegger, J., Wells, W.M., Frangi, A.F. (eds.) MICCAI 2015. LNCS, vol. 9351, pp. 234–241. Springer, Cham (2015). https://doi.org/10.1007/978-3-319-24574-4_28

14. Sirisha, U., Praveen, S.P., Srinivasu, P.N., Barsocchi, P., Bhoi, A.K.: Statistical analysis of design aspects of various YOLO-based deep learning models for object detection. Int. J. Comput. Intell. Syst. **16**, 1–29 (2023)
15. Tustison, N.J., et al.: N4ITK: improved N3 bias correction. IEEE Trans. Med. Imaging **29**, 1310–1320 (2010)
16. Wang, Y., et al.: A robust computational framework for estimating 3D bi-atrial chamber wall thickness. Comput. Biol. Med. **114** (2019)

Exploring Chemotherapy-Induced Cardiotoxicity Combining A 3D Computational Model and Preclinical Cardiac Imaging Data

Javier Villar-Valero[1(✉)], Jesus Jairo Rodríguez Padilla[2,3], Buntheng Ly[3], Juan F. Gomez[4], Mihaela Pop[2,5], Beatriz Trenor[1], and Maxime Sermesant[2,3]

[1] Centro de Investigación e Innovación en Bioingeniería, Universitat Politècnica de València, Valencia, Spain
javilva@ci2b.upv.es
[2] Centre Inria d'Université Côte d'Azur, Epione Team, Sophia Antipolis, France
[3] IHU-Liryc, Université de Bordeaux, Pessac, France
[4] Valencian International University, Valencia, Spain
[5] Sunnybrook Research Institute, Toronto, Canada

Abstract. Cardiotoxicity following anthracycline-based chemotherapy is becoming a critical clinical burden, leading in time to heart dysfunction such as arrhythmia and heart failure. The purpose of this computational study is to better understand the structure-function interaction and subtle longitudinal changes in the presence of progressive fibrosis induced by chemotherapeutic drugs such as doxorubicin, DOX. Specifically, here we investigated via simulations how the cardiac action potential wave propagates in the presence of collagenous fibrotic clusters generated by this drug. For this, we built isotropic and anisotropic 3D left ventricular (LV) models from in vivo contrast-enhanced MRI data (1.4 mm isotropic resolution) acquired in n = 2 pigs pre-DOX and post-DOX therapy (at weeks 5 in one pig, at weeks 9 in the other pig). For computations, we employed a modified biophysical Mitchell-Schaeffer model with a fast GPU-based Lattice-Boltzmann numerical method (which runs directly on MR images Cartesian grid). Overall, our results showed that several key parameters (fibrosis density, tissue conductivity, action potential duration, anisotropy) independently contribute to the perturbation of the electrical wave propagation. Future work will focus on simulating potentially lethal arrhythmia in hearts subjected to DOX therapy.

Keywords: Cardiotoxicity · Arrhythmia · Fibrosis · Computational Modeling · Simulation

1 Introduction

Drug-related treatments can have severe consequences and toxic effects on the heart muscle. Among such therapies, it has been recently shown that

O. Camara et al. (Eds.): STACOM 2024, LNCS 15448, pp. 64–74, 2025.
https://doi.org/10.1007/978-3-031-87756-8_7

chemotherapy-induced cardiotoxicity prevails as a critical clinical burden by severely altering cardiac electrical and mechanical functions [1]. One of the most widely used chemotherapeutic drugs in cancer patient is doxorubicin (DOX), which, although effective in killing cancerous cells, has been recently demonstrated to induce structural myocardial damage through a gradual deposition of collagen-based reactive fibrosis within the heart wall [2]. This leads in time to a progressive heart dysfunction (e.g. non-ischemic cardiomyopathy and abnormal heart rhythms), culminating in heart failure and potentially lethal arrhythmic events [3,4]. In particular, the abnormal propagation of electrical wave in sub-chronic phases (i.e., in the weeks after completion of DOX treatment) in the presence of diffuse fibrosis, is not well understood.

Computer simulations of electrophysiology are nowadays widely spread robust tools that can help us virtually dissect the propagation of electrical impulse through the heart [5]. In particular, 3D image-based models can predict the behaviour of action potential (AP) wave under various experimental conditions, while integrating high level of structural details [6]. Thus, such models could be adapted and exploited to study the consequences and gradual (possible irreversible) changes in heart function following chemotherapy.

In this computational study, we sought to better understand how the action potential wave propagates in the presence of DOX-induces diffuse fibrosis. We first aim to use MR images acquired in a preclinical pig model of pathology before/after DOX treatment to build 3D anatomical and morphological models of the left ventricle (LV). We then employ those 3D LV models to perform several simulation-based parametric studies. Specifically, we are interested to learn how a gradual increase in the density of DOX-induced fibrosis, along with the anisotropy and regional heterogeneities in conductivity (diffusivity) and AP duration contribute to a perturbed electrical activity within the heart under pacing conditions. The novel contribution of this work lies in its comprehensive simulation of the interplay between fibrosis and electrical activity, providing insights into the progressive and potentially irreversible changes in heart function resulting from chemotherapy.

2 Methods

2.1 Preclinical Experiment

The experiments involved n = 2 juvenile Yorkshire swine weighing ~25 kg prior to starting the doxorubicin therapy, and were carried out with approval from the research ethics board of the Animal Care Committee at Sunnybrook Research Institute Toronto, as detailed in [7]. Briefly, a 1 mg/kg weekly dose of DOX was injected intravenously throughout a four-week plan, mimicking clinical treatments in cancer patients.

Non-invasive CMR imaging was performed on a 3T MR750 scanner GE scanner using a free-breathing 3D LGE method (late gadolinium enhancement). The scans were done before the treatment (hereto referred as pre-DOX therapy time point), as well as after the completion of treatment (post-DOX) at the following time points: week 5 in pig 1, and week 9 in pig 2, respectively. The MR images were acquired at $1.4 \times 1.4 \times 1.4\,\text{mm}^3$ spatial resolution, using a conventional 3D inversion recovery fast gradient echo (IR-FGRE) pulse sequence and a bolus of 0.2 mmol/kg Magnevist (a gadolinium-based contrast agent) sensitive to collagenous fibrosis.

2.2 Left Ventricular Computational Models from CMR Images

In this work, we built isotropic (i.e., no fiber directions integrated into the domain) and anisotropic 3D left ventricular (LV) models from in vivo CMR images. For each raw cine-MRI image, an expert initially performed manual contouring of both the endocardial and epicardial surfaces of the LV, identifying the diastolic phase as the most suitable for segmentation. Once the diastolic phase was identified, a region of interest (ROI) from the posterior side of the remote myocardium was selected. Then, fibrosis was identified by applying a threshold on the intensity values on LGE images based on a clinically accepted 5SD (standard deviation) cut-off [8]. For the anisotropic cases, synthetic fiber orientations were assigned by linearly interpolating the helix angle from $-70°$ on the epicardium to $+70°$ on the endocardium [9]. Figure 1 illustrates the entire workflow for generating our computational LV models.

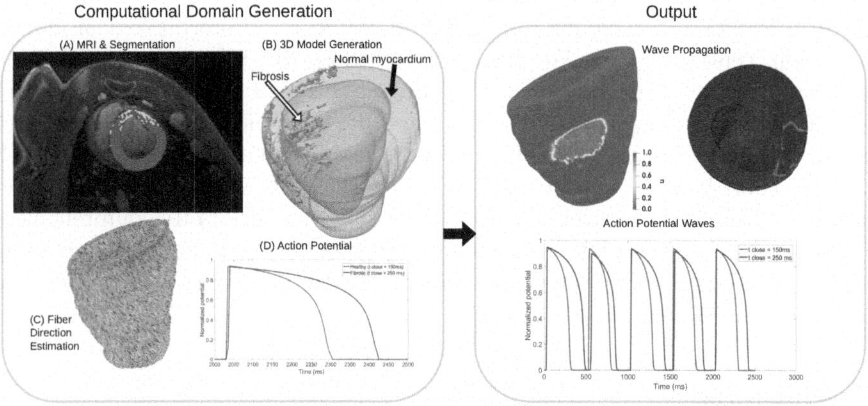

Fig. 1. Diagram of the pipeline: (left) generation of the 3D computational models from MR images with fibrosis and fiber directions integrated as well as exemplary AP waves parameterized in healthy and fibrotic zones; (right) simulation output showing activated tissue (red) within the 3D LV model and select AP waves at one voxel following a train of 5 stimuli. (Color figure online)

2.3 A Modified Mitchell-Schaeffer Model for LV Electrophysiology

To simulate action potential propagation (AP) we used a monodomain formulation of a modified Mitchell-Schaeffer (MS) model as in Djabella et al. [10]. This modification of the original MS model introduces an additional parameter, λ, that allows us to control the degree of excitability in a region of interest. The evolution equation of the system is given by:

$$\begin{aligned} \frac{\partial u}{\partial t} &= \nabla \cdot (D\nabla u) + \frac{hu(u-\lambda)(u_{max}-u)}{\tau_{in}} - \frac{u}{\tau_{out}} + J_{stim}(t), \\ \frac{\partial h}{\partial t} &= \epsilon(u,h)\left(h_{\infty}(u) - h\right), \end{aligned} \tag{1}$$

where u is the normalized AP variable, h is the gating variable which makes the gate open and close (corresponding to depolarization and repolarization phases), J_{stim} is the stimulation current at pacing location. The other parameters and their values can be found in [8]. To numerically solve this PDE system 1, we employed a fast GPU-based Lattice-Boltzmann numerical method (which runs directly on images) [11]. We validated its efficacy by comparing it to the finite element method using FEniCS on a 3D benchmark [12], demonstrating significantly faster computation times for large, high-resolution meshes, achieving a speed-up of 140× in the simulations. The results of the calibrations can be found in the Supplemental Material.

To account for tissue anisotropy, propagation speed was set to be 2.5 times slower in the direction transverse to myocardial fibers; thus, we used $d = 3.5$ for the diffusion coefficient parallel to fiber directions and $d/2.5$ for the transverse and normal directions to fibers. This gives us an estimate of a conduction velocity of 0.8 m/s along the fiber direction in healthy tissue. Under these considerations, our diffusion coefficient matrix is defined as $D = \Lambda A \Lambda^T$ with

$$A = \begin{pmatrix} d & 0 & 0 \\ 0 & \frac{d}{2.5} & 0 \\ 0 & 0 & \frac{d}{2.5} \end{pmatrix},$$

and Λ having the local basis information (fiber, transverse and normal directions).

Simulations were carried out on NVidia GTX 1080 ti (11 GB of RAM) GPU.

Stimulation Protocols and Experiments Description. For both isotropic and anisotropic cases, a series of 5 beats at 500 ms were performed from a selected point in the LV free wall. On average, the whole protocol took around 20 min per simulation. As our goal is to understand the effect of structure on electrophysiological function, the parameters concerning the shape of the APD were fixed in all our simulations: ($\tau_{open} = 120, \tau_{close} = 150, \tau_{in} = 0.3, \tau_{out} = 6, u_{max} = 1, u_{gate} = 0.13$, all in milliseconds with the exception of both u_{max} and u_{gate} which are dimensionless). In addition, a fast pacing protocol was performed with modifying the APD to assess whether this could lead to wavefront fragmentation. The protocol consisted of adding 2 pulses at 300 ms and 350 ms consecutively, with the same values for the model parameters except τ_{close}, which was set to 250 ms.

Fibrosis Modeled via Reduced Excitability. In this scenario, for the part of the domain exhibiting fibrosis we modify the excitability parameter ($\lambda = 0.2$) to decrease the propagation speed in these regions. Additionally, diffusion in both the longitudinal and transverse directions was reduced in fibrotic areas to 1/4th of the one in normal tissue [14,15].

Fibrosis Modeled as Non-conductive Tissue. In this case, we remove the voxels of MR-defined fibrosis from the ventricular domain. Consequently, no computations were performed in the fibrotic region, essentially creating clusters of tissue acting as electrical insulators.

3 Results and Discussion

We next present the simulation results obtained after varying different key parameters separately (i.e., anisotropy, fibrosis density, tissue diffusivity, and APD), in order to observe their independent influence on the perturbation of wavefront propagation and activation heterogeneity.

Figure 2 illustrates clear differences in the pattern of activation and AP wave propagation immediately following stimulation and thereafter, in the LV models from both pigs when the fiber directions were integrated (i.e. anisotropic cases) or not (isotropic cases), into the 3D computational domains. For instance, for pig 1, as the speed of wave travels 2.5 faster in the longitudinal direction (parallel to the muscle fibers), the initial activation (50 ms) is ellipsoidal in shape and, as a result, the wavefront reached the base at a later time point (200 ms) compared to the isotropic scenario where it reached the base at 150 ms. Such differences are important for realistically reproducing the AP wave spread through the heart, underscoring the need to use anisotropic model. Therefore, in the following sections, we restrain from using isotropic models.

Figure 3 presents the simulation results obtained for both pre-DOX and post-DOX at weeks 5 (pig 1) as well as pre-DOX and post-DOX at weeks 9 (pig 2). Most importantly, by week 9, fibrosis deposition had progressed substantially, which resulted in a noticeable wave perturabtion due to larger obstacles. The most significant differences were observed at 9 weeks post-DOX (pig 2), where excitability ($\lambda = 0.2$) and diffusivity (1/4 of the healthy value) were reduced.

Figure 6 shows the temporal evolution of the electrical activation of the ventricular volume, with the time steps represented in milliseconds after the last stimulus of a 5-pulse train. In the first row, the propagation of the action potential in the pre-Dox case is illustrated. We observed that in this case, the propagation proceeded normally, with a continuous wavefront, and the LV is fully activated in the absence of fibrosis (pre-DOX).

The next two rows correspond to the simulation of fibrosis treated as less excitable tissue ($\lambda = 0.2$) and as a non-conducting tissue ($d = 0$), respectively. In the simulation of fibrosis as less excitable tissue, the wavefront is observed to fragment after the stimulus ($t = 50$ ms). The propagation throughout the ventricle is more discontinuous than in the pre-DOX case, and, when the ventricle

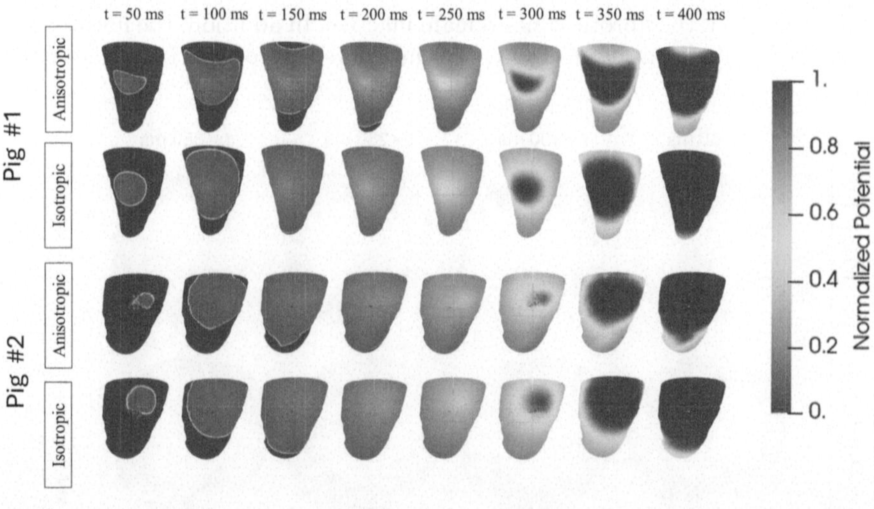

Fig. 2. AP wave propagation at different time points in the LV MRI-based models built from pig 1 (at week 5) and pig 2 (at week 9), with fiber directions (anisotropy) integrated and without fiber directions (isotropic case). Red indicates depolarized tissue and blue corresponds to the repolarized phase. (Color figure online)

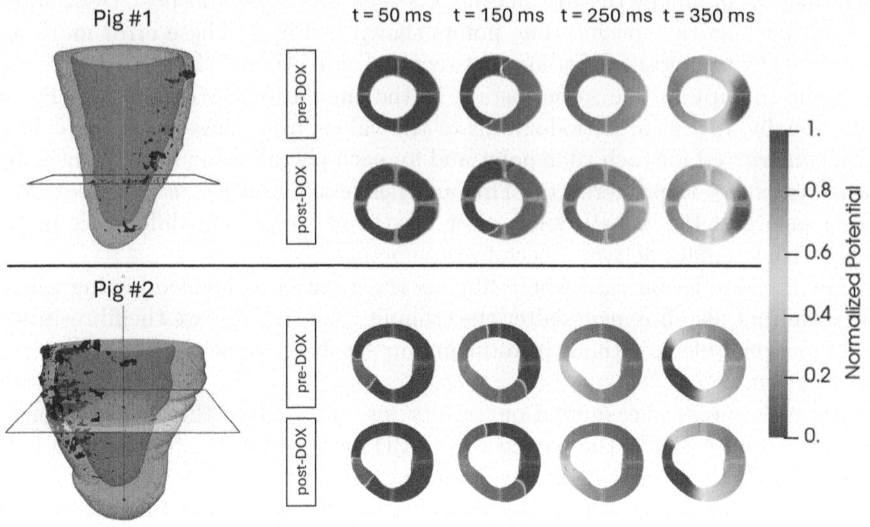

Fig. 3. The effect of fibrosis density on AP wave propagation for both pigs: (left) 3D anatomical LV models with fibrosis zone integrated; (right) simulated cardiac activation maps pre-/post-DOX presented in short-axis corresponding to the cutting planes.

is fully activated, the fibrotic zones remain inactive. In addition, the fibrotic zone repolarizes earlier than in pre-DOX, showing more heterogeneities in the APDs.

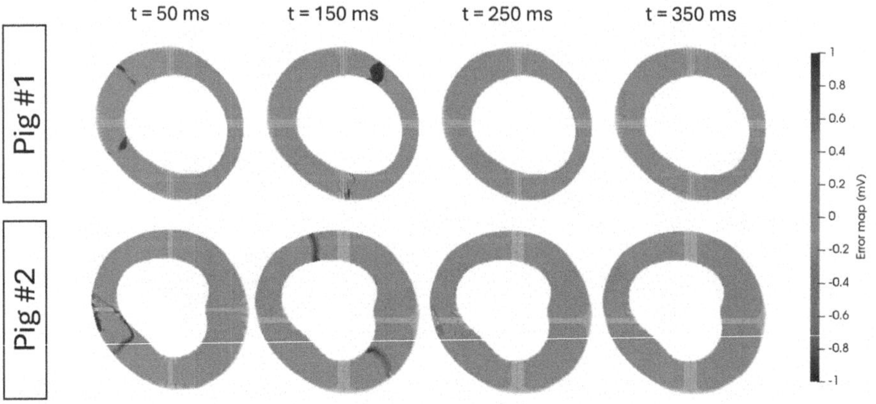

Fig. 4. Error maps showing the difference in activations between both pigs, comparing pre- and post-therapy at the time points displayed in Fig. 3.

To study the differences between cases, error maps have been computed and quantified to highlight the distinctions between pre-DOX and post-DOX states in both pigs, at the specific time points shown in Fig. 3. These error maps are calculated by taking the difference between the activation of the heart before and after the therapy at the same instant in the simulation, as depicted in Fig. 4. Additionally, in Fig. 5, periodograms of the values from these error maps have been constructed for each time point and for each pig, allowing for a quantifiable measurement of the differences. This analysis focuses on Pig #1 (1 week after treatment) and Pig #2 (9 weeks after treatment), revealing differences in the regions that remain active at certain time points.

For the simulation case where fibrosis was treated as nonconducting tissue, the wavefront also fragments after the stimulus, but in this case the fibrosis acts as a functional block without introducing further heterogeneity in the wavefront propagation.

Figure 7 shows the simulation results for pig 2 after the introduction of changes in the APD. In the second row (APD increased in the fibrotic zone), we observed that the propagation wavefront fragments into two after the stimulus (t = 100 ms). This is not seen in the first row, which corresponds to the pre-DOX simulation without fibrotic zone.

The fragmentation of the wavefront into two waves may be indicative of tissue arrhythmicity, as the introduction of these structural changes together with an early stimulus may result in a unidirectional block that triggers a reentrant arrhythmia.

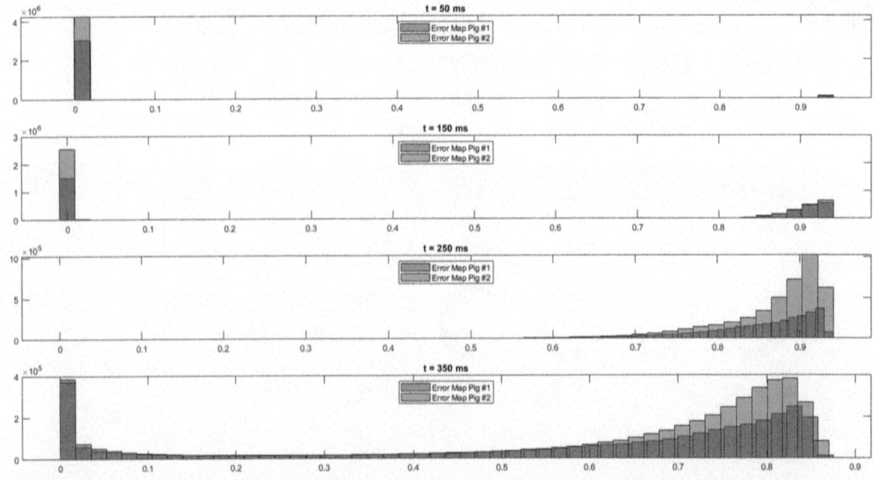

Fig. 5. Periodograms corresponding to the error maps at the four time points shown in Fig. 3.

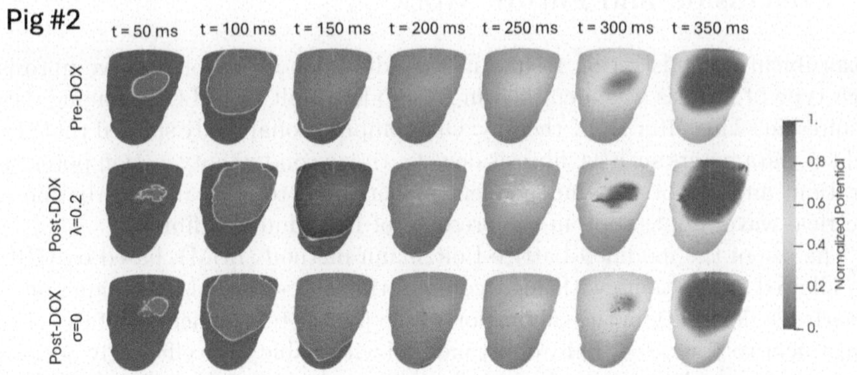

Fig. 6. Comparison of simulations obtained for pig 2: pre-DOX (top row) vs post-DOX at w9 in two cases: when the fibrotic tissue is treated with a modification of λ (middle row); and when fibrotic region treated as non-conductive tissue (bottom row).

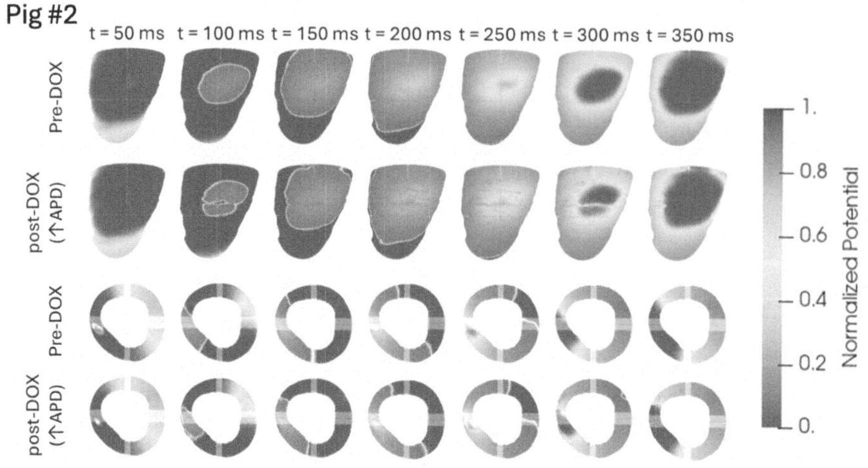

Fig. 7. Simulation of fast pacing. Case without fibrosis and w9 with increased APD in the fibrosis zone (epicardial and short-axis views).

4 Conclusion and Future Work

Doxorubicin administration results in a gradual deposition of reactive fibrosis. Such type of fibrosis was identified in LV models built from LGE imaging data acquired in swine after DOX therapy. Our computational study showed that several key parameters such as: fibrosis density, tissue conductivity, action potential duration, and anisotropy, independently contribute to a local perturbation of electrical wave propagation in the presence of DOX-induced fibrosis.

The use of the modified Lattice-Boltzmann method (LBM), based on GPU, has allowed simulations with high computational speed. This technique offers remarkable flexibility in the calibration of the parameters, which facilitates their adjustment to real cases. This robust method will enable us to efficiently perform protocols for studying arrhythmia inducibility, which in future lines will allow an exhaustive study of doxorubicin-induced cardiotoxicity. Our results highlight the ability of LBM to investigate structure-function interactions in hearts with chemotherapy-induced fibrosis, providing a valuable tool for the study of potentially lethal arrhythmias in hearts undergoing DOX therapy. We envision that our experimental-computational framework could be translated to the clinics in the future, to evaluate the longitudinal effect post-DOX (i.e., from sub-acute to chronic phases) on the electrophysiological changes and predict potential abnormal heart rhythms.

Acknowledgment. This work was supported by the SimCardioTest project, which has received funding from the European Union's Horizon 2020 research and innovation program under grant agreement No. 101016496; by the French government through the 3IA Côte d'Azur Investments in the Future project managed by the

National Research Agency (ANR) with the reference number ANR-19-P3IA-0002; by a Canadian CIHR project grant (PJT) 153212; by Grant PRE2020-091849 [MCIN/AEI/10.13039/501100011033] and "ESF Investing in your future"; Grant PID2019-104356RB-C41 [MCIN/AEI/10.13039/501100011033]; PID2022-136273OA-C33 and PID2022-140553OB-C41 [MICIU/AEI/10.13039/501100011033 and by ERDF/EU]; and by the Barcelona Supercomputing Center [IM-2021-1-0001 and IM-2021-3-0001]. The authors are grateful to the OPAL infrastructure from Université Côte d'Azur for providing resources and support.

Disclosure of Interests. The authors have no competing interests to declare that are relevant to the content of this article.

References

1. Von Hoff, D.D., Rozencweig, M., Piccart, M.: The cardiotoxicity of anticancer agents. Semin. Oncol. **9**, 23–33 (1982)
2. Tanaka, R., Umemura, M., et al.: Reactive fibrosis precedes doxorubicin-induced heart failure through sterile inflammation. ESC Heart Fail. **7**(2), 588–603 (2020)
3. Christiansen, S., Autschbach, R.: Doxorubicin in experimental and clinical heart failure. Eur. J. Cardiothorac. Surg. **30**, 611–616 (2006)
4. Herrmann, J.: Adverse cardiac effects of cancer therapies: cardiotoxicity and arrhythmia. Nat. Rev. Cardiol. **17**, 474–502 (2020)
5. Sermesant, M., Delingette, H., Ayache, N.: An electromechanical model of the heart for image analysis and simulation. IEEE Trans. Med. Imaging **25**(5), 612–625 (2006)
6. Balaban, G., Costa, C.M., et al.: 3D electrophysiological modeling of interstitial fibrosis networks and their role in ventricular arrhythmias in non-ischemic cardiomyopathy. IEEE Trans. Biomed. Eng. **67**(11), 3125–3133 (2020)
7. Lin, P., et al.: MR imaging and electrophysiological features of doxorubicin-induced fibrosis: protocol development in a small preclinical pig study with histological validation. Appl. Sci. **12**(22), 11620 (2022). https://doi.org/10.3390/app122211620
8. Karim, R., et al.: Evaluation of state-of-the-art segmentation algorithms for left ventricle infarct from late Gadolinium enhancement MR images. Med. Image Anal. **30**, 95–107 (2016)
9. Pop, M., et al.: Construction of 3D MR image-based computer models of pathologic hearts, augmented with histology and optical fluorescence imaging to characterize action potential propagation. Med. Image Anal. **16**(2), 505–523 (2012)
10. Djabella, K., Landau, M., Sorine, M.: A two-variable model of cardiac action potential with controlled pacemaker activity and ionic current interpretation. In: 2007 46th IEEE Conference on Decision and Control, pp. 5186–5191. IEEE (2007)
11. Cedilnik, N., et al.: Efficient patient-specific simulations of ventricular tachycardia based on computed tomography-defined wall thickness heterogeneity. Clin. Electrophysiol. **9**(12), 2507–2519 (2023)
12. Niederer, S.A., et al.: Verification of cardiac tissue electrophysiology simulators using an N-version benchmark. Philos. Trans. A Math. Phys. Eng. Sci. **369**(1954), 4331–4351 (2011)
13. Lopez-Perez, A., Sebastian, R., Izquierdo, M., Ruiz, R., Bishop, M., Ferrero, J.M.: Personalized cardiac computational models: from clinical data to simulation of infarct-related ventricular tachycardia. Front. Physiol. **10** (2019)

14. Gomez, J.F., Cardona, K., Martinez, L., Saiz, J., Trenor, B.: Electrophysiological and structural remodeling in heart failure modulate arrhythmogenesis. 2D simulation study. PLoS ONE **9**(7), e103273 (2014)
15. Mendonca Costa, C., Plank, G., Rinaldi, C.A., Niederer, S.A., Bishop, M.J.: Modeling the electrophysiological properties of the infarct border zone. Front. Physiol. **9** (2018)

Quantitative Comparison of Blood Flow Patterns from in Silico Simulations and 4D Flow Data Before and After Left Atrial Occlusion

Paula Casademunt[1(✉)], Xabier Morales[1], Pedro Cepas-Guillén[2], Ada Doltra[2], Jordi Mill[1], Marta Saiz-Vivó[1], Eduardo Flores-Umanzor[2], Laura Sanchis[2], Xavier Freixa[2], and Oscar Camara[1]

[1] Physense, BCNMedtech, Department of Communication and Information Technologies, Universitat Pompeu Fabra, Barcelona, Spain
paula.casademunt@upf.edu

[2] Institut Clínic Cardiovascular (ICCV), Hospital Clínic de Barcelona, IDIBAPS, Barcelona, Spain

Abstract. Atrial fibrillation (AF), the most common cardiac arrhythmia, increases the risk of thrombus, especially in the left atrial appendage (LAA). LAA occlusion (LAAO) reduces stroke risk in AF patients by isolating the LAA from the left atrium (LA). Computational fluid dynamics (CFD) simulations, offering superior resolution than current imaging techniques, are proposed to assess thrombus risk, but require validation. Emerging imaging techniques like 4D flow MRI provide in-vivo flow measurements supporting CFD credibility. Our dataset includes 4D flow magnetic resonance imaging (MRI), cardiac computational tomography (CT), and Doppler ultrasound (US) of the mitral valve for four patients (pre- and post-LAAO). We developed patient-specific CFD simulations from CT and US data, comparing blood flow patterns with 4D flow MRI. The analysis was focused on thrombus formation indicators, such as particle deposition, LAA washout, and endothelial cell activation potential (ECAP). Results from both data and simulations were within similar ranges but showed limited patient-wise agreement. Post-LAAO hemodynamic changes assessed by 4D flow MRI and CFD data agreed on reduced thrombogenic indices such as ECAP, flow stasis and vorticity. This study highlights the need for computational model validation, comparing CFD models with 4D flow MRI, being the first to quantitatively analyze 4D flow MRI data with implanted LAAO devices.

Keywords: 4D flow MRI · Computational Fluid Dynamics · Atrial Fibrillation · Left Atrial Appendage

1 Introduction

Atrial fibrillation (AF) is the most common arrhythmia in humans and significantly increases stroke risk by promoting thrombus formation in the left atrium

O. Camara et al. (Eds.): STACOM 2024, LNCS 15448, pp. 75–86, 2025.
https://doi.org/10.1007/978-3-031-87756-8_8

(LA), with AF causing 25% of acute ischemic strokes in patients over 80 [1]. More than 90% of thrombi in non-valvular AF form in the left atrial appendage (LAA) [2], a muscular ear-shaped cavity in the LA. According to Virchow's triad, blood stagnation can trigger inflammation and increase thrombus risk [3]. However, detailed in-vivo analysis of LA hemodynamics in 4D (3D plus time) in AF patients is challenging due to the spatio-temporal limitations of current clinical imaging modalities, such as transthoracic echocardiography (TTE).

As an alternative, computational fluid dynamics (CFD) models have been proposed to study blood hemodynamics in the LA. Several studies have focused on analyzing in silico flow indices in the LAA in relation to thrombogenic risk [4–6]. Furthermore, CFD models have also been used to study the effects of LAA occlusion (LAAO) [7], an intervention that deploys a device to isolate the LAA from the rest of the LA with the aim of decreasing the risk of thrombus formation. Although some efforts are being made to move towards patient-specific modeling of this intervention to aid in clinical decision-making, there is an imperative need for validation of the results of CFD analyses.

Emerging advanced imaging techniques, such as 4D (3D + t) flow magnetic resonance imaging (MRI), enable comprehensive in vivo hemodynamic studies [8], aiding the validation of CFD models. This technique provides insights into the physiology of three-dimensional structures such as LA vorticity and allows concurrent analysis of pulmonary veins (PV), which conventional TTE cannot offer. However, many studies on 4D flow MRI have focused on the aorta, with fewer examining the LA due to challenges in finding suitable acquisition settings for the heterogeneous and complex velocity fields arising in the LA. Morales *et al.* demonstrated the potential of 4D flow MRI for enhancing CFD model accuracy by using it to define boundary conditions in the LA [9].

All in all, validating CFD models is challenging but essential, following standards like V&V40 to build credibility for use in medical device regulatory processes [10]. Time-resolved three-dimensional vector fields complicate result analysis and comparisons between patients, acquisitions, or data modalities. Developing physiologically relevant markers that summarize results in a meaningful, quantitative, and informative manner is crucial. This study aims to provide methods for the quantitative assessment of hemodynamic indices, proposing 4D flow MRI data as a validation tool for CFD simulations. A reduced cohort of cardiac 4D flow MRI data and their respective CFD simulations will be used as a proof of concept.

2 Materials and Methods

In this study, we utilized a unique dataset comprising four AF patients who underwent the LAAO procedure. The dataset includes 4D flow MRI, cardiac computed tomography (CT) scans and Doppler ultrasound (US) of the mitral valve (MV) plane, collected both pre- and post-intervention. Two patients were in sinus rhythm (SR) and two had AF during the implantation. Previous work with this dataset qualitatively assessed the effect of LAA occlusion using 4D flow

MRI data [11]. In the current study, we performed CFD simulations for each case using the CT and US data. The velocity fields obtained from 4D flow MRI and CFD simulations were analyzed and processed to extract hemodynamic indices for both pre- and post-LAAO deployment. These metrics enabled a quantitative comparison between the CFD and 4D flow MRI flow fields. Specifically, we derived metrics associated with the risk of thrombus formation in the LAA, along with parameters related to vorticity, flow stasis, flow efficiency, and PV inflow trajectories.

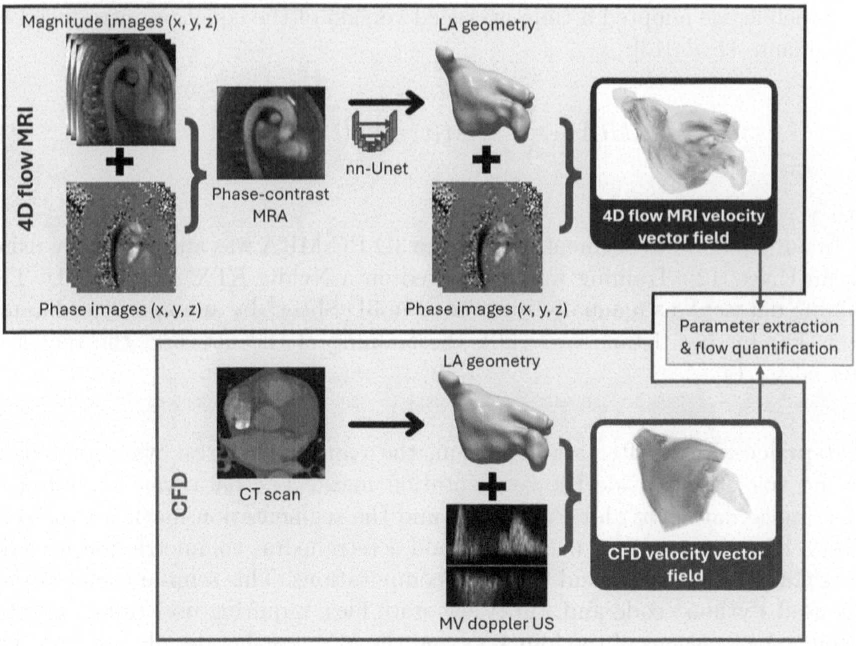

Fig. 1. Structure of the data processing pipeline. For the 4D flow magnetic resonance imaging (MRI), the magnitude data is combined with the phase (flow) data, to create a phase-contrast magnetic resonance angiogram (MRA). A segmentation of the left atria (LA) is automatically obtained using the nn-UNet [12]. The segmentation is used to delimit the region of interest, allowing subsequent data visualization and analysis of the flow within the LA. For the computational fluid dynamics (CFD) simulations, the patient-specific LA anatomy was obtained from computed tomography (CT) scans, and the flow profile at the mitral valve (MV), retrieved from Doppler ultrasounds (US) was used as a boundary condition. The resulting velocity vector fields were processed to obtain flow-related markers.

2.1 4D Flow MRI Processing Pipeline

The four analyzed 4D flow MRI cases were obtained from the Department of Cardiology at Hospital Clínic de Barcelona, Spain, with ethical approval and informed consent from all participants. The data was processed using a semi-automated open-source pipeline, depicted in the first section of Fig. 1.

Segmentation. A common approach for segmenting 4D flow MRI data is to create a 3D contrast-enhanced image called a phase-contrast magnetic resonance angiogram (PC-MRA). This image is obtained by combining the magnitude (or anatomical) data with the phase (or velocity) data. The PC-MRA highlights high blood flow areas with higher intensities using velocity data, while the magnitude image provides morphological details and reduces noise in low-velocity areas. In this pipeline, we adopted a time-averaged version of the equation introduced by Bustamante *et al.* [13]:

$$3DPC - MRA = \frac{1}{N}\sum_{t=1}^{N} M(t) \cdot (V_x^2(t) + V_y^2(t) + V_z^2(t))^{\gamma} \tag{1}$$

with $\gamma = 0.4$.

In our framework, segmentation of the 3D PC-MRA was automated by using the nn-UNet [12]. Training was completed on a Nvidia RTX A6000 GPU. The training dataset was manually segmented in 3D Slicer[1] by an experienced annotator. For further details regarding the training of the network, the reader is referred to [14].

Post-processing. After segmentation, the region of interest was isolated by zeroing velocities outside the segmentation mask. The data was converted to VTK image data (.vti) for ParaView[2] and the segmentation mask was used to build a triangular surface mesh (.stl) and a tetrahedral volumetric mesh (.vtk) using Gmsh[3] for volume and flow rate computations. This semi-automated process used Python[4] code and ParaView state files, requiring user input only for defining the locations of the four PVs and the MV. Further details can be found in [14].

2.2 Simulation Setup

CFD simulations incorporated passive movement of the LA from the longitudinal movement of the mitral valve annulus ring within a dynamic mesh approach based on the spring-based method. The ANSYS Fluent Solver 19.2 (ANSYS Inc., United States) was utilized for the simulations, using the finite volume method. Each simulation comprised three cardiac beats. Blood was characterized as an incompressible Newtonian fluid, with a density of 1060 kg/m^3 and a dynamic viscosity of 0.0035 Pa s. The time step was set to 0.01 s, with residuals for the

[1] Slicer 5.6.1 - https://www.slicer.org/.
[2] Paraview 5.11.0 - https://www.paraview.org/.
[3] Gmsh 4.11.1 - https://gmsh.info/.
[4] Python 3.12.0 - https://www.python.org/.

continuity equations set at 0.001. The boundary conditions comprised a pressure inlet at the PV and a blood flow velocity outlet at the MV. The former was obtained through catheterization from a patient with AF, while the latter was extracted from patient-specific Doppler US data, as shown in the second section of Fig. 1. More information about the simulation setup can be found in [15].

2.3 Flow Characterization

In this study, several metrics were extracted from 4D flow MRI and CFD data to assess the flow dynamics and compare the two methodologies. These metrics provided insights into the characteristics of blood flow within the LA, mostly focusing on the risk of thrombus formation.

Pathline Computation and Flow Trajectories. Mass-less particles were released at every timestep from the PVs to compute pathlines characterizing the flow entering the LA (see Fig. 2). The particles were evenly distributed, obtaining an equivalent density for PV. However, to ensure a uniform representation of each PV, the number of particles was normalized to the cross-sectional area of their PV of origin so as to account for the differences in size of the PVs. The particles were tracked throughout the cardiac cycle, and their pathlines were derived from the velocity vector field at the seed points placed in the PVs.

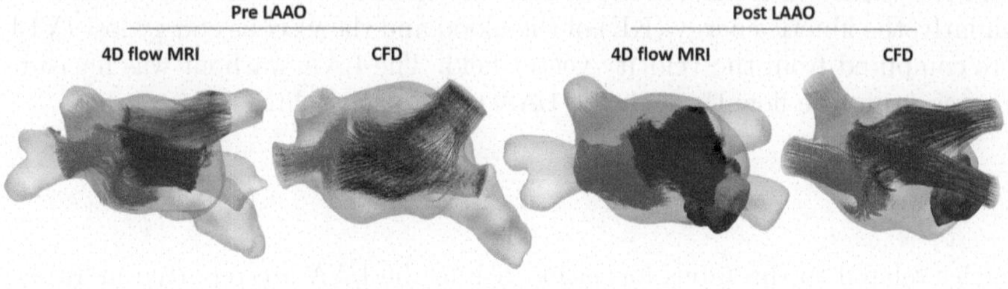

Fig. 2. Pathlines of the blood trajectory entering the left atrium (LA) through each pulmonary vein (PV), coloured by the PV of origin, for a case before and after LAAO, from CFD and 4D flow MRI data. LAAO: left atrial appendage occlusion; MRI: magnetic resonance imaging; CFD: computational fluid dynamics.

Automated Flow Collision Detection. An automated algorithm was developed to identify collision points of the flows from the four PV jets. This algorithm performed clustering of pathlines based on their PV origin (colored in Fig. 2) and computed an average flow trajectory for each PV. The algorithm first retrieved all pathlines originating from a specific PV and calculated the mean position of all particles sharing the same injection step (time of release). This provided an average trajectory with an associated standard deviation along the flow. The average trajectories were subdivided into segments of five particles, corresponding to five subsequent injection steps. The algorithm then calculated the average distance between segments of different pathlines.

This analysis was repeated iteratively, increasing the timestep, until a collision was detected or the MV opened. A collision was defined as occurring when three segments from different PV origins fell within a preset threshold distance, accounting for the standard deviation of the pathlines. If no collision was detected before the MV opened, the algorithm reset to $t = 0$ and slightly increased the threshold, allowing for a more permissive definition of collision in subsequent iterations.

Quantitative Flow-Derived Parameters. The quantitative metrics related to thrombogenicity included the LAA ostium flow, used to evaluate the washout capacity of the LAA, and the endothelial cell activation potential (ECAP). The ECAP, introduced by Di Achille *et al.* [16], was calculated as the ratio between the oscillatory shear index (OSI) and the time-averaged wall shear stress (TAWSS), focusing on hemodynamic forces near the vessel wall. As another way to measure blood stasis in the LAA, we monitored the number of particles that remained within the LAA after three complete cardiac cycles, along with the time each particle spent within the LAA cavity.

Furthermore, other extracted flow metrics include the Q-criterion$_{500}$, as a measure of vorticity, defined as the ratio of voxels with Q-criterion values above 500. Flow stasis was measured as the ratio of voxels with velocities below 0.1 m/s. Similarly, the kinetic energy (KE) of the blood and the viscous energy loss (VEL) were computed from the velocity vector field. The LAA washout was measured by integrating the flow through the LAA ostium after three cardiac cycles.

3 Results

Metrics related to thrombus formation risk in the LAA are reported in Table 1. These measurements could not be computed after LAAO implantation since the LAA is covered. The LA volume was higher in the persistent AF group, while the number of particles within the LAA after four cardiac beats was higher in the SR group, both in imaging data and simulations. The same was observed in the average residence time of the particles within the LAA and the number of particles remaining in the cavity for more than two beats. However, LAA washout results were inconsistent, with CFD reporting larger washouts in the AF group and 4D flow MRI reporting the opposite.

The reported effects of the LAAO were compared between imaging data and simulations by studying the changes in several flow-related parameters, as shown in Table 2. Particularly, the mean velocity and vorticity of the blood within the LA increased after the deployment, leading to a decrease in flow stasis. The ECAP also showed a decrease in all the conditions, except in the CFD analysis within the AF subpopulation.

Table 1. LA volume, number of particles in LAA, average time of residence of the particles in the LAA, count of particles that stayed for over two beats in the LAA, and LAA washout retrieved from pre-interventional CFD and 4D flow MRI data, for patients in SR and those in AF during the time of the data acquisition. LAA: left atrial appendage; RT: residence time; SR: sinus rhythm; AF: atrial fibrillation; MRI: magnetic resonance imaging; CFD: computational fluid mechanics.

	4D flow MRI		CFD	
	SR	AF	SR	AF
LA volume (cm^3)	106	203	106	203
N particles	493	441	334	314
Average RT	0.89	0.75	0.57	0.51
N particles > 2 beats	196	79	79	57
LAA washout (m^3)	2.1e−3	8.7e−4	1.6e−4	1e−3

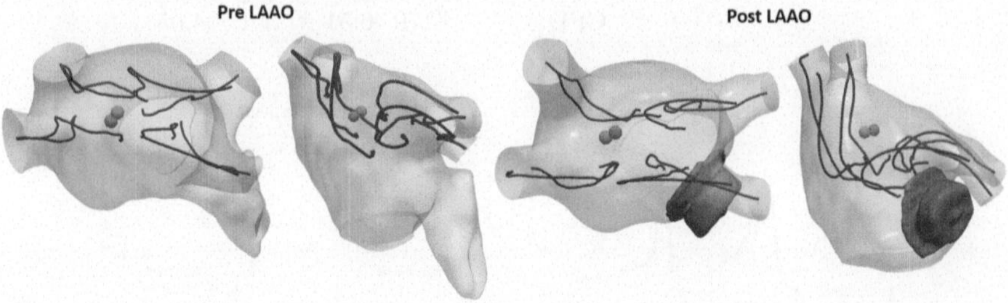

Fig. 3. Average pathline trajectory for each pulmonary vein (PV) and computed collision point, before and after the left atrial appendage occlusion (LAAO). In red, the average trajectory extracted from CFD data; in blue, the average trajectory computed from 4D flow MRI data. CFD: computational fluid dynamics; MRI: magnetic resonance imaging. (Color figure online)

The average pathline trajectory that each PV blood flow jet undergoes and the collision point are shown in Fig. 3. The overall trajectory of each blood flow jet is consistent between imaging data and simulations.

Finally, Fig. 4 shows the values of kinetic energy and viscous energy loss computed from CFD and 4D flow MRI data, before and after the deployment of the occluder. A second peak can be found both in KE and VEL from the CFD data, which considerably decreases after the intervention.

4 Discussion

Several efforts are being made to use fluid simulations to study heart flow dynamics. In particular, CFD analyses have been proposed as tools to aid in clinical decision-making, such as the deployment of the occluder in LAAO interventions [7]. To incorporate such models in the clinic, extensive verification and validation

Table 2. Mean velocity, flow stasis, ECAP and Q-crit$_{500}$ measurements pre- and post-intervention retrieved from 4D flow MRI and CFD data, for the analysed patients, divided into sinus rhythm and AF. ECAP: endothelial cell activation potential; Q-crit$_{500}$: ratio of voxels with Q-crit < 500; AF: atrial fibrillation; CFD: computational fluid mechanics; MRI: magnetic resonance imaging.

		Sinus rhythm		AF	
		Pre	Post	Pre	Post
Mean velocity (cm/s)	4D flow MRI	7.77	17.12	5.43	6.81
	CFD	10.32	17.2	10.98	12.78
Flow stasis (%)	4D flow MRI	32	11	62	32
	CFD	27	11	47	36
ECAP (1/Pa)	4D flow MRI	3.98	2.08	5.66	2.42
	CFD	1.89	0.99	1.3	1.38
Q-crit$_{500}$ (%)	4D flow MRI	3.04	2.43	0.35	0.4
	CFD	7.79	6.51	1.1	4.05

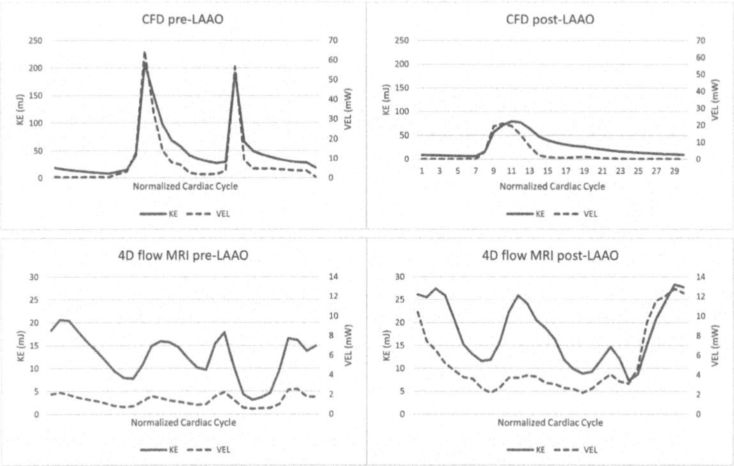

Fig. 4. Kinetic energy (KE) and viscous energy loss (VEL) before and after LAAO in a sinus rhythm patient, retrieved from 4D flow MRI and CFD data. LAAO: left atrial appendage occlusion; MRI: magnetic resonance imaging; CFD: computational fluid dynamics.

are imperative. This study proposes 4D flow MRI as a data source for validating simulations, enabling comparison of LA flow data between different modalities.

Given that most thrombi in patients with AF form in the LAA [2], considerable attention is focused on analyzing the risk of thrombus formation in this area [4–6]. Patient-specific analyses use particles as proxies to measure blood flow stasis and accumulation in the LAA [5]. Our cohort included two patients with permanent AF and two with paroxysmal AF. Higher levels of blood accumulation were observed in the paroxysmal AF group, with consistent results from both imaging data and simulations (see Table 1). Flow stasis in the LAA depends on numerous parameters, such as LAA geometry, LAA placement in relation to the PVs and LAA ostium size, among others [15]. Thus, although the persistent AF group showed more LA remodelling, other anatomical factors might be hindering LAA washout in the paroxysmal AF subgroup in our reduced cohort.

Understanding hemodynamic patterns is crucial, as certain blood flow dynamics in the LA are linked to a higher risk of thrombus formation [15]. Thus, reliable predictions of blood flow trajectories are essential. Imaging data and simulations successfully captured the flow direction from the right PVs, moving straight towards the MV due to the steep incline of the right PVs in the assessed geometry [15]. Both CFD and 4D flow MRI also captured the generation of the main vortex in the LA, indicated by the curved average pathlines of the left PVs (see Fig. 3). The computed points of collision are similar, suggesting consistent hemodynamic patterns between the two methods. Visualizing the average pathline trajectory of every PV and the point of collision simplifies vector field comparisons and reduces the dimensionality of the problem (see Fig. 2 and 3), providing a straightforward method to characterize blood flow patterns entering the cavity.

Another potential application of patient-specific fluid modeling is assessing the outcomes of the LAAO intervention [7]. This approach aims to find the optimal device choice and placement within the LAA for each patient, to avoid post-procedural complications such as device-related thrombus formation. There was good agreement on the effects of the intervention, showing improvement in all metrics and rhythmicity condition both from data and simulations. Table 2 shows larger stasis values and decreased vorticity in patients with persistent AF, which significantly improve after device deployment. This aligns with existing literature, as the absence of rhythmic contraction reduces the blood's kinetic energy and leads to the elimination of the main vortex [17].

ECAP values are notably higher in the 4D flow MRI (see Table 2), due to the relatively low resolution of the signal and the lack of a time-resolved segmentation, which leads to a mischaracterization of the actual blood flow dynamics near the cavity wall [18]. Lower velocities were consistently reported in 4D flow MRI (see Table 2) when compared to CFD data. This discrepancy is partly due to the dampening of peaks caused by the temporal averaging inherent in the acquisition protocol. The effect is even more pronounced in the AF subgroup, where the lack of signal periodicity amplifies the impact of temporal averaging [19].

While trends agreed between imaging data and simulations for most variables, the values differed significantly. These differences are not exclusively due to the CFD methodology, as there are some limitations in the current 4D flow MRI processing pipeline. Static segmentation ignoring wall motion and MV plane displacement can yield misleading results, particularly during SR when wall motion is greater. Acquisition protocols also produce unreliable data in low-velocity areas like the LAA due to a low signal-to-noise ratio [20]. Furthermore, the lack of a mass conservation constraint in 4D flow MRI suggests its velocity values may not be physiologically feasible. Incorporating physical regularizations could enhance accuracy [21]. Therefore, 4D flow MRI data should be cautiously interpreted and not considered ground truth due to these uncertainties.

Various other measurements of heart flow dynamics can be computed, including vorticity, KE, and VEL, as shown in Fig. 4. Although there is limited agreement between modalities, both report a restoration of the predominance of an early diastolic peak after the LAAO, indicative of a healthier phenotype. Additionally, CFD shows an increase in the KE/VEL ratio post-intervention, indicating a more efficient flow where more energy is used to transport the fluid with less loss through friction. The 4D flow MRI signal post-LAAO may be affected by device-related acquisition noise, which could impact the accuracy of the measurements.

Despite the inherent limitations of 4D flow MRI, it is currently the most comprehensive tool available for the in vivo characterization of cardiac fluid dynamics. Its ability to noninvasively capture patient-specific flow patterns is invaluable for both clinical assessments and as a benchmark for computational models. Therefore, it is imperative to not only acknowledge the shortcomings of 4D flow MRI but also to focus on refining CFD simulations to better match the empirical data. Enhancing the fidelity of CFD models through improved boundary conditions, incorporation of patient-specific wall motion, and more accurate representation of physiological conditions will bridge the gap between simulations and clinical data.

5 Conclusions

This study underscores the need for validating CFD simulations evaluating left atrial flow dynamics using 4D flow MRI data. By comparing a set of flow-derived measurements across methodologies in a cohort of patients who underwent LAAO, we found that despite inherent differences, both techniques identified trends correlating well with the expected intervention outcomes. Additionally, computing the average flow trajectory of each PV simplified the comparison of the reported flow dynamics. Extracting clinically relevant hemodynamic markers from both CFD and 4D flow MRI data will allow validation of the simulations, bridging the gap between advanced fluid dynamic analyses and practical clinical applications and ultimately improving patient outcomes in cardiovascular care.

Acknowledgments. This project has received funding from the European Union's Horizon 2020 research and innovation programme under grant agreement, No 101016496 (SimCardioTest) and No. 101136438 (GEMINI). We also acknowledge the support by the Spanish Ministry of Science, Innovation and Universities under the Proyectos the Generación del Conocimiento Programme (GENERALITAAT project, PIT2022-143239OB-I00). This work was also supported by the Agència de Gestió d'Ajuts Universitaris i de Recerca (grant No. 2024 FI-1 00419), and the Maria de Maeztu grant of excellence.

References

1. Alkhouli, M., Alqahtani, F., Aljohani, S., Alvi, M., Holmes, D.R.: Burden of atrial fibrillation-associated ischemic stroke in the United States. JACC: Clin. Electrophysiol. **4**(5), 618–625 (2018)
2. Alkhouli, M., Ellis, C.R., Daniels, M., Coylewright, M., Nielsen-Kudsk, J.E., Holmes, D.R.: Left atrial appendage occlusion: current advances and remaining challenges. JACC: Adv. **1**(5), 100136 (2022)
3. Watson, T., Shantsila, E., Lip, G.Y.: Mechanisms of thrombogenesis in atrial fibrillation: Virchow's triad revisited. The Lancet **373**(9658), 155–166 (2009)
4. García-Villalba, M., et al.: Demonstration of patient-specific simulations to assess left atrial appendage thrombogenesis risk. Front. Physiol. **12**, 596596 (2021)
5. Sanatkhani, S., Nedios, S., Menon, P.G., Bollmann, A., Hindricks, G., Shroff, S.G.: Subject-specific calculation of left atrial appendage blood-borne particle residence time distribution in atrial fibrillation. Front. Physiol. **5** (2021)
6. Pons, M.I., et al.: Joint analysis of morphological parameters and in silico haemodynamics of the left atrial appendage for thrombogenic risk assessment. J. Interv. Cardiol. **2022** (2022)
7. Mill, J., et al.: Patient-specific flow simulation analysis to predict device-related thrombosis in left atrial appendage occluders. REC: Interv. Cardiol. **3**(4), 278–285 (2021)
8. Markl, M., et al.: Assessment of left atrial and left atrial appendage flow and stasis in atrial fibrillation. J. Cardiovasc. Magn. Reson. **17**(Suppl 1), M3 (2015)
9. Morales, X., et al.: 4D flow magnetic resonance imaging for left atrial haemodynamic characterization and model calibration. In: Statistical Atlases and Computational Models of the Heart. M&Ms and EMIDEC Challenges: 11th International Workshop, STACOM 2020, Held in Conjunction with MICCAI 2020, Lima, Peru, 4 October 2020, Revised Selected Papers 11, pp. 156–165. Springer (2021)
10. Aldieri, A., Curreli, C., Szyszko, J.A., La Mattina, A.A., Viceconti, M.: Credibility assessment of computational models according to ASME V&V40: application to the bologna biomechanical computed tomography solution. Comput. Methods Programs Biomed. **240**, 107727 (2023)
11. Ferez, X.M., et al.: Feasibility of 4D flow cardiac magnetic resonance imaging for imaging surveillance after LAAO. JACC Cardiovasc. Interv. **17**(15), 1840–1842 (2024)
12. Isensee, F., Jaeger, P.F., Kohl, S.A., Petersen, J., Maier-Hein, K.H.: nnU-Net: a self-configuring method for deep learning-based biomedical image segmentation. Nat. Methods **18**(2), 203–211 (2021)
13. Bustamante, M., et al.: Atlas-based analysis of 4D flow CMR: automated vessel segmentation and flow quantification. J. Cardiovasc. Magn. Reson. **17**(1), 87 (2015)
14. Morales, X.: Characterization of left atrial hemodynamics with 4D flow magnetic resonance imaging, in-silico models and deep learning. Ph.D. thesis, Universitat Pompeu Fabra (2024)
15. Mill, J., et al.: The role of the pulmonary veins on left atrial flow patterns and thrombus formation. Sci. Rep. **14**(1), 5860 (2024)

16. Di Achille, P., Tellides, G., Figueroa, C.A., Humphrey, J.D.: A haemodynamic predictor of intraluminal thrombus formation in abdominal aortic aneurysms. Proc. R. Soc. Math. Phys. Eng. Sci. **470**(2172), 20140163 (2014)
17. Suwa, K., et al.: Characteristics of intra-left atrial flow dynamics and factors affecting formation of the vortex flow-analysis with phase-resolved 3-dimensional cine phase contrast magnetic resonance imaging. Circ. J. **79**(1), 144–152 (2014)
18. Petersson, S., Dyverfeldt, P., Ebbers, T.: Assessment of the accuracy of MRI wall shear stress estimation using numerical simulations. J. Magn. Reson. Imaging **36**(1), 128–138 (2012)
19. Lee, D.C., et al.: Three-dimensional left atrial blood flow characteristics in patients with atrial fibrillation assessed by 4D flow CMR. Eur. Heart J. Cardiovasc. Imaging **17**(11), 1259–1268 (2015)
20. Markl, M., Frydrychowicz, A., Kozerke, S., Hope, M., Wieben, O.: 4D flow MRI. J. Magn. Reson. Imaging **36**(5), 1015–1036 (2012)
21. Fathi, M.F., et al.: Super-resolution and denoising of 4D-flow MRI using physics-informed deep neural nets. Comput. Methods Programs Biomed. **197**, 105729 (2020)

Panoramic Anatomical Context in 3D Intracardiac Echocardiography (ICE) with 3D Registration and Geometry-Based Image Fusion

Julia Hickler[1(✉)], Lydia Neary-Zajiczek[2], Raja Sekhar Bandaru[2], Marcos Balle Sánchez[1], Christoph Hennersperger[1], and Stefan Wörz[1]

[1] LUMA Vision GmbH, Balanstraße 69b, 81541 München, Germany
{julia.hickler,stefan.woerz}@lumavision.com
[2] LUMA Vision Ltd., Beech Hill Road, Dublin D04 K5D0, Ireland
lydia.neary-zajiczek@lumavision.com
http://www.lumavision.com

Abstract. Intracardiac echocardiography (ICE) is finding increased use in interventional cardiology, but its significant learning curve for clinicians and imaging specialists in combination with increased costs act as barriers to wider use when compared with well-established modalities such as transesophageal echocardiography or fluoroscopy. Specific challenges for users include difficulty with spatial orientation, unfamiliar views of cardiac anatomy, limited field of view and the ergonomics of catheter placement. Here we present a framework for fusing 3D ultrasound (US) volumes captured in-vivo using a novel 360-degree 4D-ICE system into panoramic volumes with extended spatialinformation and improved signal-to-noise ratio to provide anatomical context for clinicians. Our approach uses 3D registration initialized with five degrees of freedom electromagnetic tracking and electrocardiogram (ECG) gating to create these panoramic volumes for visualization of key anatomical landmarks as well as positional information for the operator during complex cardiac procedures. Experimental results using in-vivo cardiac US pullback sequences of porcine specimens demonstrate the efficacy of the approach. We also compare the resulting panoramic volumes with preoperative computed tomography (CT) scans to validate identified structures.

Keywords: Intracardiac echocardiography · Ultrasound imaging · Interventional guidance · Image registration · Image fusion

1 Introduction

The use of intracardiac echocardiography (ICE) during cardiac interventions has been increasing in recent years due to rapid technological development of ultrasonic transducers that can be suitably integrated in intravenous catheters. These

J. Hickler and L. Neary-Zajiczek—These authors contributed equally to this work.

O. Camara et al. (Eds.): STACOM 2024, LNCS 15448, pp. 87–97, 2025.
https://doi.org/10.1007/978-3-031-87756-8_9

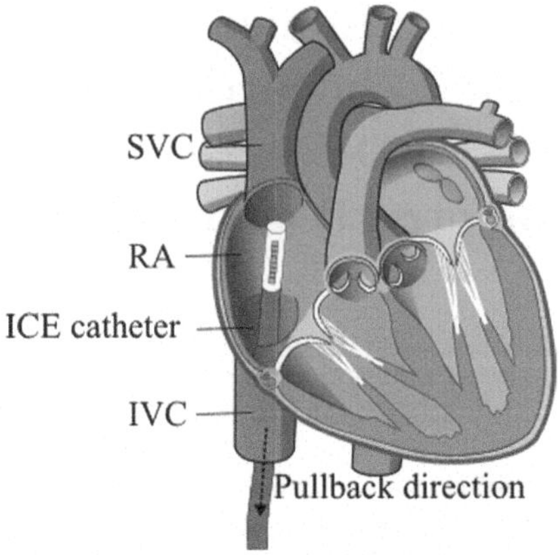

Fig. 1. Anatomy of the human heart and marked SVC-IVC pullback trajectory (modified based on image from Wikimedia Commons (Image from Stenemo (https://commons.wikimedia.org/wiki/File:Diagram_of_the_human_heart_(no_labels).svg), labels added by LUMA Vision, https://creativecommons.org/licenses/by-sa/4.0/legalcode))

devices allow for flexible and dynamic imaging of cardiac structures (see Fig. 1) without some of the key drawbacks of existing imaging modalities such as transesophageal echocardiography (TEE), transthoracic echocardiography (TTE) or contrast-enhanced fluoroscopy. In particular, removing the necessity for the general anaesthesia required for TEE, the inapplicability for certain patients (e.g., morphology, obesity) for TTE, as well as reducing radiation exposure of both patient and clinician alike as occurs with fluoroscopy are significant benefits to imaging with ICE [3]. One major disadvantage of ICE is its increased cost, as sterility requirements often necessitate single-use imaging catheters. Another disadvantage is the relative inexperience of many operators with ICE, or the unfamiliarity of interventionists with ultrasound (US) in general, often termed the "learning curve" associated with any new technique [2].

Since ICE was first described in 1981 [7], consistent technological development has produced a range of commercially available devices that consist of catheter-based US transducers and the associated processing and visualization consoles that generate the images for clinical use; such systems now offer good image quality and maneuverability, and many offer additional imaging modes such as full-colour Doppler or ultrafast US imaging [14]. Physical constraints on the size of the catheter-mounted transducer restrict the spatial resolution and therefore reduce the maximum achievable image quality, and often limit the field of view (FOV) due to the infeasibility of mounting large 2D array transducer configurations that are available for handheld TTE probes. Positioning the ICE catheter correctly to achieve the best anatomical view can also be difficult, particularly for less experienced operators [9]. The relatively recent introduction of 3D ICE with the availability of phased-array US transducers has further increased its usefulness as an imaging modality, particularly for a previously "forgotten" clinical intervention (tricuspid valve replacement or repair) that is extremely difficult to do with only TEE-based guidance [4].

Recent meta-analysis studies have concluded that ICE is safe and effective, resulting in "significantly lower fluoroscopy time, fluoroscopy dose, and shorter procedure time" for endocardial ablation of cardiac arrhythmia [8], but generally at increased cost. One study looking specifically at left atrial appendage occlusion however found that outcomes and costs were similar for ICE- and TEE-guided procedures [1]. Another study specifically examining ICE-guided transseptal punctures also confirmed the safety and effectiveness of ICE guidance as a replacement for fluoroscopy, but cautioned that this introduces additional risks when treating patients with cardiac implantable electronic devices (IEDs) such as pacemakers as the leads could be dislocated [16].

It is clear that ICE, despite its nominally higher cost than other modalities, is gaining greater acceptance by cardiac physicians due to reduction of procedure time and radiation dosage. The previously mentioned learning curve can be mitigated by providing additional anatomic context for the generated US images. Improved context and visualization would reduce the need for fluoroscopy-led guidance, the time required to interpret the images, and the risks for patients with IEDs. In this work we present our framework for providing patient-specific anatomical context for cardiac procedures based on registration and fusion of 3D ICE US image data. In particular, we consider so-called *pullback sequences* in which the ICE catheter is slowly retracted (about 10 mm/s) from the superior vena cava (SVC) to the inferior vena cava (IVC), passing through the right atrium (RA) as shown in Fig. 1. As a result, a 4D (3D+t) sequence of partly overlapping cardiac 3D US images is obtained that cover most of the heart. The goal of this work is to register and fuse all images of an ECG-gated pullback sequence into a large 3D panoramic US volume to provide anatomical context to the clinician. Compared to previous work on US registration (e.g., [5,13]), we use 4D (3D+t) US image data acquired with a novel 4D-ICE catheter with 360-degree view to reconstruct 3D panoramic volumes with extended spatial information. We also propose a novel geometry-based blending technique for image fusion that is well-tailored for 3D US image data by considering the distance of each pixel to the transducer. In contrast to [13], which is based on TTE images and where weighted blending based on overlap boundaries is used, we use 360-degree ICE images with a more realistic blending strategy for US data to reconstruct much larger 3D panoramic volumes of about $20 \times 20 \times 20\,\mathrm{cm}^3$ covering most of the heart and surrounding structures. Comparing the reconstructed panoramic US volumes with registered pre-operative CT images shows that relevant anatomical context is provided, which demonstrates the efficacy of the proposed framework.

2 Methods

First, we give a short overview of the ICE system and resulting 3D US image data (Sect. 2.1). Afterwards, we describe the proposed framework for generating

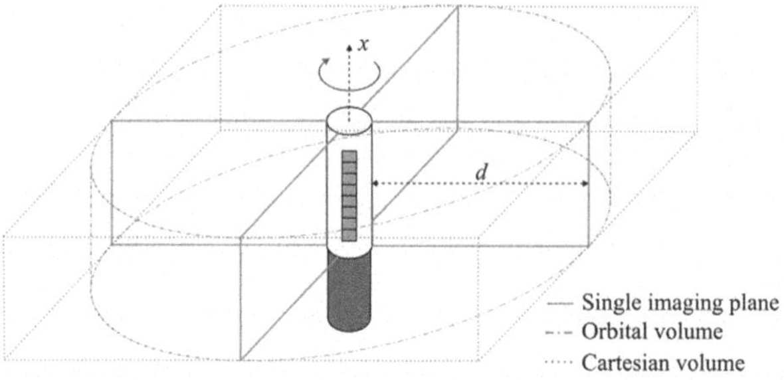

Fig. 2. Sketch of orbital geometry with imaging depth d.

3D panoramic US volumes (Sect. 2.2), which comprises of 3D image registration (Sect. 2.3) and a novel geometry-based image fusion (Sect. 2.4).

2.1 ICE System and Ultrasound Image Data

We use a novel ICE system developed by LUMA Vision[1] that comprises a 4D-ICE catheter with 360-degree view and a proprietary US processing and visualization software. The steerable catheter consists of a linear US transducer that can rotate freely around the catheter's axis inside a transparent outer sheath, enabling volume rates of up to 20 volumes per second. Figure 2 shows the orbital geometry with the rotating catheter in the center. To generate 3D-US images, high-framerate acquisitions are reconstructed with a pixel-based beamforming approach, yielding orbital B-mode imaging data. The US catheter also contains an electromagnetic (EM) tracking sensor. Additional EM tracking reference sensors are placed on the patient's chest and back to anchor the 3D position of the catheter in a patient-specific coordinate system. An ECG signal is also recorded and is synchronized with the US image data and the EM tracking data, allowing for generation of static panoramic volumes of single specific heart phases.

2.2 3D Panoramic Reconstruction

The clinical focus of this work are pullback sequences in which the ICE catheter is slowly retracted from the SVC to the IVC, passing through the RA as shown in Fig. 1. As a result, we obtain a 4D (3D+t) sequence of partly overlapping 3D US images that cover most of the heart. The goal of the subsequently described registration and fusion steps is to fuse all images of an ECG-gated pullback sequence into a large 3D panoramic US volume to provide anatomical context to the clinician.

We propose an iterative approach where EM tracking data of the very first 3D image defines the pose of the panoramic volume in the global coordinate system. The panoramic volume is initialized with the first image. Iteratively, the

[1] LUMA Vision Ltd., Dublin, Ireland, www.lumavision.com.

current image represents the moving image, which is registered to the previous fixed image (Sect. 2.3) to establish a chain of transformations to the first image. Each registered moving image is fused into the panoramic volume (Sect. 2.4).

2.3 3D Ultrasound-to-Ultrasound Image Registration

Ultrasound-to-ultrasound image registration, despite being a monomodal registration problem, is very challenging due to the unique properties of US imaging. In particular, for accurate and robust registration of US images the following US peculiarities need to be considered [5]:

- Use of similarity metrics suitable for registration where images are impacted by noise, intensity variations, depth-dependent contrast, and viewing angle dependency, all of which are present in US data from different viewing angles
- Integration of specific speckle removal steps to suppress speckle patterns in the image data, while enhancing visibility of anatomical structures
- Utilization of advanced beamforming strategies to minimize reconstruction-based artefacts such as aberrations

To address these peculiarities, we here propose and evaluate the following methods: (i) using an information-based multimodal similarity metric (mutual information) [12,15] (ii) pre-filtering the images to reduce speckle patterns using a deep-learning algorithm [10], and (iii) utilizing iMAP2 beamforming that minimizes the presence of aberrations [6].

Moreover, we use the EM tracking data to determine the relative transform between the fixed and moving images as an initial guess, and employ a conjugate gradient descent optimizer to determine the optimal geometric rigid transform between the two ECG-gated images. Accurate registration is necessary to account for any noise present in the EM tracking data in addition to systematic error introduced by breathing motion and/or imperfect ECG gating.

Optimal registration hyperparameters (e.g., similarity metric, optimizer step length) were determined using manually annotated 3D anatomical landmarks and minimizing the target registration error (TRE). Moreover, we use the TRE of a different set of manually annotated landmarks to evaluate the registration results. For registration we utilized the SimpleITK framework [11].

2.4 3D Geometry-Based Image Fusion

Given a moving image that has been transformed into the fixed image's coordinate system via registration as described above, the next step is to fuse it into the larger panoramic volume. Pixel-wise fusion strategies such as computing the pairwise minimum, maximum or average intensity can result in anatomical features becoming distorted due to the viewing angle dependence of US. In particular, "older" features can be washed out as newer images that still overlap introduce image data that has lower contrast and/or signal-to-noise-ratio as the pixels at that feature's location are further away from the imaging catheter. To

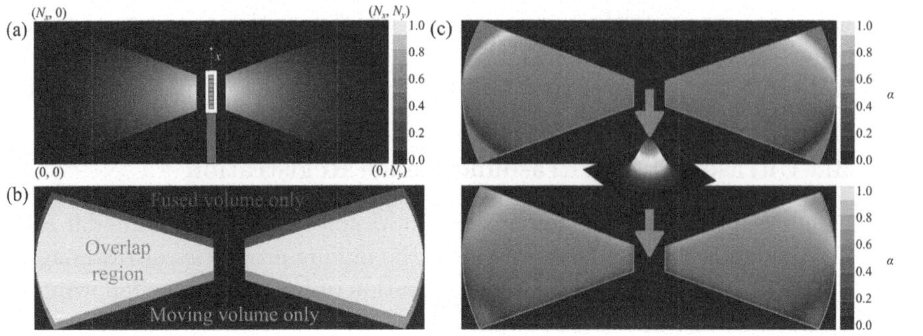

Fig. 3. (a) Cross section of 3D weight map, (b) padding strategy to bias the weight map, and (c) illustration of the effects of Gaussian blurring to reduce the presence of hard edges when fusing new image data into the panoramic volume.

address this, we have developed a novel geometry-based fusion method based on alpha blending that accounts for a pixel's position in space relative to the US transducer, which for any 3D image is well defined as shown in Fig. 2. One advantage of this weight-based blending approach is its short processing time for large US volumes while maintaining good blending results.

For a given pixel in the fused panoramic volume $I_F(i,j,k)$ that overlaps with the current moving image I_M,

$$I_F(i,j,k) := \alpha(i,j,k)\, I_F(i,j,k) + (1 - \alpha(i,j,k))\, I_M(i,j,k), \tag{1}$$

where $\alpha(i,j,k)$ is calculated from the respective image weights w_F and w_M of the fused volume and moving image as follows:

$$\alpha(i,j,k) = \frac{w_F(i,j,k)}{w_M(i,j,k) + w_F(i,j,k)}. \tag{2}$$

Here the location i, j, k refers to a 3D pixel index for an image of size $N_x \times N_y \times N_z$ (in our case $N_y = N_z$). Weight values w used for w_F and w_M are inversely proportional to the distance of a given pixel from the center of the US transducer, and can be calculated as follows:

$$\begin{gathered} r(i,j,k) = \sqrt{x^2 + y^2 + z^2}, \\ x = \frac{2}{N_x}(i-1), \quad y = \frac{1}{N_x}\left(j - \frac{N_y}{2}\right), \quad z = \frac{1}{N_x}\left(k - \frac{N_z}{2}\right) \\ w(i,j,k) = 1 - 3 \cdot r(i,j,k)/4, \end{gathered} \tag{3}$$

and the resulting image weight map w is clipped such that $0.001 \leq w \leq 1.0$. An example cross section of the weight map is shown in Fig. 3a.

After fusion, the image weights for the fused volume are updated such that $w_F(i,j,k) := \max\{w_F(i,j,k), w_M(i,j,k)\}$. Pixels with a value of 0 represent the image mask (area outside of FOV) and they are ignored during fusion. If the

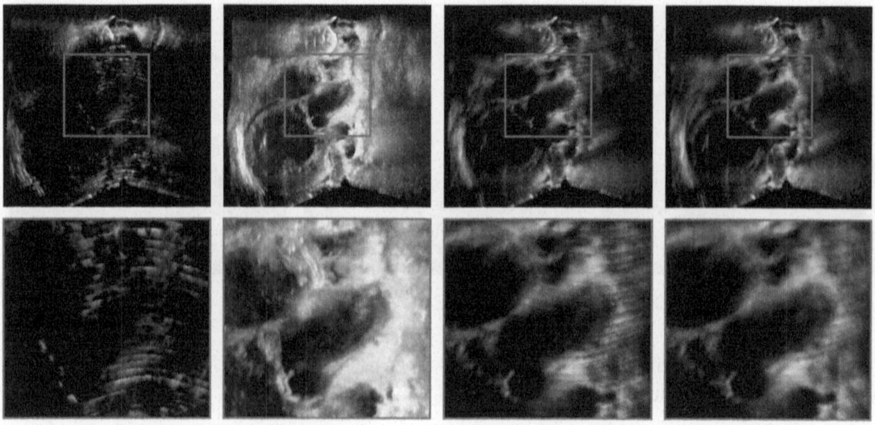
(a) (b) (c) (d)

Fig. 4. Comparison of different blending techniques: (a) no blending, (b) maximum intensity, (c) average intensity and (d) geometry-based fusion (Sect. 2.4).

fused volume contains a value of 0 at a given index, that pixel is updated with the current moving image's intensity at the same index. Pixel-wise fusion only occurs in the region where the nonzero areas of the fused volume and the moving image overlap, shown in yellow in Fig. 3b. Figure 3c (top) shows a typical map of α values that would be used to compute the fused intensities in this region according to Eqs. 1–3. Due to the strong viewing angle dependence of US images, the intensities of the fused volume and the blended or combined image can be significantly different. The edge of the α-value map as shown in Fig. 3c (top) can result in hard edges visible in the fused volume (see for example Fig. 4c). To address this, we pad the weight map immediately surrounding the overlap region with constant values, depending on which volume it is closest to. As larger alpha values prioritize the fused volume, we pad the region shown in dark blue in Fig. 3b with values of 1.0, and pad the region shown in cyan in Fig. 3b with 0.0 values to bias it towards the moving image. We then apply a Gaussian filter with a large standard deviation relative to the physical volume spacing to the α-value map to generate the map shown in Fig. 3c (bottom), which smoothly blends the combined image with the existing fused intensities and the newly-updated moving intensities.

3 Results and Validation

All US data presented here was acquired in-vivo using porcine specimens. In total, we have successfully applied the proposed framework to 37 pullback sequences acquired from seven different porcine specimens. For example, Figs. 4 and 5 show a panoramic volume generated from a typical pullback sequence. Note that this sequence was acquired with the ventilator turned off (referred to

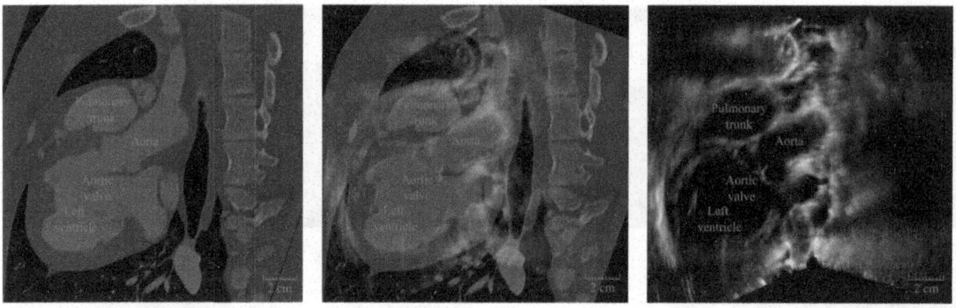

(a) CT scan (b) CT and fused US (c) Fused US

Fig. 5. Relevant anatomical features visible in (a) CT only, (b) CT with US panorama overlaid, and (c) US only.

as *breath hold* below) to minimize the impact of breathing motion. 3D US images captured at a mid-systolic ECG phase are fused using four different strategies to illustrate the efficacy of our proposed method (see Fig. 4). While the no-blending method (which never updates a fused pixel) is not able to maintain anatomical structures (Fig. 4a) and maximum intensity blending shows severe blending artefacts (Fig. 4b), average intensity blending and the proposed geometry-based fusion in Figs. 4c and 4d, respectively, are able to resolve the anatomical structures. Compared to average blending, the proposed geometry-based fusion method presents the anatomy even better, in particular, it has much less visible seams between the individual fused US volumes.

To validate our results, a pre-operative CT scan of each porcine specimen was acquired typically five days before the US imaging data was captured, and a post-operative 3D-angiogram was also recorded in which several fixed EM tracking sensors were visible. These visible reference sensors were used to manually orient the 3D-angiogram in the EM tracking data coordinate system. The pre-operative CT was then registered to the manually registered 3D-angiogram using a commercially available registration tool. For example, Fig. 5 highlights the image data and fused results for one porcine specimen. It can be seen that the CT scan and the fused panoramic US volume match very well, and that the panorama provides the spatial context that is missing in individual ICE images. The manual 3D segmentation of all four heart chambers (see Fig. 6) based on the panoramic US volume shown in Fig. 5c demonstrates how our proposed method can accurately show the anatomical context of the complete heart.

To demonstrate the efficacy of our US registration framework, we compared a fused panoramic volume generated from a *breath-hold* pullback sequence to another pullback sequence acquired immediately afterwards in the same specimen *without* breath hold, i.e. with an additional error introduced in the form of breathing motion. The breath-hold volume thus serves as an effective ground truth that can be validated using pre-operative CT data as demonstrated in Fig. 5. The ECG gating was chosen to closely match the CT scan. The initial guess for the registration was improved further by using an additional tracked

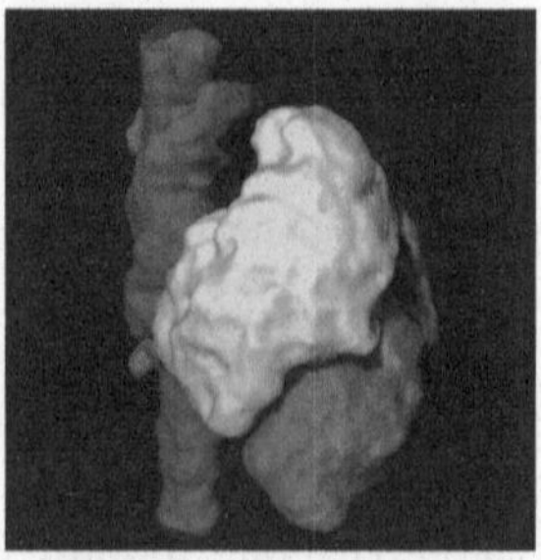

Fig. 6. Manual 4-chamber segmentation of panoramic US volume in Fig. 5c.

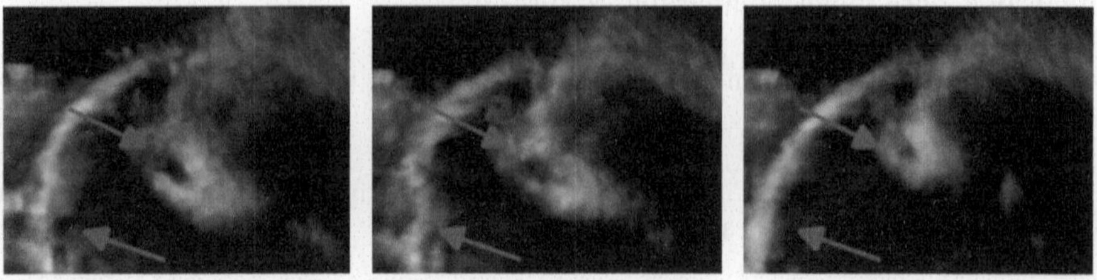

(a) TRE RMS: 1.9 mm (b) TRE RMS: 5.3 mm (c) TRE RMS: 4.1 mm

Fig. 7. Illustration of registration improvement (relevant areas highlighted by red arrows): (a) breath hold and (b) normal pullback using EM tracking only as well as (c) normal pullback with registration and breathing correction applied. (Color figure online)

EM sensor that was placed on the chest of the specimen that allowed us to estimate and correct for significant breathing motion. Observing anatomical features in the fused panoramic US volumes in Fig. 7, the registration (Fig. 7c) is able to correct some of the blurriness introduced by the breathing motion during the pullback sequence (cf. Fig. 7b). For each of the fused results shown in Fig. 7, a TRE was calculated based on manual 3D anatomical landmarks. The fused volume based on the *breath-hold* pullback has the lowest TRE of 1.9 mm (Fig. 7a). Both fusion results for the normal pullback (without breath hold) have a larger TRE with values of 5.3 mm without registration (Fig. 7b) and 4.1 mm with registration and breathing correction applied (Fig. 7c). Thus, applying the proposed registration instead of solely relying on EM tracking data achieved a notable reduction of TRE. However, the registration did not lead to a similar TRE as the breath-hold pullback. A reason for this is likely that rigid registration cannot fully correct the breathing-induced deformation of the heart.

4 Conclusion

We have presented a framework for providing patient-specific anatomical context for cardiac procedures based on registration and fusion of 3D ICE image data.

The framework uses 4D US image data acquired with a novel 4D-ICE catheter with 360-degree view as well as a novel geometry-based blending strategy to reconstruct 3D panoramic volumes. Experimental results using in-vivo cardiac US pullback sequences of porcine specimens demonstrate the efficacy of the approach, and a comparison of the reconstructed panoramic US volumes with registered pre-operative CT images shows that relevant anatomical context is provided. In future work, we will apply the proposed framework to human clinical ICE data obtained by using the same type of novel 4D-ICE catheter and perform a thorough validation. Since the device and modality are the same, the size and anatomy of porcine and human hearts are quite similar, and the framework does not involve any training step, we expect that image registration and fusion using human ICE data yields comparable results.

Acknowledgment. We thank Dr. Ivan Dudurych, MD, for providing the segmentation in Fig. 6.

Disclosure of Interests. The authors have no competing interests to declare that are relevant to the content of this article.

References

1. Alkhouli, M., Chaker, Z., Alqahtani, F., Raslan, S., Raybuck, B.: Outcomes of routine intracardiac echocardiography to guide left atrial appendage occlusion. JACC: Clin. Electrophysiol. **6**(4), 393–400 (2020). https://doi.org/10.1016/j.jacep.2019.11.014
2. Alkhouli, M., Hijazi, Z.M., Holmes, D.R., Rihal, C.S., Wiegers, S.E.: Intracardiac echocardiography in structural heart disease interventions. JACC Cardiovasc. Interv. **11**(21), 2133–2147 (2018). https://doi.org/10.1016/j.jcin.2018.06.056
3. Belani, K., Mahmood, F., Ortoleva, J.: Beyond the third dimension: intracardiac echocardiography-the next frontier for cardiac anesthesiologists. J. Cardiothorac. Vasc. Anesth. **35**(4), 979–981 (2021). https://doi.org/10.1053/j.jvca.2020.11.046
4. Chadderdon, S.M., et al.: Three-dimensional intracardiac echocardiography for tricuspid transcatheter edge-to-edge repair. Struct. Heart **6**(4), 100071 (2022). https://doi.org/10.1016/j.shj.2022.100071
5. Che, C., Mathai, T.S., Galeotti, J.: Ultrasound registration: a review. Methods **115**, 128–143 (2017). https://doi.org/10.1016/j.ymeth.2016.12.006
6. Chernyakova, T., Cohen, D., Shoham, M., Eldar, Y.C.: iMAP beamforming for high-quality high frame rate imaging. IEEE Trans. Ultrason. Ferroelectr. Freq. Control **66**(12), 1830–1844 (2019). https://doi.org/10.1109/TUFFC.2019.2933506
7. Glassman, E., Kronzon, I.: Transvenous intracardiac echocardiography. Am. J. Cardiol. **47**(6), 1255–1259 (1981). https://doi.org/10.1016/0002-9149(81)90255-1
8. Goya, M., et al.: The use of intracardiac echocardiography catheters in endocardial ablation of cardiac arrhythmia: meta-analysis of efficiency, effectiveness, and safety outcomes. J. Cardiovasc. Electrophysiol. **31**(3), 664–673 (2020). https://doi.org/10.1111/jce.14367
9. Grimsbo, M.C., et al.: Intracardiac echocardiography-guided TIPS: a primer for new operators. Semin. intervent. Radiol. **37**(4), 405–413 (2020). https://doi.org/10.1055/s-0040-1715875

10. Göbl, R., Hennersperger, C., Navab, N.: Speckle2Speckle: unsupervised learning of ultrasound speckle filtering without clean data (2022). https://doi.org/10.48550/arXiv.2208.00402
11. Lowekamp, B.C., Chen, D.T., Ibanez, L., Blezek, D.: The design of SimpleITK. Front. Neuroinform. **7** (2013). https://doi.org/10.3389/fninf.2013.00045
12. Maes, F., Collignon, A., Vandermeulen, D., Marchal, G., Suetens, P.: Multimodality image registration by maximization of mutual information. IEEE Trans. Med. Imaging **16**(2), 187–198 (1997). https://doi.org/10.1109/42.563664
13. Mao, Z., Zhao, L., Huang, S., Fan, Y., Lee, A.P.-W.: Direct 3D ultrasound fusion for transesophageal echocardiography. Comput. Biol. Med. **134**, 104502 (2021). https://doi.org/10.1016/j.compbiomed.2021.104502. https://www.sciencedirect.com/science/article/pii/S0010482521002961
14. Pham, T.H., Singh, G.D.: 3D intracardiac echocardiography for structural heart interventions. Interv. Cardiol. Clin. **13**(1), 11–17 (2024). https://doi.org/10.1016/j.iccl.2023.08.005
15. Viola, P., Wells, W.: Alignment by maximization of mutual information. In: Proceedings of IEEE International Conference on Computer Vision, pp. 16–23 (1995). https://doi.org/10.1109/ICCV.1995.466930
16. Žižek, D.: Intracardiac echocardiography-guided transseptal puncture for fluoroless catheter ablation of left-sided tachycardias. J. Interv. Card. Electrophysiol. **61**(3), 595–602 (2020). https://doi.org/10.1007/s10840-020-00858-z

Physics-Informed Neural Networks Can Accurately Model Cardiac Electrophysiology in 3D Geometries and Fibrillatory Conditions

Ching-En Chiu[1,2](✉), Aditi Roy[3], Sarah Cechnicka[4], Ashvin Gupta[5], Arieh Levy Pinto[5], Christoforos Galazis[4], Kim Christensen[6], Danilo Mandic[1], and Marta Varela[2,7]

[1] Department of Electrical and Electronic Engineering, Imperial College London, London, UK

[2] National Heart and Lung Institute, Imperial College London, London, UK
{ching-en.chiu18}@imperial.ac.uk, mvarela@sgul.ac.uk

[3] Department of Biomedical Engineering, Kings College London, London, UK

[4] Department of Computing, Imperial College London, London, UK

[5] Department of Bioengineering, Imperial College London, London, UK

[6] Department of Physics, Imperial College London, London, UK

[7] Cardiovascular and Genomics Research Institute, City St George's University of London, London, UK

Abstract. Physics-Informed Neural Networks (PINNs) are fast becoming an important tool to solve differential equations rapidly and accurately, and to identify the systems parameters that best agree with a given set of measurements. PINNs have been used for cardiac electrophysiology (EP), but only in simple 1D and 2D geometries and for sinus rhythm or single rotor dynamics. Here, we demonstrate how PINNs can be used to accurately reconstruct the propagation of cardiac action potentials in more complex geometries and dynamical regimes. These include 3D spherical geometries and spiral break-up conditions that model cardiac fibrillation, with a mean RMSE $< 5.1 \times 10^{-2}$ overall.

We also demonstrate that PINNs can be used to reliably parameterise cardiac EP models with some biological detail. We estimate the diffusion coefficient and parameters related to ion channel conductances in the Fenton-Karma model in a 2D setup, achieving a mean relative error of -0.09 ± 0.33. Our results are an important step towards the deployment of PINNs to realistic cardiac geometries and arrhythmic conditions.

Keywords: Cardiac Electrophysiology · Physics-Informed Neural Networks (PINNs) · Mathematical Modelling · Systems Biology · Parameter Identification · Atrial Fibrillation

---

Supplementary Information The online version contains supplementary material available at https://doi.org/10.1007/978-3-031-87756-8_10.

O. Camara et al. (Eds.): STACOM 2024, LNCS 15448, pp. 98–109, 2025.
https://doi.org/10.1007/978-3-031-87756-8_10

1 Introduction

Physics-Informed Neural Networks (PINNs) are a machine learning method that integrates data-driven learning with knowledge of the physical equations describing a system [8]. This domain knowledge is explicitly incorporated in the loss function of the neural network (NN). This ensures that PINNs' inferences are consistent with the physical understanding of a system and enables learning with only a small fraction of the data that conventional NNs require.

In cardiac electrophysiology (EP), PINNs have been used with eikonal models to predict arrival times of action potential in the left atrium. Sahli Costabal *et al.* initially estimated high-resolution arrival time maps using this approach [10]. Herrera *et al.* then built on it to estimate atrial fibre orientations [9]. Of most relevance to our study is the research of Herrero Martin *et al.*, which used PINNs with the monodomain equation on sparse maps of transmembrane potential to estimate EP parameters (such as the isotropic diffusion coefficient or surrogates of the action potential duration) [4]. This study was limited to 1D and 2D geometries and relied on the Aliev-Panfilov model, a simple two-variable model. In another work, PINNs were coupled with a more biologically detailed EP model, the Fenton-Karma model [3], to characterise the effects of anti-arrhythmic drugs. PINNs successfully estimated the effect of drugs on EP parameters related to the conductance of different ionic channels [2]. This work demonstrated PINNs' capability of working with experimental *in vitro* data, but was nonetheless limited to a simple 1D cable geometry.

To provide clinically useful characterisation of cardiac EP properties, PINNs will need to be deployed in 3D, using more biologically detailed EP models, and in both sinus rhythm and fibrillatory conditions. This will improve our understanding of the mechanisms of arrhythmias and help design personalised treatments that target regions with abnormal EP parameters. This is the gap this study aims to address.

Aims. We aim to use PINNs to predict the spatial-temporal propagation of cardiac action potentials:

1. Using the two-variable Aliev-Panfilov (AP) model [1],
   (a) in 3D spherical geometry, for a centrifugal wave (modelling sinus rhythm).
   (b) in 3D spherical geometry, for a single spiral wave (modelling tachycardia).
   (c) in 2D rectangular geometry, for spiral wave break-up (modelling fibrillatory conditions).
2. Using the three-variable Fenton-Karma (FK) model [3],
   (a) in 2D rectangular geometry, for a planar wave (sinus rhythm).
   (b) in 2D rectangular geometries, for a single spiral wave (tachycardia).

   For the FK model, we additionally use PINNs in inverse mode to simultaneously estimate global EP parameters, as the FK model can provide insights into more detailed EP properties. These include the diffusion coefficient, D, and parameters representing the conductance of the sodium, calcium and potassium channels. We estimate each parameter one at a time. In the following aims:

3. We use PINNs to predict the action potential propagation with one unknown EP parameter, and estimate the parameter simultaneously:
   (a) in 2D rectangular geometry, for a planar wave (sinus rhythm).
   (b) in 2D rectangular geometries, for a single spiral wave (tachycardia).

Next, we briefly introduce the AP and FK cell models used to model cardiac action potentials in the monodomain formulation. The full equations of the two models can be found in the Supplementary Materials.

Aliev-Panfilov Model. The model consists of one partial differential equation (PDE) and one ordinary differential equation (ODE), coupled together, describing fast and slow processes. The two equations describe the evolution of two variables: transmembrane potential V and a non-observable recovery variable W, related to the restitution, which enforces refractoriness properties. The variables are dimensionless: V is scaled to the $[0, 1]$ interval (AU). As in Ref [1], we used a temporal unit (TU) that corresponds to approximately 13 ms.

Fenton-Karma Model. The FK model is a more complex EP model consisting of one PDE and two ODEs, all coupled together. It has explicit formulations of three transmembrane currents: a fast inward J_{fi}, slow outward J_{so}, and slow inward J_{si}. These are analogous to the Na^+, K^+, and Ca^{2+} currents, respectively.

The three DEs describe the evolution of three variables: u, v, and w. Variable u is the dimensionless membrane potential scaled to $[0, 1]$ (corresponding to V in the AP model). Time t is in ms. v and w are latent gate variables for J_{fi} and J_{si}, respectively. There are various parameters in the model equations: τ_d, τ_o, τ_r, and τ_{si}, which are approximately inversely related to the conductance of different ion channels.

For both models, we adopt the no-flux Neumann boundary condition: $\frac{\partial V}{\partial \vec{n}} = 0$, which enforces no leakage of V outside of the domain (equivalently u in the FK model). We assume homogeneous and isotropic diffusion of the membrane potential throughout, and the diffusion coefficient D is therefore a scalar.

2 Methods

All code used in this study is available at: https://github.com/annien094/2D-3D-EP-PINNs.

2.1 Ground Truth Data Generation

Ground truth *in silico* maps of membrane potential V (or equivalently u in the FK model) in various geometries and dynamics are generated using in-house code written in Matlab. These generated data are used for the training and testing of the PINNs. 20% of the ground truth data are uniformly selected across time at each location to train the PINN solver. The remaining 80% are used to test the trained PINNs. The values of parameters used are detailed in Supplementary Table 1 &2.

3D Spherical Surface Geometry. The 3D geometry is a homogeneous spherical shell, with an inner radius of 10 cm and outer radius of 12 cm. We use the finite element method (FEM) in the Matlab PDE Toolbox to solve the AP model on a tetrahedral mesh (7670 elements, 13669 nodes, average edge length 1.61 cm). An explicit forward solver ($dt = 0.5$ TU) is used for the temporal domain.

We initialise electrical activity in two different ways. In the first case modelling sinus rhythm, the V of a small cuboid area (1 cm-side) around the north pole is raised at 0.8 AU at $t = 0$. This generates a wave that travels from the north pole down the shell with a circular wavefront. We will refer this type of wave as "centrifugal wave". In the second case modelling tachycardia, we use a cross-field protocol to generate a spiral wave: first generate a centrifugal wave; after some time, reset V for half of the sphere to zero to generate sustained spiral waves.

2D Rectangular Geometry. The 2D geometry is a square with 10 cm sides. We use central finite differences coupled with a forward Euler solver to solve the EP models. For the AP model, we use $dx = 0.1$ mm and $dt = 0.05$ TU as the spatial and temporal steps; for the FK model, we use $dx = 0.3$ mm and $dt = 0.01$ ms.

Planar waves are initiated by applying a supra-threshold external stimulus for 1 TU from one side of the rectangle. Spiral waves are created using the cross-field protocol. In the AP model simulations (Aim 1c), we observed that the generated spiral waves broke up spontaneously for the model parameters used.

2.2 PINNs Setup

We implement PINNs with the DeepXDE library [6] using a TensorFlow backend. We use a fully connected NN trained with the Adam optimiser for 150000 epochs, followed by the L-BFGS-B optimiser to facilitate convergence. Glorot initialisation and a tanh activation function are employed.

A schematic diagram of PINN's architecture is shown in Fig. 1. The inputs are space, x, and time, t, position coordinates. For the AP model, there are two output variables: V and W; for the FK model there are three: u, v, and w. In the 2D spiral break-up regime (Aim 1c), the network size is 5 layers of 60 neurons each, as we found that a larger network is needed for the more complex dynamics. For the rest of the scenarios, the network size is 4 layers of 32 neurons each. We apply a transform to the input layer: $t \rightarrow$ a vector of $\sin(kt)$ ($k = 1, 2, ...,$ up to the number of hidden layers) [6]. This helps capture the periodicity of the solutions, and we found that it leads to higher accuracies in PINN's predictions.

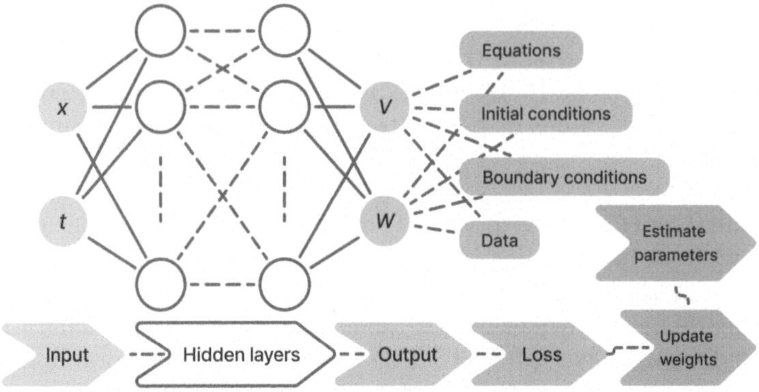

Fig. 1. The PINN architecture, using the AP model as an example for the output variables. The network takes the spatial and temporal location x and t as inputs, and outputs V and W. For the FK model, output variables are u, v, and w instead. PINN optimises its weights based on four different loss terms shown in (1). In inverse mode (Aims 3a & 3b), PINN also estimates one unknown parameter.

The PINN optimises a loss function consisting of various terms:

$$\begin{aligned}\mathcal{L} &= \mathcal{L}_{\text{eq}} + \mathcal{L}_{\text{IC}} + \mathcal{L}_{\text{BC}} + \mathcal{L}_{\text{data}} \\ &= \frac{1}{N_{\text{eq}}}\sum_{j=1}^{N_{\text{eq}}}(f_V(x_j,t_j)^2 + f_W(x_j,t_j)^2) + \frac{1}{N_{\text{IC}}}\sum_{l=1}^{N_{\text{IC}}}(V(x_l,t_0) - V_0)^2 \\ &+ \frac{1}{N_{\text{BC}}}\sum_{k=1}^{N_{\text{BC}}}(\frac{\partial V}{\partial \vec{n}}(x_k,t_k))^2 + \frac{1}{N_{\text{data}}}\sum_{i=1}^{N_{\text{data}}}(V(x_i,t_i) - V_{\text{GT}i})^2.\end{aligned} \tag{1}$$

$\mathcal{L}_{\text{eq}}$ quantifies the agreement with the EP models: AP (Supplementary Eq. 1a & 1b) or FK (Supplementary Eq. 2a–2c). Here, we use the AP model as an example, denoted by f_V and f_W. In the case of the FK model, $\mathcal{L}_{eq} = \frac{1}{N_{\text{eq}}}\sum_{j=1}^{N_{\text{eq}}}(f_u(x_j,t_j)^2 + f_v(x_j,t_j)^2 + f_w(x_j,t_j)^2)$. $\mathcal{L}_{\text{IC}}$ is the agreement with the initial conditions of the AP or FK equations. $\mathcal{L}_{\text{BC}}$ represents the agreement with the no-flux boundary condition for V, or u in FK model. $\mathcal{L}_{\text{data}}$ is the agreement with the training experimental ground truth data, denoted by V_{GT}. We only use data for the transmembrane potential V, or u in FK model, for the $\mathcal{L}_{\text{data}}$ term, since the other model variables are not experimentally measurable.

All of these agreements are quantified as mean squared errors and evaluated on the number of collocation points, N_{eq}, N_{IC}, N_{BC}, and N_{data}, detailed in Supplementary Table 3. Each term is given an equal weight, as preliminary tests showed that this led to the best outcomes.

For EP parameter estimation (Aims 3a & 3b), the unknown parameter of interest is included as an additional trainable parameter that the network will optimise as it trains with the combined loss in (1). The initial guess for each parameter is randomly taken from a uniform distribution centred at the ground truth value.

As the primary metric to assess PINNs' performance, we calculate the root mean square error (RMSE) between the ground truth and predictions for membrane potential V (or u in FK model) across all test points, which is 80% of all simulated data:

$$\mathrm{RMSE} = \sqrt{\frac{1}{N_{\mathrm{test}}} \sum_{i=1}^{N_{\mathrm{test}}} (V(x_i, t_i) - V_{\mathrm{GT}}(x_i, t_i))^2}. \tag{2}$$

We train the model using 1 RTX6000 GPU and 4 CPUs. Further details about the model hyperparameters can be found in Supplementary Table 3.

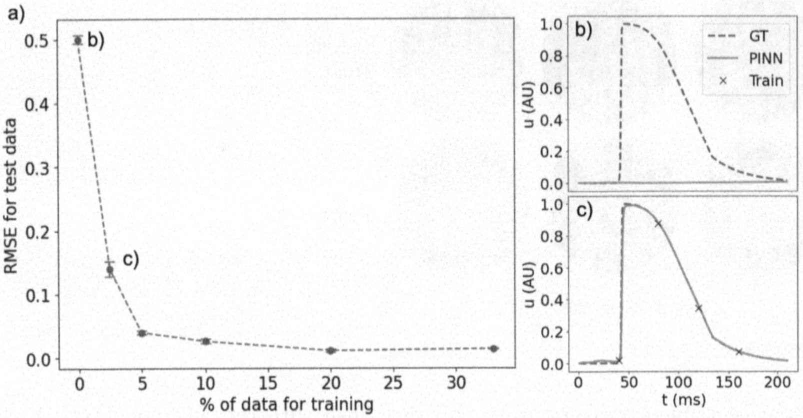

Fig. 2. The effects of training data size on PINNs' performance in forward mode. These experiments were conducted in the 2D planar wave scenario with the FK model (Aim 2a), where the total number of data points is 2.1×10^6. (a) The RMSE for test data, calculated as in (2), is plotted as a function of the percentage of training data given. (b) & (c) are examples of transmembrane potential u across time sampled at a fixed location, in the case of zero and 2.5% training data, respectively.

Choice of Train-Test Split Ratio. In forward mode, PINNs should technically be able to compute forward solutions without any training data $\mathcal{L}_{\mathrm{data}}$, but only using $\mathcal{L}_{\mathrm{eq}}$, $\mathcal{L}_{\mathrm{IC}}$, and $\mathcal{L}_{\mathrm{BC}}$, as this specifies a unique solution to the system of equations. It is with the inverse parameter estimation that data are necessary. To test this, we look at the effects of training data size on PINNs' performance in forward experiments, measured by RMSE on the test data. With zero data sample provided, PINNs struggled to learn as shown in Fig. 2b, with a poor RMSE of 0.50 ± 0.01. As the training data size increases, the RMSE first drops quickly and then plateaus (Fig. 2a). Based on these results, we keep $\mathcal{L}_{\text{data}}$ in the forward experiments and choose to use 20% of the data for training, as it is a balance between performance and size.

3 Results

3.1 Forward Solution of EP Models

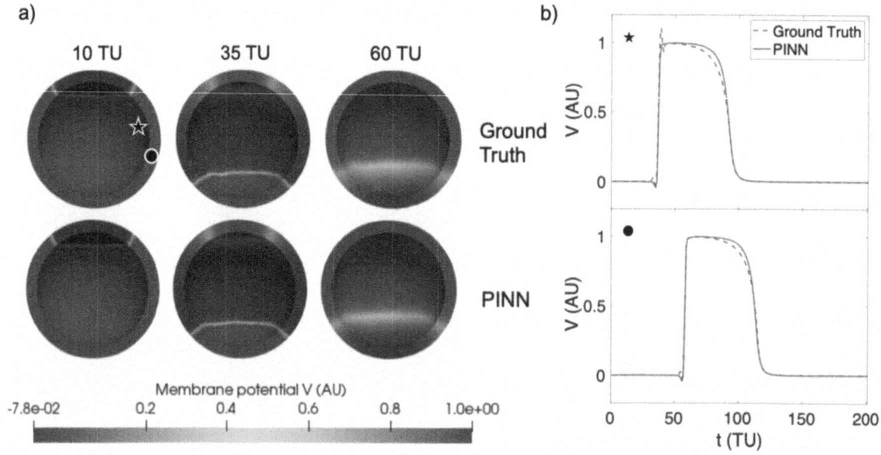

Fig. 3. A centrifugal wave on a 3D spherical shell with the AP model. (a) A visual comparison between the ground truth data from FEM solver and PINN's predictions, showing transmembrane potential V at various time points. The full movie can be found in Supplementary Video 1. (b) The time evolution of V sampled at two different locations, indicated by the star and the circle signs.

PINNs with AP Model in 3D. For a centrifugal wave on a spherical surface modelling sinus rhythm, PINNs accurately reproduced the action potential propagation with an RMSE of $(2.3 \pm 0.3) \times 10^{-2}$ on test data. A visual comparison between the ground truth data and PINNs' predictions is shown in Fig. 3. PINNs reproduces the action potentials well, with some discrepancies with the FEM solver at the wavefront and waveback (Supplementary Video 1.1).

For a spiral wave on a spherical surface, PINNs achieved an RMSE of $(3.7 \pm 0.5) \times 10^{-2}$ on test data, slightly higher than the planar wave case but still < 4% of the peak V value. Figure 4 gives the visualisation of the results. The biggest differences between the FEM ground truth data and PINN's prediction were found at the boundaries and tips of the spiral waves (Supplementary Video 2.1).

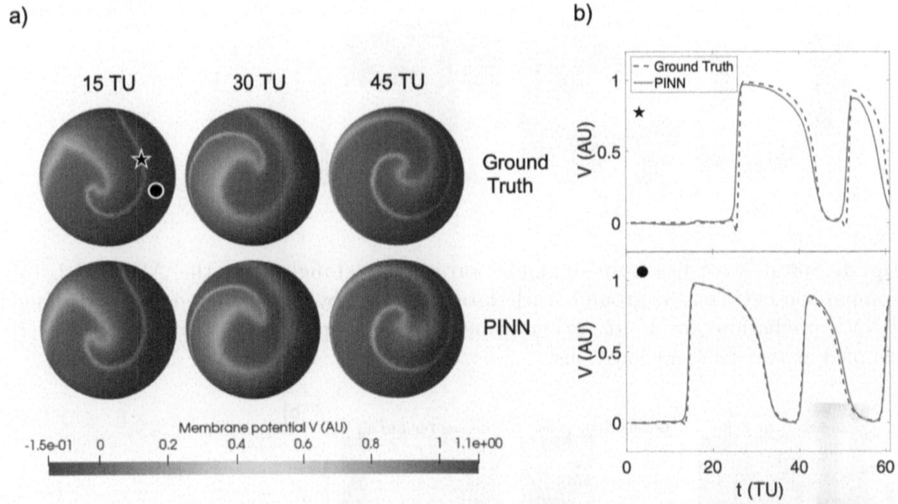

Fig. 4. A spiral wave on a 3D spherical shell with the AP model. (a) A visual comparison between ground truth data for V and PINN's prediction at various time points (full movie in Supplementary Video 2). (b) The time evolution of V sampled at two different locations.

PINNs with AP Model in Fibrillatory Conditions (2D). In the 2D spiral wave break-up regime, PINNs achieved an RMSE of $(5.1 \pm 2.6) \times 10^{-2}$ on test data (Fig. 5). The performance of PINNs in this scenario was particularly sensitive to the network initialisation, leading to a larger variance in the RMSEs.

PINNs with FK Model in 2D. Moving to the more complex FK model, for planar waves on a rectangle, PINNs achieved an excellent RMSE of $(1.1 \pm 0.2) \times 10^{-2}$. For the spiral wave, the RMSE increased to $(2.9 \pm 0.5) \times 10^{-2}$, comparable to the accuracy of PINNs using the AP model. The visualisations of these results are shown in Fig. 6 and 7.

3.2 Inverse Estimation of EP Parameters

For the FK model in 2D, we used PINNs in inverse mode to estimate the isotropic diffusion coefficient D, as well as parameters τ_d, τ_r, τ_o, and τ_{si} in Supplementary Eq. 3a–3c, which are approximately inversely related to ion channel conductances. The RMSE for u on test data in inverse mode was $(9.9 \pm 0.9) \times 10^{-3}$

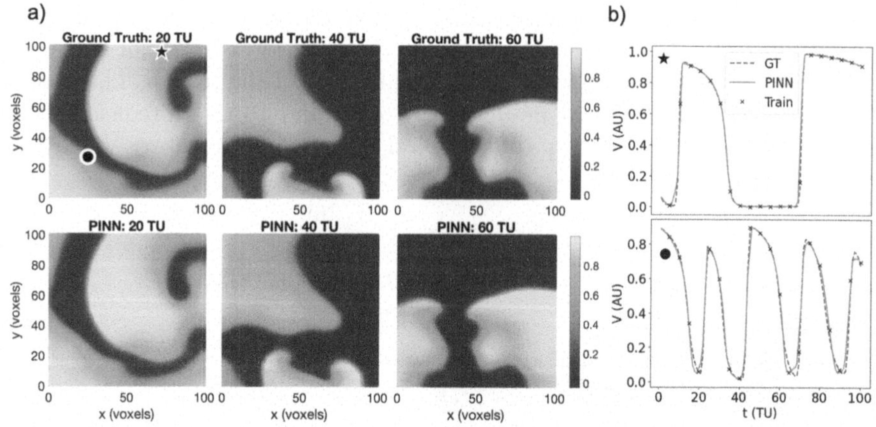

Fig. 5. Spiral wave break-up dynamics on a 2D rectangle with the AP model. (a) Comparison between the ground truth data generated by finite differences method and PINN's predictions for V at various time points (Supplementary video 3). (b) $V(t)$ sampled at two different locations.

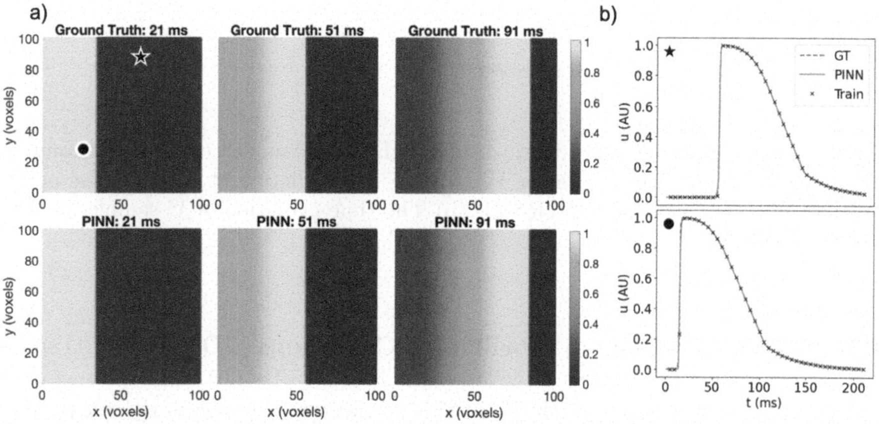

Fig. 6. A planar wave on a 2D rectangle with the FK model. (a) Comparison between the ground truth data and PINN's predictions for u at various time points (Supplementary Video 4). (b) $u(t)$ sampled at two different locations.

for planar waves and $(3.1 \pm 0.4) \times 10^{-2}$ for spiral waves, similar to the forward mode RMSEs. The parameter estimation results are summarised in Fig. 8.

PINNs were able to estimate the FK model parameters accurately in both 2D scenarios. For the Ca^{2+} and K^+ channel conductance (τ_{si} and τ_r), PINNs achieved relative errors (REs) $< 15\%$ for planar waves, and the REs were larger in the spiral wave regime. For τ_o, the REs were similar in both regimes, both

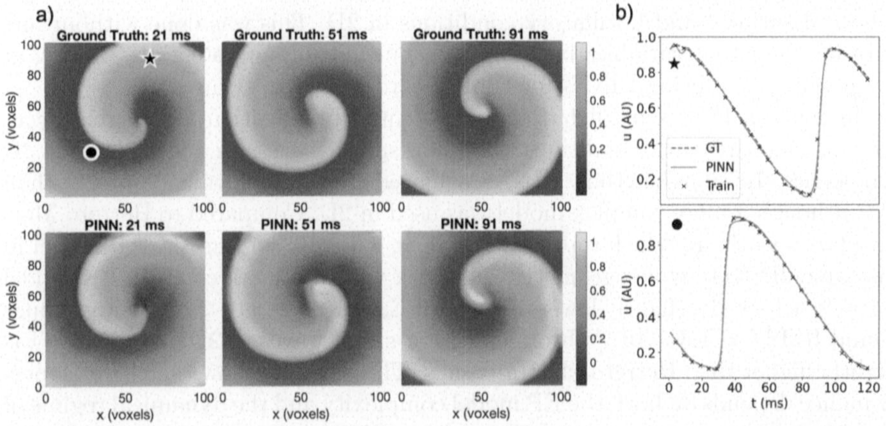

Fig. 7. A spiral wave on a 2D rectangle with the FK model. (a) Comparison between the ground truth and PINN's predictions for u at various time points (Supplementary Video 5). (b) $u(t)$ sampled at two different locations.

within 20%. For τ_d, related to the Na^+ channel, the average performance was actually better in the spiral wave regime ($< 0.5\%$). The estimation of the diffusion coefficient D in both regimes and τ_{si} for spiral wave were noticeably poorer ($> 40\%$) than all the other parameters ($< 40\%$).

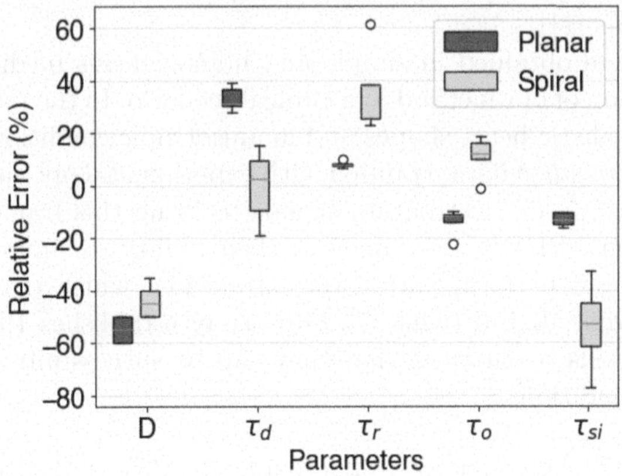

Fig. 8. Relative errors in the parameter estimates for the FK model in 2D, for both planar and spiral wave regimes.

4 Discussion

We present a PINN framework capable of accurately reconstructing cardiac action potentials in various geometries and dynamical regimes, including 3D spherical surfaces and fibrillatory conditions in 2D. This was done without any data for the latent variables but only the transmembrane potentials, hinting at the possibility of using PINNs for clinical cardiac digital twins [7].

In general, PINNs predicted action potential propagation at a higher accuracy in sinus rhythm conditions than for spiral waves, in both 2D and 3D geometries. In sinus rhythm, PINNs achieved better performance in 2D than 3D, although a more complex model was used in 2D. Compared to Herrero Martin *et al.*'s work [4], which reported RMSE $< 3.0 \times 10^{-2}$ using the AP model in the same 2D rectangular geometry, our work with the more complex FK model actually achieved a slightly lower mean RMSE, $(1.1 \pm 0.2) \times 10^{-2}$, and lower upper bound RMSE $< 1.3 \times 10^{-2}$ throughout. For spiral waves in 2D, our RMSE was slightly higher than Herrero Martin *et al.*'s. These results show that PINNs' performance depends on both the EP model complexity and the dynamical regime of the solution. Among all considered scenarios, PINNs performed the worst in the 2D fibrillatory condition, which is characterised by complex waveforms involving rapid non-periodic changes in transmembrane potential and latent variables.

PINN was additionally able to estimate, with varying degrees of accuracy (mean -0.09 ± 0.33), parameters related to ion channel conductances in the FK model in 2D geometries. This could have applications in understanding the effects of disease-induced remodelling or anti-arrhythmic drugs on different ion channels [2] in geometries more closely resembling the heart's. From our experiments, we note that a lower RMSE (2) does not necessarily correlate with a lower error in parameter estimation, which makes the refinement of parameter estimation more complicated. PINNs' convergence is also particularly difficult in inverse problems [11].

The results were obtained in simple and homogeneous media, which might limit the application of our method in a clinical scenario. In the future, we plan to train PINNs on realistic heart shapes and in anisotropic conditions. We also aim to extend beyond *in silico* data, training with experimental optical mapping data from cardiac tissue [5] or electrogram signals to bring this framework closer to clinical applications. PINNs are unique in their ability to simultaneously solve model equations and perform parameter estimation, which makes them great candidate for cardiac digital twins [7]. Our study establishes PINNs' potential for these applications by showing that they can be successfully deployed in 3D and fibrillatory conditions.

Acknowledgments. This work was supported by: St George's Hospital Charity, the NIHR Imperial Biomedical Research Centre (BRC) and the British Heart Foundation Centre of Research Excellence at Imperial College London (RE/18/4/34215). We acknowledge the computational support provided by the Imperial College Research Computing Service (DOI: 10.14469/hpc/2232).

References

1. Aliev, R.R., Panfilov, A.V.: A simple two-variable model of cardiac excitation. Chaos, Solitons Fractals **7**(3), 293–301 (1996)
2. Chiu, C.E., Pinto, A.L., Chowdhury, R.A., Christensen, K., Varela, M.: Characterisation of anti-arrhythmic drug effects on cardiac electrophysiology using physics-informed neural networks (2024). In: 2024 IEEE International Symposium on Biomedical Imaging (ISBI). pp. 1–5 (2024). https://doi.org/10.1109/ISBI56570.2024.10635234
3. Fenton, F., Karma, A.: Vortex dynamics in three-dimensional continuous myocardium with fiber rotation: filament instability and fibrillation. Chaos: Interdisc. J. Nonlinear Sci. **8**(1), 20–47 (1998)
4. Herrero Martin, C., et al.: EP-PINNs: cardiac electrophysiology characterisation using physics-informed neural networks. Front. Cardiovasc. Med. **8**, 2179 (2022). https://doi.org/10.3389/fcvm.2021.768419. https://www.frontiersin.org/articles/10.3389/fcvm.2021.768419/full
5. Lebert, J., Ravi, N., Fenton, F.H., Christoph, J.: Rotor localization and phase mapping of cardiac excitation waves using deep neural networks. Front. Physiol. **12**, 782176 (2021)
6. Lu, L., Meng, X., Mao, Z., Karniadakis, G.E.: DeepXDE: a deep learning library for solving differential equations. SIAM Rev. **63**(1), 208–228 (2021). https://doi.org/10.1137/19M1274067
7. Niederer, S.A., Lumens, J., Trayanova, N.A.: Computational models in cardiology. Nat. Rev. Cardiol. **16**(2), 100–111 (2019)
8. Raissi, M., Perdikaris, P., Karniadakis, G.E.: Physics-informed neural networks: a deep learning framework for solving forward and inverse problems involving nonlinear partial differential equations. J. Comput. Phys. **378**, 686–707 (2019). https://doi.org/10.1016/j.jcp.2018.10.045
9. Ruiz Herrera, C., Grandits, T., Plank, G., Perdikaris, P., Sahli Costabal, F., Pezzuto, S.: Physics-informed neural networks to learn cardiac fiber orientation from multiple electroanatomical maps. Eng. Comput. **38**(5), 3957–3973 (2022). https://doi.org/10.1007/s00366-022-01709-3
10. Sahli Costabal, F., Yang, Y., Perdikaris, P., Hurtado, D.E., Kuhl, E.: Physics-informed neural networks for cardiac activation mapping. Front. Phys. **8**, 42 (2020). https://doi.org/10.3389/fphy.2020.00042. https://www.frontiersin.org/article/10.3389/fphy.2020.00042/full
11. Wang, S., Sankaran, S., Wang, H., Perdikaris, P.: An expert's guide to training physics-informed neural networks. arXiv preprint arXiv:2308.08468 (2023)

Beyond the Standards: Fully-Automated Aortic Annulus Segmentation on Contrast-Free Magnetic Resonance Imaging Using a Computational Aorta Unwrapping Method

Enrique Almar-Munoz[1](✉), Mathias Pamminger[1], Christian Kremser[1], Markus Haltmeier[2], and Agnes Mayr[1]

[1] Department of Radiology, Medical University of Innsbruck, Anichstrasse 35, Innsbruck 6020, Tirol, Austria
enrique.almar@i-med.ac.at

[2] Department of Mathematics, University of Innsbruck, Technikerstrasse 13, Innsbruck 6020, Tirol, Austria

Abstract. This article presents a fully automated algorithm for segmenting the Aortic Annulus (AA) in non-contrast Cardiac Magnetic Resonance (CMR) imaging. This advancement is especially significant as it removes the need for contrast-enhanced CT coronary angiography (CCTA), the current gold standard for this task. This is crucial for patients with advanced chronic kidney disease or contrast allergies, where the use of contrast agents is contraindicated.

Selecting the appropriate valve graft during the planning of Transcatheter Aortic Valve Implantation (TAVI) relies on the precise measurement of the Aortic Annulus (AA). While numerous software tools are available for AA sizing using CCTA, there is none for CMR imaging. To address this gap, our approach utilizes two segmentation networks. The first network segments the aorta, a well-defined anatomical structure. Then, we apply an unwrapping technique to produce 2D maps of the aorta. These maps serve as input for the second segmentation network, identifying the hinge points necessary to delineate the AA structure and extract key TAVI parameters.

Comprehensive evaluations of our algorithm show a Root Mean Square Error (RMSE) of 2.671 mm in the axial direction and a Mean Absolute Error (MAE) of 1.940 mm in AA diameter. Statistical analyses reveal a correlation coefficient of 0.782 for AA diameter between the manual method and our method. A blind test further supports its reliability. Our algorithm proposes automating CMR-based AA segmentation to advance patient-friendly and low-risk imaging modalities.

Keywords: Aortic Annulus · Image-guided planning · MRI

O. Camara et al. (Eds.): STACOM 2024, LNCS 15448, pp. 110–121, 2025.
https://doi.org/10.1007/978-3-031-87756-8_11

1 Introduction

Valvular heart disease constitutes up to 20% of cardiac surgical procedures in the United States [1]. Transcatheter Aortic Valve Replacement (TAVI) is a minimally invasive procedure to replace a narrowed aortic valve, referred to as aortic valve stenosis, by utilizing catheter-based delivery. This procedure has undergone a systematic global adoption, starting with its application in elderly individuals deemed high-risk, subsequently extending to those categorized as intermediate risk, and presently encompassing low surgical risk patients [2,3].

Successful TAVI outcomes rely on precise image-based pre-procedural planning. A crucial aspect involves accurately measuring the Aortic Annulus (AA), a virtual plane at the base of the three aortic valve leaflets, known as hinge points. The AA serves as a reference point for positioning the implanted prosthetic valve. Its diameter, perimeter, and area dictate the type of graft. Radiologists manually identify the AA by navigating through correctly rotated slices until the aortic valve cusps are no longer visible. Those points are annotated as the hinge points.

Inadequate AA sizing may lead to complications such as paravalvular leaks or valve embolization, while excessive sizing can compromise prosthesis durability, instigate annulus rupture, and induce conduction abnormalities [4].

The current gold standard for this task is contrast-enhanced Cardiac Computed Tomography Angiography (CCTA) [5]. However, iodinated contrast is contraindicated in patients with advanced chronic kidney disease and contrast allergy [6,7]. Up to 80% of TAVI patients have chronic renal insufficiency [14], and using contrast agents in these patients can lead to acute kidney injury [15]. In Germany and Austria, 3% of TAVI patients require in-hospital renal replacement therapy, and 2% need chronic renal replacement therapy six months post-treatment [16], costing almost 10 million euros annually.

Cardiac magnetic resonance (CMR) imaging offers the advantage of avoiding ionizing radiation and contrast agents. In a prospective observational study, no significant differences were observed in perimeter, area, or diameter measurements between contrast-free CMR and cardiac computed tomography angiography (CCTA) [8,9]. Additionally, decisions based on MRI measurements regarding prosthesis sizing and transfemoral access did not modify TAVI strategy when compared to a CCTA-based choice [10]. A recent randomized study also demonstrated that CMR-guided TAVI was non-inferior to CCTA-guided procedures in terms of implantation success, despite the challenges CMR faces in delineating vascular calcifications [11].

Given the proven equivalence of these imaging modalities for this task, our primary objective was to develop the first fully automatic pipeline for extracting Aortic Annulus (AA) parameters using contrast-free CMR. While numerous studies have explored automatic AA estimation using CCTA, no such methods currently exist for MRI.

To achieve this, we employed a flattening technique to simulate the valve unwrapping process typically performed during physical post-mortem examinations [13] (Fig. 1). A similar approach was previously applied to the mitral valve in echocardiography [17], though their method involved creating a valve mesh, which was not feasible with our higher-noise CMR images.

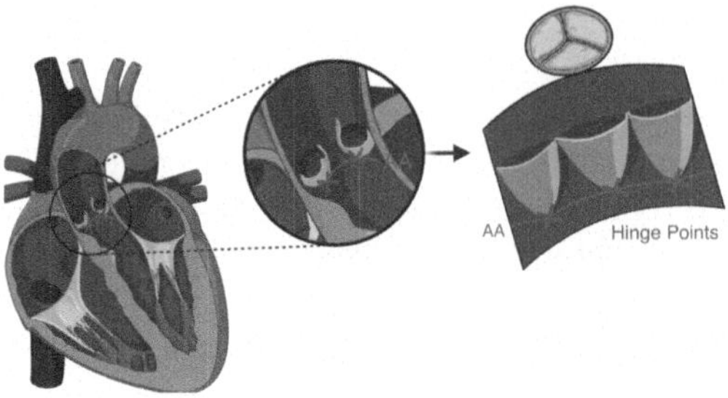

Fig. 1. Aortic valve region visualization and unwrapping simulation.

2 State-of-Art in Automatic TAVI Planning

In the literature, several automatic CCTA-based methods have been developed for AA estimation, each employing distinct techniques with varying precision. Elattar et al. [18] used a Fourier-flattening approach to detect the three hinge points, reporting AA center errors of 1.93 ± 0.90 mm, with a correlation coefficient of 0.84 for radius measurements. However, this method relies solely on the aorta's shape to find the hinge points, rather than following the medical definition of the joint between the leaflets and the aorta wall. In contrast, Santalo et al. [19] used a MeshDeformNet for 3D surface mesh generation, combined with a 3D Residual U-Net for landmark detection, achieving a high AA parameter correlation of 0.9. Saitta et al. [20] employed two 3D U-Net models for segmenting and identifying the AA, reporting a mean surface distance of 1.16 mm. Aoyama et al. [21] utilized a Spatial Configuration-Net to segment the aortic root and a 3D U-Net for detecting aortic cusps and hinge points, resulting in a perimeter error of -2.28 ± 8.78 mm. Atlas-based segmentation methods, such as those by Gao et al. [22] and Queiros et al. [23], yielded diameter errors of 0.60 ± 1.10 mm and 0.35 ± 0.70 mm, respectively. Kruger et al. [24] applied a cascade of CNNs combined with principal component analysis (PCA), achieving a mean diameter error of 1.84 ± 1.21 mm and correct implant size selection in 64% of cases.

3 Methods

3.1 Dataset

We employed a dataset of 78 navigator-gated non-contrast 3D whole-heart magnetic-resonance-angiography scans for TAVI candidates with axial slice dimensions of 320×240 mm (512×256 pixels) and the number of axial slices ranging from 72 to 96. All the scans were acquired in-house between 2015 and 2019. The dataset was partitioned into 59 training scans (4832 slices) and 19 testing scans (1504 slices). Demographic characteristics reveal a mean age of 86.83 ± 5.88 years, with a 37.29% male representation. The mean Body Mass Index (BMI) was $25.52 \pm 4.74\,\mathrm{kg\,m^{-2}}$, while the Body Surface Area (BSA) $1.75 \pm 0.20\,\mathrm{m^2}$. All individuals exhibited symptomatic severe aortic stenosis, confirmed by echocardiography.

3.2 Overview of the Method

Figure 2 shows the method pipeline. The initial step involves segmenting the aortic volume using MRI as input. Subsequently, we employed an angular projection method to simulate the aorta's unwrapping, optimizing the leaflets' visibility.

After generating 2D maps, the valve's commissures were segmented. We detected the segmentation central line and found the 2D hinge points. Later, we retrieved their corresponding 3D MRI coordinates and calculated the plane containing the three points. Ultimately, we extract the essential AA parameters.

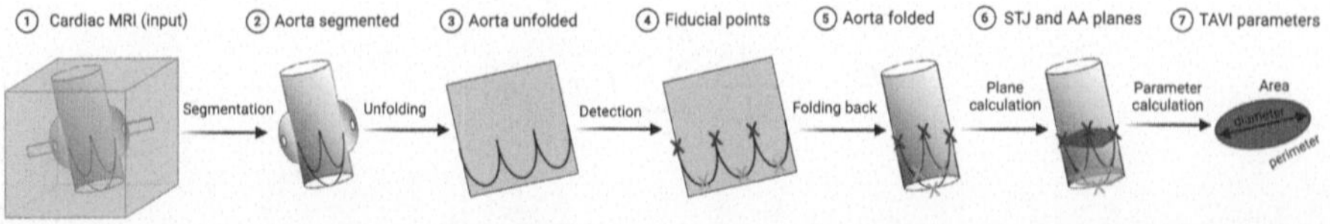

Fig. 2. Methodology's pipeline, spanning from the input CMR to the AA metrics.

3.3 Aorta Segmentation

For the aorta segmentation, we used a nnUNet [12]. Out of the 59 training scans, 30 underwent segmentation by a radiologist. Utilizing these segmentation masks, we trained the nnUNet and inferred masks for the remaining 29 scans. The training involved a 5-fold cross-validation strategy, running for 500 epochs. Early stopping was implemented to monitor the validation loss with a patience of 15.

3.4 From 3D to 2D: Aorta Unfolding

Starting with the segmented aortic volume, we performed erosion operations along the axial slices using a disk-shaped structural element with a radius of 1 pixel to remove the aortic wall components. Next, we focused on the attachments of the aortic valve leaflets by selectively removing the central part of the aorta mask while preserving a 12-pixel thickness around the aortic walls.

This pixel thickness was carefully chosen after thorough analysis to maximize the visibility of the aortic valve commissures. The thickness parameter represents a balance: reducing it too much can sharpen the image but makes it more sensitive to noise while increasing it too much can cause blurring and obscure important details.

We set the number of angular projections to $n = 100$. From the mask's center of each axial slice, we compute n projections, with the corresponding angles $I * (360/n)$, being i, the current projection. We calculated the mean of the non-zero elements among each projection line.

The unwrapped 2D map contains n columns and s (number of axial slices) rows. The coordinate system is organized such that when calculating the first angle of the initial slice, the mean projection is positioned at grid element $(0, 0)$. Subsequent projections are placed in $(0, 1)$ and so on for the first slice. Upon completing all angles for the first slice, we move to the next slice and position the mean results in the next row, specifically at $(1, i)$.

Let I be the input CMR, C_z the center of the mask for the z-th axial slice, α the angle of the i-th projection, $P_{i,z}$ the projection value at an angle α_i for slice z and p the points between the center and the image limits. The operation is summed up in Eq. 1, where $x_p = C_{z,x} + p \cdot \cos(\alpha_i)$ and $y_p = C_{z,y} + p \cdot \sin(\alpha_i)$.

$$P_{i,z} = \frac{1}{M_{i,z}} \sum_{p=1}^{M_{i,z}} I(x_p, y_p, z) \quad (1)$$

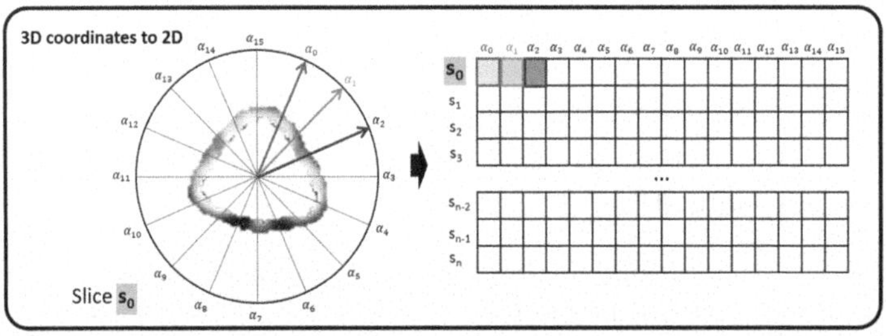

Fig. 3. Unfolded aorta map generation. The red arrow shows the mean calculation along the line for the first slice and projection. Its value is placed in $(0, 0)$. (Color figure online)

3.5 Fiducial Points Detection

Pre-processing 2D Maps: We enhanced the 2D maps by histogram equalization. As each image has dimensions of $n \times s$, with s variable for each case, we need to ensure uniform size before the model's training. Thus, we added rows at the bottom until the image reached $s = 80$. The pixel values for these rows were the average of the last 20 rows of the original image. Resize wasn't used to avoid altering the 2D-3D conversion system and to maintain isotropic pixels

Data Augmentation 2D Maps: The maps' initial column aligns with the first angular projection (0°). Starting with a different projection, such as 25 (90°), results in translated unwrapped maps along the x-axis but with the same information. For each case, four distinct unwrapped maps are generated by displacing longitudinally the data with a quarter, half, and three-quarters. To further increase variability, we introduce random contrast shifts, flips, and blurring.

2D-UNet for Commisure Segmentation: The model's architecture employs skip connections through concatenation during the up-sampling path, facilitating the retrieval of high-resolution features. The learning rate will dynamically decrease based on the validation loss. Early stopping is implemented, and validation loss is monitored. The validation split is set at 20%, batch size at 2, and epochs number at 100. The ground truth for training involved manual segmentation of the valve's commissures in the training set comprising 59 scans.

Post-processing and 2D Hinge Points Detection: From the binary predictions, we apply a modified skeletonize function to retain the central line of the segmentation. Additionally, an interpolation method ensures a value at every column in the mask. Subsequently, the algorithm identifies the three hinge points on this line by iteratively locating the absolute minimum point and removing the corresponding regions (12 columns on both the left and right sides).

3.6 From 2D to 3D: Aorta Folding

A 2D point's coordinates are denoted by axial slice and projection number (s, i). To obtain the 3D-CMR coordinates, we navigate to axial slice s and compute an angular projection with $\alpha = i \cdot (360/n)$. The final point on this line, which belongs to the aorta segmentation mask, represents the sought 3D-CMR point.

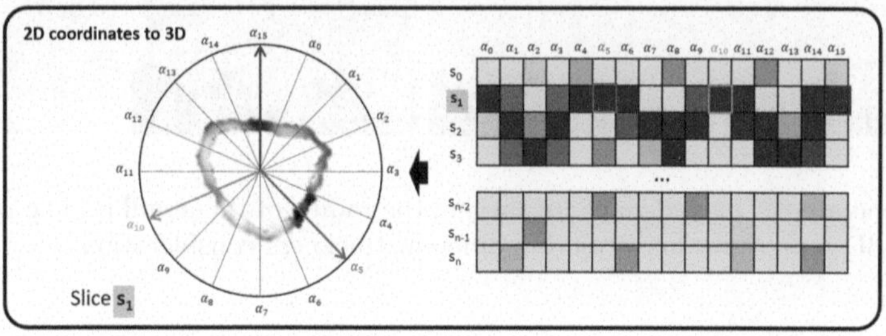

Fig. 4. Method to retrieve the 3D MRI coordinates. The red-marked pixel's row aligns with the axial CMR slice, s, and its column number, i, corresponds to the projection angle. We trace the angular projection from the center of the s slice until we reach the end of the aortic mask, obtaining the desired 3D point. (Color figure online)

3.7 Planes' Calculation

Using the 3D coordinates of the 3 hinge points, we determine the plane equation $Ax + By + Cz + D = 0$. The AA mask consists of voxels within the aorta mask that satisfy this plane equation.

4 Results

4.1 Aorta Segmentation

3D-UNet configuration achieved a Dice Score of 0.962 on the validation set for the aorta segmentation task. Post-processing techniques, specifically retaining the largest structure, further improved the Dice to 0.971.

4.2 Aorta Unfolding

Regarding unwrapped masks' generation, the computational time per CMR was 1.116 s. Figure 5 illustrates the different maps generated by varying the pixel thickness parameter. After visually inspecting the maps across the training cases, we found that a 12-pixel thickness provided the best visibility of the commissures.

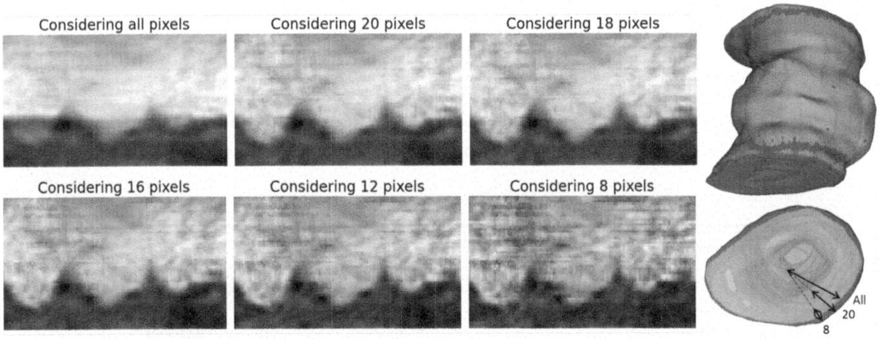

Fig. 5. Generated 2D valve maps by varying the radius of the annulus. The right part displays a 3D representation depicting different thickness considerations.

4.3 Fiducial Points Detection

When evaluating commissure segmentation, we use the Dice Score. But to ensure our metric is unbiased towards commissure thickness, we skeletonize both predicted and ground truth masks, obtaining the Error Mean Line (EML) between medium lines. Results without Data Augmentation yielded a Dice Score of 0.840 and an EML of 1.193 pixels in the axial direction. With Data Augmentation, the Dice Score increased to 0.848, and the EML decreased to 1.0152 pixels. Figure 6 visually presents aortic unwrapping maps with identified hinge points, segmented commissures and mean lines.

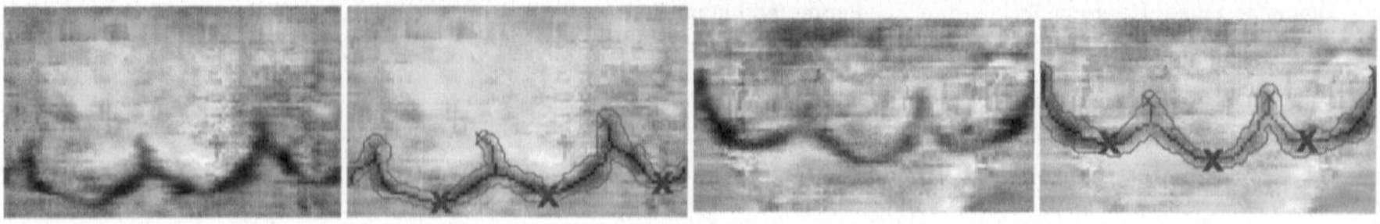

Fig. 6. Aortic valve flattening results (valve commissures, red; hinge points, blue). (Color figure online)

4.4 TAVI Parameter Estimation

We employed 19 CMR cases to test our method, categorized into 11 with average image quality and 8 with poor image quality. Figure 7 displays the results of the AA plane extraction process across different image quality scenarios.

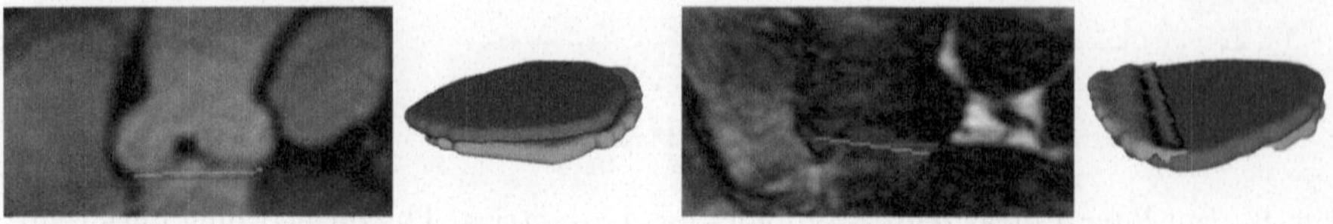

(a) Average image quality (b) Poor image quality

Fig. 7. Coronal CMR views and 3D meshes of the AA virtual planes. Green denotes our methodology, while red shows the groundtruth. (Color figure online)

Table 1 provides TAVI parameters derived from both methods, showcasing marginal metric superiority in all three metrics for our proposed methodology.

Table 1. Comparison of the AA clinical metrics.

	AA Diameter	AA Area	AA Perimeter
Groundtruth	24.845 ± 2.978 mm	487.196 ± 116.085 mm^2	78.390 ± 9.372 mm
Proposed method	26.392 ± 3.204 mm	549.616 ± 131.637 mm^2	83.296 ± 10.073 mm
Mean Absolute Error	$7.808 \pm 4.025\%$	$16.119 \pm 14.211\%$	$7.729 \pm 6.762\%$

The statistical analysis results indicate p-values of 0.1625 for diameters, 0.176 for area, and 0.1289 for perimeters (Fig. 8). These findings indicate that the two methods produce similar measurements with no significant difference. Substantial correlations are observed between the measurements of both methods, with coefficients of 0.782 for diameters, 0.768 for areas, and 0.783 for perimeters.

Technical metrics were calculated in the axial direction. The results were as follows: Mean Average Error (MAE) = 2.279 ± 0.922 mm, Root Mean Squared Error (RMSE) = 2.671 ± 1.111 mm, and Wasserstein Distance = 1.933 ± 0.920 mm.

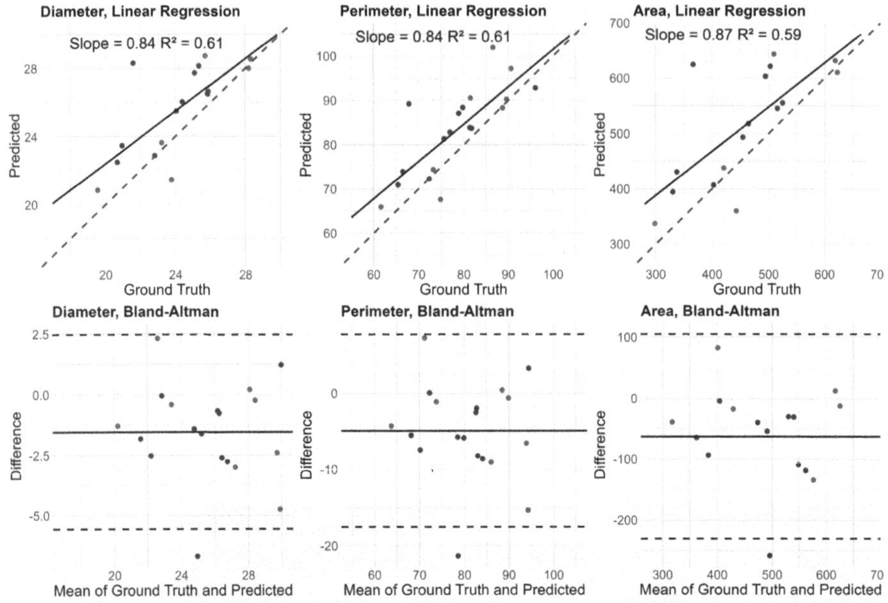

Fig. 8. *Top Row*: Linear regression plots for AA metrics. The dashed blue lines represent the line of identity, while the solid lines indicate the regression for all test data. *Bottom Row*: Bland-Altman plots. Blue dots for average quality, red for low quality. (Color figure online)

Additionally, a blind test by a second expert radiologist classified 11 cases as having similar definitions, making it challenging to discern superiority. In 4 cases, manual 3D annotation was considered better, while our method was deemed superior in 4 other cases. Regarding image quality, 2 out of the 4 cases favoring manual segmentation had low quality, while 1 out of the 4 cases favoring the automatic method exhibited low quality.

5 Discussion

This study presents a novel approach for TAVI planning using contrast-free CMR, offering a fully automated method for precise AA sizing. It introduces a simple but elegant approach for 3D-to-2D conversion, simplifying the problem. The proposed method is computationally efficient, with a total expenditure of 4.022s/CMR. Contrast-free CMR addresses the limitations of iodinated contrast administration, particularly in patients with chronic kidney disease. This is supposed a step forward to patient-friendly imaging modalities in TAVI planning.

While our initial approach involved directly segmenting the AA virtual plane, the results were inconsistent due to the low image contrast. Therefore, we adopted a more robust configuration using the two-segmentation network. The first segmentation of the aorta yielded excellent results thanks to its clear structure identification. Subsequently, we developed a technique that enhances the contrast of the leaflet commissure to facilitate the segmentation of the borders in the second network. This approach demonstrated robust performance even with low-quality scans. Moreover, employing data augmentation techniques enabled the model to perform effectively across a broader set of data.

Regarding the area of the aorta considered for the unwrapping, we found that a 12-pixel thickness was the most convenient configuration. That parameter does not adapt to the patient's ascending aorta diameter because the variation is minimal (IC 95% confidence mean diameter [23.67, 26.66]).

The Bland-Altman analysis indicates that a solitary instance among the cohort of 19 cases deviates beyond the established range of acceptability. This failure occurred due to a hinge point misplacement attributed to a confusing calcified region. The Mean Absolute Error (MAE) of 7.80% underscores the precision of our proposed methodology. Moreover, to quantify the significance of the AA Area discrepancies, it's essential to consider that each graft size has a recommended optimal AA Area interval to ensure a proper fit. For instance, the 23-mm Edwards Sapien 3 prosthesis is best suited for patients with an annular area between 338 and 430 mm^2. On average, across all graft sizes, the width of these optimal AA Area ranges is approximately 81.667 mm^2. When comparing our method, the mean area discrepancy of 16.119 mm^2 is smaller than the average range width which would require selecting a different graft size.

We conducted a test to determine how often the two methods diverged in graft size selection within our test set. This analysis was performed for two graft providers: Edwards Scientific and CoreValve. In 4 out of 19 cases, the methods differed, but in each instance, the size selected was the next closest option.

Moreover, the blind test conducted by a second expert radiologist adds an extra layer of validation to the proposed approach. Most cases were deemed comparable in quality, showcasing the algorithm's capability to rival manual segmentation. In cases where preferences were expressed, the distribution was relatively balanced between manual and automatic segmentation, indicating the subjectivity inherent in evaluating segmentation quality.

Our CMR-based automatic method yielded a correlation of 0.782 for AA diameters, which is slightly lower than the 0.84 reported by Elattar et al. [18] and the 0.9 noted by Santalo et al. [19], both using CCTA-based methods. These discrepancies could be attributed to the technology-inherent properties of CMR, which, in addition to lower spatial resolution, also entails more noise than CCTA.

We acknowledge that the dataset size is a limitation; however, the cohort size is relatively large given that CMR is not routinely used in standard clinical TAVI planning. This single-centre study includes 78 TAVI-CMR scans and is currently the sole investigation into automated CMR-TAVI planning.

6 Conclusion

In conclusion, our CMR based TAVI planning algorithm advances by automating pre-procedural assessments without iodinated contrast. The double-network architecture and the unwrapping technique effectively address the challenge of segmenting a virtual plane with limited structural information, adhering to the medical definition. Comprehensive evaluations and blind testing validate the reliability of our methodology.

References

1. Kapadia, S.R., et al.: 5-year outcomes of transcatheter aortic valve replacement compared with standard treatment for patients with inoperable aortic stenosis (PARTNER 1): a randomized controlled trial. Lancet **385**(9986), 2485–2491 (2015)
2. Fairbairn, T., et al.: Effect of transcatheter aortic valve implantation vs surgical aortic valve replacement on all-cause mortality in patients with aortic stenosis: a randomized clinical trial. Jama **327**(19), 1875–1887 (2022)
3. Deeb, G.M., et al.: 3-year outcomes in high-risk patients who underwent surgical or transcatheter aortic valve replacement. J. Am. Coll. Cardiol. **67**(22), 2565–2574 (2016)
4. Horehledova, B., et al.: Do CTA measurements of annular diameter, perimeter, and area result in different TAVI prosthesis sizes? Int. J. Cardiovasc. Imaging **34**, 1819–1829 (2018)
5. Blanke, P., et al.: Computed tomography imaging in the context of transcatheter aortic valve implantation (TAVI)/transcatheter aortic valve replacement (TAVR) an expert consensus document of the Society of Cardiovascular Computed Tomography. JACC: Cardiovasc. Imaging **12**(1), 1–24 (2019)
6. Little, S.H., Shah, D.J., Mahmarian, J.J.: Multimodality non-invasive imaging for transcatheter aortic valve implantation: a primer. Methodist Debakey Cardiovasc. J. **8**(2), 29 (2012)
7. Hawkey, M.C., et al.: Transcatheter aortic valve replacement program development: recommendations for best practice. Catheter. Cardiovasc. Interventions **84**(6), 859–867 (2014)
8. Jabbour, A., et al.: Multimodality imaging in transcatheter aortic valve implantation and post-procedural aortic regurgitation: comparison among cardiovascular magnetic resonance, cardiac computed tomography, and echocardiography. J. Am. Coll. Cardiol. **58**(21), 2165–2173 (2011)
9. Gopal, A., et al.: Noncontrast 3D CMR imaging for aortic valve annulus sizing in TAVR. Cardiovasc. Imaging **8**(3), 375–378 (2015)
10. Mooney, J., et al.: Imaging for structural heart procedures: focus on computed tomography. EuroIntervention **13**(AA), AA85–AA96 (2017)
11. Reindl, M., et al.: Cardiac magnetic resonance imaging versus computed tomography to guide transcatheter aortic valve replacement: a randomized, open-label, noninferiority trial. Circulation **148**(16), 1220–1230 (2023)
12. Isensee, F., et al.: nnU-Net: a self-configuring method for deep learning-based biomedical image segmentation. Nature Methods **18**(2), 203–211 (2021)
13. Contino, M., et al.: A geometric approach to aortic root surgical anatomy. Eur. J. Cardio-Thoracic Surg. **49**(1), 93–100 (2016)
14. Klug, G., et al.: Cardiac magnetic resonance imaging versus computed tomography to guide transcatheter aortic valve replacement: study protocol for a randomized trial (TAVR-CMR). Trials **23**(1), 726 (2022)
15. Goldfarb, S., et al.: Contrast-induced acute kidney injury: specialty-specific protocols for interventional radiology, diagnostic computed tomography radiology, and interventional cardiology. Mayo Clin. Proc. **84**(2) (2009)

16. Thongprayoon, C., et al.: AKI after transcatheter or surgical aortic valve replacement. J. Am. Soc. Nephrol. **27**(6) (2016)
17. Lichtenberg, N., et al.: Mitral valve flattening and parameter mapping for patient-specific valve diagnosis. Int. J. Comput. Assisted Radiol. Surg. **15** (2020)
18. Elattar, M., et al.: Automatic aortic root landmark detection in CTA images for preprocedural planning of transcatheter aortic valve implantation. Int. J. Cardiovasc. Imaging **32**, 501–511 (2016)
19. Santaló-Corcoy, M., et al.: TAVI-PREP: a deep learning-based tool for automated measurements extraction in TAVI planning. Diagnostics **13**(20), 3181 (2023)
20. Saitta, S., et al.: A CT-based deep learning system for automatic assessment of aortic root morphology for TAVI planning. Comput. Biol. Med. **163**, 107147 (2023)
21. Aoyama, G., et al.: Automatic aortic valve cusps segmentation from CT images based on the cascading multiple deep neural networks. J. Imaging **8**(1), 11 (2022)
22. Gao, X., et al.: Quantification of aortic annulus in computed tomography angiography: validation of a fully automatic methodology. Eur. J. Radiol. **93**, 1–8 (2017)
23. Queirós, S., et al.: Automatic 3D aortic annulus sizing by computed tomography in the planning of transcatheter aortic valve implantation. J. Cardiovasc. Comput. Tomogr. **11**(1), 25–32 (2017)
24. Krüger, N., et al.: Cascaded neural network-based CT image processing for aortic root analysis. Int. J. Comput. Assisted Radiol. Surg. **17**(3), 507–519 (2022)

Coronary Artery Calcium Scoring from Non-contrast Cardiac CT Using Deep Learning with External Validation

Hyunho Mo[1(✉)], Daniel Bos[1,2], Maryam Kavousi[2], Maarten J. G. Leening[1,2,3], and Esther E. Bron[1]

[1] Biomedical Imaging Group Rotterdam, Departments of Radiology and Nuclear Medicine, Erasmus MC University Medical Center Rotterdam, Rotterdam, The Netherlands
h.mo@erasmusmc.nl

[2] Department of Epidemiology, Erasmus MC University Medical Center Rotterdam, Rotterdam, The Netherlands

[3] Department of Cardiology, Erasmus MC University Medical Center Rotterdam, Rotterdam, The Netherlands

Abstract. Coronary artery calcium (CAC) scores are a crucial biomarker identifying asymptomatic individuals at high risk for cardiovascular disease. CAC scores using gated coronary computed tomography (CT) have been assessed by semi-automatic methods requiring human experts' contribution. We present an artificial intelligence (AI)-based automatic CAC scoring system employing image classification models after whole heart segmentation. The proposed method uses a deep learning (DL) model to detect and discard non-coronary artery calcification among all candidate calcium lesions. We developed and evaluated the DL model using a set of 435 scans publicly available for research. The model was trained on 348 scans and tested on 87 scans. The proposed method was validated internally on the test set and externally on an independent set of 1453 scans from a prospective cohort study. Computed scores by our method were compared to the manual reference standard. In our internal validation, computed scores strongly correlated with the reference scores; Pearson's r = 0.953 (95% confidence interval [CI] 0.938–0.967) and Spearman's ρ = 0.991 (95%CI 0.983–0.999); their agreement regarding Agatston risk categorization was evaluated by Cohen's κ = 0.949 (95%CI 0.924–0.975). Agreement on the external validation set was reached to Pearson's r = 0.833 (95%CI 0.791–0.875), Spearman's ρ = 0.917 (95%CI 0.909–0.926), and Cohen's κ = 0.774 (95%CI 0.764–0.783). These results demonstrate that AI has the potential to achieve reliable automatic CAC scoring on independent data from the data used for its development.

Keywords: Cardiovascular diseases · Computed Tomography (CT) · Calcium Scoring · Whole Heart Segmentation · Deep Learning

O. Camara et al. (Eds.): STACOM 2024, LNCS 15448, pp. 122–131, 2025.
https://doi.org/10.1007/978-3-031-87756-8_12

1 Introduction

Globally, cardiovascular disease (CVD) is the leading cause of mortality [1]. In Europe, more than 11 million new cases of CVD occur each year [2]. In this disease, the importance of primary prevention is well known [3] and provides an opportunity for early intervention with, for example, personalized treatment recommendation [4]. However, such primary prevention requires the identification of individuals at increased risk of CVD.

For this identification, clinicians currently use risk prediction models commonly based on atherosclerotic CVD risk factors [5] and non-contrast computed tomography (NCCT) of the heart. In particular, coronary artery calcium (CAC) scoring from gated coronary CT is one of the most important independent predictors [6]. Though its association with the outcome might improve predictive accuracy, the clinical practice for calcium scoring has a major limitation: it requires expert observation [7] that is labor-intensive and time-consuming.

Calcium scores are generally obtained by a semi-automatic process; specialized commercial software provides all the possible calcium lesions on axial CT slices, and experts then adjust them to achieve precise coronary calcium segmentation by excluding non-coronary calcium lesions [8]. Various automatic coronary calcium scoring approaches have been proposed to address this limitation. One of the earliest works proposes a supervised classification-based approach using a support vector machine to distinguish the coronary calcium from all the candidates [9]. In a different approach, coronary artery segmentation on coronary NCCT using CT angiography (CTA) atlas is followed by the k-nearest neighbors algorithm to detect calcium in the coronary arteries [10]. More recent work also has a two-stage scheme [11]; it uses two consecutive convolutional neural networks (CNNs) to detect coronary artery calcifications.

In this work, we propose an automatic coronary artery calcium scoring method that uses multi-atlas segmentation for whole heart segmentation (WHS) and a DL model as a supervised classifier for correcting false positives (FP). It was developed using 8 CTA atlas scans and a public dataset of 435 annotated CT scans. We externally validated the proposed method on an independent dataset comprising 1453 CT scans from a prospective cohort study.

The main contribution of this study is a thorough evaluation of our fully automatic scoring pipeline. We quantitatively validate the performance of the proposed method using a large set of imaging data acquired on CT scanners different from those used for collecting development data. This shows that our method combining multi-atlas segmentation with a deep learning classifier may serve as a useful automatic scoring tool, even when trained on a public dataset comprising a limited number of samples.

2 Methods

2.1 CAC Scoring Pipeline

Our pipeline aims to segment calcified regions of the coronary artery for the computation of the CAC score. As shown in Fig. 1, it consists of two consecutive

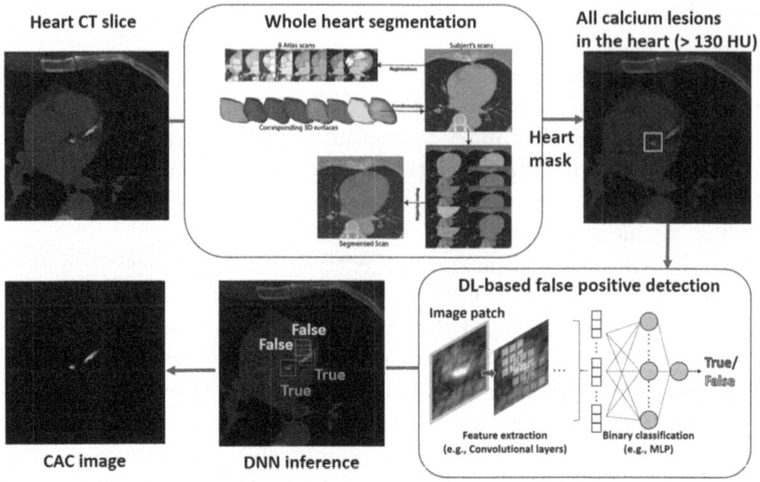

Fig. 1. Overview of the proposed CAC scoring method.

steps of (1) WHS and (2) FP detection; the first step corresponds to pinpointing the heart, and the following step mimics the correction step by experts to rule out non-coronary calcification.

For the first step, we use a multi-atlas segmentation approach [12,13] to obtain segmentation masks of the whole heart. Here, multiple atlas images are CTA scans. The cardiac atlas was drawn using student manual labeling on the 8 CTA acquired in the Erasmus MC, University Medical Center Rotterdam, The Netherlands. A radiologist then performed a correction step. The CTA images are registered using an affine registration. We use majority voting to fuse all the transformed surfaces.

Next, all the possible candidate calcium lesions in the segmented heart are extracted based on the threshold, which is set to a CT density of 130 Hounsfield units (HU) spanning a size greater than 1 mm^2 [14]. The second step is to detect FPs that denote some lesions extracted by the threshold but not coronary calcium so that those can be excluded when we compute calcium scores.

FP detection is performed by predicting the presence of CAC for fixed-size patches containing candidate calcium lesions. The patch size is set to 45×45 pixels ($\approx$250 mm^2). The patch image we defined has three channels; the first channel represents pixel values in the patch, and the remaining two channels convey the patch's x and y coordinates to localize it.

The FP detection is a binary classification task for the given patch image showing the shape and spatial location of the calcium lesion. Our method classifies the images using CNNs. We employed the backbone architecture of our model from the image classification models developed on the standard benchmark for large-scale visual recognition [15], considering that the computer vision community has already proved the generalizability of those architectures on a

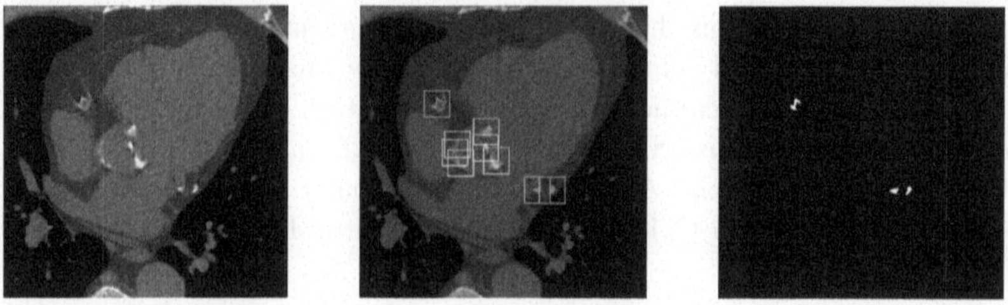

(a) (b) (c)

Fig. 2. (a) An axial CT slice of one subject from the external validation dataset, (c) showing cases in which the proposed method successfully segmented CAC based on (b) the output of the FP detection model where the yellow bounding boxes indicate FPs in the aortic valve. (Color figure online)

wide range of tasks and datasets [16]. In detail, we evaluated the four widely used image classification models to decide the backbone DL architecture of our FP detection model following the motivation highlighted in Sect. 2.1. One of the representative CNNs, VGG [16], contains merely convolutional layers and fully-connected (FC) layers. MobileNetV3 [17] is the latest MobileNet [18] that is a lightweight DNN using depthwise separable convolutions. CoAtNet [19] is a new family of models combining CNNs and Transformers [20] in the direction of achieving better trade-offs between accuracy and efficiency. For each DL model, we always chose the least complex architecture among the configurations proposed in the original paper to prevent overfitting.

After obtaining the CAC images, we computed the calcium score, Agatston score [14], following the criteria introduced in the original paper [14].

2.2 Imaging Data

Stanford AIMI COCA Dataset. To develop the FP detection model, we used the COCA dataset [21] containing de-identified annotated coronary CT scans acquired using a dual-source 128-slice multi-detector CT (MDCT) scanner (Somatom Definition FLASH, Siemens). There are 500 gated CT scans; we used 435 with at least one coronary calcium lesion. We generated 7401 patches, and 29.6% are *negative*. The age and gender of the participants are not known. We computed the Agatston score based on manual label calcifications for internal validation. The mean of the score is 488 (standard deviation [SD], 837).

Rotterdam Study. Our work was externally validated on the Rotterdam Study (RS) [22], a population-based cohort including adults (aged $\geq$ 55 years) living in Rotterdam, Netherlands. Specifically, we used gated coronary CT scans of 2524 participants who agreed to undergo a scan, out of 3229 invited participants randomly sampled from the participants visiting the research center between 2003 and 2006. In our work, the dataset used for external validation includes

1453 participants based on the following criteria: an upper age limit of 85 and a valid reference standard calcium score > 0. The scans were acquired using 16-slice or 64-slice MDCT scanners (Somatom Sensation 16 or 64, Siemens). The mean age of participants is 68.5 (SD, 5.8), and 36.9% are women. The reference standard score refers to the Agatston score manually obtained by the trained scan readers using Syngo Calcium Scoring software. The mean of the score is 299 (SD, 578).

Table 1. Summary of the comparative analysis for the FP detection method using three different DL models. The performance evaluation is based on mean ± standard deviation across 5 independent runs based on the data split described in Sect. 3.1.

DL models	Metrics			
	Accuracy	AUC	Sensitivity	Specificity
VGG [16]	**0.927 ± 0.008**	**0.965 ± 0.008**	0.943 ± 0.010	**0.888 ± 0.025**
MobileNetV3 [17]	0.901 ± 0.008	0.943 ± 0.012	**0.947 ± 0.014**	0.791 ± 0.062
CoAtNet [19]	0.923 ± 0.009	0.956 ± 0.011	0.946 ± 0.015	0.872 ± 0.032

3 Experiments and Results

3.1 Evaluation of the FP Detection Model

We randomly split the COCA participants into five non-overlapping folds and executed five independent runs of training and testing, using each fold as a test set once. For each run, a randomly selected one fold from the remaining four folds served as a validation set to monitor validation loss every training epoch and apply early stopping. The remaining three folds were used to train the model.

Based on the comparative analysis shown in Table 1, we chose VGG as the backbone of our FP detection model since it provides the best mean values as well as the smallest standard deviations in terms of three out of the four binary classification evaluation metrics. Our code specifying the computational setup and training details is available online[1].

3.2 Agreement of the Computed Score and the Reference Score

The proposed method combines the WHS and the FP detection model. In particular, we employed the VGG model as the FP detection model, which is trained on the COCA dataset under one of the train/test settings (out of five) that provides the best AUC. Our internal validation was based on one fold consisting of 87 scans that were not used for training.

Our method was externally validated on the RS data specified in Sect. 2.2 through assessing agreement between automatically and manually determined Agatston scores. The assessment is based on the three inter-rater agreement

[1] https://github.com/hyunho-mo/auto_cac.

metrics: Pearson's r for score values, Spearman's ρ for score ranks, and (linearly weighted) Cohen's κ for Agatston risk categories outlined in Fig. 4 and 6. We included both Pearson's r and Spearman's ρ to evaluate inter-rater agreement at not only the continuous level but also the ordinal level. The former is a primary metric showing the accuracy of the score predictions, and the latter is an auxiliary metric showing the capability to discriminate individuals with at least high levels of calcification in ordinal scales.

As a compared method, we computed the Agatston score for images immediately after WHS to highlight the contribution of the FP detection model.

Table 2. Results of our internal validation on the 87 scans of the COCA dataset, in terms of three agreement assessment metrics with 95% CIs through a bootstrap resampling (100 repetitions).

Metrics	Methods	
	WHS	WHS+FP detection model (The proposed method)
Pearson's r (95%CI)	0.776 (0.669–0.883)	0.953 (0.938–0.967)
Spearman's ρ (95%CI)	0.954 (0.931–0.979)	0.991 (0.983–0.999)
Cohen's κ (95%CI)	0.859 (0.826–0.893)	0.949 (0.924–0.975)

In this work. subjects with a reference CAC score of zero were excluded from validation (n > 500 for RS data) to avoid a situation where the agreement assessment is dominated by the majority of very low-risk individuals, thereby allowing a more useful evaluation for clinical implementation.

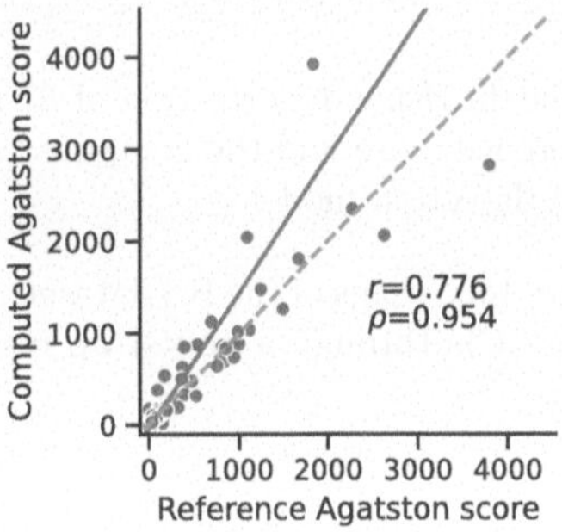

(a)

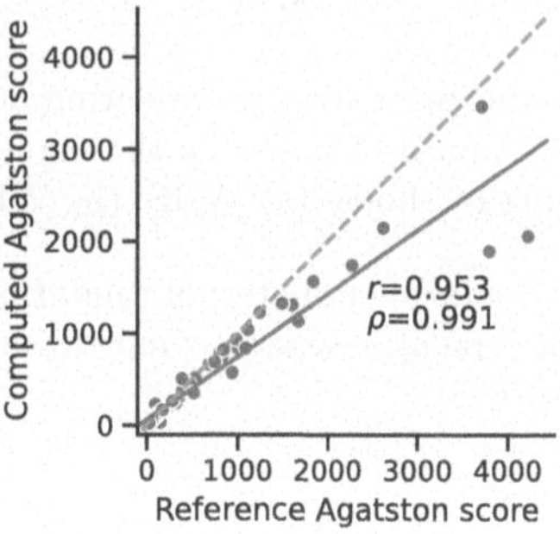

(b)

Fig. 3. Scatter plots of internal validation (Table 2) showing the correlation between the computed scores and the reference scores, based on the following methods: (a) WHS; (b) WHS+FP detection model. Note that a few subjects with too large computed scores, ≥5000, are not displayed on (a). The solid line represents a best-fit graph, and the dashed line indicates a perfect agreement.

In internal validation, as shown in Table 2, Fig. 3 and 4, the computed scores by our method showed very strong positive correlation with the reference scores. The proposed method yielded almost perfect agreement in terms of risk categorization.

The proposed method's external validation results are summarized in Table 3, Fig. 5 and 6. Although the agreement obtained here is not comparable to the agreement shown in internal validation, the proposed method enables to achieve strong positive correlation ($r = 0.833$ and $\rho = 0.917$) and a substantial agreement on risk categorization ($\kappa = 0.774$), between the two scores. In particular, the proposed method yields considerable improvement in agreement compared to the WHS method; this verifies that our DL model trained in the benchmark dataset succeeds in detecting FPs in the external validation dataset, as shown in Fig. 2.

We repeated the abovementioned experiments for a different CAC scoring method, volume score [23], while this paper focuses on reporting the results w.r.t Agatston score. Note that what we found with the results in volume score was almost the same.

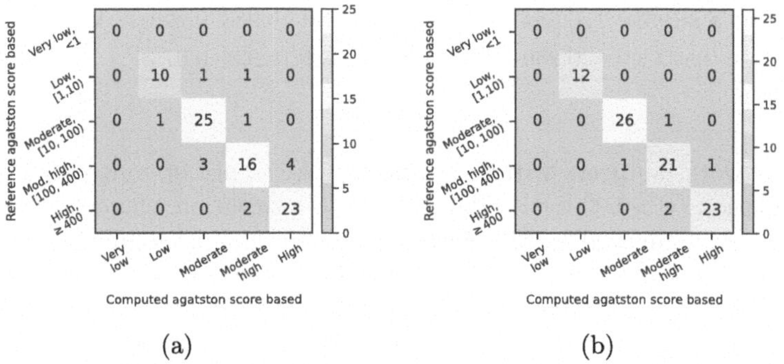

Fig. 4. Confusion matrix representing agreement in risk categorization of 87 subjects in the COCA dataset based on the manual reference score and the computed score by the following methods: (a) WHS; (b) WHS+FP detection model.

Table 3. Results of our external validation on the 1453 scans of the RS dataset, in terms of three agreement assessment metrics with 95% CIs through a bootstrap resampling (100 repetitions).

Metrics	Methods	
	WHS	WHS+FP detection model (The proposed method)
Pearson's r (95%CI)	0.500 (0.398–0.603)	0.833 (0.791–0.875)
Spearman's ρ (95%CI)	0.800 (0.788–0.813)	0.917 (0.909–0.926)
Cohen's κ (95%CI)	0.561 (0.547–0.575)	0.774 (0.764–0.783)

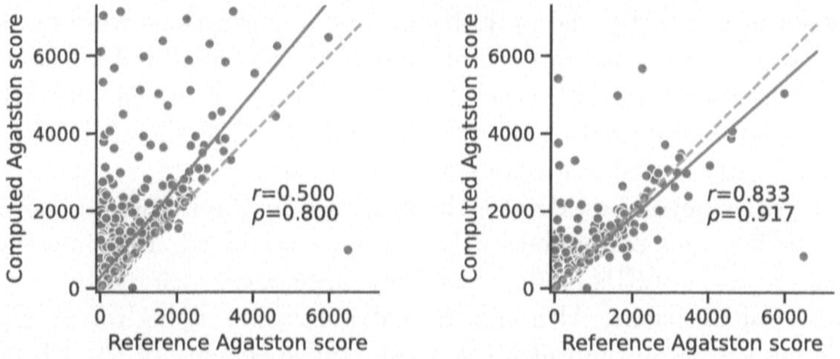

(a) (b)

Fig. 5. Scatter plots of internal validation (Table 3) showing the correlation between the computed scores and the reference scores based on the following method: (a) WHS; (b) WHS+FP detection model. Note that a few subjects with too large computed scores, ≥ 7000, are not displayed on (a). The solid line represents a best-fit graph, and the dashed line indicates a perfect agreement.

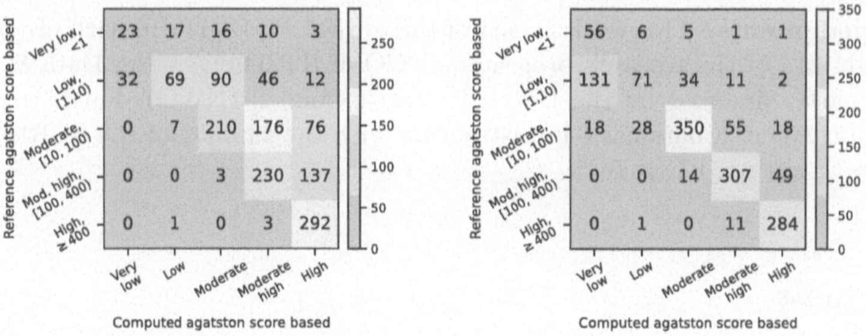

(a) (b)

Fig. 6. Confusion matrix representing agreement in risk categorization of 1453 subjects in the RS dataset based on the manual reference score and the computed score by the following methods: (a) WHS; (b) WHS+FP detection model.

4 Discussion

We developed an automatic coronary artery calcium scoring system using a deep learning-based image classification model. It has been validated on an external dataset from a large cohort study, showing strong positive correlations and substantial agreement with semi-automatic scoring.

In Fig. 3 (b), the two outliers make a best-fit graph deviate from a line representing a perfect agreement. These outliers occur when our WHS fails to yield a segmentation mask for the target subject successfully. In this case, some arteries close to the outer surface of the myocardium are excluded, and the calcium score is then underestimated. This happens also for one outlier, which has a large reference score (> 6000) in Fig. 5 (b).

As shown in Fig. 6 (b), the majority of miscategorizations were between the *very low* and *low* score categories. Our model predicts small CAC lesions on the external validation data as *negative*. Lower resolution (by 16/64-slice MDCT) of the external data compared to the training data (by 128-slice MDCT) and too few training samples of small lesions may cause this.

For categories beyond moderate, the proposed method tends to overestimate coronary calcifications compared to the reference standard. This may caused by the misclassification of FPs having relatively large sizes, such as aortic valve and mitral valve calcifications. Although miscategorization at high/very high CAC is less of a clinical issue than at lower levels, the specificity of the FP detection model still needs to be improved.

Recently, deep neural networks (DNNs) for image segmentation (e.g., U-Net [24]) have been used to achieve end-to-end CAC segmentation solely based on deep learning (DL) [7,8,25]. In future work, we will evaluate the performance of the end-to-end DL method on the dataset used in this work and further investigate the direction of improvement based on the comparison with the method proposed in this paper.

Acknowledgments. This work is part of the project MyDigiTwin with project number 628.011.213 of the research programme "COMMIT2DATA - Big Data & Health" which is partly financed by the Dutch Research Council (NWO). Furthermore, this work used the Dutch national e-infrastructure with the support of the SURF Cooperative using grant no. EINF-7675.

References

1. Roth, G.A., et al.: Global burden of cardiovascular diseases and risk factors, 1990–2019: update from the gbd 2019 study. J. Am. Coll. Cardiol. **76**(25), 2982–3021 (2020)
2. Timmis, A., et al.: European society of cardiology: cardiovascular disease statistics 2021. Eur. Heart J. **43**(8), 716–799 (2022)
3. Wang, T.J.: Assessing the role of circulating, genetic, and imaging biomarkers in cardiovascular risk prediction. Circulation **123**(5), 551–565 (2011)
4. Bos, D., Ikram, M.A.: Research aims in clinical medicine: description, identification, or explanation. World Neurosurgery **161**, 240–244 (2022)
5. Visseren, F.L., et al.: 2021 esc guidelines on cardiovascular disease prevention in clinical practice: developed by the task force for cardiovascular disease prevention in clinical practice with representatives of the European society of cardiology and 12 medical societies with the special contribution of the European association of preventive cardiology (eapc). Eur. Heart J. **42**(34), 3227–3337 (2021)
6. Budoff, M.J., et al.: Ten-year association of coronary artery calcium with atherosclerotic cardiovascular disease (ascvd) events: the multi-ethnic study of atherosclerosis (mesa). Eur. Heart J. **39**(25), 2401–2408 (2018)
7. Eng, D., et al.: Automated coronary calcium scoring using deep learning with multicenter external validation. NPJ Digit. Med. **4**(1), 88 (2021)

8. Ihdayhid, A.R., et al.: Evaluation of an artificial intelligence coronary artery calcium scoring model from computed tomography. Eur. Radiol. **33**(1), 321–329 (2023)
9. Kurkure, U., Chittajallu, D.R., Brunner, G., Le, Y.H., Kakadiaris, I.A.: A supervised classification-based method for coronary calcium detection in non-contrast ct. Int. J. Cardiovasc. Imaging **26**, 817–828 (2010)
10. Shahzad, R., et al.: Vessel specific coronary artery calcium scoring: an automatic system. Acad. Radiol. **20**(1), 1–9 (2013)
11. Lessmann, N., et al.: Automatic calcium scoring in low-dose chest ct using deep neural networks with dilated convolutions. IEEE Trans. Med. Imaging **37**(2), 615–625 (2017)
12. Kirisli, H.A., et al.: Fully automatic cardiac segmentation from 3d cta data: a multi-atlas based approach. In: Medical Imaging 2010: Image Processing, vol. 7623. SPIE (2010). 55–63
13. Shahzad, R., et al.: Automatic quantification of epicardial fat volume on non-enhanced cardiac ct scans using a multi-atlas segmentation approach. Med. Phys. **40**(9), 091910 (2013)
14. Agatston, A.S., Janowitz, W.R., Hildner, F.J., Zusmer, N.R., Viamonte, M., Jr., Detrano, R.: Quantification of coronary artery calcium using ultrafast computed tomography. J. Am. Coll. Cardiol. **15**(4), 827–832 (1990)
15. Russakovsky, O., et al.: Imagenet large scale visual recognition challenge. Int. J. Comput. Vis. **115**, 211–252 (2015)
16. Simonyan, K., Zisserman, A.: Very deep convolutional networks for large-scale image recognition. arXiv preprint arXiv:1409.1556 (2014)
17. Howard, A., et al.: Searching for mobilenetv3. In: Proceedings of the IEEE/CVF International Conference on Computer Vision, pp. 1314–1324 (2019)
18. Howard, A.G., et al.: Mobilenets: efficient convolutional neural networks for mobile vision applications. arXiv preprint arXiv:1704.04861 (2017)
19. Dai, Z., Liu, H., Le, Q.V., Tan, M.: Coatnet: marrying convolution and attention for all data sizes. Adv. Neural Inf. Process. Syst. **34**, 3965–3977 (2021)
20. Vaswani, A., et al.: Attention is all you need. Adv. Neural Inf. Process. Syst. **30** (2017)
21. Stanford University Stanford AIMI: COCA - Coronary Calcium and chest CT's (2021). http://stanfordaimi.azurewebsites.net/datasets/e8ca74dc-8dd4-4340-815a-60b41f6cb2aa
22. Ikram, M.A., et al.: The rotterdam study. design update and major findings between, et al.: and 2024. Eur. J. Epidemiol. **2024**, 1–24 (2020)
23. Hong, C., Bae, K.T., Pilgram, T.K., Suh, J., Bradley, D.: Coronary artery calcium measurement with multi-detector row ct: in vitro assessment of effect of radiation dose. Radiology **225**(3), 901–906 (2002)
24. Ronneberger, O., Fischer, P., Brox, T.: U-Net: convolutional networks for biomedical image segmentation. In: Navab, N., Hornegger, J., Wells, W.M., Frangi, A.F. (eds.) MICCAI 2015. LNCS, vol. 9351, pp. 234–241. Springer, Cham (2015). https://doi.org/10.1007/978-3-319-24574-4_28
25. Gogin, N., et al.: Automatic coronary artery calcium scoring from unenhanced-ecg-gated ct using deep learning. Diagn. Interv. Imaging **102**(11), 683–690 (2021)

Effective Approach Based on Student-Teacher Self-supervised Deep Learning for Multi-class Bi-atrial Segmentation Challenge

Moona Mazher[1], Abdul Qayyum[2(✉)], and Steven A. Niederer[2]

[1] Centre for Medical Image Computing, Department of Computer Science, University College London, London, UK

[2] National Heart and Lung Institute, Faculty of Medicine, Imperial College London, London, UK

a.qayyum@imperial.ac.uk

Abstract. Atrial fibrillation (AF), a widespread cardiac arrhythmia, is associated with significant morbidity and mortality. However, current treatments are limited due to an incomplete understanding of the atrial anatomy that drives AF. The MBAS 2024 Challenge builds on our 2018 left atrium challenge by expanding the focus to include both the left and right atria. Leveraging the largest bi-atrial LGE-MRI dataset to date, this challenge centers on advanced segmentation and biomarker identification to enhance treatment strategies for AF. The goal of this challenge is to develop automated medical semantic segmentation methods using a specialized 3-class bi-atrial dataset, ultimately aiming to improve the performance of AF treatments. We have introduced a novel two-stage process for the precise segmentation of the atrial walls and both atria. Our approach begins by utilizing self-supervised learning techniques tailored for 3D medical imaging, enabling the model to effectively capture the complex anatomical structures present in 3D medical datasets. Following this pre-training stage, the learned knowledge is transferred to a targeted 3D medical image segmentation task. In the second stage, we present an xLSTM-based model enhanced with the integration of Self-Supervised Learning (SSL) and cross-window attention mechanisms. This innovative approach has demonstrated superior performance in segmenting the atrial walls as well as the left and right blood pool cavities, offering a promising advancement in the accuracy and effectiveness of AF treatment through more detailed anatomical segmentation. Our approach achieved overall second ranked on validation leaderboard.

Keywords: Self supervised · student teacher models · xLSTM UNet · Atrium segmentation

1 Introduction

Atrial fibrillation (AF) is a prevalent cardiac arrhythmia with significant morbidity and mortality. Current treatments are hampered by limited understanding of the atrial anatomy that sustains AF. The MBAS 2024 challenge builds on our 2018 left atrium

O. Camara et al. (Eds.): STACOM 2024, LNCS 15448, pp. 132–139, 2025.
https://doi.org/10.1007/978-3-031-87756-8_13

challenge [1] by incorporating both the left and right atria, focusing on advanced segmentation and biomarker identification using the largest bi-atrial LGE-MRI dataset available. This challenge aims to develop automatic medical semantic segmentation methods using a 3-class bi-atrial dataset provided for optimal performance enhancement in AF treatment.

Medical image analysis, essential for computer-aided diagnosis, frequently encounters issues with the scarcity of labeled data, particularly for complex 3D tasks. Self-supervised learning (SSL) has emerged as a promising solution to mitigate annotation costs by leveraging large amounts of unlabeled data [2, 3]. However, SSL still requires extensive unlabeled datasets to effectively learn feature representations. The limited availability of 3D medical data due to high imaging costs and privacy concerns exacerbates this issue, with most datasets containing only a few thousand cases.

We introduce a straightforward, yet powerful self-supervised learning (SSL) approach based on the DINOv2 model to significantly reduce the need for extensive manual annotation in 3D medical image segmentation. Our method leverages readily available, unlabeled data to create robust pre-trained models that excel in downstream segmentation tasks. In the initial stage, we adapted the DINOv2 [4] model, originally designed for 2D tasks, to handle 3D medical imaging. This involves extending the model's architecture and training framework to process volumetric data, allowing it to learn rich, spatially aware representations from unlabeled 3D medical images. By doing so, we enable the model to capture the complex structures inherent in 3D medical datasets. After pre-training, we transfer the learned knowledge to a specific 3D medical image segmentation task. This is achieved by extracting the encoder from the pre-trained model and fine-tuning it alongside a task-specific decoder designed for segmentation. The pre-trained encoder provides a strong foundation, enriched with spatial features learned during the self-supervised phase, which significantly enhances the segmentation performance on the target task. We have detailed our proposed method in the following subsections.

2 Proposed Method

2.1 Dataset

The MBAS (Multi-class Bi-Atrial Segmentation Challenge) dataset was gathered from multiple centers, each adhering to a consistent protocol to ensure uniformity. The dataset comprises training and validation images with two possible resolutions: either $44 \times 640 \times 640$ or $44 \times 576 \times 576$ pixels. Each image has a voxel spacing of 2.5 mm in the slice direction and 0.625 mm in the other two spatial dimensions. This uniformity in imaging protocol and voxel spacing is crucial for maintaining consistency across the dataset, facilitating more accurate multi-class segmentation of bi-atrial structures. The challenge organizer provides 70 subjects for training, we have divided these into 65 for training and 5 for validation of proposed model. Further they provided 30 for validation without sharing ground-truth and they evaluated performance on this validation data [1].

2.2 DINOv2 to 3D Medical Imaging

The framework is implemented within a self-supervised learning (SSL) paradigm, utilizing a single online student encoder and a corresponding momentum teacher encoder. Both encoders share the same network architecture, with the teacher encoder's parameters being a momentum-based update of the student encoder's parameters. This setup enables the model to effectively learn from unlabeled data. In self-supervised learning, the use of strong data augmentations is critical for creating diverse and informative training examples. For this study, we apply a range of powerful data transformations, including flipping, scaling, Gaussian noise, Gaussian blur, adjustments to image brightness, and contrast. These transformations generate two distinct views of the input, which are then fed into a Siamese network structure. Feature Processing and Projection Both views are processed through the single encoder which extracts feature representations from the input data. These extracted features are then transformed into feature vectors using global average pooling. Following this, the feature vectors are passed through a multi-layer perceptron (MLP) projector, which is composed of four layers: The first two MLP layers each contain 2048 neurons and are followed by batch normalization (BN) and Gaussian Error Linear Unit (GELU) activation. The third layer, known as the bottleneck layer, reduces the dimensionality to 256 neurons, without BN or GELU. The final layer expands the dimensionality to K neurons and utilizes weight normalization to enhance the training process. These projections map the feature vectors into a high-dimensional latent space, where they are prepared for further comparison and learning. Momentum Teacher and Supervision Signal Simultaneously, the alternative view and vice versa) is processed by the momentum teacher encoder. The feature vectors produced by the teacher encoder are also passed through a similar MLP projector. After the MLP, the teacher's output undergoes centering and sharpening operations, followed by a softmax function. This output is then used as a supervision signal for the feature vectors generated by the student encoder.

2.3 Downstream Proposed Segmentation Model

Convolutional Neural Networks (CNNs) and Vision Transformers (ViTs) have been instrumental in advancing biomedical image segmentation. However, these models often struggle with effectively capturing long-range dependencies due to their inherent locality and computational complexity. To address these challenges, in this paper, we present xLSTM-UNet, a UNet-structured deep learning network enhanced with an efficient parallel cross-window module. This model integrates Vision-LSTM (xLSTM) as its backbone to improve medical image segmentation. xLSTM, a recent evolution of Long Short-Term Memory (LSTM) networks, has demonstrated superior performance over Transformers and State Space Models (SSMs) like Mamba in tasks such as Neural Language Processing (NLP) and image classification, as seen in Vision-LSTM (ViL) implementations [5]. We extend the success of xLSTM into the biomedical image segmentation domain by integrating it within the UNet architecture. The resulting xLSTM-UNet model leverages the local feature extraction capabilities of convolutional layers combined with xLSTM's proficiency in capturing long-range dependencies, thereby offering a powerful solution for comprehensive image analysis. To further enhance the performance of the xLSTM-UNet model, we have integrated a Cross Attention Transformer (CAT) block within the encoder module. This advanced module is designed to compute new feature representations by leveraging cross attention mechanisms between different feature sets. Specifically, the feature sets are divided into two distinct sets of windows: a base window set, and a searching window set b, both of the same size but different dimensions (Fig. 1).

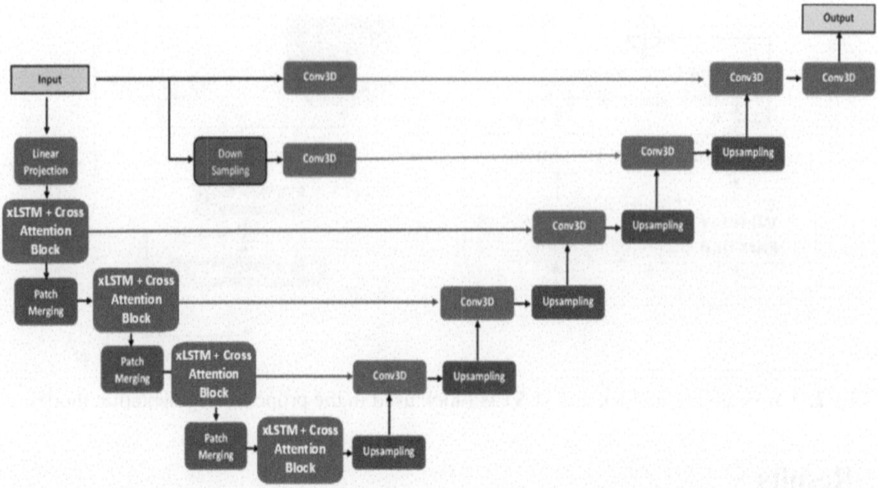

Fig. 1. Proposed based on cross window attention module.

In the CAT block, each base window is projected into a query set, while each searching window is projected into key and value sets through linear layers. The Window-based Multi-head Cross Attention (W-MCA) then calculates the cross attention between these windows, updating each base window with weighted information from its corresponding search window. This updated output is then processed through a two-layer MLP with GELU non-linearity to enhance learning, with LayerNorm (LN) applied before each W-MCA and MLP module to maintain layer stability. By incorporating the CAT block into the encoder side of the xLSTM-UNet, we enable the network to automatically identify and fuse relevant features from different regions of the input images. This facilitates more efficient and accurate feature fusion, significantly boosting the model's ability to perform complex segmentation tasks. Experimental results validate that the xLSTM-UNet, augmented with the CAT block, consistently leading CNN-based and Transformer-based models in biomedical image segmentation. The CAT block and xLSTM block are shown in Fig. 2.

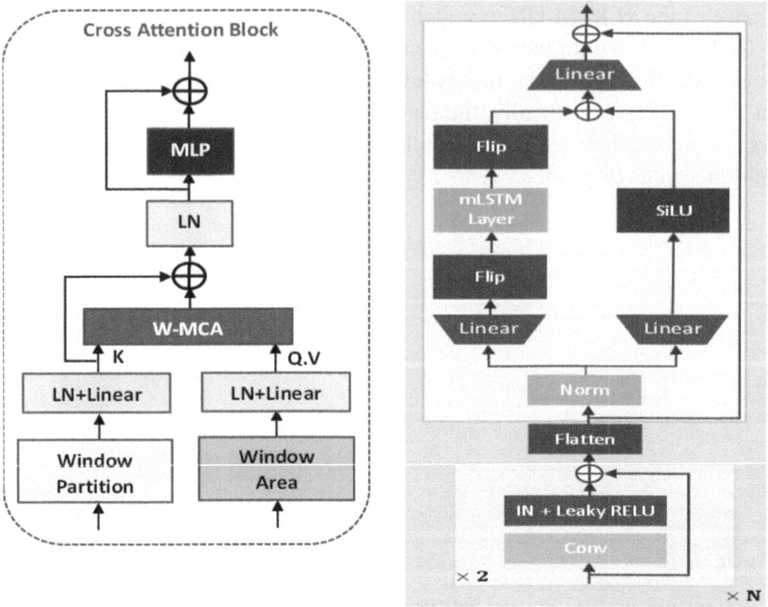

Fig. 2. Cross attention block and xLSTM block used in the proposed segmentation model.

3 Results

Our model was evaluated using five internal validation subjects, which were separated from the training set. The training set comprised 65 subjects, while the remaining 5 subjects were used for local validation to assess the performance of our model with various variations. The first and second rows correspond to two validation subjects. The second column shows the ground truth, while the third, fourth, and fifth columns display the segmentation results using our proposed models. The model incorporating the SSL (self-supervised learning) approach demonstrated superior performance compared to models without SSL. The visualization performance of proposed model is shown in Fig. 3. The average dice and HD95 produced by our proposed model achieved better performance in Arterial walls and Right and Left arterial as shown in Table 1. The HD95 for arterial wall is reasonable as compared to RA. Our proposed model achieved better performance on the leaderboard.

The average Dice score and HD95 value demonstrate that our proposed model achieves superior performance in identifying arterial walls and the right and left arterial regions, as detailed in Table 1. Specifically, the proposed model shows high Dice scores, particularly for the Left Arterial (91.03%) and Right Arterial (88.37%), indicating excellent segmentation accuracy. The HD95 values for the Wall (4.8 mm) and Left Arterial (4.81 mm) are relatively low, suggesting that our model has better boundary localization for these regions compared to the Right Arterial (7.34 mm). Overall, the proposed model performs better on the leaderboard in validation and achieved overall ranked second in validation phase, highlighting its effectiveness in segmenting arterial structures compared to existing models.

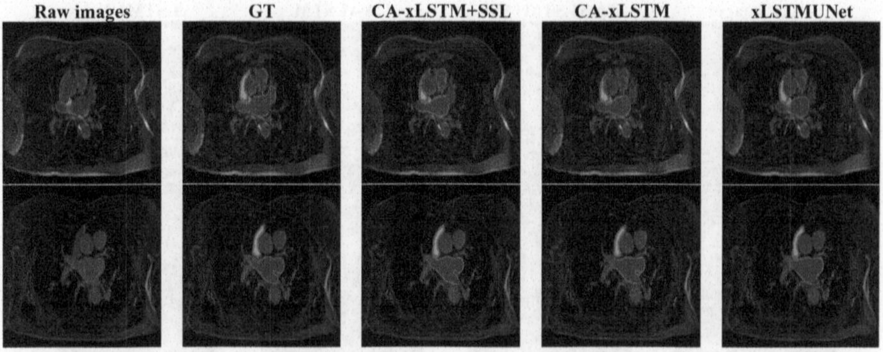

Fig. 3. Visualization results of our proposed model with different variations for subjects 35 and 41. In the images, red represents the walls, green indicates the left atrium (LA), and blue denotes the right atrium (RA).

Table 1. The results on validation dataset using proposed model on leaderboard.

	Wall		RA		LA	
	Dice	HD95	Dice	HD95	Dice	HD95
CA-xLSTM + SSL	66.36	4.8	88.37	7.34	91.03	4.81

The average Dice score and HD95 values provided in Table 1 illustrate that our proposed model significantly outperforms previous models in accurately identifying and segmenting arterial structures, including arterial walls and both right and left arterial regions.

The results are organized into three rows: the first row corresponds to subject 100, the second row to subject 85, and the third row to subject 71 from the validation set. Each row compares the performance of three different models across three columns, the first column displays the segmentation results using our CA-xLSTM + SSL model. The second column shows the results from the CA-xLSTM model without SSL. The third column presents the segmentation results obtained from the base xLSTM U-Net model as shown in Fig. 4. Among the models, the SSL-based CA-xLSTM + SSL model delivers the most accurate segmentation maps, demonstrating clear improvements over the other approaches. The base xLSTM model shows notable errors in segmenting the walls and the right atrium, indicating a less reliable performance compared to the other models.

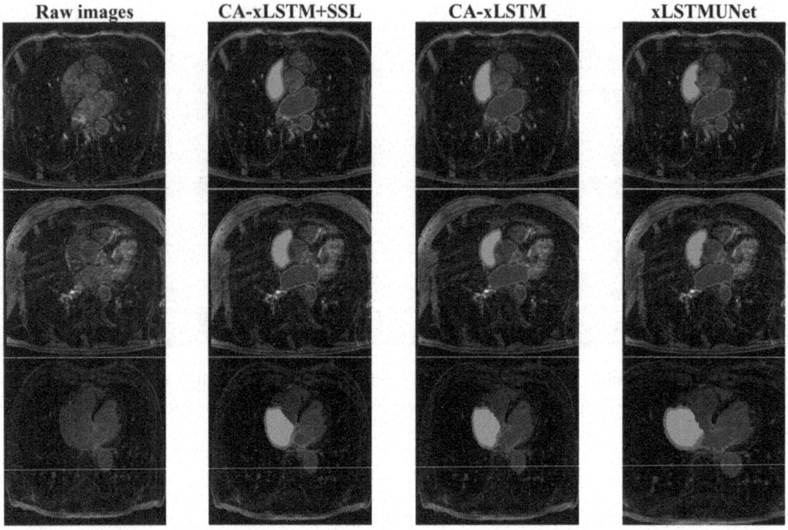

Fig. 4. Illustrates the segmentation results for the left atrium (LA), right atrium (RA), and walls using the validation dataset.

4 Discussion

The proposed model not only achieves high Dice scores and reasonable HD95 values across all regions but also ranks better on the leaderboard compared to existing models. This performance highlights the model's effectiveness in segmenting complex arterial structures, demonstrating its ability to:

- Precisely identify and segment different arterial regions, with strength in the Left and Right Arterial areas.
- Accurately localize boundaries, especially for the Wall and Left Arterial, which are crucial for detailed anatomical analysis.

In summary, the combination of high Dice scores and low HD95 values for most regions illustrates the model's robust performance and its advantage over current benchmarks in the field. This reinforces the model's potential for practical applications where precise and accurate segmentation of arterial structures is essential. HD95 measures the maximum distance between the predicted segmentation and the ground truth at the 95th percentile, providing insights into boundary localization. Lower values suggest better precision in delineating boundaries.

- Left Arterial (4.81 mm): The HD95 for the Left Arterial is comparable to that of the Wall, showing that the model performs well in localizing boundaries with minimal deviation from the ground truth.
- Right Arterial (7.34 mm): While still reasonable, the higher HD95 value for the Right Arterial indicates slightly less precision in boundary localization compared to the Wall and Left Arterial. This suggests that the model's boundary predictions for the Right Arterial may be less accurate or more variable.

- Wall (4.8 mm): The HD95 value for the Wall is relatively low, indicating that the predicted boundaries are generally close to the ground truth, with only a small maximum deviation. This suggests that the model effectively captures the Wall's shape and boundaries.

This expanded explanation provides a more comprehensive view of the metrics and their implications, helping to convey the significance of our model's performance.

5 Conclusion

In conclusion, we have developed a novel two-stage approach for the precise segmentation of atrial walls and both atria, which shows significant promise in advancing the accuracy and effectiveness of AF treatment. Our method begins with the application of self-supervised learning techniques specifically adapted for 3D medical imaging, allowing the model to accurately capture the intricate anatomical structures inherent in such datasets. This pre-training stage serves as a robust foundation for subsequent transfer to a specialized 3D medical image segmentation task. In the second stage, we introduced an xLSTM-based model, further enhanced by Self-Supervised Learning (SSL) and cross-window attention mechanisms. This approach has been shown to outperform existing methods, particularly in the segmentation of atrial walls and the left and right blood pool cavities. Our findings suggest that this advanced segmentation method could play a crucial role in improving the precision of AF treatments by providing more detailed and reliable anatomical data.

Acknowledgment. This work was supported by the Technology Missions Fund under the EPSRC Grant EP/X03870X/1, the British Heart Foundation (RG/20/4/34 803), and The Alan Turing Institute, London, UK. I am deeply grateful for their financial support and the resources provided, which were instrumental in the successful completion of this research.

References

1. Xiong, Z., et al.: A global benchmark of algorithms for segmenting the left atrium from late gadolinium-enhanced cardiac magnetic resonance imaging. Med. Image Anal. **67**, 101832 (2021)
2. Ye, Y., Zhang, J., Chen, Z., Xia, Y.: Desd: self-supervised learning with deep self-distillation for 3d medical image segmentation. In: International Conference on Medical Image Computing and Computer-Assisted Intervention, pp. 545–555. Springer Nature Switzerland, Cham (2022)
3. Mazher, M., et al.: Self-supervised spatial–temporal transformer fusion based federated framework for 4D cardiovascular image segmentation. Inf. Fusion **106**, 102256 (2024)
4. Oquab, M., et al.: Dinov2: learning robust visual features without supervision. arXiv preprint arXiv:2304.07193 (2023)
5. Alkin, B., Beck, M., Pöppel, K., Hochreiter, S., Brandstetter, J.: Vision-LSTM: xLSTM as Generic Vision Backbone. arXiv preprint arXiv:2406.04303 (2024)

Sampling-Pattern-Agnostic MRI Reconstruction Through Adaptive Consistency Enforcement with Diffusion Model

Anurag Malyala[1,2], Zhenlin Zhang[1,2], Chengyan Wang[3], and Chen Qin[1(✉)]

[1] Department of Electrical and Electronic Engineering & I-X, Imperial College London, London, UK
{a.malyala23,zhenlin.zhang23,c.qin15}@imperial.ac.uk
[2] Department of Computing, Imperial College London, London, UK
[3] Human Phenome Institute, Fudan University, Shanghai, China
wangcy@fudan.edu.cn

Abstract. Magnetic Resonance Imaging (MRI) is a powerful, non-invasive diagnostic tool; however, its clinical applicability is constrained by prolonged acquisition times. Whilst present deep learning-based approaches have demonstrated potential in expediting MRI processes, these methods usually rely on known sampling patterns and exhibit limited generalisability to novel patterns. In the paper, we propose a sampling-pattern-agnostic MRI reconstruction method via a diffusion model through adaptive consistency enforcement. Our approach effectively reconstructs high-fidelity images with varied under-sampled acquisitions, generalising across contrasts and acceleration factors regardless of sampling trajectories. We train and validate across all contrasts in the MICCAI 2024 Cardiac MRI Reconstruction Challenge (CMRxRecon) dataset for the "Random sampling CMR reconstruction" task. Evaluation results indicate that our proposed method significantly outperforms baseline methods.

Keywords: Diffusion Model · Data Consistency · MRI Reconstruction · Cardiac MRI

1 Introduction

Magnetic Resonance Imaging (MRI) is crucial in clinical diagnosis, offering high-resolution, radiation-free visualisation of internal structures. However, prolonged scanning times remain a significant limitation and can cause patient discomfort resulting in motion artefacts. While under-sampling data expedites scans, it breaks the Nyquist sampling limit, resulting in aliasing artefacts.

Supplementary Information The online version contains supplementary material available at https://doi.org/10.1007/978-3-031-87756-8_14.

O. Camara et al. (Eds.): STACOM 2024, LNCS 15448, pp. 140–151, 2025.
https://doi.org/10.1007/978-3-031-87756-8_14

Recent advancements in machine learning (ML) have enabled accurate reconstructions from under-sampled MRI acquisitions, addressing the challenge of lengthy scan times. The reconstruction of MR images is usually framed as an inverse problem traditionally solved with various techniques, including compressed sensing [9], parallel imaging, low-rank methods, etc. However, contemporary research has demonstrated that deep learning (DL)-based methods significantly outperform these conventional techniques [10], thus garnering increased attention in the field.

Deep convolutional neural networks (CNNs) are the predominant DL models for under-sampled MRI reconstruction. Schlemper et al. [17] combined data consistency with CNNs to enhance reliability and accuracy in image reconstruction. Qin et al. [14] further refined this by implementing a Convolutional Recurrent Neural Network (CRNN) to exploit temporal redundancies in dynamic MRI sequences. Guang et al. [25] suggested improving the image fidelity by applying a conditional Generative Adversarial Network (GAN) [5] as a further refinement step. Aggarwal et al. [1] proposed a model-based DL approach (MoDL) for parallel imaging data, decomposing it into a denoising and data consistency step optimised by conjugate gradient (CG) method.

The recent advancement in generative models has introduced diffusion models [7,19,20] to MRI reconstruction. Wherein two main approaches have emerged: Conditional methods as proposed by [24] and [6], which apply a conditional Denoising Diffusion Probabilistic Model (DDPM) [7] framework, and unconditional methods, which utilise posterior guidance for reconstruction-based inverse problems. The latter, exemplified by Chung et al. [3], Peng et al. [13], and Song et al. [21], are not constrained to fixed measurement models, leading to their increasing prominence in reconstruction tasks. In their recent work, Chung et al. [4] extended this to parallel imaging following the lines of [1] with Decomposed Diffusion Sampling (DDS).

Conditional methods perform well in known settings but fail to generalise to new sampling patterns, while unconditional models offer flexibility at the cost of lengthy inference and tuning. In this work, we propose "Sampling Pattern Agnostic MRI Reconstruction (SPA-MRI)", a diffusion-based framework that robustly reconstructs high-quality scans across diverse contrasts, generalising over sampling trajectories and acceleration factors. SPA-MRI achieves faster inference with adaptive hyperparameters, significantly outperforming baseline methods in evaluations.

2 Preliminaries

2.1 Diffusion Models

Diffusion models [7,19,20] form a class of generative models which learn to transform a known distribution into a target empirical distribution in a step-by-step process of noising and denoising. The noising process can be represented as an Itô stochastic differential equation (SDE) as described by Song et al. [19]:

$$d\mathbf{x} = \mathbf{f}(\mathbf{x}, t)dt + g(t)d\mathbf{w}, \tag{1}$$

where $\mathbf{x}$ is the data degraded over time $t \in [0, T]$. $\mathbf{f}(\cdot, t)$, $g(t)$ are the drift and diffusion coefficients, and $\mathbf{w}$ is the Wiener process. $\mathbf{f}(\cdot, t)$ and $g(t)$ are defined such that at $t = 0$, $x_0 \sim p_{\text{data}}$ which is the data distribution and at $t = T$, $x_T \sim p_T = \mathcal{N}(0, \mathbf{I})$, a tractable distribution. One parametrisation of this is the Variance Preserving (VP) SDE:

$$d\mathbf{x} = -\frac{1}{2}\beta(t)\mathbf{x}dt + \sqrt{\beta(t)}d\mathbf{w}, \tag{2}$$

where $\beta(t)$ is a monotonically increasing function that controls the noise scale. Song et al. [19] have demonstrated that DDPM [7] are equivalent to the setting of VP-SDE, which is used in this research.

To sample from the data distribution p_0, we solve the corresponding reverse VP-SDE [2] for Eq. (2):

$$d\mathbf{x} = [-\frac{1}{2}\beta(t)\mathbf{x} - \beta(t)\nabla_{\mathbf{x}_t} \log p_t(x_t)]dt + \sqrt{\beta(t)}d\bar{\mathbf{w}}. \tag{3}$$

In the above reverse SDE, $\nabla_{x_t} \log(p_t(x_t))$ represents the score function of the distribution, which is estimated with a neural network $s_\theta(\mathbf{x}, t)$, and $\bar{\mathbf{w}}$ is the Wiener process in reverse.

Denoising Diffusion Implicit Models. Given the iterative nature of SDEs and DDPMs, sampling is inadvertently slow. Song et al. [18] introduced a deterministic sampling strategy called Denoising Diffusion Implicit Models (DDIM), where they update the sampling rule as:

$$\mathbf{x}_{t-1} = \sqrt{\bar{\alpha}_{t-1}}\,(\mathbf{x}_0(\mathbf{x}_t)\epsilon_\theta(\mathbf{x}_t, t)) + \sqrt{1 - \bar{\alpha}_{t-1} - \eta^2\sigma^2}\epsilon_\theta(\mathbf{x}_t, t) + \eta\sigma\epsilon, \tag{4}$$

and

$$\mathbf{x}_0(\mathbf{x}_t) = \frac{\mathbf{x}_t - \sqrt{1 - \bar{\alpha}_t}\epsilon_\theta(\mathbf{x}_t, t)}{\sqrt{\bar{\alpha}_t}}, \tag{5}$$

where $\alpha_t = 1 - \beta_t$, $\bar{\alpha}_t = \prod_{s=0}^{t} \alpha_s$, and $\sigma = \beta_t(1 - \bar{\alpha}_{t-1})/(1 - \bar{\alpha}_t)$. The parameter η controls the stochasticity of the update, and $\epsilon_\theta(\mathbf{x}, t)$ is the denoising model for [7], and for DDIM, $\eta = 0$. $\mathbf{x}_{t-1}$ is the prediction of the next denoised state from $\mathbf{x}_t$ and $\mathbf{x}_0(\mathbf{x}_t)$ is the current estimate for $\mathbf{x}_0$. This deterministic sampling strategy thus allows for faster sampling from the learned prior model.[1]

2.2 Diffusion Models for Medical Inverse Problems

Reconstructing MR images from under-sampled k-space data can be stated as a linear inverse problem:

$$y = \mathcal{A}x + \epsilon, \tag{6}$$

[1] Given the formulation under Eq. (3), the score model $s_\theta(\mathbf{x}_t, t)$ is equivalent to the noise model $\epsilon_\theta(\mathbf{x}_t, t)$ as $s_\theta(\mathbf{x}_t, t) \approx -\epsilon_\theta(\mathbf{x}_t, t)/\sqrt{1 - \bar{\alpha}_t}$.

where $\mathcal{A} = \mathcal{F}\mathcal{M}s^c$ comprises three key components: the Fourier encoding matrix $\mathcal{F}$, the under-sampling mask $\mathcal{M}$, and the coil-wise sensitivities s^c. $\mathbf{y}$ represents the observed under-sampled k-space measurement, while $\mathbf{x}$ is the fully sampled image we aim to reconstruct and ϵ is the measurement noise in k-space.

To solve the inverse problem, we sample from the posterior distribution $p(\mathbf{x}|\mathbf{y})$. Using Bayes' theorem, we have $p(\mathbf{x}|\mathbf{y}) = p(\mathbf{y}|\mathbf{x})p(\mathbf{x})/p(\mathbf{y})$, where the diffusion model serves as the data prior $p(\mathbf{x})$. This approach leads to an updated conditional for the reverse VP-SDE:

$$d\mathbf{x} = [-\frac{1}{2}\beta(t)\mathbf{x} - \beta(t)(\nabla_{\mathbf{x}_t} \log p_t(\mathbf{x}_t) + \nabla_{\mathbf{x}_t} \log p(\mathbf{y}|\mathbf{x}_t))]dt + \sqrt{\beta(t)}d\mathbf{w}. \quad (7)$$

It is noteworthy that the conditional score function $\nabla_{\mathbf{x}_t} \log p(\mathbf{y}|\mathbf{x}_t)$ is generally intractable. However, a few approaches to make this tractable have been proposed. For example, Chung et al. [3] proposed an efficient approximation, expressing the unknown as $p(\mathbf{y}|\mathbf{x}_t) \approx p(\mathbf{y}|\mathbf{x}_0) = \mathbb{E}[\mathbf{x}_0(\mathbf{x}_t)]$, resulting in following approximation of the desired score called Diffusion Posterior Sampling (DPS):

$$\nabla_{x_t} \log p_t(x_t) + \nabla_{x_t} \log p_t(y|x_t) \simeq s_\theta(\mathbf{x}, t) - \frac{1}{\sigma^2}\nabla_{\mathbf{x}_t} ||\mathbf{y} - \mathcal{A}(\mathbf{x}_0(\mathbf{x}_t))||_2^2. \quad (8)$$

This approach substitutes the unknown fully-sampled image with its current approximation, which is conditional only on $\mathbf{x}_t$. Subsequent works have confirmed that this approximation enables sampling from the true posterior distribution and can retrieve the true sample [16].

In contrast to DPS, which requires model gradient calculations, Wang et al. [22] proposed the Denoising Diffusion Null-Space Model (DDNM). This method first performs unconditional sampling from $p(\mathbf{x}_t)$, then enforces consistency by back-projecting null-space errors:

$$\mathbf{x}_0(\mathbf{x}_t)' = (\mathbf{I} - \mathcal{A}^T\mathcal{A})\mathbf{x}_0(\mathbf{x}_t) + \mathcal{A}^T\mathbf{y}. \quad (9)$$

Shortcut Sampling for Fast Reconstruction. Current methods like DPS [3] and DDNM [22] generate priors in the target distribution and enforce posterior constraints. However, they overlook seed value initialisation, defaulting to Gaussian noise as default initialisation. Recognising under-sampled data as degraded versions of fully sampled data, we can improve the initialisation. Prior works addressed this by applying the diffusion process forwards using DDIM naïvely [18]:

$$\mathbf{x}_{t+1} = \sqrt{\alpha_{t+1}}(\mathbf{x}_0(\mathbf{x}_t)\epsilon_\theta(\mathbf{x}_t, t)) + \sqrt{1 - \alpha_{t+1}}\epsilon_\theta(\mathbf{x}_t, t). \quad (10)$$

While this preserves source information, the resulting samples can lack realism.

3 Methodology

Our proposed method **SPA-MRI** addresses key challenges in MRI reconstruction from under-sampled data. We mainly make three contributions: (1) an

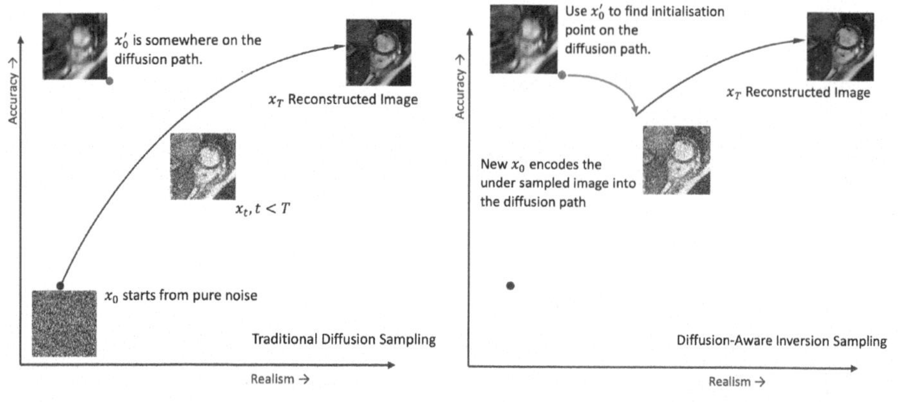

(a) Standard Diffusion Sampling (b) Diffusion Aware Inversion

Fig. 1. Conceptual visualisation of the typical diffusion sampling process (a) is compared with the enhanced diffusion-aware inversion proposed by Liu et al. (b) [8]. The inversion process comprises two stages: first, it adjusts the undersampled image x_0' to the diffusion process trajectory, and then it follows the standard diffusion process, requiring significantly fewer sampling steps.

enhanced initialisation process, (2) improved data consistency for parallel imaging, and (3) a novel frequency decomposed posterior loss. These contributions collectively enforce consistency and improve reconstruction quality and efficiency while generalising across sampling patterns. Below, we detail each component of our methodology.

Enhancement of the Forward Noising Process. To overcome the limitation of initialisation in existing approaches, inspired by Distortion Adaptive Inversion [8], we propose to follow an improved diffusion-aware forward noising process (Eq. 11) to generate the initial noise seed:

$$\mathbf{x}_{t+1} = \sqrt{\alpha_{t+1}}\,(\mathbf{x}_0(\mathbf{x}_t)\epsilon_\theta(\mathbf{x}_t, t)) + \sqrt{1 - \alpha_{t+1} - \beta_{t+1}}\epsilon_\theta(\mathbf{x}_t, t) + \sqrt{\beta_{t+1}}\mathbf{z}. \quad (11)$$

Here the initialisation is modified to include controllable perturbations $\mathbf{z} \sim \mathcal{N}(0, \mathbf{I})$ for preserving essential information. The layout and structure are shown in Fig. 1.

With this optimised initialisation, this approach can significantly reduce the effective denoising steps needed (i.e., function evaluations) while performing the reconstruction from the under-sampled data. It is also shown that the method can better preserve reconstruction realism and distribution fidelity.

Integration of Data-Consistency for Parallel Imaging. Incorporating data consistency for parallel imaging, MR is non-trivial. While methods like DPS [3], DDNM [22] and DiffuseRecon [13] work well for single-coil data, they fail to leverage parallel imaging information.

In our method, we extend DDNM's back projection method for parallel imaging. We decompose $\mathbf{x}_0(\mathbf{x}_t)$ into coil-wise estimates, perform consistency per coil, then project errors back to combined space: $\mathcal{A}^T(\mathcal{A}(\mathbf{x}_0(\mathbf{x}_t))^c - \mathbf{y}^c)$, where $c \in [0, C]$ represents coil-wise measurements. Our approach incorporates multi-coil information without the costly optimisation steps like CG, as seen in [4] and [1]. While CG methods may yield higher fidelity samples, our method achieves sufficiently good results more efficiently.

We have observed that the magnitude of the back projected errors change over time and benefits from scaling their influence, resulting in higher quality reconstructions. The scaling balances between greater influence at larger error scales for faster convergence and finer control when the errors are small to promote refined adjustments to $\mathbf{x}_0(\mathbf{x}_t)$. Therefore, we introduce an adaptive weighting schema for the consistency step:

$$\mathbf{x}_0(\mathbf{x}_t) = \mathbf{x}_0(\mathbf{x}_t) - \omega \mathcal{A}^T(\mathcal{A}(\mathbf{x}_0(\mathbf{x}_t)) - \mathbf{y}), \tag{12}$$

where $\omega = \xi * (1 + \tanh(\delta_{t+1} - \delta_t)/2)$, ξ is the base scale, and $\delta_t = ||\Delta(\mathbf{x}_0(\mathbf{x}_t), \mathbf{y})||_2$ (see Sect. 3) is the error magnitudes at time t. Using the hyperbolic tangent function, we scale the difference to $[-1, 1]$. This effectively adapts ξ to increase or decrease based on the error changes.

Frequency Decomposed Posterior Loss. Motivated by the distinct characteristics of low- and high-frequency information in MRI data, we propose a novel decomposition of the back-projection consistency measurement Δ:

$$\Delta(\mathbf{x}, \mathbf{y}) = \mathcal{A}(\mathbf{x}) - \mathbf{y}. \tag{13}$$

This approach, largely unexplored in consistency space, leverages the guaranteed accuracy of low-frequency Auto-Calibration Signal (ACS) data while focusing on reconstructing challenging high-frequency details, improving overall reconstruction quality and accuracy:

$$\Delta_{\text{d}}(\mathbf{x}_0(\mathbf{x}_t), \mathbf{y}) = \lambda_1 \Delta_{\text{low}}(\mathbf{x}_0(\mathbf{x}_t), \mathbf{y}) + \lambda_2 \Delta_{\text{high}}(\mathbf{x}_0(\mathbf{x}_t), \mathbf{y}). \tag{14}$$

Here λ_1 and λ_2 are weighting coefficients for low-frequency and high-frequency components, respectively. Specifically, $\lambda_2 > \lambda_1$, which prioritises high-frequency reconstruction. This approach is able to enhance fine detail recovery while maintaining low-frequency information integrity. Our proposed method is detailed in Algorithm 1.

Algorithm 1. SPA-MRI Sampling

for $s = 1$ to S **do** ▷ Inversion Step
  $\epsilon \leftarrow \epsilon_\theta(\mathbf{x}_s, s)$
  $\mathbf{x}_0(\mathbf{x}_s) \leftarrow \frac{\mathbf{x}_s - \sqrt{1-\alpha_s}\epsilon}{\sqrt{\alpha_s}}$
  $z \sim \mathcal{N}(0, \mathcal{I})$
  $\mathbf{x}_{s+1} = \sqrt{\alpha_{s+1}}\mathbf{x}_0(\mathbf{x}_s) + \sqrt{1 - \alpha_{s+1} - \lambda\beta_{s+1}}\epsilon + \sqrt{\lambda\beta_{s+1}}z$
end for
$\delta_T = 0$
for $t = T$ to 0 **do** ▷ Reverse Diffusion Step
  $\epsilon \leftarrow \epsilon_\theta(x_t, t)$
  $\mathbf{x}_0(\mathbf{x}_t) \leftarrow \frac{\mathbf{x}_t - \sqrt{1-\alpha_t}\epsilon_2}{\sqrt{\alpha_t}}$
  $\delta_t \leftarrow ||\Delta_d(\mathbf{x}_0(\mathbf{x}_t), \mathbf{y})||_2$
  $\mathbf{x}_0(\mathbf{x}_t) \leftarrow \mathbf{x}_0(\mathbf{x}_t) - \xi(1 + \tanh(\delta_{t+1} - \delta_t)/2)\mathcal{A}^T(\Delta_d(\mathbf{x}_0(\mathbf{x}_t), \mathbf{y}))$
  $x_{t-1} \leftarrow \sqrt{\alpha_{t-1}}x_0(x_t) + \sqrt{1 - \alpha_{t-1} - \sigma^2}\epsilon_2 + \sigma z$ ▷ $\sigma = \frac{1-\bar{\alpha}_{t-1}}{1-\bar{\alpha}_t}\beta_t$
end for
return $\hat{\mathbf{x}}_0 \leftarrow \mathbf{x}_0$

4 Experiments

Data Preprocessing. In our study, we train and evaluate on the CMRxRecon2024 challenge training dataset [23]. The dataset comprises four distinct modalities: Cardiac cine, Aorta, Mapping, and Tagging. Since the ground truth for the test set are not public yet, we implemented a stratified split, allocating 70% (981 files) for training and 30% (423 files) for testing.

Our pre-processing pipeline addresses heterogeneous image dimensions by rescaling all training data to 256×512 using PyTorch's [12] "nearest-exact" interpolation. While rescaling introduces some loss, we found this method had a negligible impact across all contrasts. Images are reverted to their respective original dimensions for the projection steps. Our model processes three consecutive frames simultaneously, using a 6-channel pseudo-real representation. Other preprocessing includes standardising to unit standard deviation and scaling to the maximum input magnitude, constraining values to the range $[-1, 1]$.

Implementation Details. We implement our model based on the Improved Denoising Diffusion Probabilistic Model (iDDPM) [11]. We train the model jointly on all modalities, using a learning rate of 10^{-4}. The noise schedule $\beta(t)$ has 4000 discretisation steps with a cosine β schedule. For inference, we use 200 DDIM steps and 25 inversion steps, with a back projection base scale ξ of 3. The centre 32×32 k-space region is used in the low frequency term in Eq. 14 with weights $\lambda_1 : \lambda_2 = 0.4 : 0.6$. A comprehensive table of all parameters is provided in the supplementary material.[2]

5 Results

We compared our results with DiffuseRecon [13] and DDNM [22] for 4, 12, and 24 acceleration factors with Gaussian, radial and uniform under-sampling patterns.

[2] Reproducibilty Note: The inference code and model weights will be made public after the conclusion of the CMRxRecon2024 challenge.

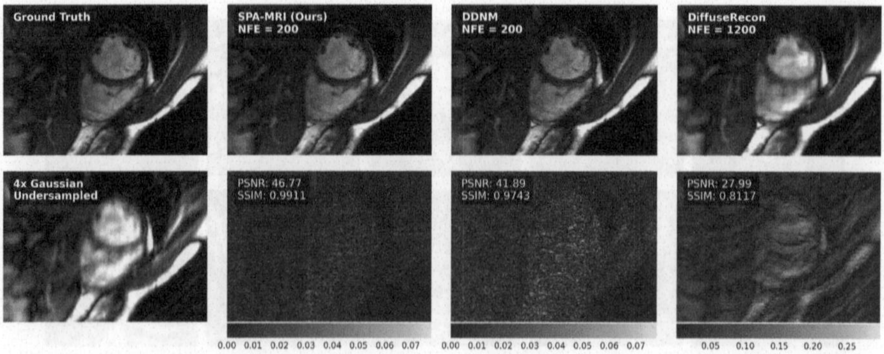

Fig. 2. Comparison of MRI reconstruction methods on 4× Gaussian undersampled Cine-SAX. Here, it shows the ground truth, zero-filled image, and reconstructions of our proposed methods, DDNM [22] and DiffuseRecon [13] along with their corresponding error maps. We also include the total Number of Function Evaluations (NFEs). For DiffuseRecon, the reconstruction is done coil-wise. The proposed approach outperforms baseline methods significantly with fewer function evaluations.

Following challenge guidelines, we evaluated central slices, initial frames for Cine, Tagging, and Aorta, and all weightings for Mapping.

Quantitative results for Gaussian under-sampling are shown in Table 1, and qualitative results are visualised in Fig. 2 and Fig. 3. Full results on radial and uniform sampling at different acceleration factors are included in the supplementary material.

For calculating metrics, we first normalise all the reconstructed samples to image space and scale them to $[0, 1]$, then use the normalised images to compute the PSNR and SSIMs. These metrics were calculated using the following formulas:

$$\text{PSNR} = 20 \cdot \log_{10}(\text{MAX}_I) - 10 \cdot \log_{10}(\text{MSE}), \tag{15}$$

where MAX_I is the maximum possible pixel intensity (1.0 in our normalized images) and MSE is the mean squared error between the ground truth and reconstructed image.

$$\text{SSIM}(x, y) = \frac{(2\mu_x\mu_y + c_1)(2\sigma_{xy} + c_2)}{(\mu_x^2 + \mu_y^2 + c_1)(\sigma_x^2 + \sigma_y^2 + c_2)}, \tag{16}$$

where μ_x and μ_y are the average pixel intensities of images x and y, σ_x^2 and σ_y^2 are the variances of x and y, σ_{xy} is the covariance of x and y, and c_1 and c_2 are constants to stabilise the division with weak denominators.

The results demonstrate that the proposed SPA-MRI method consistently outperforms baseline methods across different MRI reconstruction scenarios. From Table 1, we can observe that SPA-MRI achieves the highest PSNR and SSIM scores for all acceleration factors across all contrast types. The performance gap is particularly notable at higher acceleration factors, indicating the robustness of the frequency decomposed loss to higher accelerations.

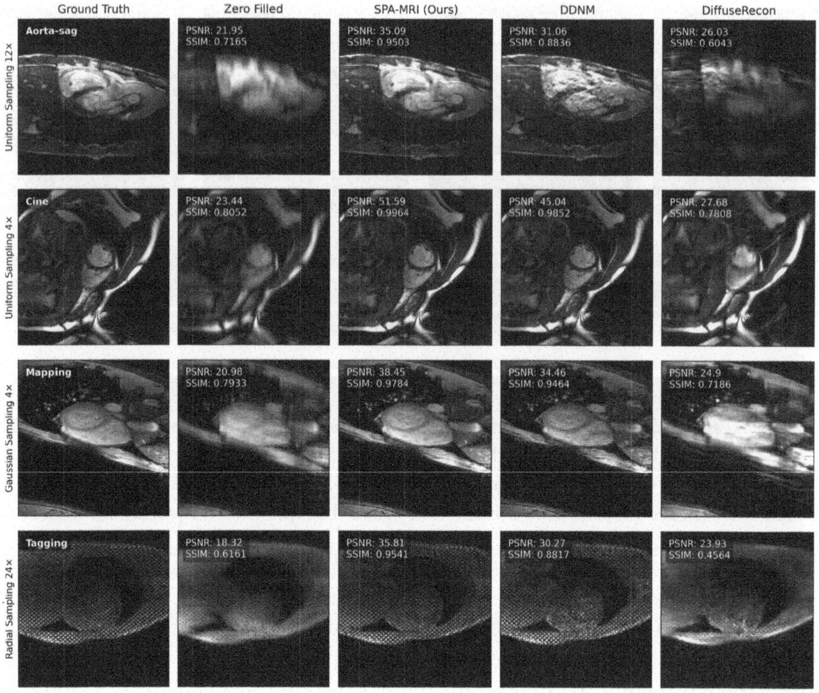

Fig. 3. Comparison of MRI reconstruction across all contrasts at varying acceleration factors and undersampling patterns. Our method consistently outperforms the competing methods across Aorta, Cine, Mapping (T1 visualised), and Tagging modalities.

Table 1. Numerical results of zero-filled, DiffuseRecon [13], DDNM [22] and our proposed methods for Gaussian undersample mask with different acceleration factors. The best results are in **bold**. Metrics were calculated on the full reconstructed images.

Acc	Contrast	Cine		Aorta		Tagging		Mapping		Total	
		PSNR ↑	SSIM ↑	PSNR ↑	SSIM ↑	PSNR ↑	SSIM ↑	PSNR ↑	SSIM ↑	PSNR ↑	SSIM ↑
4×	Zero-Filled	22.64	0.7172	21.42	0.6324	21.27	0.6697	23.16	0.7214	22.12	0.6852
	DiffuseRecon	31.28	0.7550	31.80	0.7633	30.85	0.7980	28.68	0.7605	30.65	0.7692
	DDNM	40.33	0.9695	41.85	0.9705	36.66	0.9547	36.42	0.9231	38.82	0.9545
	SPA-MRI	**44.73**	**0.9885**	**46.35**	**0.9893**	**42.45**	**0.9864**	**41.73**	**0.9818**	**43.82**	**0.9865**
12×	Zero-Filled	21.29	0.6556	20.15	0.5735	19.13	0.6011	21.81	0.674	20.60	0.6261
	DiffuseRecon	27.86	0.6084	28.07	0.6115	26.38	0.6259	25.93	0.6441	27.06	0.6225
	DDNM	33.98	0.9046	35.45	0.9074	31.76	0.8887	30.90	0.8256	33.02	0.8816
	SPA-MRI	**38.16**	**0.9647**	**39.35**	**0.9645**	**36.40**	**0.9592**	**34.70**	**0.9434**	**37.15**	**0.9580**
24×	Zero-Filled	21.06	0.6396	20.05	0.5635	18.94	0.5891	21.61	0.6669	20.42	0.6148
	DiffuseRecon	26.74	0.5496	26.59	0.5451	25.21	0.5757	25.09	0.6103	25.89	0.5702
	DDNM	31.14	0.8649	32.66	0.843	29.87	0.8348	29.89	0.7864	30.89	0.8323
	SPA-MRI	**35.02**	**0.9434**	**37.46**	**0.9517**	**33.69**	**0.9375**	**33.79**	**0.9322**	**34.99**	**0.9412**

Table 2. Comparison of total number of function Evaluations (NFEs) and average reconstruction time per slice (in seconds) using the same underlying diffusion model. Note that for SPA-MRI, the total NFEs include the inversion and the sampling steps.

Method	NFE	Time/Slice (s)
DiffuseRecon	1200	285.3
DDNM	200	44.15
SPA-MRI	200*	40.04

SPA-MRI achieves faster reconstruction times compared to DDNM while maintaining superior image quality. The slight increase in computation time relative to DiffuseRecon is offset by the significant improvement in reconstruction accuracy. Future work will focus on further optimizing the algorithm for real-time applications.

Figure 2 further visually confirms the results, where the proposed SPA-MRI produces fewer artefacts and better preserves fine details compared to DDNM and DiffuseRecon whilst having fewer inference steps. We use the same underlying UNet model [15] for sampling; thus, we compare the required sampling steps (NFEs) as well in Fig. 2, the proposed method effectively reconstructs with higher PSNR and SSIM while using the least steps.

Figure 3 consistently extends these observations across multiple MRI contrasts with varying acceleration factors and undersampling patterns. Note that the model has been trained jointly on all modalities without separate training on each of them, and the consistency enforcement mechanism has further enabled the model to deal with arbitrary undersampling patterns and acceleration factors. This generalisability is a key strength of the proposed method.

Reconstruction Time Analysis. To evaluate the computational efficiency of SPA-MRI, we conducted timing experiments on a single NVIDIA A5000 GPU. Table 2 presents the average reconstruction time per slice for SPA-MRI compared to baseline methods. The SPA-MRI method achieves high PSNR and SSIM scores while requiring significantly fewer function evaluations than DiffuseRecon and DDNM. In our results, we consider the extra sampling steps necessary for the diffusion-aware inversion process. Despite these additional steps, our enhanced seed initialisation enables us to reduce the overall sampling steps, leading to faster inference.

Limitations and Future Work. Our method achieves significantly better results than baseline methods; however, certain limitations may hinder its broader applicability. First, our approach ignores the noise components in the acquisition process (see Eq. 6), which is different from real-world scenarios; Including the noise models into the sampling process is a promising scope for future work and can potentially result in more accurate reconstructions. Besides, training a diffusion model is more computationally intensive than other traditional CNN-based methods. Our diffusion UNet [15] comprises 150 million parameters, whereas contemporary CNN-based methods typically range between 1–10 million parameters. This could limit its practicality due to the requirement for large GPUs during inference. While our inference times are competitive, additional optimization would be necessary for real-world applications. Model distillation and parameter-efficient adaption can further increase the applicability of the proposed methods.

6 Conclusion

In this paper, we proposed SPA-MRI, a diffusion-based multi-coil MRI reconstruction model that generalises across contrasts, acceleration factors, and sampling patterns. We train an unconditional diffusion model and reconstruct MR images with adaptive consistency enforcement. By combining frequency decomposed posterior loss with adaptive weighing for enforcing consistency, SPA-MRI achieves superior performance across modalities, undersampling patterns and acceleration factors, facilitating broader deployment of accelerated MRI without any downstream fine-tuning.

Acknowledgement. This work was supported by the Engineering and Physical Sciences Research Council (EPSRC) UK grant TrustMRI [EP/X039277/1]. C. Wang was supported by the Shanghai Municipal Science and Technology Major Project (No. 2023SHZD2X02A05), the National Natural Science Foundation of China (No. 62331021) and the Shanghai Sailing Program (No. 20YF1402400).

References

1. Aggarwal, H.K., Mani, M.P., Jacob, M.: Modl: model-based deep learning architecture for inverse problems. IEEE Trans. Med. Imaging **38**(2), 394–405 (2018)
2. Anderson, B.D.: Reverse-time diffusion equation models. Stochastic Processes Appl. **12**(3), 313–326 (1982)
3. Chung, H., Kim, J., Mccann, M.T., Klasky, M.L., Ye, J.C.: Diffusion posterior sampling for general noisy inverse problems. arXiv preprint arXiv:2209.14687 (2022)
4. Chung, H., Lee, S., Ye, J.C.: Decomposed diffusion sampler for accelerating large-scale inverse problems. arXiv preprint arXiv:2303.05754 (2023)
5. Goodfellow, I.J., et al.: Generative adversarial networks (2014). https://arxiv.org/abs/1406.2661
6. Güngör, A., et al.: Adaptive diffusion priors for accelerated MRI reconstruction. Med. Image Anal. **88**, 102872 (2023)
7. Ho, J., Jain, A., Abbeel, P.: Denoising diffusion probabilistic models. Adv. Neural Inf. Process. Syst. **33**, 6840–6851 (2020)
8. Liu, G., Sun, H., Li, J., Yin, F., Yang, Y.: Accelerating diffusion models for inverse problems through shortcut sampling. arXiv preprint arXiv:2305.16965 (2023)
9. ch1Lustig, M., Donoho, D., Pauly, J.: Sparse mri: the application of compressed sensing for rapid mr imaging. Magn. Resonance Med. Off. J. Int. Soc. Magn. Resonance Med. **58**(6), 1182–1195 (2007)

10. Mardani, M., et al.: Deep generative adversarial neural networks for compressive sensing mri. IEEE Trans. Med. Imaging **38**(1), 167–179 (2018)
11. Nichol, A., Dhariwal, P.: Improved denoising diffusion probabilistic models (2021). https://arxiv.org/abs/2102.09672
12. Paszke, A., et al.: Pytorch: an imperative style, high-performance deep learning library. Adv. Neural Inf. Process. Syst. **32** (2019)
13. ch1Peng, C., Guo, P., Zhou, S., Patel, V., Chellappa, R.: Towards performant and reliable undersampled mr reconstruction via diffusion model sampling. In: International Conference on Medical Image Computing and Computer-Assisted Intervention, pp. 623–633. Springer Nature, Switzerland, Cham (2022)
14. ch1Qin, C., ch1Schlemper, J., Caballero, J., Price, A., Hajnal, J., Rueckert, D.: Convolutional recurrent neural networks for dynamic mr image reconstruction. IEEE Trans. Med. Imaging **38**(1), 280–290 (2018)
15. Ronneberger, O., Fischer, P., Brox, T.: U-net: convolutional networks for biomedical image segmentation (2015). https://arxiv.org/abs/1505.04597
16. Rout, L., Raoof, N., Daras, G., Caramanis, C., Dimakis, A., Shakkottai, S.: Solving linear inverse problems provably via posterior sampling with latent diffusion models. Adv. Neural Inf. Process. Syst. **36** (2024)
17. Schlemper, J., Caballero, J., Hajnal, J., Price, A., Rueckert, D.: A deep cascade of convolutional neural networks for dynamic mr image reconstruction. IEEE Trans. Med. Imaging **37**(2), 491–503 (2017)
18. Song, J., Meng, C., Ermon, S.: Denoising diffusion implicit models. arXiv preprint arXiv:2010.02502 (2020)
19. Song, Y., Ermon, S.: Generative modeling by estimating gradients of the data distribution. Adv. Neural Inf. Process. Syst. **32** (2019)
20. Song, Y., Sohl-Dickstein, J., Kingma, D., Kumar, A., Ermon, S., Poole, B.: Score-based generative modeling through stochastic differential equations. In: International Conference on Learning Representations (2021)
21. Song, Y., Shen, L., Xing, L., Ermon, S.: Solving inverse problems in medical imaging with score-based generative models (2022). https://arxiv.org/abs/2111.08005
22. Wang, Y., Yu, J., Zhang, J.: Zero-shot image restoration using denoising diffusion null-space model. arXiv preprint arXiv:2212.00490 (2022)
23. Wang, Z., et al.: Cmrxrecon2024: a multi-modality, multi-view k-space dataset boosting universal machine learning for accelerated cardiac mri. arXiv preprint arXiv:2406.19043 (2024)
24. Xiang, T., Yue, W., Lin, Y., Yang, J., Wang, Z., Li, X.: Diffcmr: fast cardiac mri reconstruction with diffusion probabilistic models. In: International Workshop on Statistical Atlases and Computational Models of the Heart, pp. 380–389. Springer Nature Switzerland, Cham (2023)
25. Yang, G., et al.: Dagan: deep de-aliasing generative adversarial networks for fast compressed sensing mri reconstruction. IEEE Trans. Med. Imaging **37**(6), 1310–1321 (2018). https://doi.org/10.1109/TMI.2017.2785879

HyperCMR: Enhanced Multi-contrast CMR Reconstruction with Eagle Loss

Ruru Xu[1(✉)], Caner Özer[1,2,3], and Ilkay Oksuz[1]

[1] Computer Engineering Department, Istanbul Technical University, Istanbul, Turkey
xu21@itu.edu.tr
[2] Department of Artificial Intelligence and Data Engineering, Istanbul Technical University, Istanbul, Turkey
[3] Department of Applied Mathematics, University of Twente, 7522 NB Enschede, The Netherlands
https://github.com/Ruru-Xu/HyperCMR

Abstract. Accelerating image acquisition for cardiac magnetic resonance imaging (CMRI) is a critical task. CMRxRecon2024 challenge aims to set the state of the art for multi-contrast CMR reconstruction. This paper presents HyperCMR, a novel framework designed to accelerate the reconstruction of multi-contrast cardiac magnetic resonance (CMR) images. HyperCMR enhances the existing PromptMR model by incorporating advanced loss functions, notably the innovative Eagle Loss, which is specifically designed to recover missing high-frequency information in undersampled k-space. Extensive experiments conducted on the CMRxRecon2024 challenge dataset demonstrate that HyperCMR consistently outperforms the baseline across multiple evaluation metrics, achieving superior SSIM and PSNR scores.

Keywords: Cardiac MRI · Multi-Modality · Reconstruction · Deep Learning

1 Introduction

Multi-contrast cardiac magnetic resonance (CMR) imaging is a crucial tool for comprehensive cardiac assessment, providing detailed insights into cardiac structure and function across various contrast levels. Each contrast-weighted image-such as Cine, Aorta, Mapping, and Tagging-emphasizes different aspects of cardiac physiology, making them indispensable for accurate diagnosis. However, acquiring these images typically requires extended scan times, increasing patient discomfort and leading to more motion artifacts, complicating the reconstruction process [2]. Reconstructing multi-contrast CMR images from undersampled k-space data is particularly challenging because it requires the preservation of intricate details specific to each contrast, despite their distinct characteristics. For example, Cine imaging captures the dynamic motion of the heart, Aorta

O. Camara et al. (Eds.): STACOM 2024, LNCS 15448, pp. 152–163, 2025.
https://doi.org/10.1007/978-3-031-87756-8_15

imaging focuses on the vascular wall, and Mapping requires a precise depiction of myocardial tissue properties. These varying demands make it difficult to design a universal reconstruction model that performs well across multiple modalities. Traditional methods often compromise on essential details required for certain contrasts or struggle to generalize across different contrast types [1].

To address these limitations, we introduce HyperCMR, a novel framework specifically designed to handle the complexities of multi-contrast CMR reconstruction. HyperCMR builds upon existing deep learning techniques, incorporating substantial innovations to tackle the challenges inherent in multi-contrast imaging. A key enhancement is the integration of a specialized loss function, Eagle Loss [7], engineered to recover the high-frequency information in undersampled k-space.

Recent advances in deep learning have demonstrated the feasibility of reconstructing CMR images from undersampled k-space data [3,4,10,11]. However, these methods often struggle to balance global structural coherence with preserving local details across different contrast modalities. In contrast, HyperCMR not only achieves this delicate balance but also delivers superior reconstruction performance across multiple modalities, providing a robust and effective solution for multi-contrast CMR reconstruction.

2 Datasets

2.1 Data and Task Description

The CMRxRecon2024 challenge [8] provides the multi-contrast k-space data utilized in this paper. The dataset is divided into training, validation, and test sets, each containing different anatomical views and contrast settings.

The training set includes various CMR modalities, such as Cine (196 subjects), Aorta (154 subjects), Mapping (193 subjects), and Tagging (143 subjects), each with distinct imaging characteristics that highlight different aspects of cardiac anatomy and function. Additionally, the validation and test sets introduce two previously unseen contrast modalities: Flow2d and BlackBlood, further increasing the complexity of the reconstruction task. Each modality in the validation and test sets contains data from 60 subjects.

The primary objective of this challenge is to develop a contrast-general model capable of delivering high-quality image reconstruction from highly accelerated, uniformly undersampled k-space data. HyperCMR is specifically designed to address this challenge by improving the reconstruction quality across these different modalities. The inclusion of unseen contrasts in the validation and test sets underscores the importance of generalization in model performance, which is a central focus of the HyperCMR framework.

2.2 Datasets Processing and Training Idea

Based on different modalities (aorta_sag, aorta_tra, cine_lax, cine_sax, cine_lvot, T1map, T2map, tagging) and different sizes (cine_lax includes two sizes: [:, :, :,

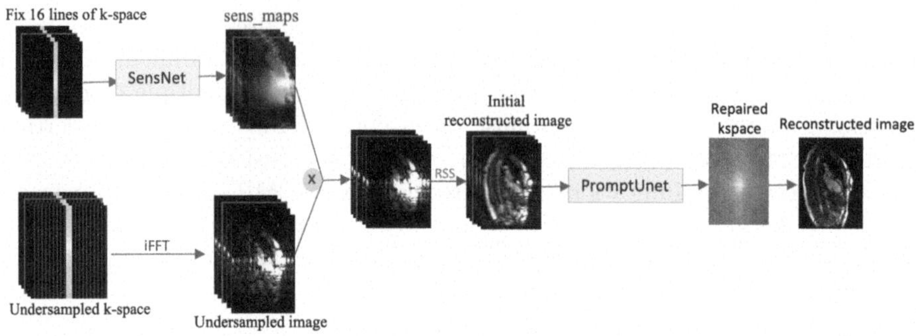

Fig. 1. Overview of the HyperCMR framework. Our pipeline includes generation of sensitivity maps and repaired k-space generation with promptUnet.

204, 448] and [:, :, :, 168, 448]. cine_sax includes three sizes: [:, :, :, 246, 512], [:, :, :, 162, 512] and [:, :, :, 204, 512]), we divide the dataset into 11 groups to facilitate us to set different batch sizes for training. The training process is shown in Algorithm 1.

Algorithm 1 Training process across modality groups

```
Require: 11 modality groups, max_epochs, model, optimizer
  for epoch in range(max_epochs) do
    total_train_loss = 0
    total_val_loss = 0
    for modality_group in [group_1, group_2, ..., group_11] do
      modality_train_loss = train_epoch(modality_group, model, optimizer, ...)
      modality_val_loss = validate_epoch(modality_group, model, ...)
      total_train_loss += modality_train_loss
      total_val_loss += modality_val_loss
    end for
    Compute total_train_loss and total_val_loss for this epoch
  end for
```

3 Methods

3.1 Enhancing PromptMR: The Development of HyperCMR

The HyperCMR framework builds upon the established PromptMR [9] model. PromptMR is particularly effective in restoring missing information in the low-frequency regions of undersampled k-space, which contributes significantly to the overall structural integrity of the reconstructed images. However, it struggles with the recovery of high-frequency details, which are crucial for preserving the finer structures and textures in CMR images. High-frequency components typically encode critical information such as sharp edges and intricate patterns, which are vital for accurate and detailed cardiac assessment. To address this limitation, HyperCMR introduces several key enhancements, utilizing a combination of loss functions specifically designed to recover high-frequency information from undersampled k-space. By optimizing the model's ability to focus on both low- and high-frequency regions, HyperCMR achieves a more balanced and comprehensive reconstruction performance.

As illustrated in Fig. 1, the HyperCMR framework leverages the temporal similarity of information by connecting slices from two adjacent time frames. With the inclusion of 10 coils, the input data format expands to 50 channels, ultimately producing a single-channel reconstructed image. The process begins by fixing 16 lines of k-space data, which are subsequently used to estimate sensitivity maps through SensNet. An initial reconstruction is performed using the inverse Fast Fourier Transform (iFFT), and the resulting image, combined with the sensitivity maps, undergoes Root Sum of Squares (RSS) calculation to yield an initial image reconstruction. This image is then refined using the PromptUnet model, which is central to both the HyperCMR and PromptMR pipelines.

3.2 Loss Functions

Our approach incorporates a set of advanced loss functions, each meticulously selected and weighted to balance the diverse aspects of CMR image reconstruction:

– **Data Fidelity Loss (k-space domain,** $\alpha_1 = 1.0$): This loss ensures that the repaired k-space data closely matches the fully-sampled ground truth, focusing on the accurate recovery of k-space information.

$$L_{fidelity} = \|k_{pred} - k_{fully}\|_2^2$$

– **SSIM Loss (image domain,** $\alpha_2 = 1.0$): This component promotes structural similarity between the reconstructed image and the ground truth, ensuring perceptual quality and structural integrity.

$$L_{recons\_ssim} = 1 - SSIM(img_{recons}, img_{fully})$$

– **Eagle Loss (image domain,** $\alpha_3 = 0.05$): Specifically designed to enhance high-frequency detail recovery, Eagle Loss plays a crucial role in edge preservation and the accurate reconstruction of fine structures in CMR images [7].

$$L_{eagle} = \text{FFT-based high-pass filter loss on variance maps}$$

– **VGG Perceptual Loss (image domain,** $\alpha_4 = 0.1$): This loss guides the network to capture higher-level perceptual features, ensuring that the reconstructed image maintains visual fidelity in comparison to the ground truth [5].

$$L_{vgg} = \sum_{l} \|VGG_l(img_{recons}) - VGG_l(img_{fully})\|_2^2$$

– **Regularization Loss (k-space domain,** $\alpha_5 = 0.01$): This loss promotes sparsity and smoothness in the repaired k-space data, helping to prevent overfitting while preserving essential details [6].

$$L_{reg} = \|k_{recons}\|_1 + \beta\|k_{recons}\|_2$$

The overall loss function is a weighted combination of these components:

$$L_{total} = \alpha_1 \cdot L_{fidelity} + \alpha_2 \cdot L_{recons\_ssim} + \alpha_3 \cdot L_{eagle} + \alpha_4 \cdot L_{vgg} + \alpha_5 \cdot L_{reg}$$

The specific weights for each loss component were determined through extensive empirical validation. We set $\alpha_1 = 1.0$ and $\alpha_2 = 1.0$ to prioritize data fidelity and structural similarity, ensuring that the reconstructed image closely resembles the ground truth in both the k-space and image domains. The weight for Eagle Loss, $\alpha_3 = 0.05$, was carefully chosen to focus on high-frequency detail recovery without overshadowing the contributions of other loss components. The VGG Perceptual Loss, weighted at $\alpha_4 = 0.1$, is crucial for maintaining visual fidelity, particularly in capturing finer image details. Finally, the Regularization Loss, set at $\alpha_5 = 0.01$, helps prevent overfitting while preserving key structural details, contributing to a robust and generalizable reconstruction model.

3.3 Optimized Eagle Loss

In our method, we improved upon the original Eagle Loss technique described in [7] by replacing the Gaussian high-pass filter with a Butterworth high-pass filter, which preserves high-frequency information crucial for multi-contrast CMR image reconstruction. Additionally, we introduced padding and other optimizations to further enhance the performance of the loss function.

As shown in Fig. 2, the effect of varying patch sizes and cutoff frequencies on the high-pass filter weights and the resulting magnitude maps within the Eagle Loss framework are as follows:

- Patch Size: This parameter determines the scale of features emphasized by the high-pass filter. Smaller patch sizes (e.g., 3) focus on finer details and higher frequencies, enhancing intricate structures but potentially amplifying noise. In contrast, larger patch sizes (e.g., 5) offer a broader context, capturing mid-level frequencies while preserving essential high-frequency information, thereby maintaining structural integrity with minimal noise amplification.
- Cutoff Frequency: The cutoff frequency in the Butterworth filter controls the threshold for emphasizing higher frequencies. A lower cutoff frequency (e.g., 0.3) allows more high-frequency components to pass through, highlighting sharp edges and fine details, but at the cost of increased noise sensitivity.

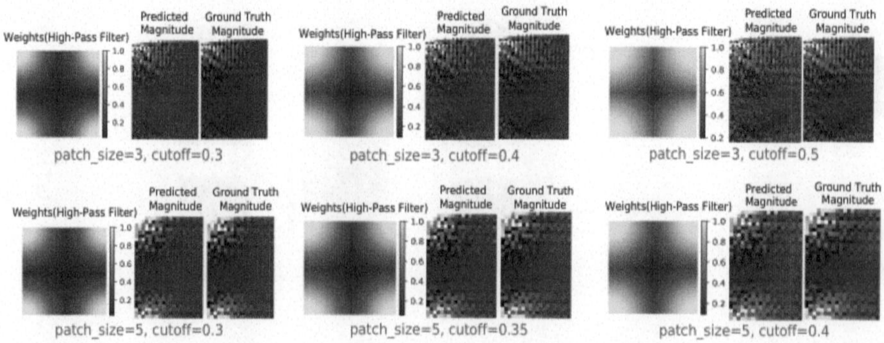

Fig. 2. Visualization of the impact of different patch sizes and cutoff frequencies on the high-pass filter and resulting magnitude maps. The selected patch size of 5 and cutoff of 0.35 provide balanced performance in capturing mid-level frequencies and spatial details in the HyperCMR framework.

Conversely, a higher cutoff frequency (e.g., 0.4 or 0.5) suppresses very high frequencies, focusing on preserving mid-level details and reducing noise.

- Selected Parameters: We selected a patch size of 5 and a cutoff frequency of 0.35 for their balanced performance. This combination effectively captures both mid-level and high-frequency details, yielding magnitude maps that closely resemble the ground truth. This balance ensures optimal structural preservation with minimal noise, which is crucial for high-fidelity reconstruction in the HyperCMR framework.

The Eagle Loss process (Fig. 3), is designed to capture the structural differences between predicted and target images by focusing on their gradient information in both the x and y directions. The process begins with gradient calculation, where the predicted and target images are convolved with predefined gradient filters (Scharr kernel) to compute the horizontal (gradient_x) and vertical (gradient_y) gradients. These gradients highlight edge-like structures, which are critical for accurate image reconstruction.

Next, for each direction (x and y), the gradient maps are divided into non-overlapping patches, and the variance of each patch is calculated. This variance map reflects the distribution of gradient magnitudes in different regions of the image. These variance maps are then passed through a 2D Fast Fourier Transform (FFT) to obtain their frequency representations. To emphasize the high-frequency components—representing fine details—the Eagle Loss applies a Butterworth high-pass filter to the FFT magnitudes.

The filtered FFT magnitudes of the predicted and target images are then compared using the L1 loss function in the frequency domain, capturing the differences in their high-frequency details. The overall Eagle Loss is the sum of the losses computed for both the x and y-direction gradients, ensuring that the model learns to preserve structural details across multiple directions. This

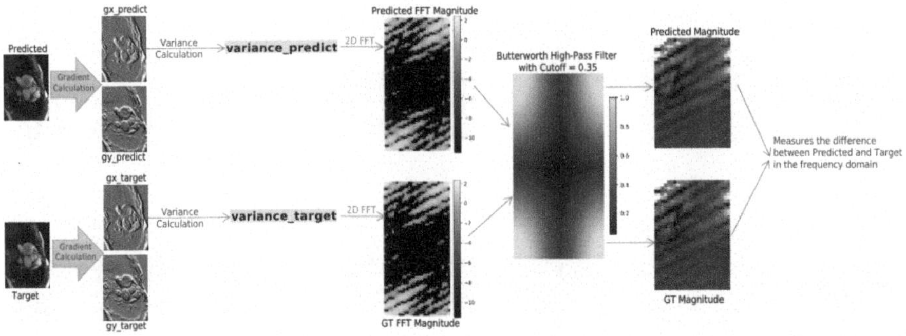

Fig. 3. The figure illustrates the detailed workflow of the Eagle Loss implementation for calculating gradients in the x-direction. The process begins by computing the gradients for both the predicted and target images, followed by calculating variance across non-overlapping patches. These variance maps are then transformed using a 2D FFT, and their magnitudes are filtered with a Butterworth high-pass filter. Finally, the L1 loss is calculated by comparing the filtered magnitudes in the frequency domain. The same process is applied to the y-direction gradients. The total Eagle Loss is computed as the sum of the loss for both x and y-direction. gradients

approach helps the model focus on recovering fine structures and details that are often lost during undersampling in MRI reconstruction.

4 Experiments and Results

4.1 Experiments

The model was implemented in PyTorch, utilizing the AdamW optimizer with an initial learning rate of 0.00009, a weight decay of 1e-4, and a StepLR learning rate scheduler with a step size of 2 and gamma of 0.95. To manage the large computational demands, gradient accumulation was employed with 8 steps.

We trained the 4x acc, 8x acc, and 10x acc models on three servers respectively. The details are as follows:

- **4x acc:** GPU: RTX 3090 24G, one epoch: 15.96 h. The batch_size of all 11 groups is set to 1.
- **8x acc:** GPU: RTX 6000 24G, one epoch: 26.13 h. The batch_size of all 11 groups is set to 1.
- **10x acc:** GPU: GV100 32G, one epoch: 20.42 h. The batch_size of the 11 groups (aorta_sag, aorta_tra, cine_lax204, cine_lax168, cine_sax246, cine_sax162, cine_sax204, cine_lvot, T1map, T2map, tagging) is set to 2, 2, 2, 2, 2, 1, 1, 2, 2, 4, 2 respectively.

Table 1. Performance across CMR modalities and acceleration factors (4×, 8×, 10×) in terms of SSIM, PSNR, and NMSE.

Modality	4×			8x			10x		
	SSIM	PSNR	NMSE	SSIM	PSNR	NMSE	SSIM	PSNR	NMSE
Tagging	0.9621	38.57	0.0118	0.9162	33.98	0.0321	0.9112	33.20	0.0380
Mapping	0.9508	37.33	0.0077	0.8880	31.87	0.0220	0.8791	30.91	0.0268
Cine	0.9391	36.54	0.0102	0.8512	30.83	0.0312	0.8121	29.13	0.0437
Aorta	0.9393	36.70	0.0137	0.8475	31.12	0.0452	0.8078	29.61	0.0609
BlackBlood	0.9426	36.14	0.0104	0.8404	30.22	0.0375	0.7842	28.30	0.0570
Flow2d	0.9770	40.98	0.0028	0.9212	33.38	0.0153	0.9121	32.38	0.0188
Overall	0.9518	37.71	0.0094	0.8774	31.90	0.0306	0.8511	30.59	0.0409

4.2 Results Analysis

To further evaluate the robustness of our model, we analyzed its performance across specific acceleration factors, including 4×, 8×, and 10×. Table 1 provides a detailed breakdown of our model's performance across various CMR modalities under these acceleration settings. Notably, our model demonstrates consistent performance across different modalities and acceleration factors, achieving an excellent balance between maintaining structural similarity and reducing noise.

Table 2 provides a detailed analysis of the impact of different loss functions on the performance of our model at an 8x acceleration factor across various CMR modalities. Three configurations were evaluated: the results from using the pre-trained PromptMR model with SSIM loss only, our model without Eagle Loss, and our model with Eagle Loss.

- Performance of Pre-trained PromptMR with SSIM Loss: The pre-trained PromptMR model, trained exclusively with SSIM loss, serves as the baseline. This model demonstrates the weakest performance across all metrics. These results indicate that while the pre-trained PromptMR model can achieve a basic level of reconstruction, it struggles with accuracy, which is crucial for precise CMR imaging.
- Effect of Removing Eagle Loss: Excluding Eagle Loss from our model still leads to a significant performance improvement over the baseline. This suggests that even without Eagle Loss, a combination of multiple loss functions contributes to more accurate reconstructions. However, the slightly higher NMSE values suggest that the model still encounters challenges in fully capturing high-frequency details.
- Impact of Including Eagle Loss: Including Eagle Loss in our model yields the best performance across all metrics and modalities. Eagle Loss effectively boosts the model's ability to recover high-frequency details and improves overall reconstruction quality.

Table 2. Result of 8x acceleration. Comparison of SSIM, PSNR, and NMSE scores for different loss functions. "PromptMR with SSIM Loss" represents the results obtained using the pre-trained PromptMR model, which was trained solely with SSIM loss. "Ours without Eagle Loss" refers to our model trained with a combination of multiple loss functions, excluding Eagle Loss. "Ours with Eagle Loss" indicates the results of our model trained with the same combination of multiple loss functions, including Eagle Loss.

Modality	PromptMR with SSIM Loss			Ours without Eagle Loss			Ours with Eagle Loss		
	SSIM	PSNR	NMSE	SSIM	PSNR	NMSE	SSIM	PSNR	NMSE
Tagging	0.7423	28.60	0.1254	0.9155	33.92	0.0326	0.9166	34.03	0.0320
Mapping	0.8878	31.86	0.0223	0.8890	31.93	0.0218	0.8941	32.18	0.0210
Cine	0.8387	30.29	0.0347	0.8497	30.69	0.0316	0.8512	30.83	0.0312
Aorta	0.8311	30.57	0.0510	0.8457	31.05	0.0458	0.8487	31.12	0.0448
BlackBlood	0.8394	30.19	0.0377	0.8422	30.27	0.0369	0.8426	30.40	0.0361
Flow2d	0.9078	32.37	0.0189	0.9205	33.20	0.0154	0.9212	33.38	0.0153
Overall	0.8412	30.65	0.0483	0.8771	31.84	0.0307	0.8791	31.99	0.0300

The ablation study demonstrates that while our model offers substantial improvements over the pre-trained PromptMR, the integration of Eagle Loss is essential for achieving state-of-the-art performance. It significantly enhances the recovery of high-frequency information in undersampled k-space, making our HyperCMR framework superior in producing high-quality, detailed reconstructions across various CMR modalities.

In addition to quantitative evaluations, we conducted a qualitative analysis by visually comparing the reconstructions from our proposed HyperCMR method against the baseline PromptMR model. Figure 4 illustrates the reconstruction quality for three different CMR modalities: aorta_tra, cine_lvot, and cine_sax, under a 10x acceleration factor. This figure demonstrates how our method effectively restores high-frequency information which is crucial for maintaining image fidelity in undersampled k-space scenarios. In the aorta_tra modality, the highlighted region reveals that our method preserves structural integrity more effectively than the PromptMR baseline. For the cine_lvot modality, our method demonstrates superior preservation of k-space textures, as indicated by the yellow arrow. Additionally, in the cine_sax modality, the superior recovery of high-frequency components further validates the efficacy of our two-stage framework in improving image fidelity and detail preservation.

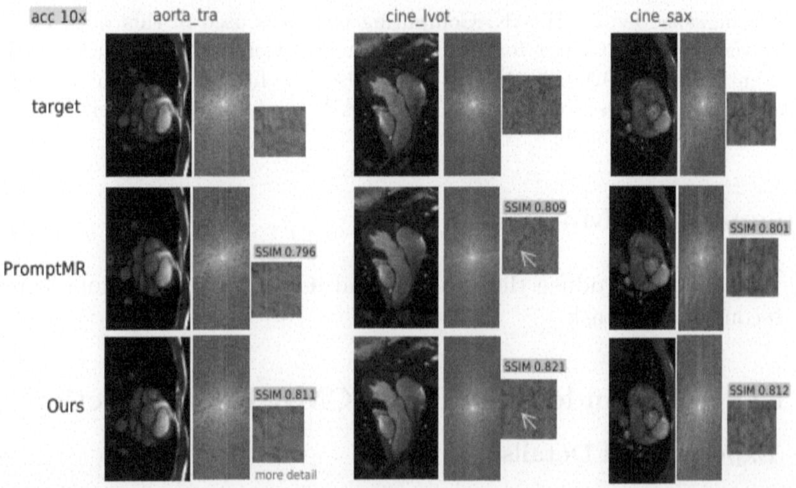

Fig. 4. Visual comparison of reconstructed images for three different CMR modalities (aorta_tra, cine_lvot, cine_sax) at 10x acceleration. The SSIM values and zoomed-in regions highlight the superior performance of our model (HyperCMR) compared to the PromptMR baseline. This figure provides a clear example of the improvements achieved by our method in preserving fine details and enhancing overall image quality, particularly in the high-frequency regions of the undersampled k-space.

5 Conclusion

In this paper, we introduce HyperCMR, a novel framework designed to enhance the reconstruction of highly accelerated multi-contrast cardiac magnetic resonance (CMR) images. Our approach builds on the PromptMR model and incorporates advanced loss functions, including a newly optimized Eagle Loss, specifically designed to recover the missing high-frequency information in the undersampled k-space. The success of HyperCMR can be attributed to several key innovations: the introduction of an advanced loss function framework, and the careful selection of patch size and cutoff frequency. These contributions have resulted in a powerful and generalizable model capable of effectively addressing the various challenges posed by multi-contrast CMR imaging.

Extensive experiments on the CMRxRecon2024 dataset demonstrate that our method achieves balanced performance across different CMR modalities and acceleration factors, significantly improving SSIM, PSNR, and NMSE scores. The experimental results confirm that our model is capable of preserving high-frequency information in the undersampled k-space, contributing to superior image reconstruction quality.

Acknowledgments. This paper has been benefitted from the 2232 International Fellowship for Outstanding Researchers Program of TUBITAK (Project No: 118C353). However, the entire responsibility of the thesis belongs to the owner. The financial support received from TUBITAK does not mean that the content of the thesis is approved in a scientific sense by TUBITAK. Computing resources used in this work were provided by the National Center for High Performance Computing of Turkey (UHeM) under grant number 4020052024. The paper also benefited from Istanbul Technical University Scientific Research Projects (ITU BAP) funds, grant numbers FHD-2024-45302 and PMA-2024-44970.

Supplementary Material

In the A, we will introduce the experimental details of the Random sampling CMR reconstruction task.

A TASK 2: Random Sampling CMR Reconstruction

A.1 Experimental Details

We performed distributed training on 4*A100 80G servers, randomly selecting half of datasets for training in each epoch. The batch_size of the 11 groups (aorta_sag, aorta_tra, cine_lax204, cine_lax168, cine_sax246, cine_sax162, cine_sax204, cine_lvot, T1map, T2map, tagging) is set to 6, 6, 4, 6, 4, 4, 2, 6, 6, 10, 4 respectively. One epoch: 23.65 h

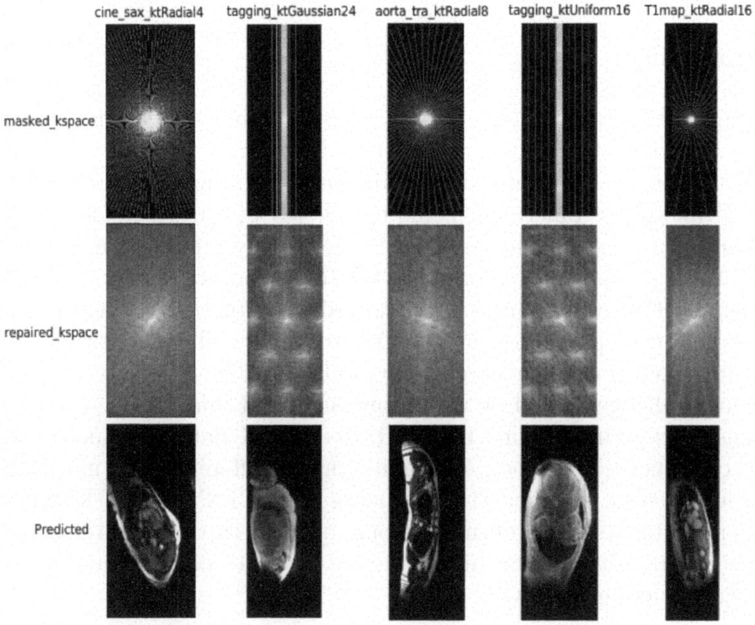

Fig. 5. Example results from Task 2. Demonstrating HyperCMR's performance in random sampling reconstruction across different CMR modalities and acceleration factors

A.2 Results

Fig. 5 showcases example results from Task 2 random sampling CMR reconstruction. The images demonstrate HyperCMR's reconstruction performance across various CMR modalities at different acceleration factors. Even at high acceleration factors, the model successfully recovers high-quality images from undersampled k-space data while preserving essential details. .

References

1. Hammernik, K., et al.: Learning a variational network for reconstruction of accelerated MRI data. Magn. Reson. Med. **79**, 3055–3071 (2018)
2. Knoll, F., et al.: Deep-learning methods for parallel magnetic resonance imaging reconstruction: a survey of the current approaches, trends, and issues. IEEE Signal Process. Mag. **37**, 128–140 (2020)
3. Lustig, M., Donoho, D., Santos, J., Pauly, J.: Compressed sensing MRI. IEEE Signal Process. Mag. **25**, 72–82 (2008)
4. Schlemper, J., Caballero, J., Hajnal, J., Price, A., Rueckert, D.: A deep cascade of convolutional neural networks for dynamic MR image reconstruction. IEEE Trans. Med. Imaging **37**, 491–503 (2017)
5. Simonyan, K., Zisserman, A.: Very deep convolutional networks for large-scale image recognition. ArXiv Preprint ArXiv:1409.1556 (2014)
6. Sun, J., Li, H., Xu, Z., et al.: Deep ADMM-Net for compressive sensing MRI. Adv. Neural Inf. Process. Syst. **29** (2016)
7. Sun, Y., et al.: EAGLE: An Edge-Aware Gradient Localization Enhanced Loss for CT Image Reconstruction. ArXiv Preprint ArXiv:2403.10695 (2024)
8. Wang, Z., et al.: CMRxRecon2024: A Multi-Modality, Multi-View K-Space Dataset Boosting Universal Machine Learning for Accelerated Cardiac MRI. ArXiv Preprint ArXiv:2406.19043 (2024)
9. Xin, B., Ye, M., Axel, L., Metaxas, D. Fill the k-space and refine the image: Prompting for dynamic and multi-contrast MRI reconstruction. In: International Workshop On Statistical Atlases And Computational Models Of The Heart, pp. 261–273 (2023)
10. Xu, R., Oksuz, I.: Segmentation-aware MRI subsampling for efficient cardiac MRI reconstruction with reinforcement learning. Image Vis. Comput. 105200 (2024)
11. Yang, Z., Shen, D., Chan, K., Huang, J.: Attention-based MultiOffset deep learning reconstruction of chemical exchange saturation transfer (AMO-CEST) MRI. IEEE J. Biomed. Health Inform. (2024)

A Multi-contrast Cardiac MRI Reconstruction Method Using an Advanced Unrolled Network Architecture

Yajing Zhang(✉), Yanxin Huang, Zhixin Xu, and Jin Qi(✉)

University of Electronic Science and Technology of China, Chengdu, China
jqichina@hotmail.com

Abstract. This paper introduces a deep learning-based approach for reconstructing multi-contrast cardiac magnetic resonance images (CMRI) from highly undersampled k-space data. Traditional methods, such as parallel imaging and compressed sensing, often suffer from slow reconstruction times and limited acceleration rates. To address these challenges, we propose an advanced unrolled network architecture that separately optimizes low-frequency and high-frequency components, leveraging diverse information from k-space, the image domain, and temporal sequences. This method demonstrates robust performance across various imaging sequences and anatomical views, including long-axis, short-axis, and aortic views. The proposed approach, evaluated on the CMRxRecon challenge dataset, effectively improves image quality while maintaining clinical interpretability. Our results show improvements in PSNR, SSIM metrics compared to traditional methods, highlighting the potential of this approach to enhance CMRI reconstruction and expand its clinical applicability.

Keywords: Cardiac MRI (CMRI) · Image Reconstruction · undersampled k-space · Unrolled Networks

1 Introduction

Cardiac magnetic resonance imaging (CMRI) is widely recognized as the most accurate and reliable non-invasive technique for evaluating the heart's anatomy [1] functionality, and pathology. It plays a crucial role in the clinical diagnosis and management of various cardiovascular conditions [2], such as coronary artery disease, heart failure, and atrial fibrillation, as well as in postoperative rehabilitation. However, a significant drawback of MRI is its slow imaging speed [3], which can cause patient discomfort and introduce motion artifacts into the images. To minimize the adverse effects of physiological movements, such as breathing and heartbeats, data acquisition is often limited to shorter windows (e.g., during end-diastole), resulting in high undersampling of k-space data and making image reconstruction more challenging.

CMRI inherently provides multiple contrasts [4], including cardiac cine, mapping, tagging, phase-contrast, and dark-blood imaging [5]. It also involves capturing images from various anatomical perspectives, such as long-axis (2-chamber, 3-chamber, and

O. Camara et al. (Eds.): STACOM 2024, LNCS 15448, pp. 164–171, 2025.
https://doi.org/10.1007/978-3-031-87756-8_16

4-chamber views), short-axis, outflow tract, and aortic views (cross-sectional and sagittal). The undersampling patterns for accelerated imaging are diverse, ranging from uniform undersampling to variable-density sampling, which further complicates the reconstruction process.

Traditional methods for reconstructing undersampled images primarily involve parallel imaging and compressed sensing algorithms [6–8]. While these classical methods are theoretically sound and provide excellent interpretability, they often require careful hyperparameter tuning, are limited by lower feasible acceleration rates, and involve lengthy reconstruction times due to iterative calculations.

Deep learning (DL) techniques offer an alternative approach, enabling the reconstruction of MR images from highly undersampled k-space data within a clinically acceptable timeframe [9–11]. Among DL-based approaches, unrolled networks have emerged as a powerful and effective architecture. These networks integrate imaging models into the framework and leverage large datasets to learn image priors, offering strong interpretability. However, conventional DL-based reconstruction models often require specific training and deployment tailored to each imaging scenario (e.g., imaging sequence, view, or device vendor), which restricts their clinical utility in real-world settings.

Therefore, there is a strong demand to develop more effective and generalizable deep learning methods. This paper introduces a novel approach based on an advanced unrolled structure, which applies separate optimization techniques to the low-frequency and high-frequency components of images according to their distinct characteristics. The proposed method learns and integrates diverse information from k-space, the image domain, and temporal sequences to improve the network's robustness across various data types [12].

2 Methods

Following the approach of SOTA (state-of-the-art) models [13], we divided the training task into two training stages. In the first stage, we use the SOTA method to recover multichannel data from undersampled k-space to the undersampled image domain. Figure 1 illustrates the processing flow of stage one. First, the images are transformed from k-space to the image domain, followed by down-sampling using a convolutional network. Each down-sampling block contains a down-sampling operation and a channel attention module. The output of stage one is obtained by fusing the up-sampled data with the information extracted from the Prompt module.

In the second stage, we recover the undersampled images to fully sampled images [14]. The model structure is shown in Fig. 2. It consists of three modules: an input module (IM), a feature extraction module, and an output module (OM). The IM handles early

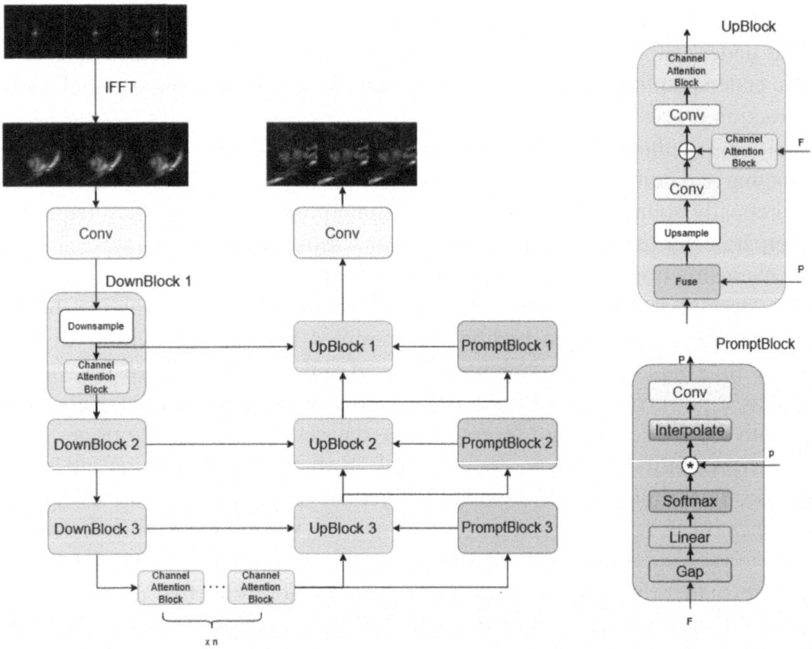

Fig. 1. Overview of the first-stage model architecture. Inverse Fast Fourier Transform (IFFT) is applied to map the signal to the image domain before it is input into the model. Green modules represent upsampling, blue modules indicate downsampling, and pink modules denote Prompt components. Channel attention is applied during downsampling and feature concatenation, while Prompt features are integrated during upsampling for better k-space reconstruction. (Color figure online)

visual processing, extracting low-level features from the input image and mapping them from the input image space to a high-dimensional feature space represented by finite elements. It employs a two-dimensional convolutional layer (Conv2D) to map the input image into shallow features, which are then fed into the subsequent feature extraction module. The feature extraction module consists of six residual Swin Transformer blocks (RSTBs) and one Spatial Frequency Block (SFB). Each RSTB is composed of six Swin Transformer layers (STLs) and an SFB arranged in a residual structure. STLs utilize windowed multi-head self-attention (W-MSA) and shifted windowed multi-head self-attention (SW-MSA) for feature extraction. To overcome the limitations of local receptive fields in STLs and convolutional layers, RSTBs incorporate a globally-aware SFB.

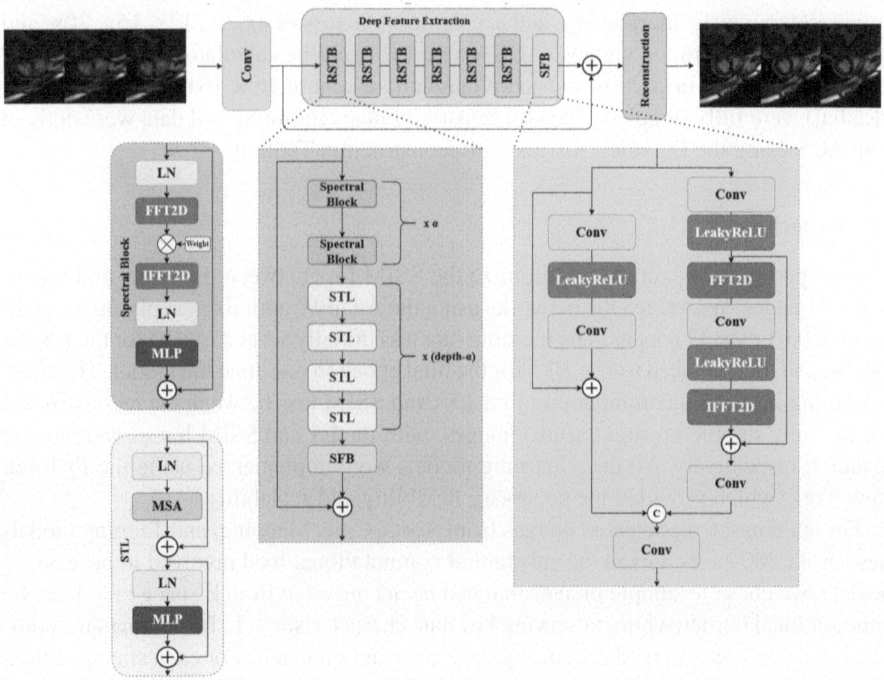

Fig. 2. Overview of the second-stage model architecture. The second stage is primarily used for refining the image, utilizing cascaded RSTB modules to restore the image. The RSTB (Residual Swin Transformer Block), primarily composed of Spectral Block, STL (Swin Transformer Layer), and SFB (Spectral Fusion Block). The final result is obtained after this refinement.

3 Experiments and Results

3.1 Datasets

The CMRxRecon challenge dataset was utilized to develop this method. In total, 330 volunteers were scanned using a 3T MRI scanner (MAGNETOM Vida, Siemens Healthineers, Germany). Image acquisition followed the guidelines for CMR exams outlined in prior publications. The dataset contains multi-contrast k-space data, including cardiac cine, T1/T2 mapping, and tagging, among others. It also features images of various anatomical views, such as long-axis (LAX, including 2-chamber, 3-chamber, and 4-chamber), short-axis (SAX), left ventricle outflow tract (LVOT), and aortic views (both transversal and sagittal). Cine, phase-contrast, and tagging were acquired using the TrueFISP sequence, while T1/T2 mapping was performed using the FLASH sequence. For T1/T2 mapping, signals were collected at the end of diastole with ECG triggering [15]. Typically, 5–15 slices were acquired for each contrast. The cardiac cycle was divided into 12–25 phases with a temporal resolution of approximately 50 ms. The typical geometric parameters consisted of a spatial resolution of $1.5 \times 1.5\ \text{mm}^2$, a slice thickness of 8.0 mm, and a slice gap of 4.0 mm. Various undersampling scenarios employed different k-space trajectories, including uniform, Gaussian, and pseudo-radial undersampling with temporal/parametric interleaving, and acceleration factors of 4x, 8x, 12x, 16x, 20x, and 24x, with autocalibration signals (ACS) excluded from the calculations. For ACS, the central 16 lines (Ny in ktUniform and ktGaussian) or central 16 × 16 regions (Nx*Ny in ktRadial) were fully sampled. The coil sensitivity maps for multi-coil data were derived from ACS using the Berkeley Advanced Reconstruction Toolbox.

3.2 Experiments

In our experiment, we aimed to minimize the SSIM loss between the target and reconstructed images. We trained the networks using the AdamW optimizer, applying a weight decay of 0.01 over 12 epochs. The learning rate was initially set at 2×10^{-4} for the first 11 epochs, and then reduced to 2×10^{-5} for the final epoch to fine-tune the model. The overall training loss was a combination of L1 loss and SSIM loss between the reconstructed and the fully sampled (ground truth) images, with the L1 and SSIM losses weighted at 10 and 1, respectively. All deep learning models were implemented using the PyTorch framework, which provided the necessary flexibility and scalability.

For our dataset, we selected images from Aorta, Cine, Mapping, and Tagging modalities across 200 cases. Given the substantial computational load involved in processing the data, we chose to sample images that had been converted from k-space to reduce the computational burden while preserving key data characteristics. To further manage computational costs, we merged data along the z-axis and time axis. For each slice position, we selected the five neighboring slices as the final input for training, thus maintaining relevant spatial context. We set the batch size to one, ensuring a manageable load during training. A single model, capable of handling all image contrasts and acceleration factors, was trained over 12 epochs on an NVIDIA A100-PCIE GPU with 40 GB of memory.

To assess the performance of our method, we employed three evaluation metrics: normalized mean square error (NMSE), peak signal-to-noise ratio (PSNR), and structural similarity index (SSIM), which were used to compare the reconstructed image sequences with the ground truth. Additionally, we benchmarked our results against three commonly used reconstruction methods: Zero-filling (ZF), Grappa, and Sense. The performance metrics for these methods were calculated and compared with our approach, highlighting the effectiveness of our method in producing high-quality image reconstructions.

3.3 Results

The reconstructed images under various sampling patterns are presented in Fig. 3, emphasizing the substantial improvements achieved by our method in comparison to the traditional GRAPPA algorithm. For a comprehensive visual assessment, we selected a specific slice and corresponding time frame from one case for each contrast type. These side-by-side comparisons offer a clear and detailed evaluation of the performance across different sampling strategies.

Regardless of the sampling method employed, our algorithm consistently outperforms the traditional approach, particularly in handling undersampled images with varying contrast levels. This improvement is evident not only in the enhanced clarity of fine anatomical details but also in the significant reduction of image artifacts, even in the most complex regions. Our method excels in preserving intricate features, such as the myocardium and small vessels, which are essential for accurate clinical interpretation, while the GRAPPA algorithm struggles to maintain such fidelity.

The results highlight the robustness of our algorithm in managing diverse contrast levels and sampling strategies, showcasing its effectiveness in reducing artifacts and improving image sharpness. Whether in cases of extreme undersampling or images with complex contrast variations, our approach consistently delivers superior visual detail, with fewer artifacts and more reliable anatomical reconstruction compared to the traditional method.

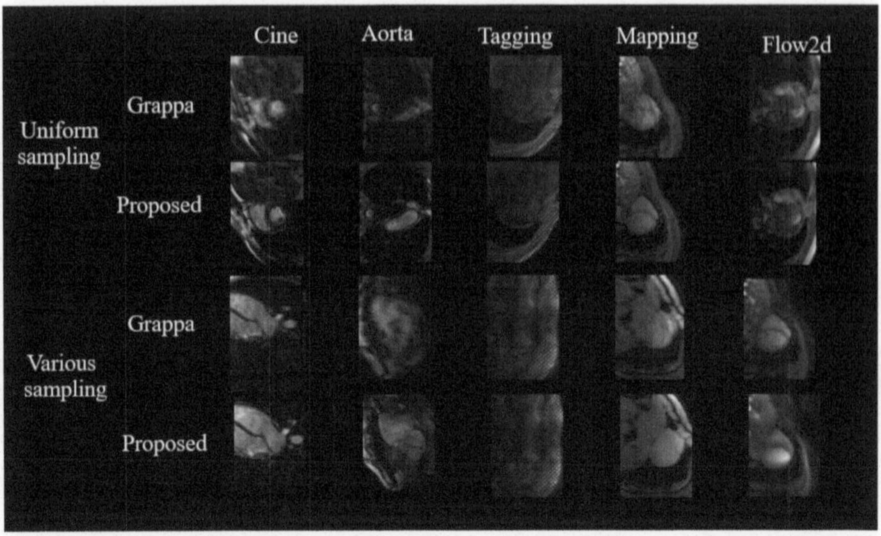

Fig. 3. Reconstruction results of multi-contrast cardiac images (Cine, Aorta, Tagging, and Flow2D) from Grappa and our proposed methods.

Table 1 provides a comprehensive quantitative evaluation of the reconstructed images across different contrast levels, detailing the performance of our method in comparison to other reconstruction techniques. The metrics evaluated include normalized mean square error (NMSE), peak signal-to-noise ratio (PSNR), and structural similarity index (SSIM), all of which provide critical insights into the accuracy and quality of the reconstructions. Notably, our method consistently demonstrates superior performance in terms of both PSNR and SSIM, highlighting its ability to produce high-quality reconstructions with greater structural fidelity and precision.

In addition to the general comparison, the table also presents results under various sampling patterns, including uniform and variable mask sampling strategies. The results reveal a common trend: reconstructions under variable mask sampling tend to yield lower performance compared to uniform sampling. This is a well-known challenge in image reconstruction, as variable mask sampling introduces greater complexity and makes it harder to accurately recover fine details and maintain image integrity. Despite this difficulty, our method outperforms traditional algorithms across both uniform and variable sampling conditions, demonstrating its robustness in handling different sampling strategies.

The ability of our approach to excel under the more challenging variable mask sampling scenario further validates its versatility and effectiveness. While traditional methods struggle to achieve consistent results under such conditions, our algorithm delivers superior performance, producing images with higher PSNR and SSIM values. This consistent improvement across both uniform and variable sampling patterns underscores the adaptability of our method and its potential to provide high-quality reconstructions in diverse imaging scenarios.

Table 1. The evaluation metrics results of different reconstruction methods

	Method	Metric		
		PSNR	SSIM	NMSE
Uniform sampling	ZF	20.5536	0.5026	0.238
	Grappa	28.3721	0.7351	0.0921
	Sense	27.1099	0.7041	0.1125
	Ours	30.1032	0.8077	0.1297
Various sampling	ZF	19.4182	0.461	0.2853
	Grappa	24.9678	0.6449	0.1554
	Sense	24.478	0.62	0.1734
	Ours	28.2563	0.7239	0.2099

4 Discussion

In this study, we proposed a universal approach for multi-contrast cardiac magnetic resonance (CMR) image reconstruction, utilizing a prompt-based architecture designed to process subsampled multi-contrast k-space data with varying acceleration factors and trajectories. Our method successfully reconstructs high-quality, clinically interpretable images by leveraging both k-space and image-domain features through the integration of physical and SSIM loss functions. Additionally, by incorporating the swin transformer blocks, we demonstrated the method's effectiveness across multiple CMR contrasts and views, including long-axis (LAX), short-axis (SAX), and left ventricular outflow tract (LVOT) in Cine imaging, as well as T1 and T2 mapping.

Moreover, the quantitative evaluation further supports our findings, as our method consistently outperformed other reconstruction techniques across key metrics, such as PSNR and SSIM. This demonstrates the robustness and versatility of our approach across different imaging contrasts and acceleration factors, ensuring reliable reconstructions across a wide range of clinical scenarios.

Future research will focus on optimizing the framework to enhance computational efficiency, particularly in terms of speed, while maintaining the high reconstruction quality. Efforts will be made to further refine the model architecture, reducing the computational burden without compromising on image accuracy, enabling faster processing times that are crucial for real-time clinical workflows. Additionally, expanding this method to handle more complex imaging challenges, such as higher undersampling rates and more diverse k-space trajectories, will be essential in advancing CMR image reconstruction techniques.

References

1. Wang, C., et al.: Recommendation for cardiac magnetic resonance imaging-based phenotypic study: imaging part. Phenomics **1**(4), 151–170 (2021)
2. Quarta, G., et al.: Cardiovascular magnetic resonance imaging in hypertrophic cardiomyopathy: the importance of clinical context. **19**(6), 601–610 (2018)
3. Fagiry, M.A., et al.: The recent advances, drawbacks, and the future directions of CMRI in the diagnosis of IHD. **11**(1), 14958 (2021)
4. Wang, C., et al.: CMRxRecon: an open cardiac MRI dataset for the competition of accelerated image reconstruction, 2023
5. Wang, Z., et al.: CMRxRecon2024: A Multi-Modality, Multi-View K-Space Dataset Boosting Universal Machine Learning for Accelerated Cardiac MRI, 2024
6. Pruessmann, K.P., et al.: SENSE: sensitivity encoding for fast MRI. **42**(5), 952–962 (1999)
7. Griswold, M.A., et al.: Generalized autocalibrating partially parallel acquisitions (GRAPPA). **47**(6), 1202–1210 (2002)
8. Bernstein, M.A., Fain, S.B.: Effect of windowing and zero-filled reconstruction of MRI data on spatial resolution and acquisition strategy. **14**(3), 270–280 (2001). S.J.J.J.o.m.r.i.a.o.j.o.t.i.s.f.m.r.i.m. Riederer
9. Qin, C., et al.: Convolutional recurrent neural networks for dynamic MR image reconstruction. **38**(1), 280–290 (2018)
10. Lyu, J., et al., STADNet: Spatial-Temporal Attention-Guided Dual-Path Network for cardiac cine MRI super-resolution. **94**, 103142 (2024)
11. Lyu, J., et al.: Region-focused multi-view transformer-based generative adversarial network for cardiac cine MRI reconstruction. **85**, 102760 (2023)
12. Ouyang, C., et al.: Generalizing Deep Learning MRI Reconstruction across Different Domains, 2019
13. Lyu, J., et al.: The state-of-the-art in Cardiac MRI Reconstruction: Results of the CMRxRecon Challenge in MICCAI 2023. 2024
14. Huang, J., et al.: Swin transformer for fast MRI. **493**, 281–304 (2022)
15. Wang, C., et al., Black blood myocardial T 2 mapping **81**(1), 153–166 (2019)

Implicit Neural Representations for Registration of Left Ventricle Myocardium During a Cardiac Cycle

Mathias Micheelsen Lowes[1](✉), Jonas Jalili Pedersen[2], Bjørn S. Hansen[1], Klaus Fuglsang Kofoed[2], Maxime Sermesant[3], and Rasmus R. Paulsen[1]

[1] Department of Applied Mathematics and Computer Science, Technical University of Denmark, Kongens Lyngby, Denmark
mmilo@dtu.dk

[2] The Heart Center, Rigshospitalet, University of Copenhagen, Copenhagen, Denmark

[3] Inria Université Côte d'Aur, Nice, France

Abstract. Understanding the movement of the left ventricle myocardium (LVmyo) during the cardiac cycle is essential for assessing cardiac function. One way to model this movement is through a series of deformable image registrations (DIRs) of the LVmyo. Traditional deep learning methods for DIRs, such as those based on convolutional neural networks, often require substantial memory and computational resources. In contrast, implicit neural representations (INRs) offer an efficient approach by operating on any number of continuous points. This study extends the use of INRs for DIR to cardiac computed tomography (CT), focusing on LVmyo registration. To enhance the precision of the registration around the LVmyo, we incorporate the signed distance field of the LVmyo with the Hounsfield Unit values from the CT frames. This guides the registration of the LVmyo, while keeping the tissue information from the CT frames. Our framework demonstrates high registration accuracy and provides a robust method for temporal registration that facilitates further analysis of LVmyo motion.

Keywords: Cardiac CT · Deformable Image Registration · Implicit Neural Representations · Left Ventricle Myocardium · Signed Distance Fields · Deep Learning

1 Introduction

The movement of the left ventricle myocardium (LVmyo) during the cardiac cycle is complex and difficult to accurately model. Understanding and modelling the movement of the LVmyo is important for assessing cardiac function and diagnosing cardiac diseases. Modelling the movement of the LVmyo can be viewed as a deformable image registration (DIR) task requiring several registrations over a single cardiac cycle.

O. Camara et al. (Eds.): STACOM 2024, LNCS 15448, pp. 172–182, 2025.
https://doi.org/10.1007/978-3-031-87756-8_17

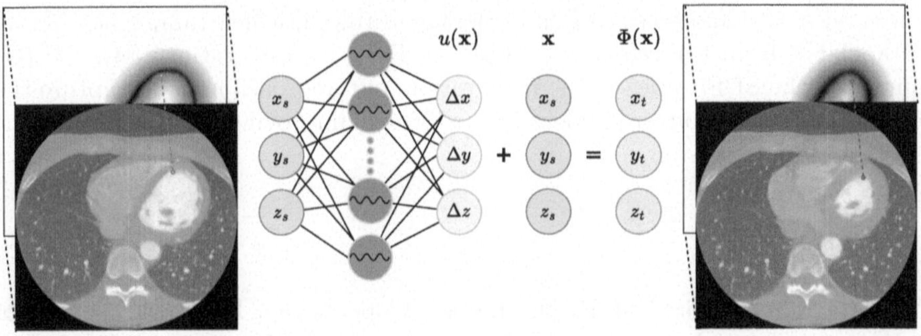

Fig. 1. Registration process for a single point. The MLP learns the deformation $u(\mathbf{x})$ by sampling values from the CT frame and the SDF in the source domain at $\mathbf{x}$ and in the target domain at $\Phi(\mathbf{x})$.

DIR involves establishing a correspondence between a source and a target image by deforming the source to align with the target. Often, the source image is mapped to a deformable vector field such that the location of each point is given by $\Phi(\mathbf{x}) = u(\mathbf{x}) + \mathbf{x}$, where $\mathbf{x}$ is the position in the source image, see Fig. 1. Applying Φ to the coordinates of the source image aligns it to the target image.

Early approaches for DIR were mainly based on B-splines [8,9,17]. Through optimization, the parameters of the B-splines are calculated for the specific registration. Instead of being imposed to parameter optimization for each registration, deep learning approaches can be trained on a dataset and then generalize to new data. Many deep learning approaches for DIR are based on convolutional neural networks (CNNs) [1,6,11,14,18]. CNNs offer the advantage of rapid inference post-training. However, they operate within the discrete voxel space of images, resulting in substantial memory demands. Consequently, downsampling the images is often necessary, which often compromise the registration accuracy.

Implicit neural representations (INRs) address several limitations of using CNNs for DIR, primarily by operating on continuous points. This eliminates the restriction to a specific grid and reduces the memory requirements significantly. An INR leverages a neural network to approximate any signal or function. In the context of registration, the INR represents the function Φ, mapping a continuous input point to a continuous output point. Wolterink et al. [16] demonstrated the ability of an INR to parameterize a DIR between two 3D chest CT images, outperforming other learning based methods on the DIR-LAB dataset [3] without the need for training data. Several later works also explore and build on the use of INRs for registration [2,4,5,12,13]. Some of these works incorporate CNNs into the INR framework, achieving marginal performance improvement while losing the simplicity of INRs and increasing the computational requirements.

Building on the framework of Wolterink et al. [16], we extend the use of INR for DIR to the domain of cardiac CT with a specific focus on the registration of the LVmyo. Our work contributes by adding the signed distance field (SDF) of the LVmyo to the registration process. The SDF provide geometric information about the shape and position of the myocardium, which cannot be directly extrapolated from CT scans only using an INR. By integrating both the HU values and the SDF values in the registration, our approach aims to capture the detailed and continuous movement of the LVmyo, improving the precision of the registration at each step.

2 Data

The data were acquired at Rigshospitalet, Copenhagen, Denmark, as part of the ongoing Copenhagen General Population Study. It consists of 100 randomly selected participants from the study, who meet the requirements: individuals > 40 years, non-pregnant and normal kidney function. The participants underwent retrospective ECG-gated cardiac computed tomography angiography, allowing for continuous scanning of the heart over a cardiac cycle, resulting in one temporal scan per participant. Each temporal scan consists of 20 volumetric frames denoted $f_0, \ldots, f_{19}$, these are also divided into a percentage representation of the whole interval. The frame at 0% (f_0) represents the end-diastole (ED) phase (maximum filling of the left ventricle), while the end-systole (ES) phase (maximum emptying of the left ventricle) usually occurs around 35–40% of the series. The reconstructed spatial resolution of each frame is $0.5 \times 0.5 \times 2$ mm in the x-, y-, and z-directions, respectively.

In Fig. 1, axial slices of two frames belonging to the same participant are displayed. The slice on the left is from the ED phase, while the slice on the right is from the ES phase.

For a single participant, four expert-annotated landmarks are annotated on each CT frame in the cardiac cycle. These landmarks are defined from the coronary arteries as follows:

1. The start of the left anterior descending artery (LAD).
2. The start of the circumflex artery (CX).
3. The point where the first obtuse marginal (OM) branches off the right circumflex artery.
4. The point where the last diagonal branches off the LAD.

Landmarks 1, 3 and 4 are illustrated in Fig. 2. Due to the high axial spacing of 2 mm, noise may be present in the landmarks since the exact point of the coronary artery could fall between two slices.

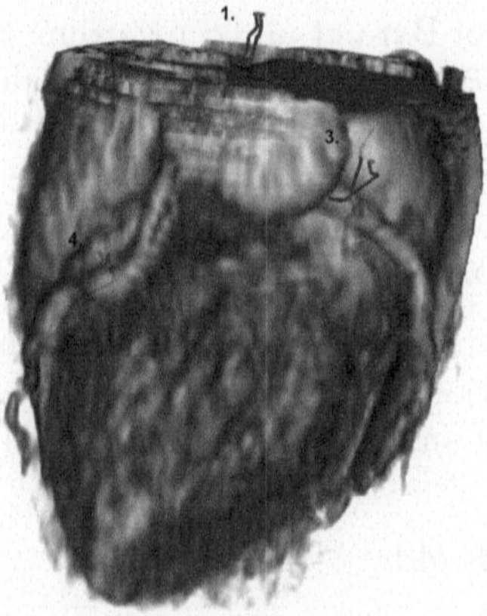

Fig. 2. 3D rendering of the CT frame in the ES-phase. The landmarks 1, 3 and 4 are rendered as lines representing the movement over a cardiac cycle, they are colored blue at 0% with a gradient to red at 95%. (Color figure online)

3 Methods

The LVmyo registration in a cardiac cycle can be seen as a series of pairwise DIRs, each between a source image $I_S : \Omega \subset \mathbb{R}^3 \to \mathbb{R}$ and a target image $I_T : \Omega \subset \mathbb{R}^3 \to \mathbb{R}$. Each DIR aims to find a mapping $\Phi : \mathbb{R}^3 \to \mathbb{R}^3$ in the form of $\Phi(\mathbf{x}) = u(\mathbf{x}) + \mathbf{x}$, such that $(I_S \circ \Phi)(\mathbf{x}) = I_T(\mathbf{x})$ for all $\mathbf{x} \in \Omega$. The optimal registration between two images is found by optimization as

$$\Phi^* = \arg\min_{\Phi} \{ \mathcal{L}_{\text{sim}}(I_S \circ \Phi, I_T) + \lambda \mathcal{L}_{\text{reg}}(\Phi) \}, \tag{1}$$

where $\mathcal{L}_{\text{sim}}$ and $\mathcal{L}_{\text{reg}}$ are similarity and regularization losses, respectively. The similarity loss used for this project is the normalized cross correlation loss. As for the regularization loss we use the symmetric Jacobian determinant regularization defined by van Harten et al. [5] as

$$\mathcal{L}_{\text{sjac}}(\Phi) = \min \left\{ \frac{(\det \nabla\Phi - 1)^2}{|\det \nabla\Phi|}, \tau \right\}, \tag{2}$$

with $\nabla\Phi$ as the Jacobian matrix of Φ evaluated at point $\mathbf{x}$ and τ as a hyperparameter to clip the regularization penalty. As the van Harten et al. [5] suggest, we use $\tau = 10$ in our work.

3.1 Implicit Neural Representations

The INR of a DIR is achieved using a multi-layer perceptron (MLP) to parameterize the deformation field. The MLP takes a continuous coordinate $\mathbf{x}$ as input and outputs the corresponding deformation vector for the given point. Thus, the INR represents the deformation $\Phi(\mathbf{x}) = u(\mathbf{x}) + \mathbf{x}$.

Based on the findings of Byra et al. [2] regarding the activation functions of INRs, we adopted the `SIREN` [10] architecture for modeling the MLP. The `SIREN` model utilizes sinusoidal activation functions in the network. This defines the i'th layer of the neural network as

$$\phi_i(\mathbf{x}_i) = \sin\left(\omega\left(\mathbf{W}_i\mathbf{x}_i + \mathbf{b}_i\right)\right), \tag{3}$$

where $\mathbf{W}_i$ and $\mathbf{b}_i$ are the trainable parameters of a linear layer, $\mathbf{x}_i$ is the input vector, and ω is a modulation hyper-parameter. We use $\omega = 30$ as Byra et al. [2] suggest. A visualization of an INR with one hidden layer is shown in Fig. 1.

3.2 Signed Distance Fields

The SDFs utilized in the registration process are derived from segmentation of the LVmyo using the TotalSegmentator [15] framework. The SDFs provide more meaningful value over the whole image domain compared to the raw LVmyo segmentations. They are generated using the physical spacing of the images to account for the differences in spacing over the different image dimensions.

For the registration process, the SDFs are defined similarly to the source and target images as $S_S : \Omega \subset \mathbb{R}^3 \rightarrow \mathbb{R}$ and $S_T : \Omega \subset \mathbb{R}^3 \rightarrow \mathbb{R}$. As the domain of the SDFs are the same domain as the images, the mapping between the source and target domains Φ is also applied to the SDFs.

The optimal registration Eq. (1) is modified to include the SDFs

$$\Phi^* = \arg\min_{\Phi}\left\{(1-\alpha)\mathcal{L}_{\text{sim}}(I_S \circ \Phi, I_T) + \alpha\mathcal{L}_{\text{sim}}(S_S \circ \Phi, S_T) + \lambda\mathcal{L}_{\text{reg}}(\Phi)\right\}. \tag{4}$$

The weighting between the SDFs and the images in the training process is controlled by the hyperparameter α.

4 Experiments

Two approaches are used to register the LVmyo over a cardiac cycle: sequential and non-sequential registration. In the sequential approach, each frame is registered to the next (f_0 to f_1, f_1 to f_2, etc.). In the non-sequential approach, f_0 is directly registered to all subsequent frames (f_0 to f_1, f_0 to f_2, etc.). For comparability, both methods register frames from 0 to 19 without looping back.

Both approaches are primarily evaluated using the LVmyo segmentations of the target image and the transformed segmentations of the source image. This is done with the metrics; Dice Similarity Coefficient (DSC) and the 95% Hausdorff Distance (HD95). For the single patient with annotated landmarks, we also evaluate the target registration error (TRE) in Euclidian distance.

4.1 Experimental Details

Each INR used for registration is parameterized with 5 hidden layers containing 256 neurons per layer. The first network mapping from f_0 to f_1 is trained for 2000 epochs, subsequent networks are trained for 1000 epochs, each using the previously trained network as initialization. The Training is conducted using the Adam optimizer [7] with a learning rate of 10^{-5} and a batch size of 10,000 points per epoch. The points are randomly sampled within a dilated mask of the whole heart, acquired from TotalSegmentator [15]. For the loss Eq. (4) we use $\lambda = 0.05$ for controlling the regularization and $\alpha = \{0.0, 0.8, 1.0\}$ for experimentation with the weighting between SDF and CT scan in the loss. The coordinate systems of both the SDFs and the CT frames are scaled to the interval of $[-1, 1]^3$ following the procedure of Wolternik et al. [16]. All experiments were performed on an NVIDIA RTX A4000 GPU, where the training time for the first network in the sequence is ~1 min and ~ 30 s for subsequent networks.[1]

4.2 Results

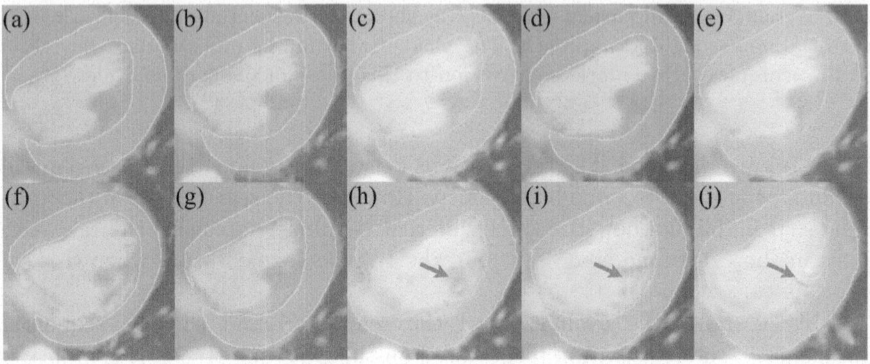

Fig. 3. Axial slices of a registration example with the sequential approach in the top row (a) - (e) and the non-sequential approach in the bottom row (f) - (j). (a) and (b) shows the source and target frames which are before the ES-phase and the ES-phase it self, respectively. (c) - (e) shows the transformed CT scan and LVmyo segmentation after the registration from (a) to (b) using $\alpha = 0.0$, $\alpha = 0.8$ and $\alpha = 1.0$, respectively. (f) and (g) shows the source and target frames which are before the ED-phase and the ES-phase. Again (h) - (j) shows the transformation using $\alpha = 0.0$, $\alpha = 0.8$ and $\alpha = 1.0$, respectively. The arrows highlights trabeculated area of the LV.

Figure 3 illustrates a single registration with both the sequential and non-sequential approaches across varying α-values. The difference in the registration for the two approaches is from the source images (a) and (f), as the target images

[1] Our code is publicly available at: https://github.com/MMLowes/INR_reg_LVMyo.

Table 1. Evaluation of the two approaches with three different α-values. All values are averaged over all time steps and the DSC and HD95 are also averaged over all patients.

α-value	Sequential			Non-sequential		
	0.0	0.8	1.0	0.0	0.8	1.0
DSC [%]	94.79	97.78	**98.31**	91.18	96.90	97.28
HD95 [mm]	1.147	0.4364	**0.3901**	2.445	0.5212	0.4987
TRE [mm]	11.40	8.703	11.03	**4.739**	6.273	8.347

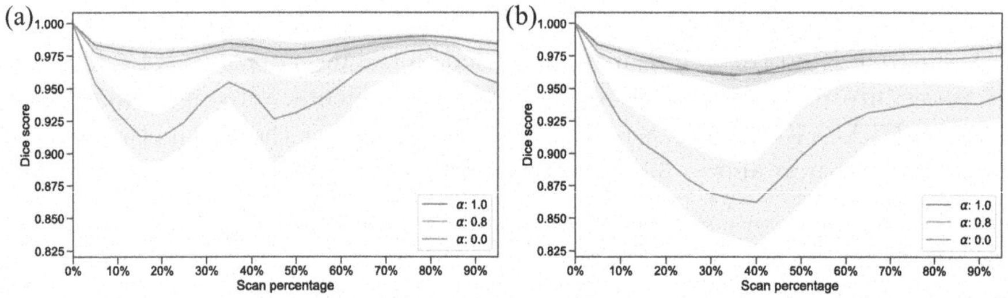

Fig. 4. Evaluation between the LVmyo segmentation and the moved LVmyo segmentation averaged over all patients in DSC. (a) shows the sequential approach, while (b) shows the non-sequential approach. The shaded area marks one standard deviation, and the scan percentage marks registration to this phase of the cardiac cycle.

(b) and (g) are the same. The registrations between (a) and (b) in (c) - (e) are all very similar and of the same high qualtiy. However, the registration of (f) to (g) in (h) - (j) highlights how the three different α-values emphasize the raw voxel values in the CT frame when $\alpha = 0.0$, the LVmyo segmentation when $\alpha = 1.0$ and a mixture when $\alpha = 0.8$.

Table 1 shows the evaluation of the sequential and the non-sequential approaches to the registration task. Based on the DSC and HD95, it is indicated that training with $\alpha = 1.0$ yields the best performance for both approaches. However, the TRE of the method with $\alpha = 1.0$ is worse in performance than when using $\alpha = 0.8$. The values for DSC and HD95 in Table 1 are averaged over both time and patients. The same values for DSC are illustrated in Fig. 4 as a function of the scan percentage, where the values are only averaged over patients. This shows that in the non-sequential approach the registration accuracy is lower closer to the ES-phase of the cardiac cycle, mainly when using $\alpha = 0.0$. The performance when using $\alpha = 0.8$ and $\alpha = 1.0$ are quite similar, there are no clear places where $\alpha = 1.0$ outperforms $\alpha = 0.8$ significantly.

From the patient with annotated landmarks, the movement of landmark 4 is visualized in Fig. 5. It is apparent that modeling the movement of the landmark with $\alpha = 1.0$ differs from the annotated movement of the landmark. Meanwhile, modeling the movement with $\alpha = 0.0$ and $\alpha = 0.8$ follows the actual movement better.

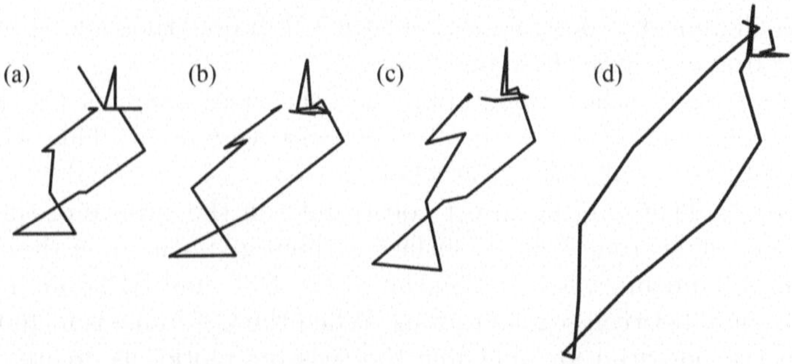

Fig. 5. Movement of landmark 4 through a cardiac cycle illustrated in the sagittal plane. (a) is the annotated movement, while (b), (c) and (d) shows the movement using the non-sequential approach with $\alpha = 0.0$, $\alpha = 0.8$ and $\alpha = 1.0$, respectively.

5 Discussion

The sequential and the non-sequential approaches differ in how they handle registration. In the sequential approach, each registration step only needs to capture a small deformation. In contrast, the non-sequential approach handles increasingly larger deformations as the registration progresses closer to the ES-phase of the cardiac cycle, as the movement is relative to the ED-phase. Consequently, we expect the sequential approach to achieve a higher DSC, as the registration should be easier for the INR to parameterize. However, the performance differences are not as substantial as anticipated, except for the models without the SDF. A difference of less than 0.5% in DSC and 0.05 mm in HD95 between the sequential and the non-sequential approaches is a minor improvement. Indicating that the inclusion of SDF effectively addresses the challenge of larger deformations in the non-sequential approach.

The advantage of the non-sequential approach is seen in the landmark tracking. Modeling the movement of a set of points over a cardiac cycle with the sequential displacement of the points leads to a build-up of errors, which the non-sequential approach avoids as the displacement of the points are from the first frame each time. This build up of errors is what gives the high TRE in Table 1. Although, this comparison is not strictly fair and only with landmarks from a single patient, it shows the major differences in the two approaches and why the sequential is not strictly better.

In general, the inclusion of SDFs in the registration makes a difference in the registration quality. This is apparent in Fig. 4, where both registrations with $\alpha > 0$ outperform the registration with $\alpha = 0.0$, that is, without the SDFs. Including the original CT in the registration ($\alpha = 0.8$) results in a slight drop in performance. However, this can potentially be attributed to the choice of evaluating only the LVmyo. The SDFs are based directly on the LVmyo segmentations, so aligning the SDFs perfectly gives the best score in both DSC and

HD95. This alignment is possible without the CT frame, thus adding it will not improve performance with these metrics.

It is known from echocardiography that the contraction of the LVmyo is a complex movement that also involves a twisting motion. This will not be faithfully captured if only using the segmentation masks or the derived SDF in the registration. There are no direct constraints on the surface of the LVmyo, thus points can slide around on the surface without giving a larger similarity loss $\mathcal{L}_{\text{sim}}$. If this is happening, it is not shown in the DSC and HD95 metrics as the evaluation is only between segmentations. When the CT frame is included in the registration the potential for modeling the twisting motion is greater, as there are more changes in the HU values on the surface. An example of this is what Fig. 5 shows. The movement of this landmark is easier to accurately model when including HU values in the registration. This indicates that accurate modeling of the LVmyo is not feasible from only the SDF.

In Fig. 3, examples of registrations are shown. Upon visual inspection of the non-sequential approach (f)-(j), the registration using solely SDF (j) achives the most accurate transformation of the LVmyo segmentation. However, the trabeculated area of the LV does not look as realistic as the other registrations (c) and (d), see arrows in figure. Using only the CT frame provides the best registration of the trabeculated area, but a worse registration of LVmyo segmentations. The combination of CT frame and SDF appears realistic in the trabeculated area of the LV, and the transformed LVmyo segmentation fits the target segmentation quite well.

We have previously utilized the `elastix` [8] framework for this type of registration. The performance of `elastix` lies around 60% DSC on the LVmyo, when including the SDF in the registration from the ED-phase to the ES-phase. Another DIR method tried is the VoxelMorph [1] framework which achieves a performance of 80% DSC in the same registration setup mapping from the ED-phase to the ES-phase when including the SDF as an extra feature of the input images. Both `elastix` and VoxelMorph have shorter inference times than the INR registration framework at ~19 s and ~0.4 s per image pair. However, the VoxelMorph framework needs training for ~18 h on our setup, where the INRs are optimized during inference. Inspired by VoxelMorph we implemented the dice loss on TotalSegmentator segmentations to substitute the SDF. This yielded no significant improvement to using only HU values for the registration.

6 Conclusion

We show that implicit neural representations are capable of modeling the movement of the left ventricle myocardium over a cardiac cycle. Additionally, the inclusion of the signed distance field in the registration process guides the registration to be more precise in the area of interest, specifically the left ventricle. Both the sequential and the non-sequential approaches show promising results and have potential for further development. Future work could include using both methods simultaneously to regularize each other as a form of cycle-consistency learning over a whole cardiac cycle.

References

1. Balakrishnan, G., Zhao, A., Sabuncu, M., Guttag, J., Dalca, A.V.: Voxelmorph: a learning framework for deformable medical image registration. IEEE TMI: Trans. Med. Imaging **38**, 1788–1800 (2019)
2. Byra, M., Poon, C., Rachmadi, M.F., Schlachter, M., Skibbe, H.: Exploring the performance of implicit neural representations for brain image registration. Sci. Rep. **13**(1) (2023). https://doi.org/10.1038/s41598-023-44517-5
3. Castillo, R., et al.: A framework for evaluation of deformable image registration spatial accuracy using large landmark point sets. Phys. Med. Biol. **54**(7), 1849–1870 (2009). https://doi.org/10.1088/0031-9155/54/7/001, epub 2009 Mar 5
4. van Harten, L., Herten, R.L.M.V., Stoker, J., Isgum, I.: Deformable image registration with geometry-informed implicit neural representations. In: Medical Imaging with Deep Learning (2023). https://openreview.net/forum?id=Pj9vtDIzSCE
5. van Harten, L.D., Stoker, J., Išgum, I.: Robust deformable image registration using cycle-consistent implicit representations. IEEE Trans. Med. Imaging **43**(2), 784–793 (2024). https://doi.org/10.1109/TMI.2023.3321425
6. Heinrich, M.P., Hansen, L.: Voxelmorph++ going beyond the cranial vault with keypoint supervision and multi-channel instance optimisation. In: Biomedical Image Registration: 10th International Workshop. WBIR 2022, Munich, 10–12 Germany 2022, Proceedings, pp. 85–95. Springer, Cham (2022)
7. Kingma, D.P., Ba, J.: Adam: a method for stochastic optimization. CoRR arXiv:1412.6980, https://api.semanticscholar.org/CorpusID:6628106 (2014)
8. Klein, S., Staring, M., Murphy, K., Viergever, M.A., Pluim, J.P.: elastix: a toolbox for intensity-based medical image registration. IEEE Trans. Med. Imaging **29**(1), 196–205 (2010). https://doi.org/10.1109/TMI.2009.2035616, epub 2009 Nov 17
9. Rueckert, D., Sonoda, L.I., Hayes, C., Hill, D.L., Leach, M.O., Hawkes, D.J.: Nonrigid registration using free-form deformations: application to breast mr images. IEEE Trans. Med. Imaging **18**(8), 712–721 (1999). https://doi.org/10.1109/42.796284
10. Sitzmann, V., Martel, J.N., Bergman, A.W., Lindell, D.B., Wetzstein, G.: Implicit neural representations with periodic activation functions. In: Proceedings of NeurIPS (2020)
11. Sokooti, H., de Vos, B., Berendsen, F., Lelieveldt, B.P., Išgum, I., Staring, M.: Nonrigid image registration using multi-scale 3d convolutional neural networks. In: International Conference on Medical Image Computing and Computer-Assisted Intervention, pp. 232–239. Springer, Cham, September 2017
12. Sun, S., Han, K., Kong, D., You, C., Xie, X.: Mirnf: medical image registration via neural fields. ArXiv **abs/2206.03111**, https://api.semanticscholar.org/CorpusID:249431729 (2022)
13. Tian, L., Sengupta, S., Greer, H., Estépar, R.S.J., Niethammer, M.: Nephi: neural deformation fields for approximately diffeomorphic medical image registration. ArXiv **abs/2309.07322**, https://api.semanticscholar.org/CorpusID:261823163 (2023)
14. de Vos, B.D., Berendsen, F.F., Viergever, M.A., Sokooti, H., Staring, M., Igum, I.: A deep learning framework for unsupervised affine and deformable image registration. Med. Image Anal. **52**, 128–143 (2019). https://doi.org/10.1016/j.media.2018.11.010,
15. Wasserthal, J., et al.: TotalSegmentator: robust segmentation of 104 anatomic structures in CT images. Radiol. Artif. Intell. **5**(5) (2023). https://doi.org/10.1148/ryai.230024

16. Wolterink, J.M., Zwienenberg, J.C., Brune, C.: Implicit neural representations for deformable image registration. In: Medical Imaging with Deep Learning 2022 (2022)
17. Wu, Z., Lan, T., Wang, J., Ding, Y., Qin, Z.: Medical image registration using b-spline transform. Int. J. Simul. Syst. Sci. Technol. **17**, 1.1–1.6 (2016). https://doi.org/10.5013/IJSSST.a.17.48.01
18. Yang, X., Kwitt, R., Styner, M., Niethammer, M.: Quicksilver: fast predictive image registration - a deep learning approach. Neuroimage **158**, 378–396 (2017). https://doi.org/10.1016/j.neuroimage.2017.07.008

Deep Multi-contrast Cardiac MRI Reconstruction via vSHARP with Auxiliary Refinement Network

George Yiasemis[1,2(✉)], Nikita Moriakov[1,2], Jan-Jakob Sonke[1,2], and Jonas Teuwen[1,2,3]

[1] Netherlands Cancer Institute, Amsterdam, The Netherlands
{g.yiasemis,n.moriakov,j.sonke,j.teuwen}@nki.nl
[2] University of Amsterdam, Amsterdam, The Netherlands
[3] Radboud University Medical Center, Nijmegen, The Netherlands

Abstract. Cardiac MRI (CMRI) is a cornerstone imaging modality that provides in-depth insights into cardiac structure and function. Multi-contrast CMRI (MCCMRI), which acquires sequences with varying contrast weightings, significantly enhances diagnostic capabilities by capturing a wide range of cardiac tissue characteristics. However, MCCMRI is often constrained by lengthy acquisition times and susceptibility to motion artifacts. To mitigate these challenges, accelerated imaging techniques that use k-space undersampling via different sampling schemes at acceleration factors have been developed to shorten scan durations. In this context, we propose a deep learning-based reconstruction method for 2D dynamic multi-contrast, multi-scheme, and multi-acceleration MRI. Our approach integrates the state-of-the-art vSHARP model, which utilizes half-quadratic variable splitting and ADMM optimization, with a Variational Network serving as an Auxiliary Refinement Network (ARN) to better adapt to the diverse nature of MCCMRI data. Specifically, the subsampled k-space data is fed into the ARN, which produces an initial prediction for the denoising step used by vSHARP. This, along with the subsampled k-space, is then used by vSHARP to generate high-quality 2D sequence predictions. Our method outperforms traditional reconstruction techniques and other vSHARP-based models.

Keywords: Deep Cardiac MRI Reconstruction · MRI Reconstruction via vSHARP · Multi-contrast MRI Reconstruction

1 Introduction

Cardiac Magnetic Resonance Imaging (CMRI) is a cornerstone clinical imaging modality that provides detailed 4D (3D + dynamic) images of the heart, essential for assessing and evaluating cardiovascular health, structure, and function. Multi-contrast CMRI (MCCMRI) enhances this capability by acquiring multiple imaging sequences with varying contrast weightings, such as cardiac cine,

O. Camara et al. (Eds.): STACOM 2024, LNCS 15448, pp. 183–192, 2025.
https://doi.org/10.1007/978-3-031-87756-8_18

mapping, tagging, phase-contrast, and dark-blood [16]. These sequences capture different anatomical views, including long-axis, short-axis, outflow tract, and aortic, offering a comprehensive assessment of the heart.

Despite the clinical importance of MCCMRI, it is often hindered by extensive acquisition times, which increases susceptibility to periodic or aperiodic body motion and causes patient discomfort. A viable solution is to reduce these lengthy acquisition times by subsampling the k-space during the MCCMRI acquisition. This approach accelerates the process by scanning fewer k-space data points than required by the Nyquist sampling theorem [6,11].

This is typically achieved by applying a variety of subsampling patterns up to a specific acceleration factor. Common techniques include Cartesian trajectories such as equidistant sampling (each column sampled at a uniform distance from the next and previous), random-column sampling, variable density point sampling, pseudo-radial sampling, etc. [21,22]. While these patterns are often applied uniformly across all time frames of the dynamic acquisition for ease of implementation, distinct patterns can also be applied to each time frame [14], exploiting potential cross-frame complementary information.

Although subsampling the k-space accelerates MCCMRI, it also degrades the quality of the reconstructed images, rendering them less useful for clinical practice. The Cardiac MRI Reconstruction (CMRxRecon) challenge [4] held at MICCAI in 2023 demonstrated that deep learning (DL)-based approaches represent the state-of-the-art (SOTA) compared to traditional methods.

In this work, we build upon the method that achieved one of the highest reconstruction quality submissions (2nd place in the challenge with minimal differences from the winning team) and demonstrated high reconstruction speed (more than 5$\times$ faster than the winning team) [19]. Specifically, the authors used the variable Splitting Half-quadratic ADMM algorithm for Reconstruction of inverse-Problems (vSHARP) [20], which was adapted for 2D dynamic CMRI reconstruction. They demonstrated that a unified model was able to quickly (2–15 s) and effectively reconstruct 4D CMRI volumes of varying contrast (Cine, T1, T2), views (short or long-axis), and acceleration factors (4$\times$, 8$\times$, 10$\times$).

Building on vSHARP, we introduce an enhanced version with an Auxiliary Refinement Network (ARN) for multi-contrast, multi-view, multi-scheme, and multi-acceleration 2D dynamic reconstruction. Specifically, we employ a Variational Network [12] as the ARN, which generates an initial state for the auxiliary variable in vSHARP's ADMM denoising step, using the subsampled k-space measurements as input to inform the reconstruction process of the specific type of input data. We outline 2D dynamic MRI reconstruction in Sect. 2, introduce our proposed methods in Sect. 3, present our experiments in Sect. 4, and conclude with a discussion of our results in Sect. 5.

2 Background

2.1 Accelerated Dynamic CMRI Reconstruction

In the context of accelerated 2D dynamic CMRI reconstruction, the objective is to produce a 2D dynamic image $\mathbf{x}^* \in \mathbb{C}^{n \times n_f}$ from a 2D dynamic subsampled multi-coil k-space $\mathbf{y} \in \mathbb{C}^{n \times n_c \times n_f}$, where n denotes the spatial dimension of each time frame, n_c the number of acquisition coils, and n_f the total number of time frames in the dynamic acquisition. This can be formulated as a solution to a regularized least-squares minimization problem:

$$\mathbf{x}^* = \underset{\mathbf{x} \in \mathbb{C}^{n \times n_f}}{\operatorname{argmin}} \frac{1}{2} ||\mathcal{T}_{\mathbf{U},\mathbf{S}}(\mathbf{x}) - \mathbf{y}||_2^2 + \rho \mathcal{G}(\mathbf{x}), \tag{1}$$

where $\mathcal{T}_{\mathbf{U},\mathbf{S}}$ denotes the forward operator, which transforms the 2D dynamic image $\mathbf{x} \in \mathbb{C}^{n \times n_f}$ into separate coil 2D dynamic images using sensitivity maps $\mathbf{S} \in \mathbb{C}^{n \times n_c \times n_f}$, transforms them to the k-space domain by applying the two-dimensional Fourier transform $\mathcal{F}$, and subsamples them using a binary subsampling pattern $\mathbf{U}$. Moreover, $\mathcal{G} : \mathbb{C}^{n \times n_f} \to \mathbb{R}$ represents an arbitrary functional assumed to impose prior reconstruction information [3]. Equation 1 can be solved numerically using iterative methods [17] such as gradient descent or ADMM.

2.2 Accelerated Dynamic CMRI Reconstruction with vSHARP

Artificial intelligence-based methods address (1) using deep neural networks. Although many approaches exist for dynamic CMRI reconstruction [2,7,13], the 3D vSHARP [19] forms the basis of our work due to its SOTA performance in CMRxRecon 2023. vSHARP approaches a solution to (1) by applying half-quadratic variable splitting (HQVS) followed by ADMM unrolling in N iterations. Each iteration consists of three steps:

$$\mathbf{z}^{(j+1)} = \underset{\mathbf{z} \in \mathbb{C}^{n \times n_f}}{\operatorname{argmin}} \mathcal{G}(\mathbf{z}) + \frac{\rho}{2}||\mathbf{x}^{(j)} - \mathbf{z} + \frac{\mathbf{u}^{(j)}}{\rho}||_2^2 := \mathcal{D}_{\theta_j}(\mathbf{x}^{(j)}, \mathbf{z}^{(j)}, \frac{\mathbf{u}^{(j)}}{\rho}), \tag{2a}$$

$$\mathbf{x}^{(j+1)} = \underset{\mathbf{x} \in \mathbb{C}^{n \times n_f}}{\operatorname{argmin}} \frac{1}{2} ||\mathcal{T}_{\mathbf{U},\mathbf{S}}(\mathbf{x}) - \tilde{\mathbf{y}}^k||_2^2 + \frac{\rho}{2}||\mathbf{x} - \mathbf{z}^{(j+1)} + \frac{\mathbf{u}^{(j)}}{\rho}||_2^2, \tag{2b}$$

$$\mathbf{u}^{(j+1)} = \mathbf{u}^{(j)} + \rho(\mathbf{x}^{(j+1)} - \mathbf{z}^{(j+1)}), \quad j = 0, \cdots, N-1. \tag{2c}$$

As shown in (2), vSHARP performs alternating optimization with respect to three variables: $\mathbf{z}$, the auxiliary variable derived from HQVS; $\mathbf{x}$, the image prediction; and $\mathbf{u}$, the Lagrange multipliers. At each iteration, (2a) is estimated using a DL-based denoiser denoted as $\mathcal{D}_{\theta j}$, specifically a 3D U-Net [23], while (2b) is solved numerically via a gradient descent unrolled scheme over $N_{\mathbf{x}}$ steps. Furthermore, vSHARP initializes $\mathbf{z}^{(0)}, \mathbf{x}^{(0)}$ with SENSE reconstruction [9], and for $\mathbf{u}^{(0)}$, it employs a DL-based module $\mathcal{I}_\psi$ using $\mathbf{x}^{(0)}$:

$$\mathbf{z}^{(0)} = \mathbf{x}^{(0)} = \sum_{k=1}^{n_c} \mathbf{S}^{k^*} \mathcal{F}^{-1}(\mathbf{y}^k), \quad \mathbf{u}^{(0)} = \mathcal{I}_\psi(\mathbf{x}^{(0)}). \tag{3}$$

Consistent with [19], sensitivity maps are estimated using the autocalibration region of the subsampled data and refined via a 2D U-Net [10].

3 Methods

3.1 Deep Learning Framework

Auxiliary Refinement Network. To adapt to the diverse nature of our data, which vary in contrast, subsampling scheme, and acceleration factor, we incorporate an auxiliary refinement network (ARN), denoted as $\mathcal{A}_\omega$. This network generates an initial image from the subsampled data, which serves as the initialization for the auxiliary variable in vSHARP's ADMM, replacing the SENSE reconstruction in (3) for $\mathbf{z}^{(0)}$:

$$\mathbf{z}^{(0)} = \sum_{k=1}^{n_c} \mathbf{S}^{k^*} \mathcal{F}^{-1}(\mathbf{r}^k), \quad \mathbf{r} = \mathrm{DC}_{\mathbf{U},\mathbf{y}} \circ \mathcal{A}_\omega(\mathbf{y}, \mathbf{S}), \tag{4}$$

where $\mathcal{A}_\omega$ can be any algorithm that inputs k-space data and complementary sensitivity maps, and outputs refined k-space data, and DC denotes the data consistency operator (defined in Sect. 3.3). We opted for a 3D extension of the Variational Network [12], utilizing 3D U-Nets as regularizers with four scales and 12 channels in the first scale, and implemented 8 cascades.

Reconstruction Model. Our experiments utilize vSHARP as the reconstruction method, as detailed in Sect. 2.2, with $N = 12$ optimization iterations. We employ 3D U-Nets with four scales and 32 channels in the first scale for (2a), and perform $N_\mathbf{x} = 6$ gradient descent steps for (2b). Sensitivity map estimation is performed using a 2D U-Net with four scales and 32 channels in the first scale.

3.2 Implementation and Training Details

Our experiments, including model implementation, data handling, and subsampling, were conducted using the Deep Image Reconstruction Toolkit [18], on NVIDIA A100 or H100 GPUs.

Optimization. We trained our models using the Adam optimizer, starting with a learning rate of 1.6e–4, which was linearly increased to 5e–4 over the first 2k iterations, and subsequently decreased by 5% every 50k iterations. Due to the large input k-space data size and the gradients computed during training, we utilized mixed precision training [5] to allow for larger model implementations. To ensure training stability, we clipped all gradients' norms at 10.0 [8].

Data Augmentations. During training, we employed random augmentations. Specifically, we applied horizontal or vertical flips with a 50% probability, as well as time-reversal with a 50% probability, which involved reversing the time sequence. These augmentations were performed in the image domain by transforming the k-space data to the image domain, applying the augmentations, and then transforming it back to the k-space domain.

3.3 Data Consistency

Our reconstruction model, vSHARP, generates a series of image predictions from each ADMM optimization iteration $\{\mathbf{x}^{(j)}\}_{j=1}^{N}$. To ensure each prediction is consistent with the initial measurements $\mathbf{y}$, we enforce hard data consistency:

$$\hat{\mathbf{x}}^{(j)} = \sum_{k=1}^{n_c} \mathbf{S}^{k^*} \mathcal{F}^{-1}((\hat{\mathbf{y}}^{(j)})^k), \quad \hat{\mathbf{y}}^{(j)} = \mathrm{DC}_{\mathbf{U},\mathbf{y}}\big(\mathbf{S}^1 \mathcal{F}(\mathbf{x}^{(j)}), \cdots, \mathbf{S}^{n_c} \mathcal{F}(\mathbf{x}^{(j)})\big), \quad (5)$$

where $\mathrm{DC}_{\mathbf{M}}$ is the hard data consistency operator defined as:

$$\mathrm{DC}_{\mathbf{U},\mathbf{y}}(\mathbf{w}) = \mathbf{U}(\mathbf{y}) + (\mathbb{1} - \mathbf{U})(\mathbf{w}). \quad (6)$$

3.4 Loss Function

Given that our training data is fully sampled, we have access to ground truth images and k-space data, enabling us to train our models in a supervised manner using a dual-domain loss strategy. This involves computing the loss in both the image and frequency domains. Specifically, assuming $\mathbf{x}^*$ and $\mathbf{y}^*$ are the ground truth root-sum-of-squares image and fully sampled k-space, we define the following loss function:

$$\begin{aligned} &\sum_{t=1}^{n_f} \Big[\mathcal{L}_{\mathrm{SSIM}}(|\mathbf{z}_t^{(0)}|, \mathbf{x}_t^*) + \big|\big||\mathbf{z}_t^{(0)}| - \mathbf{x}_t^*\big|\big|_1\Big] + \frac{||\mathbf{r} - \mathbf{y}^*||_1}{||\mathbf{y}^*|||_1} + \mathcal{L}_{\mathrm{SSIM3D}}(|\mathbf{z}^{(0)}|, \mathbf{x}^*) \\ &\quad + \sum_{j=1}^{N} 10^{\frac{(j-N)}{(N-1)}} \Big(\sum_{t=1}^{n_f} \Big[\mathcal{L}_{\mathrm{SSIM}}(|\hat{\mathbf{x}}_t^{(j)}|, \mathbf{x}_t^*) + \big|\big||\hat{\mathbf{x}}_t^{(j)}| - \mathbf{x}_t^*\big|\big|_1\Big] \quad (7) \\ &\quad + \frac{||\hat{\mathbf{y}}^{(j)} - \mathbf{y}^*||_1}{||\mathbf{y}^*|||_1} + \mathcal{L}_{\mathrm{SSIM3D}}(|\hat{\mathbf{x}}^{(j)}|, \mathbf{x}^*)\Big), \end{aligned}$$

where $\mathbf{z}^{(0)}$ and $\mathbf{r}$ as in Sect. 3.1, and $\{\hat{\mathbf{x}}^{(j)}\}_{j=1}^{N}$ and $\{\hat{\mathbf{y}}^{(j)}\}_{j=1}^{N}$ are as defined in Sect. 3.3. $\mathcal{L}_{\mathrm{SSIM}}$ and $\mathcal{L}_{\mathrm{SSIM3D}}$ are as defined in the literature [19].

3.5 Evaluation Metrics

For evaluation, we employed three common image reconstruction fidelity metrics: structural similarity index measure (SSIM), peak-signal-to-noise ratio (PSNR), and normalized mean squared error (NMSE).

4 Experiments

4.1 Data

The dataset used for our experiments consists of 1,404 multi-contrast 5D (3D + time + coils) volumes of k-space data, including cardiac cine, T1/T2 mapping, tagging, phase-contrast (flow2d), and dark-blood imaging [16]. These were acquired using a 3T MRI scanner with protocols outlined in previous studies [15]. The data encompasses various anatomical views, such as long-axis (LAX) views (2, 3, and 4-chamber), short-axis (SAX), left ventricular outflow tract (LVOT), and aortic (transversal and sagittal) views. The training set includes fully-sampled cine, aorta, mapping, and tagging data, while the inference set includes subsampled data and two additional unseen contrasts (flow2d and black-blood).

4.2 Experimental Setups

We consider two different experimental setups:

(**A**) Multi-contrast with equispaced (same mask applied to all time-frames) sampling at accelerations $R = 4, 8, 10$, evaluated on seen and unseen modalities (1,614 volumes in total).

(**B**) Multi-contrast with kt (interleaved masks applied across time-frames) multi-scheme (equispaced, Gaussian 1D, radial) sampling, at $R = 4-24$, evaluated on seen modalities (1,281 volumes in total).

For each experimental setup, we trained separate models responsible for reconstructing inference data within each setup, accommodating any contrast, view, acceleration, and subsampling scheme (applicable only for **B**).

4.3 Comparisons

We compared the following methods:

1. Zero-filled reconstruction (no optimization).
2. Two traditional reconstruction algorithms: GRAPPA [1], SENSE [9].
3. Proposed method (vSHARP without ARN module as outlined in Sect. 3.1).
4. Variants of the proposed method:
   (a) vSHARP with an ARN configuration with 8 channels in the first scale of each regularizer, denoted as (S).
   (b) vSHARP with $N = 14$, $N_{\mathbf{x}} = 8$, denoted as (L) with an ARN as outlined in Sect. 3.1. For training this model, k-space "cropping" augmentations following [19] were performed to regulate memory during training.

   Note that for these methods in experimental setup **A**, distinct models for each acceleration factor were trained.

4.4 Results

Tables 1 and 2 summarize the results of our experiments for setups **A** and **B**, respectively. Sample reconstructions at 10× acceleration from setup **A** are illustrated in Fig. 1. We also present the average inference times for reconstructing a 5D k-space volume using both the original vSHARP and our proposed method in Table 3 acquired on the same machine (NVIDIA A100).

As shown from the results, our proposed method consistently outperformed all other techniques across both experimental setups. It achieved the best scores

Table 1. Experimental setup **A** results using different methods. Metrics are averaged across contrasts and acceleration factors

Method	Average Metrics		
	SSIM	PSNR	NMSE
Zero-Filled	0.5026	20.5536	0.238
GRAPPA	0.7351	28.3721	0.0921
SENSE	0.7041	27.1099	0.1125
vSHARP w/o ARN	0.8789	32.7229	0.0321
vSHARP with ARN (S)	0.8937	33.5504	0.0259
vSHARP (L) with ARN	0.8911	33.4788	0.0264
vSHARP with ARN (proposed)	**0.9014**	**33.9011**	**0.0242**

Table 2. Experimental setup **B** results using different methods. Metrics are averaged across contrasts and acceleration factors

Method	Average Metrics		
	SSIM	PSNR	NMSE
Zero-Filled	0.4610	19.4182	0.2853
GRAPPA	0.6449	24.9678	0.1554
SENSE	0.6200	24.4780	0.1734
vSHARP w/o ARN	0.8421	30.6929	0.0428
vSHARP with ARN (S)	0.8639	31.6883	0.0408
vSHARP (L) with ARN	0.8555	31.3105	0.0448
vSHARP with ARN (proposed)	**0.8810**	**32.4700**	**0.0346**

Table 3. Average inference time per 5D volume (in seconds)

Experimental Setup	Inference Time (s)	
	vSHARP w/o ARN	vSHARP with ARN (proposed)
A	7.06	7.78
B	9.01	10.40

in all metrics, including SSIM, PSNR, and NMSE, surpassing traditional methods such as GRAPPA and SENSE, as well as other variations of the vSHARP method. Interestingly, the models trained for individual acceleration factors in setup **A** did not surpass our unified model trained across all acceleration factors. In addition, Fig. 1 shows that vSHARP based reconstructions are superior compared to traditional methods. Also, we can see that all vSHARP variants are able to reconstruct well the two unseen modalities.

In terms of reconstruction times, as expected our proposed method shows an increase of around 0.7 to 1.4 s, as evident from Table 3.

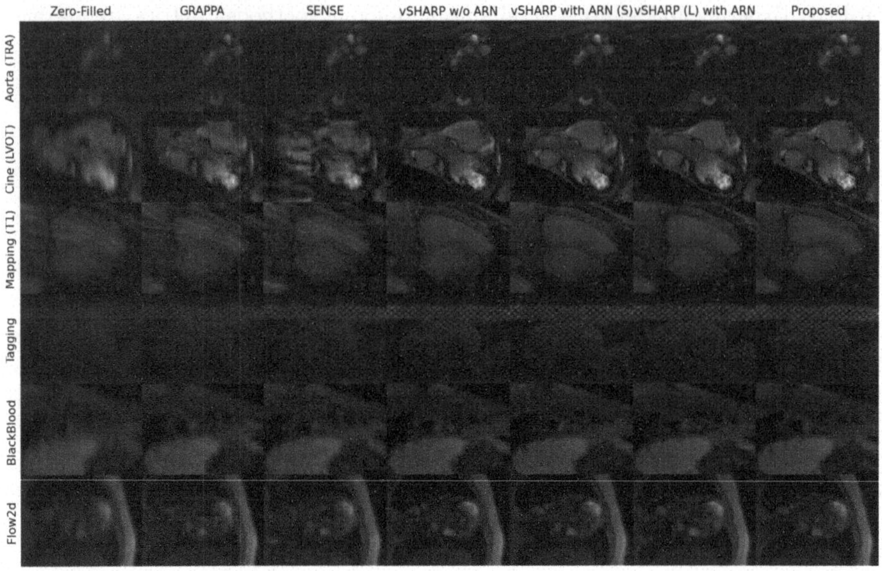

Fig. 1. Sample reconstructions from the inference set in Experimental Setup **A**

5 Discussion and Conclusion

This study introduces a new deep learning-based reconstruction method that enhances the vSHARP algorithm with an Auxiliary Refinement Network (ARN) for accelerated multi-contrast cardiac MRI. Integrating the ARN into the vSHARP framework improves reconstruction quality, outperforming traditional methods and other vSHARP variants. The method's effectiveness across a wide range of imaging sequences, views, and acceleration factors demonstrates its practical utility in clinical settings, where the use of multiple specialized models may not be feasible. However, it should be noted that the use of ARN introduces an increase in reconstruction time.

Our experiments focused exclusively on a single type of ARN-the Variational Network, a deep learning-based approach. Alternative ARNs, including traditional techniques like GRAPPA or SENSE, or other deep learning-based methods, could be investigated, though non-DL approaches may require significantly more computation time. Additionally, we employed the ARN to inform the reconstruction model about input types (contrast, view, subsampling scheme, acceleration factor) via auxiliary variable initialization. Other conditioning methods, such as using a multi-class label input to a trained embedding space model, could also be explored.

Future work should focus on incorporating additional contrasts, subsampling schemes, and anatomical regions into the training data to further improve the model's robustness and generalizability.

References

1. Griswold, M.A., et al.: Generalized autocalibrating partially parallel acquisitions (grappa). Magn. Reson. Med. **47**(6), 1202–1210 (2002). https://doi.org/10.1002/mrm.10171
2. Küstner, T., et al.: Cinenet: deep learning-based 3d cardiac cine mri reconstruction with multi-coil complex-valued 4d spatio-temporal convolutions. Sci. Rep. **10**(1) (2020). https://doi.org/10.1038/s41598-020-70551-8
3. Lustig, M., Donoho, D., Pauly, J.M.: Sparse mri: the application of compressed sensing for rapid mr imaging. Magn. Reson. Med. **58**(6), 1182–1195 (2007). https://doi.org/10.1002/mrm.21391
4. Lyu, J., et al.: The state-of-the-art in cardiac mri reconstruction: results of the cmrxrecon challenge in miccai 2023. arXiv preprint arXiv:2404.01082 (2024)
5. Micikevicius, P., et al.: Mixed precision training (2018). arxiv:1710.03740
6. Nyquist, H.: Certain topics in telegraph transmission theory. Trans. Am. Inst. Electr. Eng. **47**, 617–644 (1928). https://api.semanticscholar.org/CorpusID:8632488
7. Oscanoa, J.A., et al.: Deep learning-based reconstruction for cardiac mri: a review. Bioengineering **10**(3), 334 (2023). https://doi.org/10.3390/bioengineering10030334
8. Pascanu, R., Mikolov, T., Bengio, Y.: On the difficulty of training recurrent neural networks. In: Dasgupta, S., McAllester, D. (eds.) Proceedings of the 30th International Conference on Machine Learning. Proceedings of Machine Learning Research, vol. 28, pp. 1310–1318. PMLR, Atlanta, Georgia, USA, 17–19 June 2013. https://proceedings.mlr.press/v28/pascanu13.html
9. Pruessmann, K.P., Weiger, M., Scheidegger, M.B., Boesiger, P.: Sense: sensitivity encoding for fast mri. Magn. Reson. Med. Official J. Int. Soc. Magn. Reson. Med. **42**(5), 952–962 (1999)
10. Ronneberger, O., Fischer, P., Brox, T.: U-net: convolutional networks for biomedical image segmentation. In: Navab, N., Hornegger, J., Wells, W., Frangi, A. (eds.) Medical Image Computing and Computer-Assisted Intervention - MICCAI 2015, MICCAI 2015, LNCS, vol. 9351, pp. 234–241. Springer, Cham (2015). https://doi.org/10.1007/978-3-319-24574-4_28
11. Shannon, C.E.: Communication in the presence of noise. Proc. IRE **37**(1), 10–21 (1949)
12. Sriram, A., et al.: End-to-end variational networks for accelerated MRI reconstruction. In: Martel, A.L., (eds.) MICCAI 2020. LNCS, vol. 12262, pp. 64–73. Springer, Cham (2020). https://doi.org/10.1007/978-3-030-59713-9_7
13. Sun, J., Wang, C., Guo, L., Fang, Y., Huang, J., Qiu, B.: An unrolled neural network for accelerated dynamic mri based on second-order half-quadratic splitting model. Magn. Reson. Imaging **113**, 110218 (2024). https://doi.org/10.1016/j.mri.2024.110218
14. Tsao, J., Boesiger, P., Pruessmann, K.P.: k-t blast and k-t sense: dynamic mri with high frame rate exploiting spatiotemporal correlations. Magn. Reson. Med. **50** (2003). https://api.semanticscholar.org/CorpusID:9806611
15. Wang, C., et al.: Recommendation for cardiac magnetic resonance imaging-based phenotypic study: imaging part. Phenomics **1**(4), 151–170 (2021). https://doi.org/10.1007/s43657-021-00018-x
16. Wang, Z., et al.: Cmrxrecon2024: a multi-modality, multi-view k-space dataset boosting universal machine learning for accelerated cardiac mri. arXiv:2406.19043 (2024)

17. Willoughby, R.A.: Solutions of ill-posed problems (a. n. tikhonov and v. y. arsenin). SIAM Rev. **21**(2), 266–267 (1979). https://doi.org/10.1137/1021044
18. Yiasemis, G., Moriakov, N., Karkalousos, D., Caan, M., Teuwen, J.: Direct: deep image reconstruction toolkit. J. Open Source Softw. **7**(73), 4278 (2022)
19. Yiasemis, G., Moriakov, N., Sonke, J.J., Teuwen, J.: Deep cardiac mri reconstruction with admm. In: International Workshop on Statistical Atlases and Computational Models of the Heart, pp. 479–490. Springer (2023)
20. Yiasemis, G., Moriakov, N., Sonke, J.J., Teuwen, J.: vsharp: variable splitting half-quadratic admm algorithm for reconstruction of inverse-problems. Magn. Reson. Imaging 110266 (2024)
21. Yiasemis, G., Sánchez, C., Sonke, J.J., Teuwen, J.: On retrospective k-space subsampling schemes for deep mri reconstruction. Magn. Reson. Imaging **107**, 33–46 (2024). https://doi.org/10.1016/j.mri.2023.12.012
22. Yiasemis, G., Zhang, C., Sánchez, C.I., Sonke, J.J., Teuwen, J.: Deep mri reconstruction with radial subsampling. In: Medical Imaging 2022: Physics of Medical Imaging, vol. 12031, pp. 801–810. SPIE (2022)
23. Özgün Çiçek, Abdulkadir, A., Lienkamp, S.S., Brox, T., Ronneberger, O.: 3d u-net: learning dense volumetric segmentation from sparse annotation (2016). arxiv: 1606.06650

Multi-model Ensemble Approach for Accurate Bi-Atrial Segmentation in LGE-MRI of Atrial Fibrillation Patients

Lucas Beveridge[1(✉)] and Le Zhang[2]

[1] School of Computer Science, College of Engineering and Physical Sciences, University of Birmingham, Birmingham, UK
lxb032@student.bham.ac.uk

[2] School of Engineering, College of Engineering and Physical Sciences, University of Birmingham, Birmingham, UK

Abstract. Atrial fibrillation (AF) is the most prevalent form of cardiac arrhythmia and is associated with increased morbidity and mortality. The effectiveness of current clinical interventions for AF is often limited by an incomplete understanding of the atrial anatomical structures that sustain this arrhythmia. Late Gadolinium-Enhanced MRI (LGE-MRI) has emerged as a critical imaging modality for assessing atrial fibrosis and scarring, which are essential markers for predicting the success of ablation procedures in AF patients. The Multi-class Bi-Atrial Segmentation (MBAS) challenge at MICCAI 2024 aims to enhance the segmentation of both left and right atriums and their walls using a comprehensive dataset of 200 multi-center 3D LGE-MRIs, labeled by experts. This work presents an ensemble approach that integrates multiple machine learning models, including Unet, ResNet, EfficientNet and VGG, to perform automatic bi-atrial segmentation from LGE-MRI data. The ensemble model was evaluated using the Dice Similarity Coefficient (DSC) and 95% Hausdorff distance (HD95) on the left & right atrium wall, right atrium cavity, and left atrium cavity. On the internal testing dataset, the model achieved a DSC of 88.41%, 98.48%, 98.45% and an HD95 of 1.07, 0.95, 0.64 respectively. This demonstrates the effectiveness of the ensemble model in improving segmentation accuracy. The approach contributes to advancing the understanding of AF and supports the development of more targeted and effective ablation strategies.

Keywords: Atrial Fibrillation · Late Gandolinium-Enhanced MRI · Multi-class Bi-Atrial Segmentation · Ensemble Model

1 Introduction

Atrial fibrillation (AF) is the most prevalent form of cardiac arrhythmia, posing significant risks due to its association with increased morbidity and mortality rates. Despite advancements in treatment, the effectiveness of current clinical

O. Camara et al. (Eds.): STACOM 2024, LNCS 15448, pp. 193–199, 2025.
https://doi.org/10.1007/978-3-031-87756-8_19

interventions for AF remain limited, largely due to an incomplete understanding of the atrial anatomical structures that contribute to the perpetuation of AF. Recent developments in medical imaging, particularly Late Gadolinium-Enhanced MRI (LGE-MRI), have facilitated the study of atrial fibrosis and scarring, which are critical markers for predicting the success rates of ablation procedures in AF patients. The extent and distribution of atrial fibrosis, as observed in LGE-MRI scans, offer vital insights into patient-specific treatment planning and the potential for improving clinical outcomes.

At the 27th International Conference Medical Image Computing and Computer Assisted Intervention (MICCAI) 2024, the Multi-class Bi-Atrial Segmentation (MBAS) challenge aims to advance the field of atrial analysis by expanding the scope to include both the left and right atriums and their respective walls [12]. Building on the foundational work of the 2018 Left Atrium Challenge [11], this competition focuses on leveraging multi-class machine learning models to enhance the segmentation of atrial structures and the identification of key biomarkers, such as atrial volume and fibrosis, from LGE-MRI scans. With the largest dataset of its kind, comprising 200 multi-center 3D LGE-MRIs labeled by experts, the MBAS 2024 challenge offers a robust platform for developing and evaluating artificial intelligence driven approaches to cardiac analysis.

This work presents an ensemble method that combines several machine learning (ML) models to address the challenge of automatic bi-atrial segmentation from LGE-MRI data. Our approach aims to optimize performance on the provided 3-class dataset by leveraging the complementary strengths of these architectures, inspired by the approach of Zheng et al. [13]. We evaluate our ensemble model using the Dice Similarity Coefficient (DSC) and the 95% Hausdorff distance (HD95), reflecting the assessment of the MBAS challenge. By improving the segmentation of atrial structures, our method contributes to the broader goal of refining ablation strategies and enhancing patient-specific treatment plans for atrial fibrillation.

2 Methods

2.1 Dataset

The data used to train the ML models was sourced from the MBAS challenge organizers. It consists of 200 multi-center 3D LGE-MRIs, each scan labelled by three experts. The MBAS challenge organizers themselves split these 200 scans into 70 training, 30 validation and 100 testing. The validation and testing sets were used to externally assess the performance of participants' submissions, and so the only labelled data available to participants were the 70 training scans. These scans had dimensions of either $40 \times 640 \times 640$ or $40 \times 576 \times 576$, and consisted of three classes (or four if the background is included as a class): left & right atrium wall, right atrium cavity, and left atrium cavity.

The 70 labelled training scans were subsequently split into internal training, validation and testing datasets in a ratio of 84:8:8. Each 3D scan was treated as a series of 2D images stacked in the depth dimension, giving a total of 3080 2D images either of size 640×640 or 576×576. This gave 2587 training images, 246 validation images, and 247 testing images. In line with the approach of Zheng et al. [13], three different splits were created; the testing dataset was fixed while the training and validation datasets were different for each split. These splits were created randomly across the entire domain of 2D images. This meant that slices of a single 3D scan could potentially be split across training, validation and testing datasets, with the goal of improving generalization.

Regularization was also introduced in the form of data augmentation. This was performed at runtime as a model was being trained. The augmentations applied were horizontal flip, vertical flip, and rotation by an angle, randomly chosen as one of 45°, 135°, 225°, or 315°. Each augmentation had a 10% probability of being applied to an image, and only one augmentation would be applied to an image.

2.2 Model Architectures

Like Zheng et al. [13], four CNN architectures were used: UNet [6], ResNet [2], EfficientNet [10], and VGG [7]. All models take a single-channel 2D LGE-MRI slice as input, and output a four-channel image of the same size (either 640×640 or 576×576). Each output channel represents the logits of each pixel belonging to a particular class: background, wall, right atrium and left atrium respectively. The final activation layer is an *argmax*, converting the four-channel logits to a single-channel integer label. All architectures apart from VGG included regularization in the form of dropout [9] at a probability of 20%.

UNet. UNets were first proposed by Ronneberger et al. [6] for biomedical image segmentation. They consist of an encoder-decoder architecture with skip connections in-between. The architecture used in this case comprised four downsampling layers, a bottleneck, and four upsampling layers. Each downsampling layer consisted of a convolution with kernel size 4×4, stride 2 and padding 1, batch normalization [3], dropout and a LeakyReLU activation with negative slope 0.2. Each upsampling layer mirrored the corresponding downsampling layer, but replaced the convolution with a transposed convolution of the same parameters.

ResNet. ResNets were introduced by He et al. [2] to address the vanishing gradient problem and facilitate the training of very deep neural networks. As the standard ResNet architecture is designed for image classification, it must be adapted for segmentation. This was done by using a ResNet34 [2] as the encoder of a UNet. The decoder consists of upsampling layers followed by convolutions of size 3×3, batch normalisation, dropout, and ReLU activations.

EfficientNet. EfficientNets were introduced by Tan and Le [10], utilizing a compound scaling method that uniformly scales all dimensions of depth, width, and resolution, making it computationally efficient while maintaining high performance. EfficientNetB0 was used as the encoder of a UNet, with the same decoder architecture used for the ResNet.

VGG. VGG is a classification architecture first proposed by Zisserman [7], with a deep but simple convolutional structure consisting of small 3×3 kernels. As with the ResNet and EfficientNet, the VGG architecture used in this approach integrates a VGG16 encoder into a UNet for segmentation. The implementation follows that of Zhou [14], where the decoder consists of transposed convolutions of size 2×2 and stride 2, followed by regular convolutions, batch normalisation, and ReLU activation.

2.3 Training and Testing

Each architecture (UNet, ResNet, EfficientNet and VGG) was trained on each of the three splits, giving a total of 12 models. Each model was trained for 300 epochs using the Adam optimizer [4] and a OneCycle learning rate scheduler [8]. The initial learning rate was 1×10^{-4}, which increased to a maximum of 1×10^{-2}, and then annealed to a minimum of 1×10^{-6}. A joint loss function comprising cross entropy and the Dice metric was used.

Once all models were trained, the Ensemble model was created and trained for 30 epochs with the same optimizer and learning rate scheduler as above. The loss function was a weighted combination of cross entropy, Dice metric, and log HD95 in the ratio 0.2 : 0.4 : 0.4. The ensemble model wrapped all sub-models with four 1×1 convolutions, one for each class. This is effectively a per-class weighted combination of each of the sub-models, and so optimizing the Ensemble is equivalent to finding the weight for each sub-model that gives the best Dice and HD95 metrics per class. The final activation of the Ensemble model is $argmax$, giving a single-channel, integer label. All models were evaluated on the internal testing dataset, comparing their outputs to corresponding ground truth labels. Dice Score and HD95 were used as evaluation metrics.

All models were implemented and trained in PyTorch [5], using elements from the Medical Open Network for AI (MONAI) [1].

3 Results

Table 1 shows the Dice and HD95 scores for every model on each class. In general, the UNet achieves the worst results, consistently achieving the lowest Dice scores and highest HD95 distances in all three classes. This is most likely due to the UNet having the smallest number of parameters amongst all models, therefore forming the simplest model of the LGE-MRI data and its heart atria. This reasoning is reinforced by the EfficientNet and ResNet achieving better results than the UNet; both have deeper architectures and are therefore able to capture the data distribution more accurately. ResNet attains narrowly superior results over EfficientNet, which could be due to its smaller model size [10]. However, the difference in scores is small enough that it could instead be due to minor differences such as random weight initialization. Surprisingly, VGG achieves the greatest results of all the sub-models. This is unexpected as the VGG architecture predates ResNets and EfficientNets; both were proposed as an advancement in training speed and accuracy [2,7,10]. This could be because VGGs have a higher complexity than ResNets [2], allowing them to adapt to the complicated problem of multi-class segmentation more easily.

The Ensemble model achieves the best results, with the highest Dice score and HD95 distance in each class by a considerable amount. This reflects the strength of a model ensemble, able to combine the strengths of multiple, individual models to attain superior results. Figure 1 shows the outputs of each model on three MRI scans from the internal testing dataset, along with their ground truth labels. It shows that each model has small inaccuracies, such as the atrial wall being too thick, or the edges of each atrium not being the right shape, and that the Ensemble is able to mitigate these inaccuracies through collective strength.

Table 1. Dice and HD95 scores for the left & right atrium wall (Wall), right atrium (RA) and left atrium (LA) of all twelve models - four architectures (UNet, ResNet, EfficientNet and VGG) and three dataset splits (A, B and C). The best Dice and HD95 score for each class is in bold.

		Wall		**RA**		**LA**	
Architecture	**Split**	Dice	HD95	Dice	HD95	Dice	HD95
UNet	A	64.36	8.18	89.53	6.75	91.61	4.77
	B	65.33	8.34	89.96	11.80	91.97	4.33
	C	65.20	6.92	91.24	4.58	91.21	3.75
ResNet	A	74.88	3.92	93.80	2.32	94.89	2.40
	B	73.45	4.07	94.11	2.73	94.05	2.94
	C	74.80	3.08	95.02	2.98	94.15	1.94
EfficientNet	A	72.18	4.52	93.80	3.25	94.05	2.40
	B	70.96	3.80	94.69	2.39	93.70	2.83
	C	71.99	3.30	94.75	2.67	93.47	2.65
VGG	A	75.27	3.61	93.11	2.42	94.45	2.50
	B	75.31	3.22	94.37	2.38	92.88	2.47
	C	75.16	3.05	94.39	2.76	94.05	2.11
Ensemble	N/A	**88.41**	**1.07**	**98.48**	**0.95**	**98.45**	**0.64**

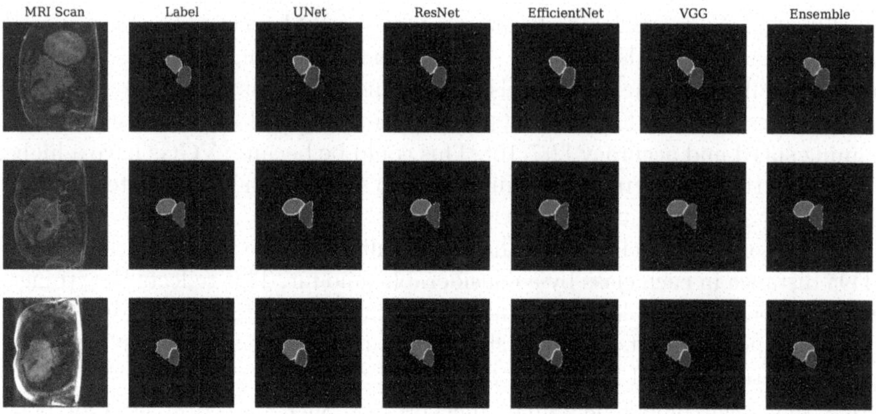

Fig. 1. Outputs of each model (UNet, ResNet, EfficientNet, VGG and Ensemble) for three LGE-MRI inputs (MRI Scan), compared to their respective ground truths (Label). The models UNet, ResNet, EfficientNet and VGG were all trained on dataset split A.

4 Discussion and Conclusion

The results of this study demonstrate the effectiveness of the ensemble approach in multi-class bi-atrial segmentation from LGE-MRI data, with superior performance compared to individual models such as UNet, ResNet, EfficientNet, and VGG. The ensemble model, leveraging the complementary strengths of each architecture, achieved the highest Dice and HD95 scores across all atrial regions, significantly outperforming any single model. This reflects the power of ensemble learning, where model diversity helps to reduce errors and improve generalization. The unexpected performance of VGG, outperforming more modern architectures like ResNet and EfficientNet, suggests that model complexity may play a role in better capturing the intricate features of LGE-MRI data, although further investigation into specific factors like initialization and training dynamics could provide additional insights.

Ultimately, this ensemble-based method offers a robust solution for automatic multi-class segmentation of the left and right atrium from LGE-MRI data, achieving high accuracy and precision as evidenced by the Dice and HD95 metrics. The success of the ensemble model in this context suggests it could be an important tool for advancing atrial fibrillation treatment strategies by providing more reliable segmentation of atrial structures. Future work could explore further refinements, such as hyperparameter tuning or integration of other architectures, to push the boundaries of segmentation accuracy even further.

References

1. Cardoso, M.J., et al.: MONAI: an open-source framework for deep learning in healthcare, 2022. https://doi.org/10.48550/arXiv.2211.02701
2. He, K., Zhang, X., Ren, S., Sun, J.: Deep residual learning for image recognition. In: Proceedings of the IEEE Conference on Computer Vision and Pattern Recognition. IEEE Computer Society, pp. 770–778, 2016. https://doi.org/10.1109/CVPR.2016.90
3. Ioffe, S., Szegedy, C.: Batch normalization: accelerating deep network training by reducing internal covariate shift. In: Proceedings of the 32nd International Conference on Machine Learning, vol. 37. PMLR, pp. 448–456, 2015. https://doi.org/10.48550/arXiv.1502.03167
4. Kingma, D.P., Ba, J.: Adam: a method for stochastic optimization. In: 3rd International Conference on Learning Representations, 2015. https://doi.org/10.48550/arXiv.1412.6980
5. Paszke, A., et al.: PyTorch: An Imperative Style, High-Performance Deep Learning. Library (2019). https://doi.org/10.48550/arXiv.1912.01703
6. O. Ronneberger, P.F., Brox, T.: U-Net: convolutional networks for biomedical image segmentation. In: Medical Image Computing and Computer-Assisted Intervention, pp. 234–241. Springer International Publishing, 2015. https://doi.org/10.1007/978-3-319-24574-4_28
7. Simonyan, K., Zisserman, A.: Very deep convolutional networks for large-scale image recognition. In: International Conference on Learning Representations, 2015. https://doi.org/10.48550/arXiv.1409.1556
8. Smith, L.N., Topin, N.: Super-Convergence: Very Fast Training of Neural Networks Using Large Learning Rates, 2018. https://doi.org/10.48550/arXiv.1708.07120
9. Srivastava, N., Hinton, G., Krizhevsky, A., Sutskever, I., Salakhutdinov, R.: Dropout: a simple way to prevent neural networks from overfitting. J. Mach. Learn. Res. **15**(56), 1929–1958 (2014). http://jmlr.org/papers/v15/srivastava14a.html
10. Tan, M., Le, Q.: EfficientNet: rethinking model scaling for convolutional neural networks. In: Proceedings of the 36th International Conference on Machine Learning, vol. 97. PMLR, pp. 6105–6114, 2019. https://doi.org/10.48550/arXiv.1905.11946
11. Xiong, Z., et al.: A global benchmark of algorithms for segmenting the left atrium from late gadolinium-enhanced cardiac magnetic resonance imaging. Med. Image Anal. **67**, 101832 (2021). https://doi.org/10.1016/j.media.2020.101832
12. Zhao, J., Xu, F., Feng, F.: MICCAI MBAS 2024 - Multi-class Bi-Atrial Segmentation Challenge, 2024. https://codalab.lisn.upsaclay.fr/competitions/18516
13. Zheng, Y., Wu, F., Papież, B.W.: An ensemble method to automatically grade diabetic retinopathy with optical coherence tomography angiography images. In: Mitosis Domain Generalization and Diabetic Retinopathy Analysis, pp. 46–58. Springer Nature, Switzerland, 2023. https://doi.org/10.1007/978-3-031-33658-4_6
14. Zhou, Z.: PyTorch-Unet: Unet with VGG encoder, 2019. https://github.com/zhoudaxia233/PyTorch-Unet/blob/02faa36d185cc631fd6fc8a70d4b9b1755e95e76/vgg_unet.py

Two-Stage nnU-Net for Automatic Multi-class Bi-atrial Segmentation from LGE-MRIs

Y. On[1(✉)], C. Galazis[2], C. Chiu[3], and M. Varela[1,4]

[1] National Heart and Lung Institute, Imperial College London, London, UK
yu.on16@imperial.ac.uk
[2] Department of Computing, Imperial College London, London, UK
[3] Department of Electrical and Electronic Engineering, Imperial College London, London, UK
[4] Cardiovascular and Genomics Research Institute, City St George's University of London, London, UK

Abstract. Late gadolinium enhancement magnetic resonance imaging (LGE-MRI) is used to visualise atrial fibrosis and scars, providing important information for personalised atrial fibrillation (AF) treatments. Since manual analysis and delineations of these images can be both labour-intensive and subject to variability, we develop an automatic pipeline to perform segmentation of the left atrial (LA) cavity, the right atrial (RA) cavity, and the wall of both atria on LGE-MRI. Our method is based on a two-stage nnU-Net architecture, combining 2D and 3D convolutional networks, and incorporates adaptive histogram equalisation to improve tissue contrast in the input images and morphological operations on the output segmentation maps. We achieve Dice similarity coefficients of 0.92 ± 0.03, 0.93 ± 0.03, 0.71 ± 0.05 and 95% Hausdorff distances of (3.89 ± 6.67) mm, (4.42 ± 1.66) mm and (3.94 ± 1.83) mm for LA, RA, and wall, respectively. The accurate delineation of the LA, RA and the myocardial wall is the first step in analysing atrial structure in cardiovascular patients, especially those with AF. This can allow clinicians to provide adequate and personalised treatment plans in a timely manner.

Keywords: Late Gadolinium Enhancement MRI · nnU-Net · Left/Right Atrium · Atrial Wall · Contrast Limited Adaptive Histogram Equalisation · Atrial Segmentation

1 Introduction

Atrial fibrillation (AF) is characterised by irregular and rapid electrical activity originating from the atrial myocardium. It is the most common form of cardiac arrhythmia, affecting around 37 million people worldwide [1]. One common treatment option is catheter ablation, in which specific areas of the heart are

O. Camara et al. (Eds.): STACOM 2024, LNCS 15448, pp. 200–208, 2025.
https://doi.org/10.1007/978-3-031-87756-8_20

ablated (scarred) to block the propagation of abnormal electrical signals. There is some evidence that the degree and distribution of atrial fibrosis is an important determinant of AF treatment success [2]. Moreover, it has been suggested that ablating atrial fibrotic regions may lead to better outcomes in some AF patients [3].

Late gadolinium enhancement (LGE) is a technique commonly used in cardiac magnetic resonance imaging (MRI) to study myocardial scars and fibrosis. Due to the different washout kinetics of the administered gadolinium-based contrast agents, regions of myocardial injury show enhanced intensities compared to healthy tissues >10 min after administration [2]. LGE-MRI has been proposed as a useful tool for AF treatment planning. However, clinical studies are mostly based on time-consuming manual analysis, and reliable automatic methods for LGE-MRI analysis are highly desired.

To this end, we introduce an automatic two-stage segmentation method for LGE-MRI of the atria, as part of the 2024 Multi-class Bi-Atrial Segmentation (MBAS) Challenge under MICCAI.

2 Methods

The proposed segmentation pipeline consists of four main steps: region of interest (ROI) extraction, image histogram equalisation, segmentation, and morphological post-processing, as illustrated in Fig. 1. ROI extraction is performed in the first stage using a lightweight 2D nnU-Net. In the second stage, a 3D nnU-Net is used for segmentation. Aligning with the competition's rules, all neural networks are trained on the provided dataset and no pre-trained weights are used.

2.1 Data

The multi-centre dataset provided consists of 70 3D bi-atrial LGE-MRI scans for training, 30 for validation, and 100 for the final test phase of the competition. The training set comes with ground truth (GT) segmentation masks for three classes: left atrial (LA) cavity, right atrial (RA) cavity, and left and right atrial wall. The images have a voxel spacing of $0.625 \times 0.625 \times 2.5\ \text{mm}^3$ and an image matrix size of either $640 \times 640 \times 44$ or $576 \times 576 \times 44$. While the test set images maintain the same voxel spacing, they may have different matrix sizes. This dataset expands upon a previous challenge [4], which focused exclusively on the LA, with the ultimate goal of improving ablation outcomes for patients with AF.

2.2 Pre-processing

Model Architecture: Image Cropping. To reduce computational complexity, standardise image size, and minimise runtime in segmenting the LA and RA, we crop the images to include only the heart and surrounding structures. The steps we take are outlined below.

First, we merge the three original segmentation masks into a single binary image. Then, the images are downsampled using linear interpolation, with an

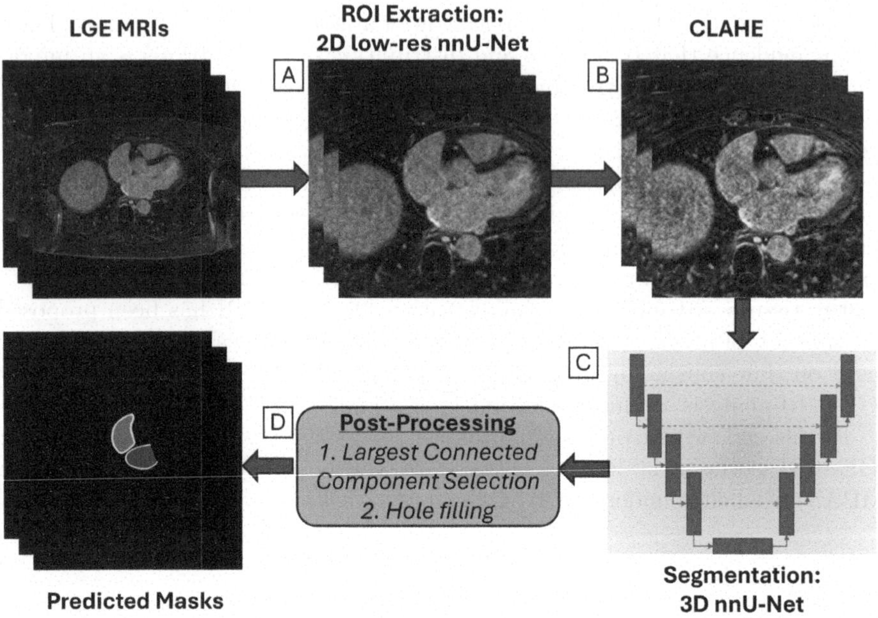

Fig. 1. The proposed segmentation pipeline. 3D LGE-MRIs are inputs to a **(A)** 2D nnU-Net to extract a region of interest (ROI) encapsulating the heart. **(B)** CLAHE is applied to the ROIs to improve the contrast of the images, which are the inputs to a **(C)** 3D nnU-Net that segments the LA, RA, and myocardial wall. The predicted masks are **(D)** post-processed with largest connected component selection and hole filling. *CLAHE: Contrast limited adaptive histogram equalisation; LA: Left atrium; RA: right atrium*

in-plane reduction factor of 4 and a through-plane reduction factor of 2, to reduce inference time. The segmentation maps are similarly downsampled using a nearest-neighbour scheme.

We use a modified three-level 2D convolutional nnU-Net [5] (version 2.4.1) implemented in PyTorch, for slices acquired perpendicular to the long axis. The convolutional layers use a feature map size of 16, 32, and 64 at each level, a 3×3 kernel with strides of 1×1, 2×2, and 2×2, respectively, resulting in a 120k-parameter model. All other hyperparameters are retained as set by the default nnU-Net configuration. The network is optimised for 250 epochs. For inference, we select the weights that result in the best validation loss, which is the combined Dice and cross-entropy loss.

For ROI extraction, we calculate the 3D centroid on the predicted binary segmentation map and crop the image using a $320 \times 320 \times 44$ voxel box, ensuring surrounding context is preserved. Voxels extending beyond the image boundaries are padded with a value of 0.

Contrast Limited Adaptive Histogram Equalisation. Even for human experts, it can be difficult to delineate the myocardium wall from the LA or RA in poor-contrast regions. To tackle this problem, Contrast Limited Adaptive Histogram Equalisation (CLAHE) [6,7] is applied as a pre-processing step to improve tissue contrast in the images. Unlike global histogram equalisation [8], which calculates the histogram of pixel intensities and performs equalisation globally on the cumulative density function, CLAHE applies the equalisation on a local scale. Briefly, the input image is divided into a $H \times W \times D$ grid, in which the size of $H/W/D$ is 1/8 of the corresponding dimension of the image, and the histogram is calculated separately in each grid.

2.3 Segmentation

Model Architecture: Segmentation. To segment the images into the three atrial categories, we use the 3D convolutional nnU-Net model [5] with a total of 3M trainable parameters. To improve generalisability, we apply data augmentations during training, which consist of spatial transformations (e.g. rotation, transpose, mirroring), blurring, sharpening, additive white noise, brightening, and gamma correction. Details on specific parameters can be found in the nnU-Net's "nnUNetTrainerDA5" configuration. The model is optimised for 1,000 epochs. The predicted segmentation maps are then resized back to their original dimensions using nearest-neighbour resampling.

2.4 Post-processing

To improve the quality of the segmentation maps, two morphological operations are employed as a post-processing step using SimpleITK [9]: largest connected component selection and greyscale hole filling. Both operations are configured for a 3D face connectivity of 6.

2.5 Experimental Setup

We evaluate the models using the Dice similarity coefficient (DSC) and 95% Hausdorff distance (HD95, in mm). The HD95 represents the HD calculated based on the closest 95% of points between two sets. During the development phase, we perform a 5-fold validation on the training set. For the submission to the validation and test phases of the competition, we optimise the weights by randomly splitting the 70 training cases into 65 for training and 5 validation.

The neural networks are trained on a single Nvidia RTX 6000 GPU, and the remaining pre-processing and post-processing steps are executed on a 3XS Intel Core i7 (10700K, 3.8 GHz, 8 Core) CPU. The approximate inference time for a single case is 19.5 s. This includes 7.7 s, 0.5 s, 11 s, and 0.3 s, for ROI extraction, CLAHE, segmentation, and post-processing steps, respectively.

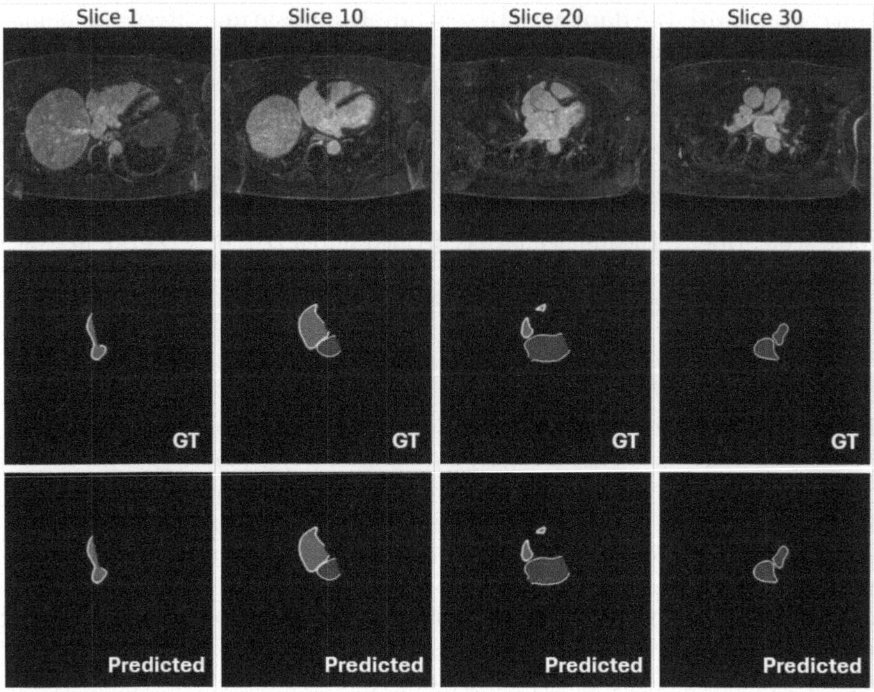

Fig. 2. A representative example case comparing the ground truth (GT) and predicted masks at every 10 slices. **(Top)** LGE image (from the training set); **(Middle)** GT segmentation masks; **(Bottom)** final predicted masks after post-processing. Key: left atrium (blue), right atrium (red), and myocardial wall (yellow). (Color figure online)

3 Results

Our proposed two-stage nnU-Net-based pipeline demonstrates high accuracy in segmenting the LA and RA. The 5-fold cross validation results on the 70-subject training set are: a DSC of 0.92 ± 0.03, 0.93 ± 0.03 and 0.71 ± 0.05, and HD95 of (3.89 ± 6.67) mm, (4.42 ± 1.66) mm and (3.94 ± 1.83) mm for the LA, RA and myocardial wall, respectively. Overall, we see accurate segmentation of the atria, including their appendages. A representative case can be viewed in Fig. 2.

Furthermore, we compare our pipeline with four other variants: 1) a 2D nnU-Net for segmentation, 2) without CLAHE, 3) without post-processing, and 4) without both CLAHE and post-processing. As summarised in Fig. 3, our proposed pipeline overall outperforms all other variants, with the exception of a drop in performance on the wall's HD95. Specifically, the use of both pre-processing and post-processing yields DSC improvements of 0.8%, 0.58% and 4.3%, and HD95 changes of −32.71%, −5.22% and 16.74% for the LA, RA and wall, respectively. While the 2D nnU-Net variant is advantageous in terms of the execution time, it exhibits a large performance drop compared to its 3D counterpart, especially for the wall, due to limited spatial information.

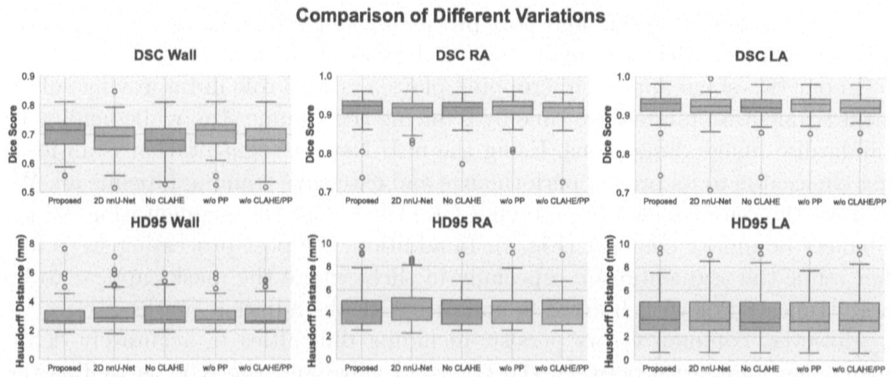

Fig. 3. The box plots show the Dice similarity coefficient (DSC) **(Top)** and 95% Hausdorff distance (HD95) **(Bottom)** for four other variations of our proposed two-stage model. In each plot, from left to right we show: our proposed method, 2D nnU-Net for segmentation, without CLAHE, without post-processing (PP), and without both CLAHE and PP.

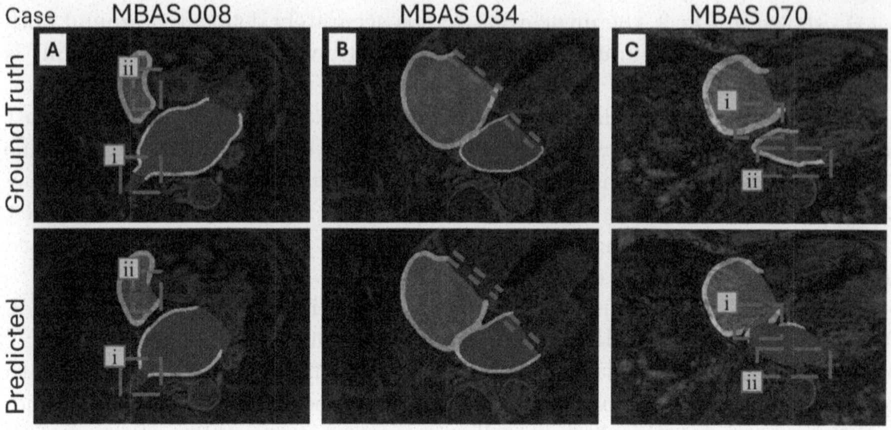

Fig. 4. Examples of prediction errors from three different cases, with each column representing a separate case. **Top**: ground truth segmentations; **Bottom**: predicted segmentations. Marked regions in green highlight prediction errors. (Color figure online)

4 Discussion

The proposed pipeline accurately delineates the LA, RA, and myocardial wall from the LGE-MRIs. Our results compare well with reported literature metrics (LA DSC: 0.93 ± 0.02 [4]; RA DSC: $0.90 \pm n/a$ [10]) for neural network-based LGE-MRI segmentations. This performance is notable given the computational time and hardware restrictions imposed by the competition: inference times no longer than 10 min for 100 test cases on a single Nvidia Tesla V100 GPU and exclusion of additional training data or pre-trained models. We pragmatically focus on the pre-processing and post-processing steps to enhance the segmentation performance with our small training dataset of 70 cases.

In our two-stage approach, cropping plays a crucial role in improving subsequent segmentation inference time by reducing the volume size, while helping to standardise image dimensions. Using the nnU-Net for segmentation is an ideal first choice due to its proven performance and extensive training pipeline [5]. We improve the performance by applying CLAHE before the segmentation, which enhances boundary contrast (Fig. 3). In addition, the post-processing steps further refine the segmentation, especially in slices where the masks appear to be more truncated or with blurred boundaries with the adjacent masks.

However, common errors persist, including difficulties in accurately delineating the boundaries between the tricuspid valve and the RA, as well as the mitral valve and the LA, as shown in Fig. 4B. Additionally, structures like the pulmonary veins, superior vena cava, and inferior vena cava are occasionally trimmed (see pulmonary vein example in Fig. 4Ai) or overextended in comparison to the GT. These issues arise due to the lack of distinct features that the model can rely on for precise segmentation. Additionally, these errors are further exacerbated by unavoidable inconsistencies in manual annotation.

As shown in Fig. 2, the myocardial wall is accurately delineated around both the RA and LA, with uniform thickness. The RA wall ($\sim 3.54 \pm 0.19$ mm) is thicker than the LA wall ($\sim 2.31 \pm 0.23$ mm), consistent with the GT. However, small segments of the wall are occasionally missed, either due to low contrast neighbouring tissue being misinterpreted as atrium openings (Fig. 4Aii), or due to high levels of noise and blurring (Fig. 4Cii). Additionally, the model may struggle to accurately identify the thin septal wall separating the LA and RA blood pools, due to its inherently low contrast (Fig. 4Ci). This issue is worsened in images with high noise or ringing artefacts.

Omitting CLAHE results in the largest decline in the wall's DSC to 0.64 ± 0.56 and an increase in HD95 to (5.04 ± 5.15) mm. This highlights CLAHE's ability to enhance boundary contrast at important regions of the wall, with only negligible computational cost of 0.5 s per cropped case. On the other hand, omitting post-processing has a smaller impact on the overall performance, as expected. Nevertheless, it plays a key role in addressing more challenging cases where noise and artefacts are prominent, preventing large errors in boundary identification and refining segmentations.

The combined results indicate that while CLAHE is essential for improving wall segmentation accuracy, post-processing serves as an important corrective step, especially for outlier cases with high noise or imaging artefacts. However, we note that each of these steps results in small changes, either positive or negative, for the LA and RA segmentations. This might suggest additional benefits in improving the neural network architecture.

We aim to improve our two-stage nnU-Net model through a multi-task approach [11], as prior research has shown it can improve robustness and generalisability [12]. Our current model performs poorly on low-quality images, so we plan to develop a joint nnU-Net that integrates segmentation, de-noising, and inpainting to improve both image quality and segmentation accuracy. In addition, we also plan to further improve post-processing steps by using contour correction and wall dilation. Unfortunately, these steps would unavoidably increase inference time, making them unsuitable under the challenge's constraints.

Atrial segmentation is a crucial step that enables quantitative analysis of atrial structure and function. This leads to important applications in prognosis, diagnosis, and treatment. However, the LA and RA are naturally challenging to segment, especially when compared with the left ventricle. This is primarily due to several factors: 1) the complex geometries of the LA and RA, 2) variability in atrial size and shape among individuals, and 3) thin and inconsistent wall thickness [13–17]. Therefore, continuous advancements of automated segmentation methods are needed to further improve the LA and RA analysis.

5 Conclusions

For the MBAS challenge, we propose an automatic pipeline to accurately segment the LA, RA, and myocardial wall from LGE-MRIs. Our segmentation is based on the 3D convolutional nnU-Net and incorporates important preprocessing steps: 2D nnU-Net-based image cropping, CLAHE-based image equalisation, and morphological post-processing steps.

Despite the challenges posed by image quality variations and the complex anatomical geometries of the atria, our approach achieves a competitive DSC and HD95 with a very quick inference time (average 19.5 s). Our pipeline has the potential to support clinical decision-making and improve the planning of treatments like catheter ablation in AF patients.

Acknowledgments. This work was supported by: St George's Hospital Charity, the NIHR Imperial Biomedical Research Centre (BRC) and the British Heart Foundation Centre of Research Excellence at Imperial College London (RE/18/4/34215). We acknowledge the computational support provided by the Imperial College Research Computing Service (DOI: https://doi.org/10.14469/hpc/2232).

References

1. Lippi, G., Sanchis-Gomar, F., Cervellin, G.: Global epidemiology of atrial fibrillation: an increasing epidemic and public health challenge. Int. J. Stroke **16**(2), 217–221 (2021)
2. Li, L., Zimmer, V.A., Schnabel, J.A., Zhuang, X.: Medical image analysis on left atrial LGE MRI for atrial fibrillation studies: a review. Med. Image Anal. **77**, 102360 (2022)
3. Marrouche, N.F., et al.: Effect of MRI-guided fibrosis ablation vs conventional catheter ablation on atrial arrhythmia recurrence in patients with persistent atrial fibrillation: the DECAAF II randomized clinical trial. JAMA **327**(23), 2296–2305 (2022)

4. Xiong, Z., et al.: A global benchmark of algorithms for segmenting the left atrium from late gadolinium-enhanced cardiac magnetic resonance imaging. Med. Image Anal. **67**, 101832 (2021)
5. Isensee, F., et al.: nnU-Net: a self-configuring method for deep learning-based biomedical image segmentation. Nat. Methods **18**(2) (2021)
6. Reza, A.M.: Realization of the contrast limited adaptive histogram equalization (CLAHE) for real-time image enhancement. J. VLSI Sig. Proc. Syst. Sig., Image, Video Technol. **38**(1), 35–44 (2004)
7. Stimper, V., Bauer, S., Ernstorfer, R., et al.: Multidimensional contrast limited adaptive histogram equalization. IEEE Access **7**, 165437–165447 (2019)
8. Abdullah-Al-Wadud, M., Kabir, M.H., Dewan, M., Chae, O.: A dynamic histogram equalization for image contrast enhancement. IEEE Trans. Consum. Electron. **53**(2), 593–600 (2007)
9. Lowekamp, B.C., Chen, D.T., Ibáñez, L., Blezek, D.: The design of simpleITK. Front. Neuroinform. **7**(DEC) (2013)
10. Zhu, J., Bai, J., Zhou, Z., et al.: RAS dataset: a 3D cardiac LGE-MRI dataset for segmentation of right atrial cavity. Sci. Data **11**(1), 401 (2024)
11. Zhao, Y., Wang, X., Che, T., Bao, G., Li, S.: Multi-task deep learning for medical image computing and analysis: a review. Comput. Biol. Med. **153**, 106496 (2023)
12. Zhang, Yu., Yang, Q.: A survey on multi-task learning. IEEE Trans. Knowl. Data Eng. **34**(12), 5586–5609 (2021)
13. Barbero, U., Ho, S.Y.: Anatomy of the atria. Herzschrittmachertherapie + Elektrophysiologie **28**(4) (2017)
14. Ho, S.Y., Sánchez-Quintana, D.: The importance of atrial structure and fibers (2009)
15. Ho, S.Y., Cabrera, J.A., Sanchez-Quintana, D.: Left atrial anatomy revisited. Circ.: Arrhythmia Electrophysiol. **5**(1) (2012)
16. Varela, M., et al.:3D high-resolution atrial wall thickness maps using black-blood PSIR. J. Cardiovasc. Magn. Reson. **17** (2015)
17. Varela, M., et al.: Novel MRI technique enables non-invasive measurement of atrial wall thickness. IEEE Trans. Med. Imag. **36** (2017)

An Ensemble of 3D Residual Encoder UNet Models for Solving Multi-class Bi-atrial Segmentation Challenge

Alexander Zolotarev(✉), Kiane Johnson, Abbas Khan, Gregory Slabaugh, and Caroline Roney

Queen Mary University of London, London, UK
{a.zolotarev,c.roney}@qmul.ac.uk

Abstract. Atrial fibrillation is the most common heart rhythm disorder and is linked to an elevated risk of stroke. Selecting the appropriate treatment for AF remains a significant challenge, but biophysical simulations offer a promising solution by enabling the creation of patient-specific cardiac digital twins for in-silico testing of various treatments. To achieve this, it is essential to have an accurate and efficient tool for segmenting individual atrial structures, including atrial cavities and walls.

We proposed a method using a residual encoder nnUNet for segmenting LGE-MRI scans. Our approach excelled in the validation stage of the MBAS challenge, achieving Dice scores of 91.25 for the left atrium cavity, 88.44 for the right atrium cavity, and 67.16 for the atrial walls.

Keywords: Atrial segmentation · Late gadolinium-enhanced MRI · Atrial fibrillation

1 Introduction

Atrial fibrillation (AF) is the most prevalent form of cardiac arrhythmia and is linked to significant morbidity and mortality [4]. Current treatments for AF are often ineffective, largely due to a limited understanding of the atrial anatomy that sustains AF within the human atria. Recently, Late gadolinium-enhanced MRI (LGE-MRI) has become a prominent tool for studying fibrosis and scarring in the atria, and clinical studies on AF patients have shown that the extent and pattern of atrial fibrosis identified by LGE-MRIs can reliably predict ablation outcomes [5]. Therefore, direct analysis of atrial structure in AF patients is crucial for advancing understanding and personalized treatment of AF.

2 Methodology

Building on the previous 2018 left atrium challenge [6], this expanded challenge now includes both the left and right atria and their walls. It emphasizes multi-class machine learning from LGE-MRIs to improve ablation outcomes for AF

O. Camara et al. (Eds.): STACOM 2024, LNCS 15448, pp. 209–213, 2025.
https://doi.org/10.1007/978-3-031-87756-8_21

patients by testing methods for segmentation and biomarker identification (e.g., atrium volume and fibrosis) across 200 multi-center 3D LGE-MRIs—the largest dataset in this field, with each scan precisely labelled by three experts.

The dataset was separated into 70 3D LGE-MRI cases for training, 30 for validation, and an additional 100 designated for the final test phase. There are 3 segmentation targets: atrial walls, and right (RA) and left (LA) atrial cavities. The performance was evaluated by Dice Similarity Coefficient (DSC) and 95% Hausdorff distance (HD95) metrics.

We utilized nnUNet framework [2], which automates the configuration of UNet for the given dataset. All experiments were run using standard 3D nnUNet 5-fold cross-validation on the 70 training cases. The resulting 5 models were used in an ensemble to predict 30 cases in the validation stage. We tested if adding residual encoders to the original architecture changes the performance metrics (Fig. 1).

Residual Encoder UNet is the extension of plain UNet which incorporates residual blocks [1] within the architecture. Residual connections may help improve the model's performance by reducing the vanishing gradient problem.

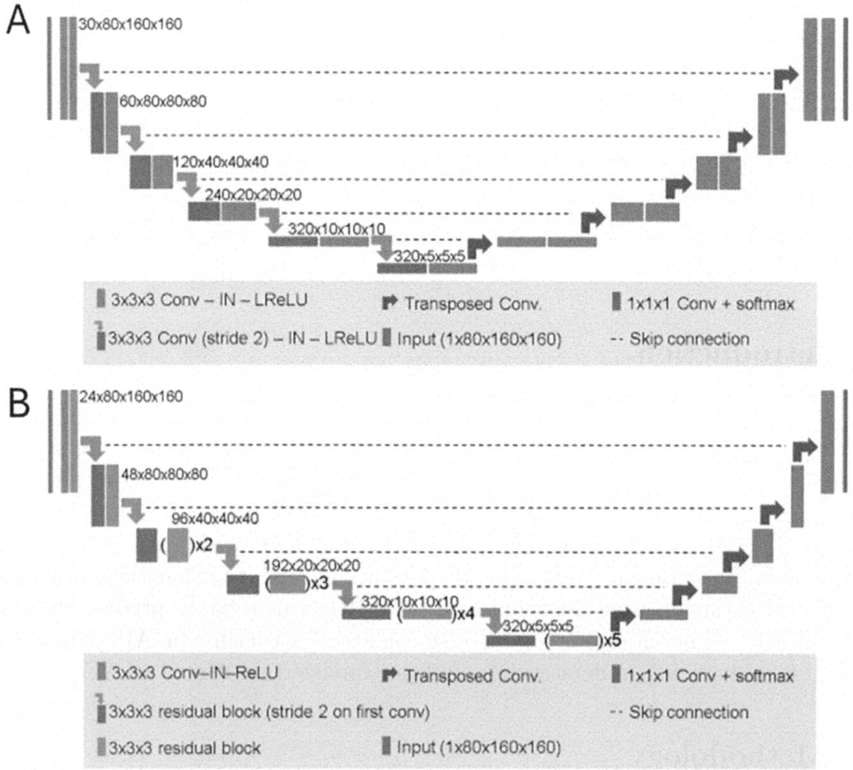

Fig. 1. The 3D UNet (A) and 3D Residual Encoder UNet (B) were compared in terms of best Dice scores and Hausdorff distances for the Multi-Class Bi-Atrial Segmentation Challenge. Adapted from [3]

The training was conducted on NVIDIA GeForce RTX 3080 video card for 1000 epochs.

3 Results

Our approach secured **first** place in the validation stage of the Multi-Class Bi-Atrial Segmentation Challenge. The organizers have also provided the raw ranks for each of the 6 metrics (Table 1). Our ensemble achieved the best Dice scores across all teams (for all 3 classes). However, 95% Hausdorff distances (the lower, the better) calculated for our solution are not so successful (from second place for HD95 of atrial walls to seventh for HD95 of the right atrial cavity).

Table 1. The performance of solutions developed by different teams in the validation stage. R - Rank.

Team	R	Atrial Walls				Right Atrial Cavity				Left Atrial Cavity			
		Dice	R	HD95	R	Dice	R	HD95	R	Dice	R	HD95	R
QMUL	1	67.16	1	4.24	2	88.44	1	7.23	7	91.25	1	4.79	3
PEDRETS	2	66.55	2	4.68	6	88.25	4	7.38	10	91.14	3	4.81	6
SJTU	3	65.02	9	3.88	1	88.38	2	6	1	90.48	12	4.82	7
CEMRG	4	66.36	5	4.8	8	88.37	3	7.34	8	91.03	4	4.81	5
AI3MED	5	66.39	4	4.53	4	88.06	5	7.68	13	90.99	5	4.91	9

Notably, we have shown that the Residual Encoder UNet outperforms the plain UNet in five out of six performance metrics (all metrics for left and right atrial cavities and 95% Hausdorff distance for atrial walls). Only the Dice score for atrial walls showed the opposite dynamics - 67.53 for plain UNet versus 67.16 for residual encoder UNet (Table 2).

Table 2. The performance of different ensemble architectures in the validation stage.

	Atrial Walls		Right Atrium		Left Atrium	
	Dice	HD95	Dice	HD95	Dice	HD95
3D UNet	**67.53**	5.01	88.17	7.64	91.22	4.83
3D ResEnc UNet	67.16	**4.24**	**88.44**	**7.23**	**91.25**	**4.79**

4 Discussion

We have shown that our solution achieved the high metrics for the task of atrial structure segmentation. The Dice scores between predictions and ground truth masks provided by organizers are 88.44 and 91.25 for right and left atrial cavities respectively. However, the metric for the atrial wall is significantly lower, Dice score is equal to 67.16. It indicates the most complicated part of this challenge.

We can easily see this on the visualisation test. Figure 2 shows four LGE-MRI scans of case number 10 from the validation set, with prediction masks on them. There are clear and correct masks of left and right atrial cavities on scan number 1, with masks of atrial walls surrounding both cavities from all sides. We can count this prediction as ideal, with high Dice score for all 3 classes.

For comparison, scan number 2 shows the atrial wall partly missed around the left atrium. This is an example of a scan with a high Dice score for left and right atrial cavities and a lower Dice score for atrial walls.

However, there are some scans with mistakes for cavity masks, usually for the left atrium. Scan number 3 shows the left atrial cavity both inside and outside atrial walls, and open tips of atrial wall masks. This mistake can be caused by the regions with almost the same intensity within and outside the atrial cavity without a clear border between them. Another complicated case can be from the first (such as scan number 4) or last scans with masks where the model struggled to find relevant anatomical structures.

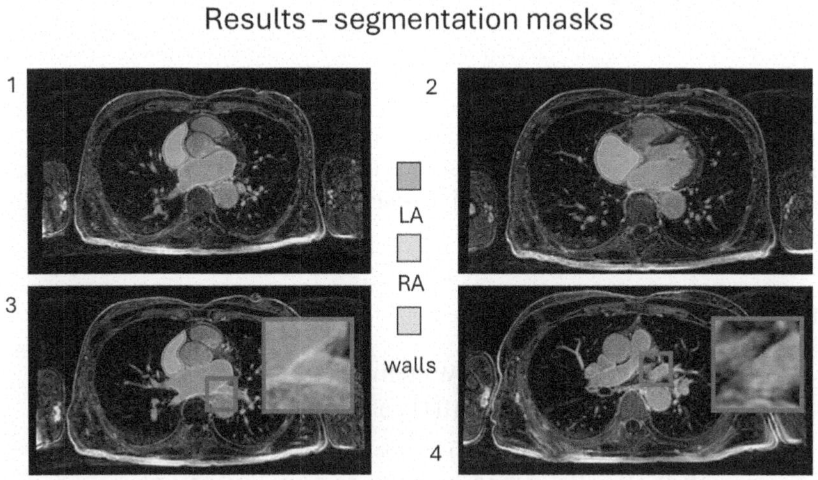

Fig. 2. The examples of successful and unsuccessful prediction masks from the same LGE-MRI case. LA - Left Atrium, RA - Right Atrium.

5 Conclusion

Our team achieved first place in the validation stage of the Multi-Class Bi-Atrial Segmentation Challenge by applying the ensemble of 5 Residual Encoder UNet models. This advancement in AI and clinical methodologies marks a paradigm shift in cardiac imaging and has the potential to impact other medical domains, ultimately aiming to refine ablation strategies for treating persistent AF.

Acknowledgments. This work was funded by NIHR Barts BRC (NIHR203330) and the UKRI Future Leaders Fellowship (MR/W004720/1) grants.

Disclosure of Interests. The authors have no competing interests to declare that are relevant to the content of this article.

References

1. He, K., Zhang, X., Ren, S., Sun, J.: Deep residual learning for image recognition. In: 2016 IEEE Conference on Computer Vision and Pattern Recognition (CVPR), pp. 770–778 (2015)
2. Isensee, F., et al.: nnU-Net: a self-configuring method for deep learning-based biomedical image segmentation. Nat. Methods **18**, 203–211 (2020)
3. Isensee, F., Maier-Hein, K.H.: An attempt at beating the 3D U-Net (2019). https://arxiv.org/abs/1908.02182
4. Joglar, J.A., et al.: 2023 ACC/AHA/ACCP/HRS guideline for the diagnosis and management of atrial fibrillation: a report of the American college of cardiology/American heart association joint committee on clinical practice guidelines. J. Am. Coll. Cardiol. **83**(1), 109–279 (2024). https://doi.org/10.1016/j.jacc.2023.08.017
5. Marrouche, N.F., et al.: Effect of MRI-guided fibrosis ablation vs conventional catheter ablation on atrial arrhythmia recurrence in patients with persistent atrial fibrillation: the DECAAF II randomized clinical trial. JAMA **327**(23), 2296–2305 (2022). https://doi.org/10.1001/jama.2022.8831
6. Xiong, Z., et al.: A global benchmark of algorithms for segmenting the left atrium from late gadolinium-enhanced cardiac magnetic resonance imaging. Med. Image Anal. **67**, 101832 (2021). https://doi.org/10.1016/j.media.2020.101832

Evaluating Convolution, Attention, and Mamba Based U-Net Models for Multi-class Bi-Atrial Segmentation from LGE-MRI

Claas Thesing[1,2], Alfonso Bueno-Orovio[1], and Abhirup Banerjee[2](✉)

[1] Department of Computer Science, University of Oxford, Oxford, UK
{claas.thesing,alfonso.bueno}@cs.ox.ac.uk
[2] Institute of Biomedical Engineering, Department of Engineering Science, University of Oxford, Oxford, UK
abhirup.banerjee@eng.ox.ac.uk

Abstract. Atrial Fibrillation (AF) is the most prevalent cardiac arrhythmia, associated with a significantly increased risk of stroke and heart failure. Thus, it is critical to analyse the progression of AF, which requires accurate segmentation of the atria in medical images. This study addresses the challenge of automating this segmentation process for Late Gadolinium-Enhanced Magnetic Resonance Imaging (LGE-MRI) scans through a multi-class segmentation approach targeting the left and right atria and atrial walls. Instead of introducing a novel architecture, we evaluate existing U-Net-based models employing different backbone architectures, including convolutional layers, attention-based layers, and Mamba blocks, and assess their performance over the Multi-class Bi-Atrial Segmentation (MBAS) Challenge dataset. Additionally, we compare the impact of different preprocessing techniques on the segmentation performance. Achieving the third rank in the MBAS Challenge with a convolutional nnU-Net model, we demonstrate its superior performance compared to other models. Our findings aim to inform the development of more efficient and accurate segmentation methods in the clinical evaluation of AF.

Keywords: Bi-Atrial Segmentation · LGE-MRI · nnU-Net · Attention layer · Mamba · U-Net

1 Introduction

Atrial Fibrillation (AF) is the most common cardiac arrhythmia, impacting about 0.51% of the global population [11]. It is associated with an increased risk of stroke and heart failure, making it a critical area of focus in cardiovascular medicine [19]. To analyse the progression of AF, Late Gadolinium-Enhanced Magnetic Resonance Imaging (LGE-MRI) has emerged as the most widely-used imaging procedure [7]. Unlike standard MRI, LGE-MRI uses the

O. Camara et al. (Eds.): STACOM 2024, LNCS 15448, pp. 214–225, 2025.
https://doi.org/10.1007/978-3-031-87756-8_22

contrast agent gadolinium to enhance the contrast between different tissues in the heart. This allows to better characterise the extent of fibrosis and scar tissues in the atria [20], which can be useful for predicting the success rate of ablation therapy for patients with AF [2,22,27].

Despite the advantages of LGE-MRI, the interpretation of these medical images requires an accurate segmentation of the left atrium (LA) and right atrium (RA), as well as of the atrial wall. The traditional approach of manual segmentation is labour-intensive, time-consuming, and prone to inter-observer variability. These drawbacks underscore the need for automated segmentation methods that can provide reliable and reproducible results at scale.

In the last decade, there has been significant progress in developing deep-learning-based methods to automate image segmentation tasks. Among these methods, the U-Net architecture [18] has emerged as a particularly effective model for medical image segmentation. The U-Net architecture is characterised by its encoder-decoder structure with skip connections and originally was implemented using 2D convolutions. However, since then, several extensions and improvements have been proposed, including the usage of 3D convolutions to capture spatial information in 3D medical images more accurately [1]. Furthermore, the U-Net architecture has been combined with more advanced models. For example, UNETR [6] and SWIN-UNETR [5] replace specific convolutional blocks with transformer layers to capture the global context of the image more effectively. U-Mamba [16], on the other hand, uses a Mamba block to reduce the quadratic time complexity of the attention mechanism in the transformer layers during inference. Since all these methods come with their own set of hyperparameters to tune, Isensee et al. [8] designed nnU-Net, a framework that uses the original convolutional implementation but automates the hyperparameter search based on key properties of the dataset.

Previous work in the field of atrial segmentation primarily focused on the LA only [25]. For the 2018 Atrial Segmentation Challenge, different methods were proposed to segment the LA. These include combining different models for segmenting the region of interest or developing novel losses [23,24,26].

In this study, we consider a multi-class segmentation problem for the automated segmentation of both the LA and RA along with the atrial walls. Instead of developing a completely new architecture and further extending the multitude of already existing models, we focus on comparing existing U-Net based models with different backbone architectures such as convolutional layers, attention-based layers, and Mamba blocks and investigate their performance on the Multi-class Bi-Atrial Segmentation (MBAS) Challenge dataset. Moreover, we vary the preprocessing technique to analyse its impact on the segmentation performance.

2 Methods

2.1 Dataset and Preprocessing

This study utilises the MICCAI MBAS Challenge dataset, which includes a total of 100 3D LGE-MRI scans during the training phase. The organisers of the challenge released 70 scans and their corresponding ground-truth segmentation masks for training and 30 scans for testing. For the latter, the ground-truth segmentation masks were not provided, but the evaluation of some model predictions could be requested up to three times. The final evaluation was performed on a separate test set of 100 scans with all the data hidden from the participants such that the test code had to be submitted. For the internal evaluation, we split the 70 instances into training and testing following an 80%/20% split.

The data was annotated by three experts and contains the manual segmentation of the right atrium cavity, the left atrium cavity, and the left and right atrial walls. Each MRI has a resolution of $0.625 \times 0.625 \times 2.5\,\text{mm}^3$ and contains either $576 \times 576 \times 44$ or $640 \times 640 \times 44$ voxels. The axes of the volume are oriented to match the alignment of the patient's body in the MRI scanner.

We use two independent preprocessing techniques to determine their suitability for the task of bi-atrial segmentation: Z-Score normalisation and Contrastive Limited Adaptive Histogram Equalisation (CLAHE) [17]. While Z-Score normalisation standardises the intensity values of the images to have a mean of zero and a standard deviation of one, CLAHE enhances the contrast of the images by equalising the histogram of local regions [21].

2.2 Algorithms

We evaluated the performance of the following U-Net-shaped models: nnU-Net [8], nnU-Net ResEnc [9], UNETR [6], Swin UNETR [5], U-MambaBot as well as U-MambaEnc [16]. A simplified overview of the architecture can be found in Fig. 1. For a detailed description of the models, we refer the reader to the original papers. Essentially, these models consist of an encoder-decoder structure with skip connections linking encoder and decoder blocks across different scale levels. The encoder is responsible for extracting features from the input volume, while the decoder generates the final segmentation mask. Although all models use convolutional layers in the decoder to better capture local information, they differ in the layers used within their encoder blocks: nnU-Net employs convolutional layers, UNETR and Swin UNETR utilise attention-based layers, and U-MambaBot and U-MambaEnc use Mamba blocks [4] to extract global features. For each model, we considered both 2D and 3D configurations, where the 3D configuration processes the entire MRI scan at once, while the 2D configuration processes each slice of the volume individually.

In contrast to nnU-Net, nnU-Net ResEnc also includes skip connections in the encoder blocks to improve the gradient flow. The difference between UNETR and Swin UNETR lies in the version of the transformer layer used in the encoder blocks. While UNETR uses a standard Vision Transformer (ViT) [3], Swin UNETR leverages the SwinTransformer [13] for a more efficient hierarchical design. On the other hand, U-MambaEnc differs from U-MambaBot in the placement of the Mamba block. In U-MambaEnc, the Mamba block is placed in every encoder block, whereas in U-MambaBot it is only placed in the last bottleneck layer. An overview of the different layer structures is provided in Fig. 2.

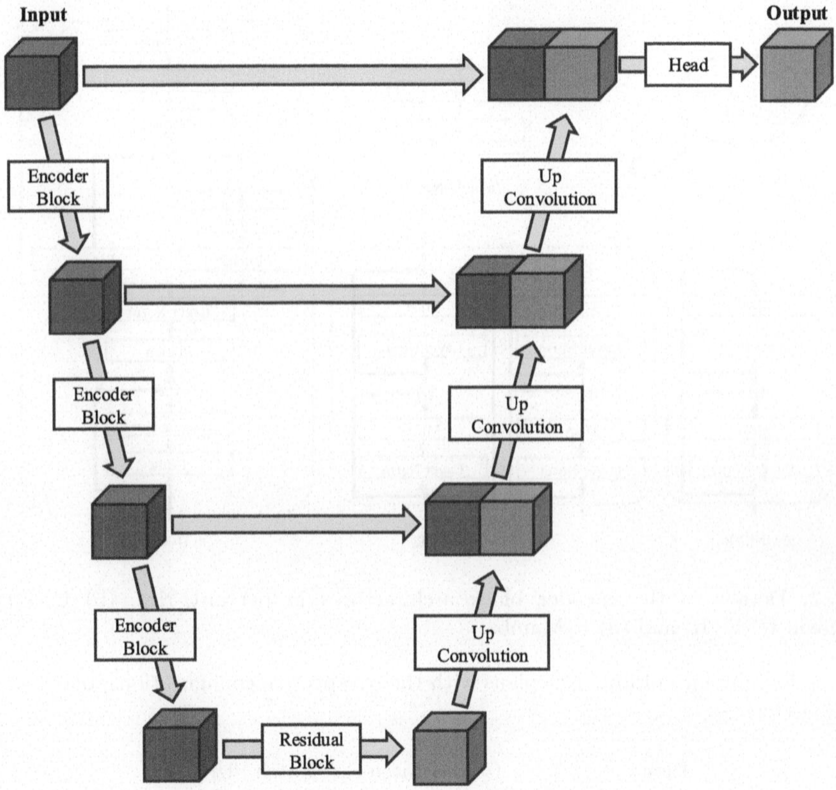

Fig. 1. Simplified structure of the considered U-Net architecture.

2.3 Training Procedure

We trained each model using the nnU-Net framework [8] to ensure a comparable training setup across all models. Further, we configured hyperparameters such as batch size and patch size following the experiment planner of the framework. This planner automatically adjusts the hyperparameters based on key properties of the dataset, e.g., the resolution and size of the MRI scans. Next to the default planner, we used the planner for the residual encoder that aims to fit the model into 16 GB of VRAM. UNETR and Swin UNETR manually overwrite the patch size of the plan to match the architectural constraint that all axes of the patch size must be divisible by 32. While planning to use the 16 GB plan in most cases, we had to switch to the default plan or reduce the batch size for some models that required more VRAM than anticipated and caused out-of-memory errors. The patch and batch sizes for the different plans are summarised in Table 1. Table 2 outlines which plan was used to train each model.

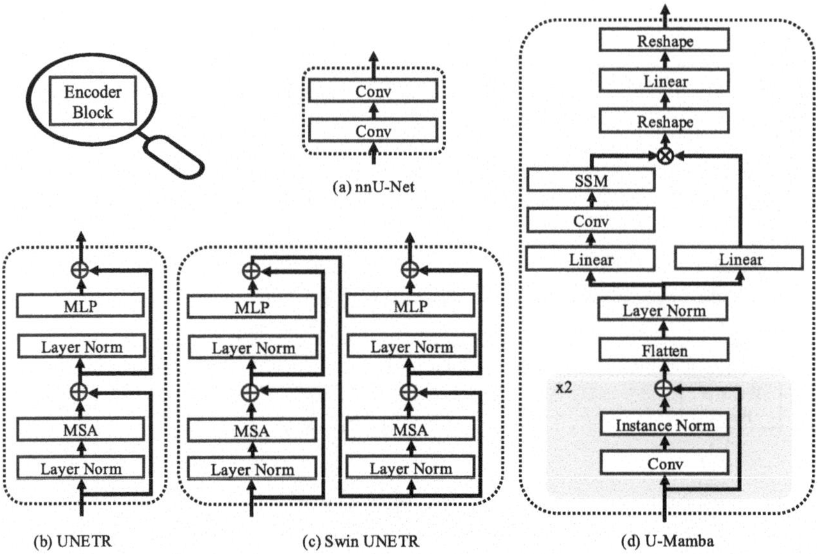

Fig. 2. Details of the encoder block architectures in (a) nnU-Net, (b) UNETR, (c) Swin UNETR, and (d) U-Mamba.

Table 1. Summary of nnU-Net plans with their respective configurations, batch sizes, and patch sizes.

Plan	Config	Batch Size	Patch Size
default	2D	8	640×640
default	3D	2	$20 \times 320 \times 256$
ResEnc 16 GB	2D	16	640×640
ResEnc 16 GB	3D	2	$28 \times 320 \times 320$
UNETR-3D	3D	2	$32 \times 320 \times 256$
Swin UNETR-3D	3D	2	$32 \times 256 \times 224$

To increase the variability of the training set and improve the generalisation of the model, we used the data augmentation pipeline integrated in the nnU-Net framework [8]. This includes random rotations, random scaling, random elastic deformations, random flipping, and gamma correction. Additionally, we padded the image appropriately, if the image was smaller than the patch size.

We followed the recommendations of the original papers to implement different optimisers and learning rate schedules. Thus, we trained the nnU-Net und U-Mamba models with stochastic gradient descent, while UNETR and Swin UNETR use AdamW [14]. The initial learning rates were set to 10^{-2}, 10^{-2}, 10^{-4}, and 8×10^{-4} for nnU-Net, U-Mamba, UNETR, and Swin UNETR, respectively. All models but Swin UNETR were trained with a polynomial learning rate scheduler [12], while Swin UNETR used a cosine learning rate scheduler [15]. The

Table 2. Summary of models and their corresponding training plans.

Model	Config	Plan
nnU-Net	2D	default
nnU-Net	3D	default
nnU-Net ResEnc	2D	ResEnc16 GB
nnU-Net ResEnc	3D	ResEnc16 GB
U-MambaBot	2D	ResEnc16 GB
U-MambaBot	3D	default
U-MambaEnc	2D	ResEnc16 GB
U-MambaEnc	3D	default
UNETR	2D	default
UNETR	3D	UNETR-3D
Swin UNETR	2D	default
Swin UNETR	3D	Swin UNETR-3D

models were trained for at most 500 epochs, and early stopping was employed based on the validation loss with a patience of 30 epochs. Following Isensee et al. [8], we used the compound loss of the commonly used cross-entropy loss for classification tasks and the Dice loss, which is averaged over all classes C, for training:

$$\mathcal{L}_{\text{total}} = \frac{1}{|C|} \sum_{c \in C} \mathcal{L}^c_{\text{Dice}} + \mathcal{L}_{\text{CE}},$$

where

$$\mathcal{L}^c_{\text{Dice}} = -\frac{2 \sum_i p^c_i g^c_i}{\sum_i p^c_i + \sum_i g^c_i}$$

and

$$\mathcal{L}_{\text{CE}} = -\sum_{c \in C} \sum_i g^c_i \log(p^c_i).$$

Here, p^c_i denotes the predicted probability of class c for voxel i and $g^c_i \in \{0, 1\}$ the corresponding binary ground truth label, i.e. $g^c_i = 1$ if voxel i belongs to class c and $g^c_i = 0$ otherwise.

For submission to the challenge, we used the best-performing model from the validation set and additionally post-processed the predictions by keeping only the largest connected component for each class.

2.4 Implementation Details

We integrated all models, preprocessing, and post-processing steps into the nnU-Net framework [8] to ensure a fair comparison in the model setup and a consistent training and preprocessing pipeline. The framework is implemented in Python and uses PyTorch as the main deep learning library. All models were trained on an NVIDIA Tesla V100 32 GB GPU.

2.5 Evaluation Metrics

To assess the performance of our models, we employ two widely recognised metrics in the field of medical image segmentation: the Dice Similarity Score (DSC) and the Hausdorff Distance (HD). Given the multi-class nature of our task, the metrics are computed on a per-class basis and then averaged across all classes. Formally, for a given class c, the DSC is defined as

$$\mathrm{DSC}_c = \frac{2|P_c \cap G_c|}{|P_c| + |G_c|},$$

where P_c and G_c are the predicted and ground truth binary voxels for class c, respectively. Essentially, the DSC measures the overlap between the predicted and ground truth segmentation masks. The HD, on the other hand, is defined as

$$\mathrm{HD}_c = \max\left(\max_{p\in P_c}\min_{g\in G_c} d(p,g), \max_{g\in G_c}\min_{p\in P_c} d(p,g)\right),$$

where $d(p,g)$ is the Euclidean distance between predicted voxel p and ground truth voxel g. The HD measures the maximum distance between the predicted and ground truth segmentation masks. Instead of reporting the HD directly, we report the 95th percentile of the HD to mitigate the effect of outliers.

3 Results

The quantitative results for all models on our internal validation set are presented in Tables 3 and 4 for Z-score and CLAHE normalisation, respectively. The training process for U-MambaEnc suffered from numerical instability and did not converge. Therefore, we exclude the corresponding results from the analysis.

We observe that in case of applying CLAHE normalisation, the performance of the 3D models is generally better than the 2D models with respect to DSC. However, this is not always the case for HD95, especially for the UNETR and Swin UNETR models. For example, the 3D Swin UNETR model has a significantly higher HD95 for every class with a mean of 18.51 compared to the 2D configuration with a mean of 8.93. In contrast, when considering Z-Score normalisation, the 3D models generally perform worse in terms of HD95 compared to the 2D models, while no clear trend can be observed for the DSC.

Comparing the individual classes, we detect the highest DSC for the LA, generally closely followed by the RA. The atrial wall has the lowest DSC for all models. In contrast, the models tend to achieve a slightly better HD95 for the atrial wall compared to the atrial cavities.

The best performing model in terms of the mean DSC is the 3D MambaBot model with a score of 0.8497 using Z-Score normalisation, followed by the 3D nnU-Net ResEnc and 3D MambaBot model with a DSC of 0.8485 each using CLAHE normalisation. However, since the 3D nnU-Net ResEnc achieves the best mean HD95 of 4.09, compared to 5.04 and 4.39 for the 3D MambaBot

Table 3. Segmentation results when applying Z-score normalisation. The best results are highlighted in **bold** and the second best results are underlined. Higher values are better for the DSC, and lower values are better for the HD.

Model	Config	Wall		RA		LA		Mean	
		DSC	HD95	DSC	HD95	DSC	HD95	DSC	HD95
nnU-Net	2D	0.6936	3.89	0.9098	7.30	0.9162	4.76	0.8399	5.31
nnU-Net	3D	0.6989	4.94	0.8812	7.97	0.9198	**4.37**	0.8333	5.76
nnU-Net ResEnc	2D	0.6775	4.42	0.8986	6.18	0.9108	4.62	0.8289	5.07
nnU-Net ResEnc	3D	0.7012	3.84	**0.9142**	**4.18**	0.9230	7.95	0.8461	5.32
MambaBot	2D	0.7002	4.06	0.9072	6.19	0.9187	4.88	0.8420	**5.04**
MambaBot	3D	**0.7124**	**3.21**	0.9136	4.44	**0.9232**	11.52	**0.8497**	6.39
MambaEnc	2D	0.6048	6.337	0.88352	9.48	0.8843	7.46	0.7748	7.77
UNETR	2D	0.5812	16.36	0.7753	32.30	0.8334	14.21	0.7300	20.96
UNETR	3D	0.5516	40.90	0.7670	75.12	0.8453	46.89	0.7212	54.30
Swin UNETR	2D	0.6428	6.06	0.8696	8.37	0.8932	9.65	0.8019	8.03
Swin UNETR	3D	0.6445	16.40	0.8616	18.89	0.8859	20.42	0.7973	18.57

models, we submitted this model for the final test phase. To further boost the performance, we additionally applied the post-processing steps of the nnU-Net framework [8] and kept only the largest connected component for each class. The submitted model achieved the third place in the MBAS challenge. Both transformer-based models, UNETR and Swin UNETR, performed remarkably worse than the other models.

Further to the quantitative metrics, we also provide qualitative results in Fig. 3 by visualising the predicted segmentation masks. Due to space constraints, we only include the results for a selected subset of models when applying CLAHE normalisation for a randomly selected input volume. While the segmentation mask of the 3D nnU-Net ResEnc and 3D MambaBot models closely resemble the ground truth, the transformer-based models show a few failure cases. For example, in slice 20 the 3D Swin UNETR model fails to segment the fine details of the atrial wall. Even more severe, the 2D UNETR model predicts areas of the LA to be part of the RA in slice 31. Additionally, its predictions are generally less smooth and accurate compared to the other models.

4 Discussion

One notable observation in our study is that the performance across the considered models remain less sensitive to the preprocessing technique used. While the DSC tends to be higher when applying Z-Score normalisation, HD95 is lower when using CLAHE normalisation. This indicates that enhancing the contrast in low-contrast MRI scans does not necessarily improve the segmentation per-

Table 4. Segmentation results when applying CLAHE normalisation. The best results are highlighted in **bold** and the second best results are underlined. Higher values are better for the DSC, and lower values are better for the HD.

Model	Config	Wall		RA		LA		Mean	
		DSC	HD95	DSC	HD95	DSC	HD95	DSC	HD95
nnU-Net	2D	0.6853	4.16	0.8975	5.62	0.9116	5.29	0.8315	5.02
nnU-Net	3D	0.6997	23.13	0.9059	15.91	0.9112	23.38	0.8389	20.80
nnU-Net ResEnc	2D	0.6863	3.80	**0.9139**	6.06	0.9180	4.57	0.8394	4.81
nnU-Net ResEnc	3D	0.7093	**3.22**	0.9129	**4.62**	**0.9232**	**4.42**	**0.8485**	**4.09**
MambaBot	2D	0.6900	3.98	0.9130	5.13	0.9133	5.08	0.8388	4.73
MambaBot	3D	**0.7119**	3.47	0.9126	4.83	0.9210	4.89	**0.8485**	4.39
MambaEnc	2D	0.6000	7.29	0.8296	9.46	0.8792	11.82	0.7695	9.53
UNETR	2D	0.5630	13.98	0.7563	40.28	0.8011	15.57	0.7068	23.28
UNETR	3D	0.5415	51.34	0.7706	73.82	0.8407	47.94	0.7176	57.70
Swin UNETR	2D	0.6403	6.55	0.8688	10.59	0.8849	9.66	0.7980	8.93
Swin UNETR	3D	0.6509	13.61	0.8589	14.98	0.8964	26.94	0.8021	18.51

formance. Note that this observation is in contrast to the findings of Xiong et al. [25], although they focused on the segmentation of the LA only.

The better performance of the 3D models compared to their 2D counterparts can be attributed to the additional spatial context that is available to the models. The 3D models can leverage the spatial information in the Z-direction, which is especially useful for the lower and upper slices in the MRI stack, where the atria is not as prominent as in the middle slices.

Segmenting the atrial wall proved more challenging than segmenting the atrial chambers. The cause for this is most likely the thin structure of the wall compared to the larger cavities of the LA and RA. For example, the wall of the LA has an average thickness of 1.48 mm [10], such that on average at most three consecutive voxels belong to the wall along its short axis. For all considered architectures, this makes it more difficult for the encoder to extract a resonable feature, whereas the decoder fails to resolve that level of detail.

The competitive performance of the original nnU-Net based on convolutional layers align with the results of Isensee et al. [9]. While the other tested models claim to better capture long-range dependencies via transformers or structured state space sequence models, we do not observe a clear benefit of this compared to using localised convolutions. One reason for the general poor performance of the transformer-based models could be the limited size of the dataset, since transformer tend to require large-scale datasets.

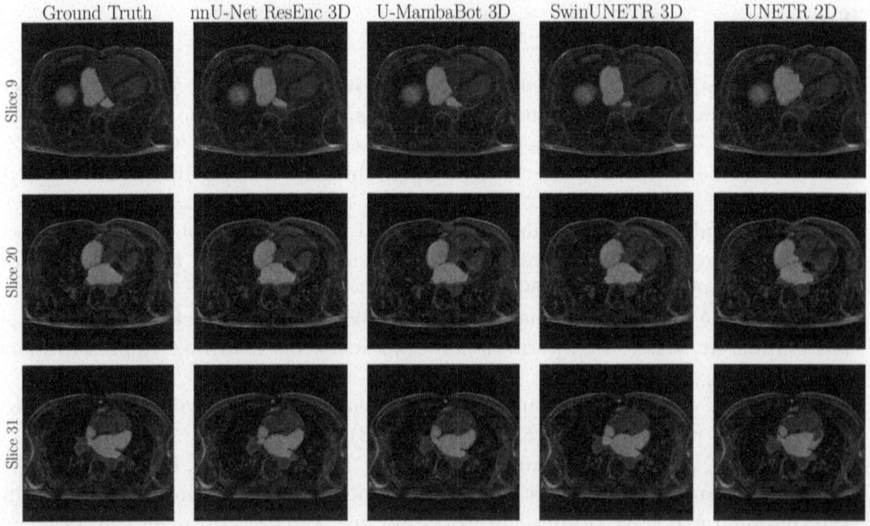

Fig. 3. Segmentation results for a subset of the models for different slices of a randomly selected volume. Green pixels correspond to the LA, orange ones to the RA, and blue ones to the atrial wall. (Color figure online)

5 Conclusion

In this study, we evaluated several U-Net-based models with different architectural encoders, including convolutional layers, attention-based layers, and Mamba blocks, for the segmentation of the left and right atrium as well as for the atrial wall from LGE-MRI scans. We tested each method using two different preprocessing techniques: Z-Score and CLAHE normalisation. Our results indicate that there is no clear benefit of using one preprocessing method over the other. Moreover, despite recent advancements exploring more complex techniques for image segmentation, the 3D nnU-Net based on convolutional layers continues to deliver the best performance. Externally validated on 100 test instances, this model achieved the third rank in the MBAS challenge, demonstrating its promising potential in studying atrial fibrillation and developing effective treatments against it.

Acknowledgments. This study was supported by a DAAD Scholarship to CT, a UK Research and Innovation grant (10110728) to ABO, and a Royal Society University Research Fellowship (URF\R1\221314) to AB. For the purpose of open access, the authors have applied a Creative Commons attribution (CC BY) licence to any Author Accepted Manuscript version arising.

Disclosure of Interests. The authors have no competing interests to declare that are relevant to the content of this article.

References

1. Çiçek, Ö., Abdulkadir, A., Lienkamp, S.S., Brox, T., Ronneberger, O.: 3D U-Net: learning dense volumetric segmentation from sparse annotation. In: Ourselin, S., Joskowicz, L., Sabuncu, M.R., Unal, G., Wells, W. (eds.) MICCAI 2016. LNCS, vol. 9901, pp. 424–432. Springer, Cham (2016). https://doi.org/10.1007/978-3-319-46723-8_49
2. Dasí, A., et al.: In silico trials guide optimal stratification of atrial fibrillation patients to catheter ablation and pharmacological medication: the i-stratification study. Europace **26**(6), euae150 (2024)
3. Dosovitskiy, A., et al.: An image is worth 16x16 words: transformers for image recognition at scale. In: International Conference on Learning Representations (2021)
4. Gu, A., Dao, T.: Mamba: Linear-time sequence modeling with selective state spaces. arXiv preprint: arXiv:2312.00752 (2023)
5. Hatamizadeh, A., Nath, V., Tang, Y., Yang, D., Roth, H.R., Xu, D.: Swin UNETR: swin transformers for semantic segmentation of brain tumors in MRI images. In: International MICCAI Brainlesion workshop, pp. 272–284. Springer (2021)
6. Hatamizadeh, A., et al.: UNETR: transformers for 3D medical image segmentation. In: Proceedings of the IEEE/CVF Winter Conference on Applications of Computer Vision (WACV), pp. 574–584 (2022)
7. Hopman, L.H.G.A., et al.: Right atrial function and fibrosis in relation to successful atrial fibrillation ablation. Eur. Heart J.- Cardiovasc. Imaging **24**(3), 336–345 (2022)
8. Isensee, F., Jaeger, P.F., Kohl, S.A., Petersen, J., Maier-Hein, K.H.: nnU-Net: a self-configuring method for deep learning-based biomedical image segmentation. Nat. Methods **18**(2), 203–211 (2021)
9. Isensee, F., et al.: nnU-Net revisited: a call for rigorous validation in 3D medical image segmentation. arXiv preprint: arXiv:2404.09556 (2024)
10. Li, L., Li, L., Yang, D., Nong, S., Luo, C., Gui, C.: Analysis of the thickness characteristics of the left atrial posterior wall and its correlation with the low and no voltage areas of the left atrial posterior wall in patients with atrial fibrillation. J. Cardiothorac. Surg. **19**(1), 187 (2024)
11. Lippi, G., Sanchis-Gomar, F., Cervellin, G.: Global epidemiology of atrial fibrillation: an increasing epidemic and public health challenge. Int. J. Stroke **16**(2), 217–221 (2021)
12. Liu, W.: ParseNet: Looking wider to see better. arXiv preprint: arXiv:1506.04579 (2015)
13. Liu, Z., et al.: Swin transformer: hierarchical vision transformer using shifted windows. In: Proceedings of the IEEE/CVF International Conference on Computer Vision, pp. 10012–10022 (2021)
14. Loshchilov, I.: Decoupled weight decay regularization. arXiv preprint: arXiv:1711.05101 (2017)
15. Loshchilov, I., Hutter, F.: SGDR: stochastic gradient descent with warm restarts. arXiv preprint: arXiv:1608.03983 (2016)
16. Ma, J., Li, F., Wang, B.: U-Mamba: enhancing long-range dependency for biomedical image segmentation. arXiv preprint: arXiv:2401.04722 (2024)
17. Pizer, S.M., et al.: Adaptive histogram equalization and its variations. Comput. Vis., Graph. Image Process. **39**(3), 355–368 (1987)

18. Ronneberger, O., Fischer, P., Brox, T.: U-Net: convolutional networks for biomedical image segmentation. In: Navab, N., Hornegger, J., Wells, W.M., Frangi, A.F. (eds.) MICCAI 2015. LNCS, vol. 9351, pp. 234–241. Springer, Cham (2015). https://doi.org/10.1007/978-3-319-24574-4_28
19. Sharp, A.J., Betts, T.R., Banerjee, A.: Leveraging 3D atrial geometry for the evaluation of atrial fibrillation: a comprehensive review. J. Clin. Med. **13**(15) (2024)
20. Siebermair, J., Kholmovski, E.G., Marrouche, N.: Assessment of left atrial fibrosis by late gadolinium enhancement magnetic resonance imaging: methodology and clinical implications. JACC: Clin. Electrophysiol. **3**(8), 791–802 (2017)
21. Stimper, V., Bauer, S., Ernstorfer, R., Schölkopf, B., Xian, R.P.: Multidimensional contrast limited adaptive histogram equalization. IEEE Access **7**, 165437–165447 (2019)
22. Takagi, T., et al.: Impact of right atrial structural remodeling on recurrence after ablation for atrial fibrillation. J. Arrhythmia **37**(3), 597–606 (2021)
23. Vesal, S., Ravikumar, N., Maier, A.: Dilated convolutions in neural networks for left atrial segmentation in 3D gadolinium enhanced-MRI. In: Statistical Atlases and Computational Models of the Heart. Atrial Segmentation and LV Quantification Challenges, pp. 319–328. Springer (2019)
24. Xia, Q., Yao, Y., Hu, Z., Hao, A.: Automatic 3D atrial segmentation from GE-MRIs using volumetric fully convolutional networks. In: Statistical Atlases and Computational Models of the Heart. Atrial Segmentation and LV Quantification Challenges, pp. 211–220. Springer (2019)
25. Xiong, Z., et al.: A global benchmark of algorithms for segmenting the left atrium from late gadolinium-enhanced cardiac magnetic resonance imaging. Med. Image Anal. **67**, 101832 (2021)
26. Yang, X., et al.: Combating uncertainty with novel losses for automatic left atrium segmentation. In: Statistical Atlases and Computational Models of the Heart. Atrial Segmentation and LV Quantification Challenges, pp. 246–254. Springer (2019)
27. Zhao, J., et al.: Three-dimensional integrated functional, structural, and computational mapping to define the structural "fingerprints" of heart-specific atrial fibrillation drivers in human heart ex vivo. J. Am. Heart Assoc. **6**(8), e005922 (2017)

On the Foundation Model for Cardiac MRI Reconstruction

Chi Zhang(✉), Michael Loecher, Cagan Alkan, Mahmut Yurt, Shreyas S. Vasanawala, and Daniel B. Ennis

Stanford University, Stanford, CA, USA
zcglhf@stanford.edu

Abstract. In recent years, machine learning (ML) based reconstruction has been widely investigated and employed in cardiac magnetic resonance (CMR) imaging. ML-based reconstructions can deliver clinically acceptable image quality under substantially accelerated scans. ML-based reconstruction, however, also requires substantial data and computational time to train the neural network, which is often optimized for a fixed acceleration rate or image contrast. In practice, imaging parameters are often tuned to best suit the diagnosis, which may differ from the training data. This can result in degraded image quality, and multiple trained networks are needed to fulfill the clinical demands. In this study, we propose a foundation model that uses adaptive unrolling, channel-shifting, and Pattern and Contrast-Prompt-UNet (PCP-UNet) to tackle the problem. In particular, the undersampled data goes through a different number of unrolled iterations according to its acceleration rate. Channel-shifting improves reconstructed data quality. The PCP-UNet is equipped with an image contrast and sampling pattern prompt. *In vivo* CMR experiments were performed using mixed combinations of image contrasts, acceleration rates, and (under)sampling patterns. The proposed foundation model has significantly improved image quality for a wide range of CMR protocols and outperforms the conventional ML-based method.

Keywords: Cardiac MR · Image reconstruction · Machine learning

1 Introduction

Cardiac magnetic resonance (CMR) imaging is a widely applied clinical tool for the diagnosis of various cardiac diseases. CMR exams are time-consuming and research currently focuses on ways to reduce exam times. Most approaches to faster CMR exams rely on undersampling the acquired k-space data and using advanced reconstruction methods to generate images of acceptable clinical quality. In particular, recent advances in machine learning (ML) based reconstruction [1–5,8] have enabled substantial increases in acceleration rates (e.g. 10–20x) for CMR, which extends its availability to patients, as well as enabling advanced CMR techniques [9]. These approaches, however, are generally bespoke for each

O. Camara et al. (Eds.): STACOM 2024, LNCS 15448, pp. 226–235, 2025.
https://doi.org/10.1007/978-3-031-87756-8_23

acquisition protocol. ML-based reconstruction approaches also require substantial amounts of data and compute time to be trained, but once trained provide very fast reconstruction times that outperform conventional methods (e.g. compressed sensing).

Existing ML-based reconstructions can be split into two major categories: Scan-specific and data-driven approaches. In the former, neural networks learn to interpolate k-space from the fully sampled k-space center (i.e. calibration lines) of the image to be reconstructed [5–7]. In the latter, a neural network is trained using a large imaging database [1–4]. Restricted by the limited number of calibration lines, scan-specific approaches are typically forced to employ compact network designs, and perform poorly when reconstructing images acquired with high acceleration rates. In contrast, when sufficient data is available for training, the data-driven approaches leverage more advanced network designs, including transformers [10] and diffusion models [11]. The data-driven approach can pushing the acceleration rates even higher while maintaining satisfactory image quality.

To optimize image quality and minimize the risk of distribution shifts, protocol-specific learning is commonly adapted in data-driven approaches [1,2,9], which requires that the images for used training and inference are matched in image contrast, sampling patterns, and acceleration rates. Any mistmatch between training and inference will typically lead to a significant degradation in image quality [1]. This is a major practical issue of the data-driven approaches because the imaging protocol and parameters are generally refined for each CMR exam. In a typical CMR exam, images of multiple views and contrast are acquired to assist diagnosis. Further adjustments in sampling pattern and acceleration rates may also be needed [12–14]. One could deploy multiple networks matched to all exam possibilities, but this adds substantial complexity and is not generally feasible.

In this work, we propose a universal foundation model for machine learning-based CMR reconstruction that allows reconstruction of mixed contrast, views, sampling patterns, and acceleration rates. To do so we employ an unrolled network [1] architecture to solve a regularized reconstruction problem. We inherit the analytical conclusion from compressed sensing [15], to handle different acceleration rates by using different numbers of unrolled steps. We adapt the Prompt-UNet [16] for mixed image contrasts and views, and extend it with additional sampling embedding, which we refereed as Pattern and Contrast-Prompt-UNet (PCP-UNet) to optimize image quality of different sampling patterns. To better reconstruct various sampling patterns, we also apply a novel channel-shift technique to achieve a wider receptive field in the PCP-UNet with minor computational overhead.

Experiments were conducted using *in vivo* CMR data from the CMRxRecon 2024 challenge. The proposed method outperformed a simple unrolled network with a fixed number of unrolled steps using either UNet or Prompt-UNet. The proposed PCP-UNet outperformed Prompt-UNet using fixed or adaptive unrolls. Statistical results showed improved structural similarity index measure (SSIM) were obtained using PCP-Unet, which is suitable for handling reconstruction of images with a variety of image contrast, views, sampling patterns, and acceleration rates.

2 Methodology

2.1 Regularized Reconstruction and Unrolling

Let $\mathbf{E}_{\Omega}$ be the forward acquisition model associated with sampling pattern $\mathbf{\Omega}$, $\mathbf{y}$ denotes the acquired k-space data, and the regularization term is $\mathcal{R}(\mathbf{x})$. The regularized reconstruction problem is modeled as:

$$\arg\min_{\mathbf{x}} \|\mathbf{E}_{\Omega}\mathbf{x} - \mathbf{y}\|_2^2 + \mathcal{R}(\mathbf{x}) \tag{1}$$

A common approach to solve (1) is model unrolling via variable splitting [1,15], which introduces an auxiliary variable $\mathbf{u}$ to the original problem:

$$\arg\min_{\mathbf{x}} \|\mathbf{E}_{\Omega}\mathbf{x} - \mathbf{y}\|_2^2 + \mathcal{R}(\mathbf{u}) \quad s.t. \quad \mathbf{x} = \mathbf{u} \tag{2}$$

Equation 2 can be solved by iteratively solving the sub-problems:

$$\mathbf{u}^{(i)} = \arg\min_{\mathbf{u}} \mu\|\mathbf{x}^{(i-1)} - \mathbf{u}\|_2^2 + \mathcal{R}(\mathbf{u}) \tag{3}$$

$$\mathbf{x}^{(i)} = \arg\min_{\mathbf{x}} \|\mathbf{E}_{\Omega}\mathbf{x} - \mathbf{y}\|_2^2 + \mu\|\mathbf{u}^{(i)} - \mathbf{x}\|_2^2 \tag{4}$$

where Eq. 3 is often solved using a neural network [1]. Equation 4 is the data-consistency (DC) term, which is solved via *e.g.* gradient descent [2,9,16] or the conjugate gradient method [1].

2.2 Adaptive Unrolling

The acceleration rate impacts the condition number of $\mathbf{E}_{\Omega}$. This suggests the number of unrolled iterations (UI) required for convergence is different for different acceleration rates. This is also implied by previous works, where highly accelerated acquisitions were reconstructed using more UI to improve image quality [16]. To reconstruct images for a wide range of acceleration rates, we propose to use adaptive unrolling that performs fewer (more) UI for relatively low (high) acceleration rates. Figure 1 depicts the proposed unrolled network. Specifically, our unrolled cascade is implemented using up to sixteeen UI. Each UI has its own neural network and conjugate gradient solver, whose parameters are independent from the other UI. An input image enters the unrolled cascade from a specific UI according to its acceleration rate. As such, images with the highest acceleration rate go through all sixteen UI, and images with lower acceleration rates will be reconstructed via the latter UI only. As a result, each UI has its own neural network and conjugate gradient solver optimized for inputs with specific level of artifacts.

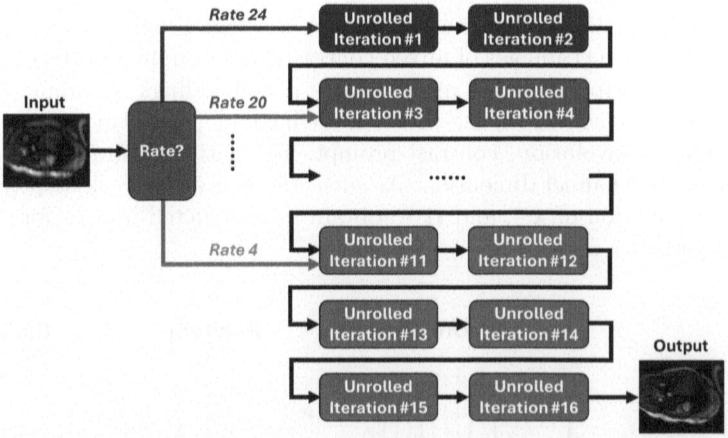

Fig. 1. Adaptive unrolling. Undersampled measurements are fed to a specific unrolled iteration according to the acceleration rate. Images of higher acceleration rate proceed through more unrolled iterations than those of lower acceleration rates. Each unrolled iteration has its own network regularizer and conjugate gradient with parameters independently determined relative to the other unrolled iterations.

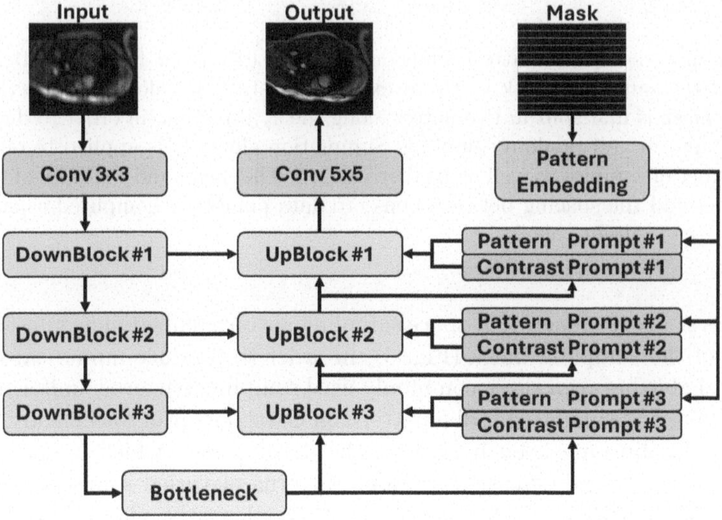

Fig. 2. Pattern and Contrast-Prompt (PCP) UNet. PCP-UNet inherits the Prompt-UNet concept with additional k-space undersampling pattern prompt modules that are similar to the contrast prompt modules in Prompt-UNet.

2.3 Pattern and Contrast-Prompt (PCP) UNet

Based on Prompt-UNet [16], we extend the contrast prompt with a pattern-prompt to reconstruct images of mixed contrasts and sampling patterns (Fig. 2). The pattern prompt modules receive pattern embeddings as input, followed by attention and convolutions, which is defined in [16]. Feature maps from image domain convolution, contrast-prompt, and pattern-prompt are concatenated along the channel dimension. As such, the neural network step (Eq. 3) is solved by a function of $\mathbf{x}^{(i)}$ and $\boldsymbol{\Omega}$ to obtain reconstructed images for different sampling patterns.

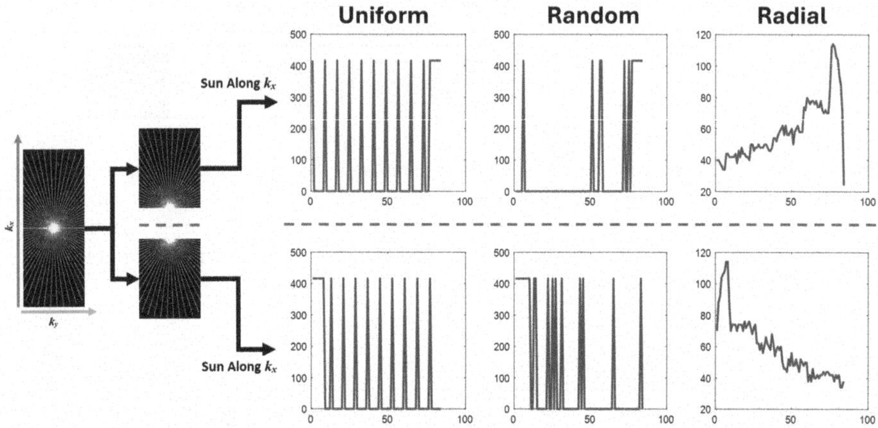

Fig. 3. k-space sampling pattern embeddings were estimated from statistical information of the sampling mask, with an example of statistics along k_x direction. The sampling mask is first split in two halves along the k_x direction, in order to distinguish between uniform and random sampling. Summation along k_x is applied to obtain the total number of samples as well as its distribution. The mean and variance of the sample number and the spacing between non-zero data points are computed as statistical feature of the sampling mask.

In this study we generate k-space sampling pattern embeddings via statistical features of the sampling mask (Fig. 3), in order to produce numerical features distinguishable between three commonly used sampling patterns: uniform, Gaussian random, and simulated radial pattern on Cartesian grid. Specifically, we first split the sampling mask by half along the k_x direction, which is intentionally aimed to differentiate Gaussian random patterns of varied sampling densities. Then we accumulate the total sample numbers along the phase and frequency encoding directions of each half of the sampling mask. This is important for distinguishing between uniform and random sampling. We further compute the statistics of the spacing between samples along each direction. These embedded features can be used to distinguish between sampling patterns and also reflects the sampling density of the sampling mask.

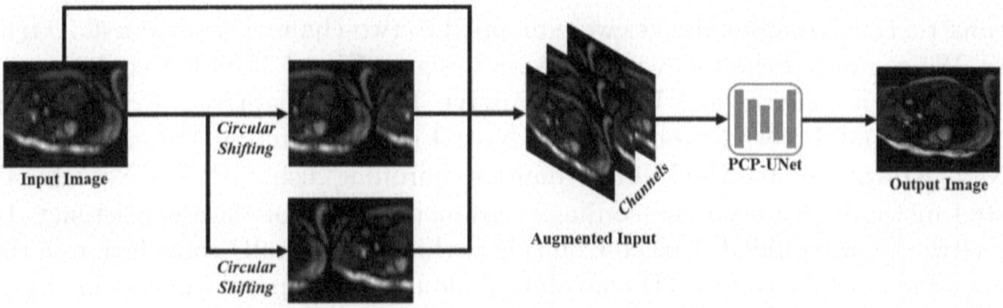

Fig. 4. Enlarging receptive field via channel-shifting. Circular shifting along the subsampled direction(s) is performed to produce shifted replicas of the input image. The replicas are subsequently concatenated with the input along channel dimension to produce an augmented input, which is fed into a regular CNN with additional channels in its input layer.

2.4 Enlarging Receptive Field via Channel-Shift

Various sampling patterns result in different point-spread functions (PSF) in the image domain. Notably, the PSF can span the entire FOV due to the presence of a fully sampled center, as well as unevenly distributed sampling positions. This indicates a global receptive field is necessary to better distinguish between PSFs of different sampling patterns, which can be leveraged to improve image reconstruction quality. To efficiently enlarge the receptive field, we propose to augment the input image with circular-shifted replicas, which are then concatenated as additional image channels. The augmented input is fed into PCP-UNet with additional input channels (Fig. 4). With a sufficient number of shifted replicas, this channel-shifting can enable having aglobal receptive field, while accepting arbitrary input sizes. Channel-shifting has a minor computational overhead equivalent to adding an additional convolutional layer with multiple channels. Channel-shifting has no additional memory consumption when the number of input channel is not greater than the maximum number of channels in the hidden layers.

2.5 *In Vivo* CMR Dataset and Experiment Details

We performed our experiments using the CMRxRecon dataset [17], which contains fully sampled cine, aorta, tagging, and mapping (T1 and T2) CMR images from various views including short-axis, two, three, four-chambers, and long-axis views. Typically 5–10 slices and 12–25 cardiac phases were acquired. We split multi-slice acquisitions into single-slice 2D+time (2D+t) images and re-assembled the multiple slice after reconstruction. We used 150,480 2D+t images for training, and 4,158 images for validation.

Each fully sampled image was retrospectively undersampled using uniform, Gaussian random and pseudo radial patterns at undersampling rates of 4, 8, 12, 16, 20 and 24. Coil sensitivity profiles were estimated using ESPIRiT [18]. In reconstruction, complex images were mapped to two-channel real-valued 3D tensors. We compared the following network designs: **Fixed UNet**: Fixed amount of UI using a conventional UNet; **Adaptive UNet**: Adaptive unrolling using a conventional UNet; **Fixed PCP-UNet**: Fixed amount of UP using PCP-UNet; **Adaptive PCP-UNet**: Adaptive unrolling using PCP-UNet. In all tested methods, we used the conjugate gradient solver for data consistency. In all networks, we applied 2.5D convolutions, which perform 2D convolution in the image domain followed by 1D convolution along the temporal dimension.

3 Results

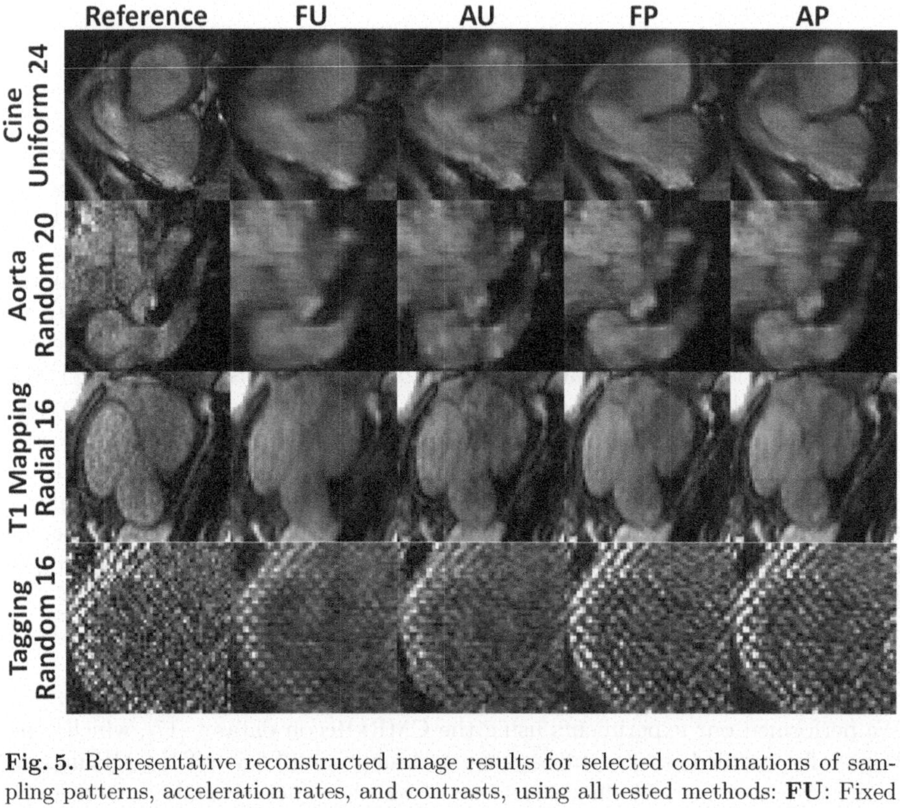

Fig. 5. Representative reconstructed image results for selected combinations of sampling patterns, acceleration rates, and contrasts, using all tested methods: **FU**: Fixed amount of UI with a UNet regularizer; **AU**: Adaptive amount of UI with a UNet regularizer; **FP**: Fixed amount of UI with the PCP-UNet regularizer; **AP**: Adaptive amount of UI with the PCP-UNet regularizer.

For evaluation, we computed SSIM of single-slice 2D+t images with respect to fully sampled references, over a cropped central area covering 1/2 of k_x and 2/3 of k_y to remove background signals.

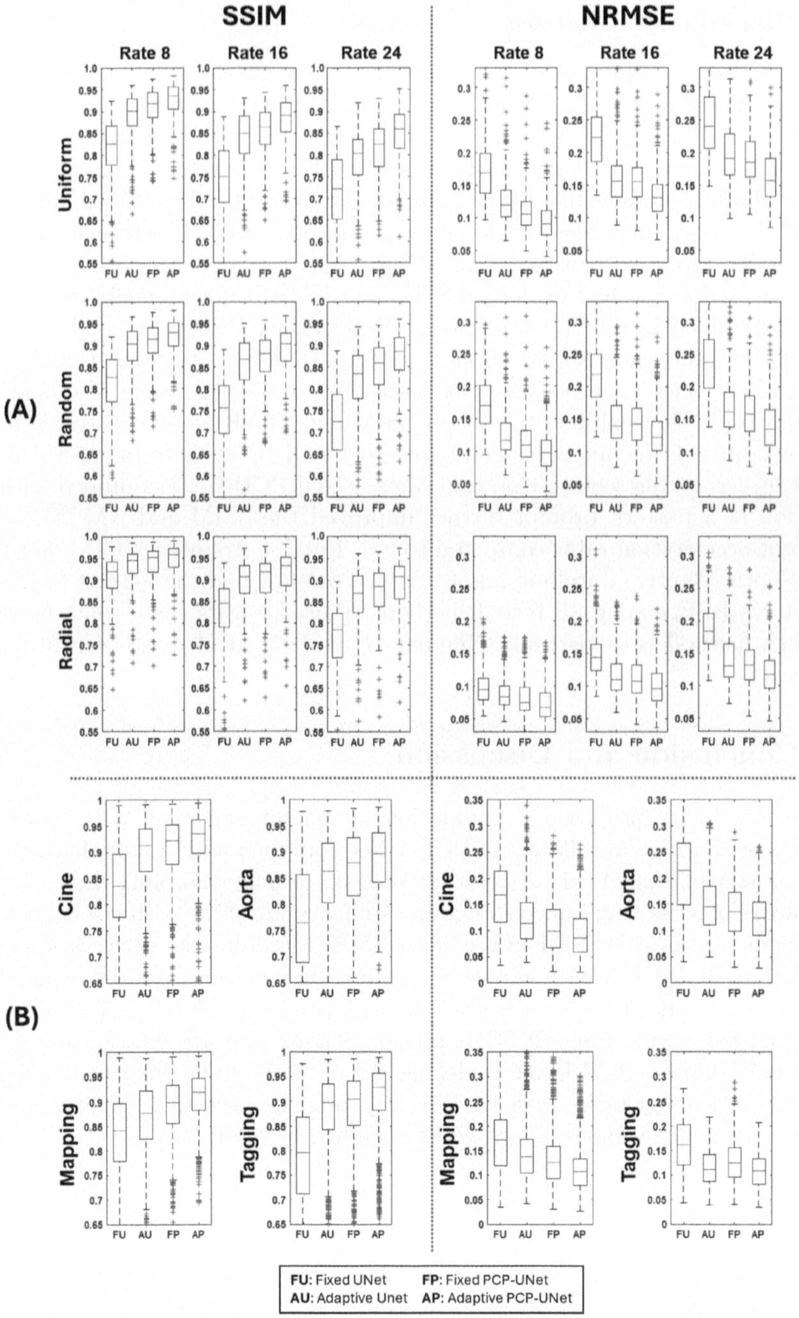

Fig. 6. SSIM and NMRSE statistics for each reconstruction method (Fixed UNet, Adaptive UNet, Fixed PCP-UNet and Adaptive PCP-UNet) with respect to: (A) the k-space undersampling rate (8x, 16x, or 24x) and sampling pattern (uniform, random and radial), and (B) image contrast. AP generally outperforms (higher SSIM, lower NMRSE) the other methods.

3.1 Experimental Results

Figure 5 depicts representative reconstructed images for each reconstruction method, with selected combinations of sampling patterns, rates, and contrasts. In all cases, Fixed UNet shows visual blurring and aliasing artifacts. Adaptive UNet shows increased sharpness compared to Fixed UNet, but it still exhibits visible aliasing artifacts due to the high acceleration rate. Both Fixed and Adaptive PCP-UNet shows less visible aliasing. Adaptive PCP-UNet better recovers fine details compared to Fixed PCP-UNet.

Figure 6 depicts the boxplots of SSIM and NMRSE for all tested reconstruction methods (Fixed UNet, Adaptive UNet, Fixed PCP-UNet and Adaptive PCP-UNet) and each undersampling pattern (uniform, random, radial) for a range of acceleration rates (8x, 16x, and 24x), and for a range of image contrasts. Fixed UNet exhibits the lowest SSIM and PSNR among all tested methods. Adaptive UNet improves compared to Fixed UNet in terms of SSIM and NMRSE across different acceleration rates. Fixed PCP-UNet equipped with the addition of a pattern prompt further improved the SSIM and NMRSE across different acceleration rates compared to FU. Finally, Adaptive PCP-UNet combined both adaptive unrolling and a pattern prompt had the highest SSIM and lowest NMRSE among all tested methods. Paired t-tests show that the SSIM for each method is statistically different ($P = 0.05$) at all patterns, rates, and contrasts.

4 Conclusion and Discussion

In this study we proposed a foundation model for general CMR reconstruction using adaptive unrolling and PCP-UNet equipped with a new channel-shift technique. We applied this to a variety of k-space undersampling rates, k-space sampling patterns, and image contrasts. Adaptive unrolling with a regular UNet significantly improved SSIM and reduced NMRSE for different acceleration rates. Employing PCP-UNet with adaptive unrolling further improved the image quality across different image contrasts. Our Adaptive PCP-UNet approach works very well for a wide range if CMR image contrast and acceleration rates. The proposed Adaptive PCP-UNet model has the potential to be fine-tuned for a specific reconstruction task. Transfer learning could be used to apply our method to other image anatomies, which will be studied in the future.

References

1. Hammernik, K., et al.: Learning a variational network for reconstruction of accelerated MRI data. Magn. Reson. Med. **79**(6), 3055–3071 (2018)
2. Aggarwal, H.K., Mani, M.P., Jacob, M.: MoDL: model-based deep learning architecture for inverse problems. IEEE Trans. Med. Imaging **38**(2), 394–405 (2018)
3. Knoll, F., et al.: Deep-learning methods for parallel magnetic resonance imaging reconstruction: a survey of the current approaches, trends, and issues. IEEE Signal Process. Mag. **37**(1), 128–140 (2020)

4. Schlemper, J., Caballero, J., Hajnal, J.V., Price, A.N., Rueckert, D.: A deep cascade of convolutional neural networks for dynamic MR image reconstruction. IEEE Trans. Med. Imaging **37**(2), 491–503 (2017)
5. AkçSakaya, M., Moeller, S., Weingärtner, S., Uğurbil, K.: Scan-specific robust artificial-neural-networks for k-space interpolation (RAKI) reconstruction: database-free deep learning for fast imaging. Magn. Reson. Med. **81**(1), 439–453 (2019)
6. Hosseini, S., et al.: Accelerated coronary MRI with sRAKI: a database-free self-consistent neural network k-space reconstruction for arbitrary undersampling. PLoS ONE **15**(2), e0229418 (2020)
7. Zhang, C., Moeller, S., Demirel, O.B., Uğurbil, K., Akçakaya, M.: Residual RAKI: a hybrid linear and non-linear approach for scan-specific k-space deep learning. Neuroimage **256**, 119248 (2022)
8. Oscanoa, J.A., et al.: Deep learning-based reconstruction for cardiac MRI: a review. Bioengineering **10**(3), 334 (2023)
9. Oscanoa, J.A., Middione, M.J., Syed, A.B., Sandino, C.M., Vasanawala, S.S., Ennis, D.B.: Accelerated two-dimensional phase-contrast for cardiovascular MRI using deep learning-based reconstruction with complex difference estimation. Magn. Reson. Med. **89**(1), 356–369 (2023)
10. Huang, J., et al.: Swin transformer for fast MRI. Neurocomputing **493**, 281–304 (2022)
11. Chung, H., Ye, J.C.: Score-based diffusion models for accelerated MRI. Med. Image Anal. **80**, 102479 (2022)
12. Rajiah, P.S., François, C.J., Leiner, T.: Cardiac MRI: state of the art. Radiology **307**(3), e223008 (2023)
13. Curtis, A.D., Cheng, H.: Primer and historical review on rapid cardiac CINE MRI. J. Magn. Reson. Imaging **55**(2), 373–388 (2022)
14. Lotz, J., Meier, C., Leppert, A., Galanski, M.: Cardiovascular flow measurement with phase-contrast MR imaging: basic facts and implementation. Radiographics **22**(3), 651–671 (2002)
15. Lustig, M., Donoho, D., Pauly, J.M.: Sparse MRI: the application of compressed sensing for rapid MR imaging. Magn. Reson. Med. Official J. Int. Soc. Magn. Reson. Med. **58**(6), 1182–1195 (2007)
16. Xin, B., Ye, M., Axel, L., Metaxas, D.N.: Fill the k-space and refine the image: prompting for dynamic and multi-contrast MRI reconstruction. In: International Workshop on Statistical Atlases and Computational Models of the Heart, pp. 261–273. Springer Nature Switzerland, Cham (2023)
17. Wang, C., Lyu, J., Wang, S.: CMRxRecon: an open cardiac MRI dataset for the competition of accelerated image reconstruction. arXiv preprint: arXiv:2309.10836 2023
18. Uecker, M., et al.: ESPIRiT-an eigenvalue approach to autocalibrating parallel MRI: where SENSE meets GRAPPA. Magn. Reson. Med. **71**(3), 990–1001 (2014)

Multi-loss 3D Segmentation for Enhanced Bi-atrial Segmentation

Enrique Almar-Munoz[1(✉)], Markus Tiefenthaler[1], Nohemi Sofia Leon Contreras[2], Adria Aguilar-Minguez[3], Christian Kremser[1], Markus Haltmeier[4], and Agnes Mayr[1]

[1] Department of Radiology, Medical University of Innsbruck, Anichstrasse 35, 6020 Innsbruck, Tirol, Austria
enrique.almar@i-med.ac.at
[2] Pupil Labs GmbH, Sanderstrasse 28, 12047 Berlin, Germany
[3] Department of Computer Engineering, Polytechnic University of Valencia, Cami de Vera, 46022 Valencia, Valencia, Spain
[4] Department of Mathematics, University of Innsbruck, Technikerstrasse 13, 6020 Innsbruck, Tirol, Austria

Abstract. Introduction: Atrial fibrillation (AF) treatment is often impeded by limited anatomical insights. Late Gadolinium-Enhanced MRI (LGE-MRI) has emerged as a valuable tool for predicting the success of ablation procedures. In this study, we present a deep learning approach for enhanced segmenting atrial chambers and walls using 3D LGE-MRI, developed as part of our participation in the MICCAI Multi-class Bi-Atrial Segmentation (MBAS) 2024 Challenge.

Methods: We employed a preprocessing pipeline to standardize intensity values and enhance image contrast. Multiple state-of-the-art models were utilized for segmentation: both 2D and 3D nnU-Net architectures, Swin-UNETR, MedNeXt, and U-Mamba. To optimize segmentation performance, we experiment with various loss functions, including Dice Loss, TopK Loss, Boundary Loss, and Hausdorff Distance Loss.

Results: In the validation phase of the challenge, our approach achieved the second-place ranking. Utilizing a 3D nnU-Net architecture trained with a Dice Loss function, we obtained Dice Scores of 66.55% for the Atrial Wall, 88.25% for the Right Atrium, and 91.14% for the Left Atrium.

Conclusion: The MedNext-B model achieved the best performance on our internal validation set using a combined loss function incorporating DC, HD95, TopK, CE, and Boundary losses. However, the difference in performance compared to using only DC and CE losses was not statistically significant. To meet the inference time requirements, we used data cropping.

Keywords: Cardiac Atria · Deep Learning · Late Gadolinium-Enhanced MRI

O. Camara et al. (Eds.): STACOM 2024, LNCS 15448, pp. 236–244, 2025.
https://doi.org/10.1007/978-3-031-87756-8_24

1 Introduction

Atrial fibrillation (AF) represents the most prevalent sustained cardiac arrhythmia globally, impacting millions of individuals and posing significant health risks such as stroke, heart failure, and other cardiovascular complications [1]. Characterized by rapid and erratic atrial contractions, AF is associated with the formation of blood clots and subsequent thromboembolic events [2]. Despite advancements in treatment modalities—including pharmacotherapy, cardioversion, and catheter ablation—current strategies remain suboptimal, largely due to a limited understanding of the atrial structural abnormalities that perpetuate the arrhythmia [3].

Late gadolinium enhancement magnetic resonance imaging (LGE-MRI) has emerged as a pivotal diagnostic tool for visualizing atrial fibrosis and scarring, which are critical contributors to the pathophysiology of AF [4]. The quantification and spatial distribution of atrial fibrosis, as imaged via LGE-MRI, are key determinants of the success of catheter ablation procedures. Accurate delineation of atrial anatomy and fibrotic tissue distribution can significantly influence patient selection for ablation, optimize procedural strategies, and enhance both risk stratification and monitoring of treatment outcomes [5].

This paper presents our approach to the MICCAI Multi-class Bi-Atrial Segmentation 2024 Challenge. By using the largest LGE-MRI dataset, the objective of the challenge is to achieve the highest possible segmentation accuracy, which is crucial for selecting patients who are likely to benefit from catheter ablation, planning more effective ablation strategies, enhancing risk assessment, and monitoring treatment outcomes over time.

2 Methods

2.1 Dataset

We utilized exclusively the bi-atrial LGE-MRI dataset provided by the challenge, which comprises a total of 70 3D LGE-MRI scans for training, 30 scans for validation, and 100 scans for the final testing phase. Due to the unavailability of ground truth data for the validation and test sets, we partitioned the training dataset into two subsets: 60 scans for training and 10 for internal validation. After identifying the optimal model through this internal validation process, the final model was trained using the complete set of 70 training scans.

The training and validation sets consist of 43 scans with dimensions of $608 \times 608 \times 44$ pixels and 57 scans with dimensions of $640 \times 640 \times 44$ pixels. The grayscale intensity values range from 0 to 4095. The average grayscale levels for the training set is 103.72 ± 158.82.

Figure 1 illustrates the three views of MRI scans with the corresponding ground truth labels, as well as a 3D rendering of these labels. Among the training cases, background voxels account for $98.24 \pm 0.585\%$ of the total. For the non-background voxels, the aortic walls constitute $24.18 \pm 3.61\%$, the left atrium $36.53 \pm 5.32\%$, and the right atrium $39.29 \pm 4.99\%$. There is a significant class imbalance between background and non-background voxels. However, among the non-background classes (1–3), the differences in voxel populations are small.

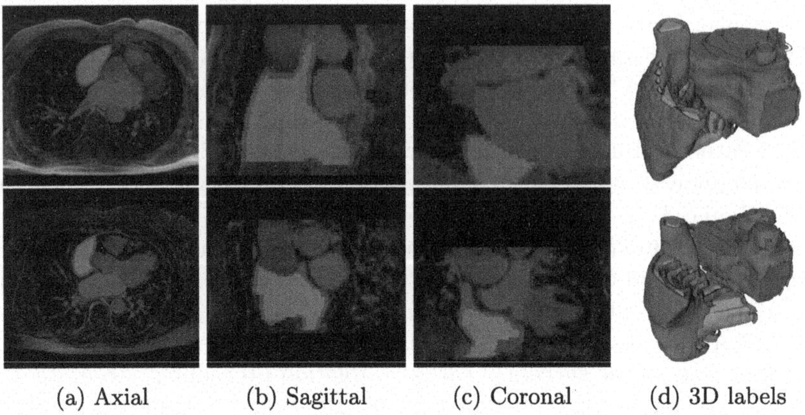

Fig. 1. Overlay of manual labels on MRI views. Red denotes the Atrium Walls, green indicates the Right Atrium Cavity, and blue shows the Left Atrium Cavity. (Color figure online)

2.2 Overview of the Methodology

Initially, we evaluated a range of state-of-the-art 2D and 3D segmentation models to identify the architecture that delivered the best performance. After selecting the optimal model, we explored various loss functions and their combinations to further enhance segmentation accuracy. Latly, to address any potential limitations in inference time, as specified by the challenge constraints, we investigated strategies such as data cropping and preprocessing techniques to optimize computational efficiency.

2.3 Pre-processing

Image preprocessing was performed to standardize intensity values and enhance contrast across the dataset. Given that the intensity values spanned a range from 0 to 4095, with a mean intensity of 101.15 indicative of a long-tailed distribution, contrast enhancement using Contrast Limited Adaptive Histogram Equalization (CLAHE) was applied. Following CLAHE, the intensity values were normalized to a scale of 0 to 256. This systematic preprocessing was applied uniformly to the entire dataset, ensuring consistency in data quality.

Figure 2 shows the difference in image visualization and histogram before and after applying the proposed preprocessing.

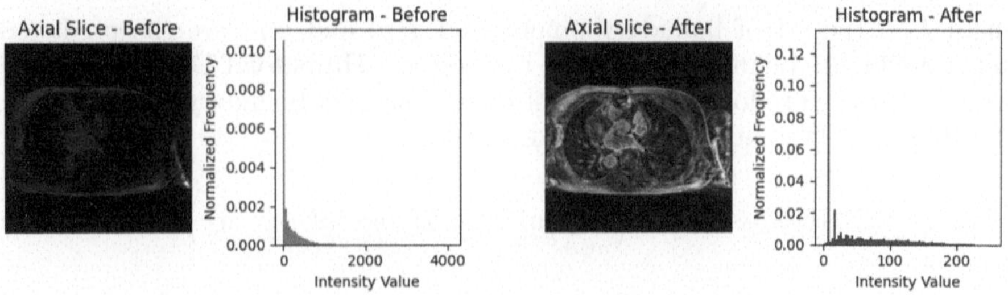

Fig. 2. Effect of the preprocessing on image appearance and intensity distribution. The left panels display the central axial slice and histogram before preprocessing, while the right panels present the same slice and histogram after it.

2.4 Models

In this study, several advanced deep-learning architectures were employed. The **2D and 3D nnU-Net** models were utilized due to their robust performance in medical image segmentation tasks. The nnU-Net automatically configures the U-Net architecture by determining the optimal hyperparameters and preprocessing steps based on the dataset [6]. The **Swin-UNETR** is a transformer-based model that integrates Swin Transformers as the encoder with a UNet-like decoder, leveraging self-attention mechanisms to capture long-range dependencies in volumetric medical images [7]. **MedNeXt** introduces a novel approach by combining convolutional neural networks (CNNs) with transformer blocks, creating a hybrid architecture that aims to benefit from both local feature extraction and global context understanding [9]. Due to the time constraint, we will use the little version of it. Finally, **U-Mamba** is a deep learning model designed for multi-scale medical image analysis, incorporating a U-Net backbone with multi-scale attention modules to enhance feature extraction at different scales [11].

After identifying the best model for our configuration, we experimented with various loss functions to enhance its performance. Specifically, we tested **Dice Loss**, which is particularly effective for handling class imbalances. We also evaluated **TopK Loss**, which focuses on penalizing incorrect predictions, aiming to refine the model's sensitivity to the most critical regions. The TopK Loss is given by Eq. 1:

$$\text{TopK Loss} = \frac{1}{N}\sum_{i=1}^{N} \max\left(0, 1 - \text{softmax}(y_i)\right) \tag{1}$$

where N is the number of top predictions considered, and y_i represents the scores for each prediction. **Boundary Loss** was also incorporated to address the accuracy of object boundaries, enhancing the model's ability to delineate the atrial walls more precisely. Boundary Loss can be expressed by Eq. 2:

$$\text{Boundary Loss} = \frac{1}{|P|}\sum_{p\in P} \left\|\nabla f(p) - \nabla g(p)\right\|^2 \tag{2}$$

where P is the set of boundary points, and $f(p)$ and $g(p)$ are the predicted and ground truth boundary gradients, respectively. **Hausdorff Distance (HD) Loss** was utilized to focus on the worst-case scenario of boundary discrepancies. The Hausdorff Distance Loss is provided in Eq. 3:

$$\text{HD Loss} = \max\left(\sup_{a\in A}\inf_{b\in B}\|a-b\|, \sup_{b\in B}\inf_{a\in A}\|a-b\|\right) \tag{3}$$

where A and B are the sets of boundary points from the predicted and ground truth segments, respectively, and $\|\cdot\|$ denotes the Euclidean distance.

3 Training

We trained the models using an NVIDIA A40 GPU and an NVIDIA A100 GPU. The patch size was set to $20 \times 320 \times 256$ in the 3D configuration and 512×512 in the 2D configurations. The batch size was always set to 2. We utilized the AdamW optimizer with an initial learning rate of 1e-3. The model was trained for $1,000$ epochs using the best configuration.

4 Results

4.1 Segmentation Results Employing Different Models

Table 1 shows the results of different network architectures in the internal validation set. The two challenge metrics (Dice Similarity Coefficient, DSC; 5% Hausdorff Distance, HD95) and Volume Similarity (VS) are compared in the table for the different segmentation classes.

Among atrial wall (AW), right atrium cavity (RAC), and left atrium cavity (LAC) structures, UMambaBot and MedNext-B consistently outperform other models in terms of DSC and HD95, with MedNext-B demonstrating the best HD95 for AW and LAC. Volume Similarity (VS) values are generally negative or close to zero, with minimal variation between models. The SwinUNETR model performs significantly worse in HD95, especially for the atrial wall and cavities, with values over 100, suggesting less precise boundary delineation.

Figure 3 shows the 3D render of the given groundtruth and the output of the outperforming model.

4.2 Segmentation Results Employing Different Loss Functions

Table 2 presents a comparison of performance metrics for various combinations of loss functions. The goal of using different combinations of these loss functions is to assess their impact on the accuracy and boundary delineation of the model's segmentation results.

Table 1. Performance metrics for various models on bi-atrial segmentation. Metrics include Dice Similarity Coefficient (DSC), 95% Hausdorff Distance (HD95), and Volume Similarity (VS). The abbreviations are as follows: RAC denotes Right Atrium Cavity, LAC denotes Left Atrium Cavity, and AW denotes Atriums Wall. The best results for each metric and class are highlighted in bold.

Model	DSC	HD95	VS
nnUNet2D	**0.996** (±0.002)	3.810 (±1.710)	
AW	0.702 (±0.054)	3.074 (±1.803)	−0.043 (±0.078)
RAC	0.903 (±0.040)	3.645 (±2.148)	−0.090 (±0.143)
LAC	0.936 (±0.012)	2.413 (±0.616)	−0.023 (±0.041)
nnUNet3D	**0.996** (±0.002)	3.932 (±1.386)	
AW	0.715 (±0.044)	2.911 (±1.351)	0.006 (±0.064)
RAC	0.912 (±0.014)	3.631 (±1.421)	−0.078 (±0.096)
LAC	0.931 (±0.022)	3.019 (±1.920)	−0.043 (±0.064)
SwinUNETR	0.991 (±0.007)	133.242 (±90.854)	
AW	0.597 (±0.069)	66.368 (±79.355)	0.033 (±0.205)
RAC	0.773 (±0.176)	112.963 (±109.062)	−0.070 (±0.210)
LAC	0.837 (±0.108)	93.143 (±92.377)	0.081 (±0.276)
UMambaBot	**0.996** (±0.001)	3.794 (±1.359)	
AW	**0.718** (±0.042)	2.861 (±1.383)	0.006 (±0.060)
RAC	**0.918** (±0.013)	**3.088** (±1.123)	−0.063 (±0.079)
LAC	0.936 (±0.017)	2.538 (±1.287)	−0.041 (±0.056)
MedNext-B	**0.996** (±0.002)	**3.632** (±1.535)	
AW	0.710 (±0.048)	**2.767** (±1.497)	**0.005** (±0.064)
RAC	0.908 (±0.035)	3.438 (±1.941)	**-0.071** (±0.122)
LAC	**0.941** (±0.013)	**2.214** (±0.706)	−0.037 (±0.040)

Table 2. Performance of different loss function combinations in terms of DSC and HD95. The best results for each metric are highlighted in bold.

Loss Functions used	DSC	HD95
DC and CE *losses*	0.856	2.878
DC and HD95 *losses*	0.850	2.754
DC, HD95, TopK, CE and Boundary *losses*	**0.857**	**2.879**

The highest DSC score is obtained by the most comprehensive setup that includes DC, HD95, TopK, CE, and Boundary losses, reaching 0.857. However, the reduced combination of DC and CE losses shows almost the same performance.

The paired t-test for DSC ($t(3) = -0.11$, $p = 0.92$) and HD95 ($t(3) = -0.005$, $p = 0.996$) did not show significant differences between the loss functions. Nevertheless, these configurations were validated in a set of only 10 scans.

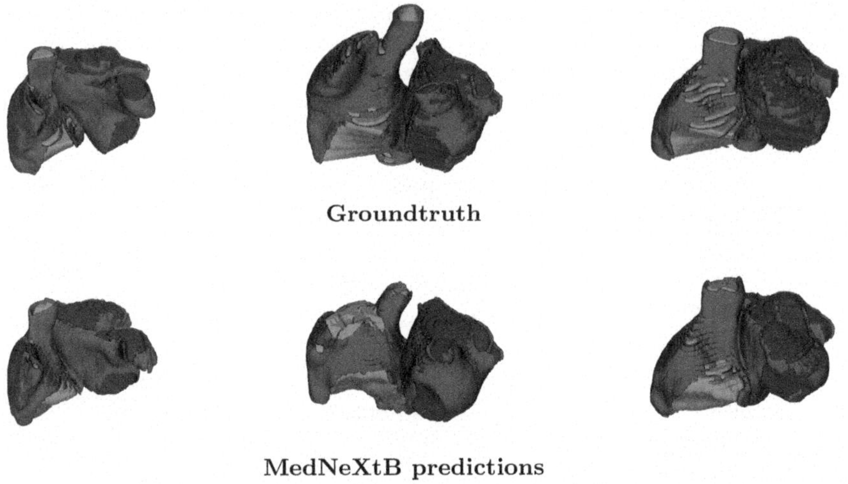

Fig. 3. 3D comparison between the labels and MedNect-B output. Red shows the Atrium Walls, green the Right Atrium Cavity, and blue the Left Atrium Cavity. (Color figure online)

4.3 Cropping Inclusion

The challenge imposes a constraint of a maximum inference time of 10 min for processing 100 cases. As such, we present the results of several models evaluated on both the original and cropped data. The following table summarizes the inference times and challenge metrics for each model. Notably, excessive inference times exceeding acceptable limits are highlighted in red for emphasis (Table 3).

Table 3. Inference time **per case** and segmentation metrics for the original data shape and the cropped version. The DSC and HD95 shown are for the average of class 1 (AW), class 2 (RAD), and class 3 (LAC).

	Original Data			Cropped Data		
Model	Time(s)	DSC	HD95	Time(s)	DSC	HD95
nnUNet2D	4	0.847	**3.044**	1.2	0.797	6.582
nnUNet3D	12	**0.853**	3.187	4	0.852	3.107
SwinUNETR	39	0.736	90.825	13	0.780	11.270
UMambaBot	28	0.857	2.829	9	0.853	**2.842**
MedNext-B	45	0.853	2.806	14	**0.856**	2.878

The best configuration that meets the time requirements without cropping is the nnUNet3D, yielding a DSC of 0.853. With cropping, we can use the MedNext-B model, which increases the internal validation DSC to 0.856.

4.4 Pre-processing Inclusion

After applying the previously explained pre-processing techniques, the results showed no improvement in performance metrics. Specifically, using MedNext-B, the DSC decreased from 0.857 to 0.841, while the HD95 increased from 2.879 to 2.995. One possible explanation for this decline in performance could be the normalization process applied to the initial data, which was originally in the range of 0 to 4095 and subsequently normalized to a range of 0 to 256.

5 Discussion

Our findings demonstrate that models such as UMambaBot and MedNext-B consistently delivered the best results, achieving high DSC scores and low HD95 values for critical structures. These metrics are essential for evaluating the precision of segmentation, especially in terms of capturing the boundaries of the atrial structures, which directly impact the success of ablation procedures. In contrast, SwinUNETR, while a novel transformer-based model, showed significantly poorer performance in boundary delineation, particularly in HD95, suggesting that traditional convolutional models, with or without hybrid configurations, still outperform attention-based models in this task.

The use of multiple loss functions slightly enhances model performance. The addition of boundary-specific loss functions such as Boundary Loss and HD95 was particularly beneficial for models dealing with fine atrial structures, suggesting that focusing on boundary accuracy is crucial for precise segmentation. Despite these improvements, the paired t-test showed no statistically significant differences between loss function combinations, which may be attributed to the limited sample size of the internal validation set. Larger validation sets in future studies would provide more robust comparisons.

Regarding preprocessing, applying CLAHE for contrast enhancement and normalization mitigated the long-tailed intensity distribution but decreased the model performance.

The analysis of cropped versus original data also revealed interesting insights. While cropping improved inference times significantly, allowing models to meet the challenge's time constraints, the segmentation accuracy kept almost invariant. This is due to the big data augmentation that we use, specifically due to the zoom augmentation.

Despite these promising results, several limitations should be acknowledged. First, the dataset used for validation was relatively small, limiting the generalizability of our findings. Additionally, the absence of ground truth for the test set in the challenge prevents a definitive evaluation of the models' performance in real-world clinical settings.

6 Conclusion

In conclusion, our study demonstrates the potential of advanced deep learning models for improving bi-atrial segmentation in LGE-MRI scans. Integrating multiple loss functions, robust preprocessing, and innovative model architectures resulted in significant improvements in segmentation accuracy. Our MedNext-B and UMambaBot models outperformed others, particularly in atrial wall and cavity segmentation, with a combination of Dice, Hausdorff, TopK, and boundary losses proving most effective. Inference time was optimized using cropping strategies.

References

1. Lippi, G., Sanchis-Gomar, F., Cervellin, G.: Global epidemiology of atrial fibrillation: an increasing epidemic and public health challenge. Int. J. Stroke **16**(2), 217–221 (2021)
2. Hart, R.G., Halperin, J.L.: Atrial fibrillation and thromboembolism: a decade of progress in stroke prevention. Ann. Intern. Med. **131**(9), 688–695 (1999)
3. Morin, D.P., et al.: The state of the art: atrial fibrillation epidemiology, prevention, and treatment. Mayo Clinic Proc. **91**(12) (2016)
4. Li, L.,et al.: edical image analysis on left atrial LGE MRI for atrial fibrillation studies: a review. Med. Image Analy. **77**, 102360 (2022)
5. McGann, C., et al.: Atrial fibrillation ablation outcome is predicted by left atrial remodeling on MRI. Circ.: Arrhythmia Electrophysiol. **7**(1), 23–30 (2014)
6. Isensee, F., et al.: nnU-Net: self-adapting framework for U-Net-based medical image segmentation. arXiv preprint: arXiv:1809.10486 (2018)
7. Hatamizadeh, A., et al.: Swin unetr: Swin transformers for semantic segmentation of brain tumors in MRI images. In: International MICCAI Brainlesion workshop. Springer International Publishing, Cham (2021)
8. Myronenko, A.: 3D MRI brain tumor segmentation using autoencoder regularization. In: Brainlesion: Glioma, Multiple Sclerosis, Stroke and Traumatic Brain Injuries: 4th International Workshop, BrainLes 2018, Held in Conjunction with MICCAI 2018, Granada, Spain, September 16, 2018, Revised Selected Papers, Part II 4. Springer International Publishing (2019)
9. Roy, S., et al.: MedNeXt: transformer-driven scaling of convnets for medical image segmentation. In: International Conference on Medical Image Computing and Computer-Assisted Intervention. Springer Nature Switzerland, Cham (2023)
10. Liao, W., et al.: LightM-UNet: mamba assists in lightweight UNet for medical image segmentation. arXiv preprint: arXiv:2403.05246 (2024)
11. Ma, J., Li, F., Wang, B.: U-mamba: Enhancing long-range dependency for biomedical image segmentation. arXiv preprint: arXiv:2401.04722 (2024)

Classification of Mitral Regurgitation from Cardiac Cine MRI Using Clinically-Interpretable Morphological Features

Y. On[1(✉)], K. Vimalesvaran[1], S. Zaman[2], M. Shun-Shin[1], J. Howard[1], N. Linton[1], G. Cole[2], A. A. Bharath[1], and M. Varela[1,3]

[1] Imperial College London, Exhibition Road, London SW7 2AZ, UK
yu.on16@imperial.ac.uk

[2] Imperial College Healthcare NHS Trust, Du Cane Road, London W12 0HS, UK

[3] Cardiovascular and Genomics Research Institute, City St George's University of London, London SW17 0RE, UK

Abstract. The assessment of mitral regurgitation (MR) using cardiac MRI, particularly Cine MRI, is a promising technique due to its wide availability. However, some of the temporal information available in clinical Cine MRI may not be fully utilised, as it requires detailed temporal analysis across different cardiac views. We propose a new approach to identify MR which automatically extracts 4-dimensional (3D + Time) morphological features from the reconstructed mitral annulus (MA) using Cine long-axis (LAX) views MRI.

Our feature extraction involves locating the MA insertion points to derive the reconstructed MA geometry and displacements, resulting in a total of 187 candidate features. We identify the 25 most relevant mitral valve features using minimum-redundancy maximum-relevance (MRMR) feature selection technique. We then apply linear discriminant analysis (LDA) and random forest (RF) model to determine the presence of MR. Both LDA and RF demonstrate good performance, with accuracies of 0.72 ± 0.05 and 0.73 ± 0.09, respectively, in a 5-fold cross-validation analysis.

This approach will be incorporated in an automatic tool to identify valvular diseases from Cine MRI by integrating both handcrafted and deep features. Our tool will facilitate the diagnosis of valvular disease from conventional cardiac MRI scans with no additional scanning or image analysis penalty.

All code is made available on an open-source basis at: https://github.com/HenryOn2021/MA_Morphological_Features.

Keywords: Cardiac MRI · Feature Engineering · Mitral Regurgitation · Cine MRI · Linear Discriminant Analysis · Random Forests · Disease Classification

O. Camara et al. (Eds.): STACOM 2024, LNCS 15448, pp. 245–256, 2025.
https://doi.org/10.1007/978-3-031-87756-8_25

1 Introduction

Mitral Regurgitation. Mitral regurgitation (MR) is one of the most common cardiac conditions: it currently affects 31% of the population in Europe [1], and its incidence is rising due to increases in life expectancy. MR occurs when the mitral valve (MV) leaflets do not close properly, leading to the reflow of blood from the left ventricle (LV) to the left atrium (LA) during systole, which can be categorised as primary or secondary. In the former case, it is usually caused by degenerative MV and patients can be asymptomatic with normal LV function [2]; whereas the latter case involves the deformation of the mitral annulus (MA) caused by LV dysfunction [3].

According to clinical guidelines [1,4], transthoracic echocardiography (TTE) is still used as the first choice for diagnosis of MR. However, it lacks reproducibility, subjectivity, and is highly dependent on operator skills [5–7], especially for the detection of mitral regurgitant flow [1]. The assessment of MR severity is challenging even for experienced clinicians, as it comprise of qualitative, semi-quantitative, and quantitative measurements [8]. Therefore, computed tomography (CT) and cardiac magnetic resonance imaging (cMRI) [9,10] are increasingly used to characterise MR for functional analysis and evaluation of severity. There are, however, few tools to help identify MR from Cine MRI. This blocks the clinical potential of these images for MR diagnosis and management.

Cardiac Cine MRI. Cine (or dynamic) MRI is one of the most commonly acquired cardiac MRI sequences, used to assess the function and morphology of the heart chambers for virtually all clinical cardiac MRI indications [10]. Conventional Cine MRI usually employs a 2D balanced steady-state free precession (bSSFP) readout reconstructed into approximately 30 or 50 cardiac phases. Cine MRI is acquired in both short-axis stacks (full coverage of the LV) and single-slice long-axis (2-, 3-, and 4-chamber) views [11].

The MV leaflets are thin and protein-rich, making them hard to image using most types of MRI. Regurgitant jets can be observed as a region of reduced blood signal in bSSFP; however, they are often not clearly visible [12]. In contrast, the MA insertions can be reliably identified in Cine MRI [10,13], opening the possibility of studying MR annulus morphology and dynamics using Cine MRI. Some past studies have studied MR patients using Cine MRI. Xiao *et al.* [14] developed a semi-supervised neural network pipeline for MR classification using Cine MRI. However, despite being fully automated, it lacked clinical interpretability as we cannot explain the model's decision-making process (known as the 'black-box' problem). Wu *et al.* [15] developed a semi-automatic approach to extract dynamic features from reconstructed MA using Cine MRI to identify diastolic dysfunction, which showed comparative results with TTE measurements. However, they did not include the rich morphological features from the reconstructed MA in their study. Ennis *et al.* [2] used feature tracking to approximate the MA points and extract dynamic and morphological features from the reconstructed MA at end-diastole and end-systole phases. They however focused on evaluating treatment outcomes in patients with degenerative MV disease, not in MR identification. Leng *et al.* [16] also used feature tracking to extract both morphological and dynamics features from reconstructed MA using Cine MRI, and showed MR patients have significantly reduced mitral dynamics and mild annular deformation. However, to date, no studies propose optimally combining a spectrum of morphological features across full cardiac cycles for Cine-based MR classification.

Feature Selection. Feature selection methods can be used to reduce high-dimensional data to a smaller subset of features, which can be used as inputs to different machine-learning problems. Minimal redundancy and maximal relevance (MRMR) is a filter-based feature selection method which ranks features based on two components - relevance and redundancy [17,18]. In an iterative approach, MRMR uses F-statistics or mutual information to identify the features that correlate the most (are most relevant) with a given binary label (e.g., a clinical diagnosis) [19,20]. The feature is added to an optimal feature set, in which Pearson correlation [21] or mutual information is used to remove highly correlated features with lower relevance (to minimise redundancy).

Machine Learning. We propose to use Linear Discriminant Analysis (LDA) [22] and Random Forest (RF) [23] to evaluate the performance of the imaging features for MR detection. LDA looks for a linear decision boundary in feature space by fitting the class-conditional densities to the data using Bayes's rule [22]. RF is an ensemble learning algorithm that combines the outcomes of multiple decision trees to make predictions. A random subset of the features is trained in each decision tree to reduce over-fitting and improve accuracy.

Aim. We propose a machine-learning approach to detect mitral regurgitation using imaging features extracted from long-axis (2-chamber, 3-chamber and 4-chamber) bSSFP Cine MRI.

2 Methods

Patient Demographic. 187 subjects referred for clinical cardiac MRI, consisting of 98 with no MR (mean age 54 ± 17 years, 43% female) and 89 with MR (63 ± 16 years, 38% female), were imaged under ethical approval in this retrospective study. Clinical diagnostic reports based on TTE were used to divide patients into each cohort. Among the 89 MR cases, 73% have mild severity, 17% moderate, and 9% are severe. The patient demographics and volumetric characteristics are shown in **Supplementary Table 1**.

Image Acquisition. Conventional breath-held, gated bSSFP Cine CMR [24] was performed using 1.5 or 3T Siemens (Erlangen, Germany) Prisma or Aera MRI scanners in three different hospitals. Long-axis views were acquired and reconstructed to 30 cardiac phases at a spatial resolution of $0.7 - 2.1 \times 0.7 - 2.1 \times 5.0\ mm^3$. The cardiac phases were ordered such that phase 0 corresponds to the electrocardiography R-peak (end-diastole).

Data Analysis. Two mitral valve insertion points are manually labelled at every 5th cardiac phase (CP, out of 30) in each of the 2-, 3-, and 4-chamber long-axis views (see Fig. 1A-C). The inter-operator error in the positioning of mitral valve insertion points is assessed 20 randomly-selected patients. The coordinates of each of these 36 (= 6 points × 6 cardiac frames) points are converted to patient (world) coordinates using DICOM header information and used to study the morphology and dimensions of the mitral annulus (MA).

For each labelled CP, the 6 MA points are used to find the best-fit plane in 3D space and the best-fit ellipse on that plane (see Fig. 1D and E). From Fig. 1D, the MA points depict a close approximation of ellipse, which is proposed as a simple morphological approximation of the saddle-shaped MA. The fitted ellipses are co-registered to the common centroid at $(0, 0, 0)$ across patients, and the semi-major axis of the ellipse is aligned with the x-axis at CP_0 (see **Supplementary Fig. 1**). The best-fit plane is identified using singular value decomposition (SVD) [25] and the ellipse is fitted by projecting the 3D points to the 2D plane using Rodrigues' rotation [26]. At each CP, we extract the following features from the fitted ellipse: area, perimeter, semi-minor axis length (b), semi-major axis length (a), eccentricity, $b : a$ ratio, MA heights (the sum of the most superior and the most inferior points perpendicular to the best-fit plane), and the (x,y,z) components of the normal plane vectors. We additionally estimate changes across consecutive cardiac phases in: the angle of the MA plane with a horizontal plane (tilt) and the in-plane rotation angle (θ) of the ellipse's semi-major axis with respect to the x-axis. The magnitude and individual (x-, y-, z-) axis displacements of the 3D MA point positions across the cardiac cycle are also measured (see Fig. 1D, E).

Data Selection. In total, 187 candidate features are extracted. The comprehensive list of features and the number of values measured from individual features are outlined in **Supplementary Table 2**. We select the most salient features by estimating individual MRMR coefficients between each feature and MR/no MR clinical labels [17,18]. The top K features with the highest relevance and lowest redundancy are selected as the input vectors for the classification stage. To decide how many features to keep for the subsequent analyses, we tested the K = 5, 10, 25, 50 features and decided on K = 25, which produced the best cross-validation (CV) results.

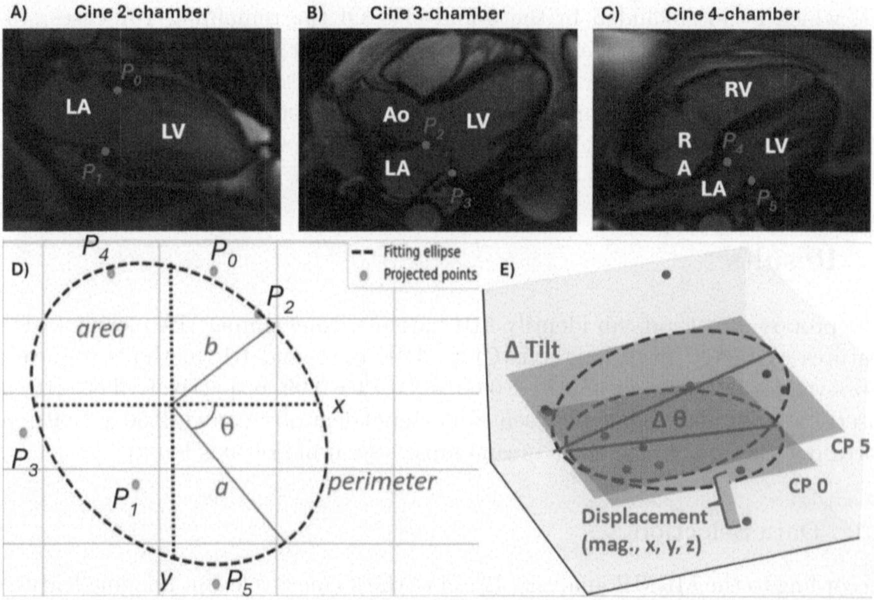

Fig. 1. **(A-C)** The mitral annulus (MA) insertion points are labelled (red circles) in: A) 2-chamber, B) 3-chamber and C) 4-chamber long-axis Cine views. **(D)** Example of 6 MA points in scanner coordinates (blue dots) projected to 2D space and the corresponding best-fit ellipse (black dashed line). Some of the measured ellipse properties: semi-minor axis (of length b), semi-major axis (of length a), in-plane rotation angle θ, area, and perimeter are highlighted in red. **(E)** The same set of points in 3D space, alongside the best-fit ellipse and the best-fit plane (yellow surface) estimated with SVD. Two example phases are used to illustrate some of the features being extracted in 3D space. *LA: left atrium, LV: left ventricle, Ao: aorta, RA: right atrium, RV: right ventricle; P_0: mitral anterior (2-ch); P_1: mitral posterior (2-ch); P_2: mitral septal (3-ch); P_3: mitral free wall (3-ch); P_4: mitral septal (4-ch); P_5: mitral free wall (4-ch); Mag.: magnitude; CP: cardiac phase; Δ: change in.* (Color figure online)

Machine Learning Classification. We have selected two machine learning models to classify the optimal feature set ($K = 25$) into one of the two classes (No MR and MR): linear discriminant analysis (LDA) [22] and random forest (RF) [23]. To optimise the RF model, we perform a grid search for the number of estimators, the number of features to consider when looking for the best split, maximum tree depth, and maximum leaf nodes using the training and validation set with the optimal feature set. The best RF configuration for the dataset is listed in **Supplementary Table 4**.

Performance Assessment. To mitigate the risk of overfitting and provide a reliable evaluation of the models, we perform stratified 5-fold cross-validation (CV) with a constant random seed (for reproducibility) and data shuffling [27]. From the 187 cases, 10 cases from each class are randomly selected as a test set, which is not included in the 5-fold CV. Of the remaining 167 cases, 134 cases are used for training and 33 for validation in each fold. The test set is used for independent evaluation of the models to minimise data leakage. We assess the classification performance using the following metrics: accuracy, specificity, sensitivity, F1 score, and Area Under the Receiver Operating Characteristics Curve (ROC AUC) [28].

3 Results

The proposed method can identify MR patients using simple MA morphological features extracted from long-axis Cine MRI. LDA and RF methods achieve a cross-validation accuracy of 0.72 ± 0.05 and 0.73 ± 0.09, respectively. The error in insertion point placements between two independent observers is 2.50 ± 1.99 mm, corresponding to 10.5% of the mean ellipse's semi-major axis length.

3.1 Data Selection

According to the MRMR analysis, 12 out of the 25 most relevant imaging features are showed in Table 1, rank in descending order using the MRMR F-statistics relevance. From the LDA column, the ellipse perimeter at $CP_5 \& CP_{10}$, the semi-major axis length (a) and the semi-minor axis length (b) at CP_{10} have high absolute importance coefficients; whereas the magnitude displacement (u_{mag}) of *P2* between CP_{20-15}, displacement of *P5* along x-axis between CP_{15-10}, MA height, and the semi-minor axis length (b) at CP_5 have high coefficient values in the RF model. The remaining 13 of 25 features are listed in **Supplementary Table 3**.

3.2 Performance Assessment

Table 2 summarises the evaluation metrics of the two models in classifying 'No MR' and 'MR' cohorts with the hand-engineered features. From the 5-fold CV and test set results, RF performs marginally better than LDA for all metrics.

In Fig. 2A and B, the 4 most important LDA features ($perimeter_{CP_{10}}$, $perimeter_{CP_5}$, $a_{CP_{10}}$, and $b_{CP_{10}}$) already show some separability between the 2 classes when considered in pairs. In Fig. 3A-D, the plots illustrate the change in the 4 feature values with the highest RF coefficients across the full cardiac cycle, which further depict the class separation in the selected features at the specific CP (blue box).

Table 1. 12 out of the selected features ($K = 25$) with the highest MRMR relevance (F-statistics), the corresponding LDA and RF models' coefficients, and the $mean \pm std$ values of the 2 cohorts. The 4 features with the highest absolute model coefficients are highlighted in **bold**. All distance measurements are in **mm**. $u_{x/y/z}$: displacement of x-/y-/z-coordinate; u_{mag}: displacement magnitude of x-,y-,z-coordinates; b: semi-minor axis length; a: semi-major axis length; CP: cardiac phase.

Feature	F-stat	LDA	RF	Value (No MR)	Value (MR)
$u_{mag} \cdot P_3 \cdot CP_{20-15}$	19.3	−0.32	0.06	5.73 ± 3.57	3.66 ± 2.76
$u_{mag} \cdot P_2 \cdot CP_{20-15}$	18.0	−0.89	**0.07**	3.92 ± 2.18	2.70 ± 1.67
$perimeter_{CP_{10}}$	16.1	**3.58**	0.03	129.4 ± 20.3	145.4 ± 33.0
$area_{CP_{10}}$	15.2	0.34	0.03	1345.7 ± 438.2	1734.5 ± 868.6
$height_{CP_5}$	15.1	0.39	**0.08**	5.99 ± 3.59	9.00 ± 6.64
$perimeter_{CP_5}$	14.9	**1.01**	0.05	130.8 ± 29.6	151.6 ± 43.0
$u_{mag} \cdot P_0 \cdot CP_{20-15}$	14.9	0.26	0.05	5.06 ± 4.01	3.19 ± 2.24
$u_x \cdot P_5 \cdot CP_{15-10}$	14.2	0.57	**0.07**	-0.97 ± 2.26	0.52 ± 3.08
$u_{mag} \cdot P_1 \cdot CP_{20-15}$	14.1	−0.07	0.04	4.93 ± 3.29	3.31 ± 2.47
$a_{CP_{10}}$	13.9	**−2.12**	0.03	21.6 ± 2.95	24.1 ± 5.91
b_{CP_5}	13.2	−0.92	**0.07**	19.0 ± 5.90	22.7 ± 7.63
$b_{CP_{10}}$	11.1	**−1.76**	0.03	19.5 ± 4.25	21.9 ± 5.73

Table 2. Classification performance of LDA and RF with 5-fold CV (validation set) and test set.

Model	Accuracy	Specificity	Sensitivity	F1-Score	AUC
LDA (Validation set)	0.72 ± 0.05	0.77 ± 0.08	0.66 ± 0.07	0.69 ± 0.06	0.71 ± 0.05
RF (Validation set)	0.73 ± 0.09	0.77 ± 0.09	0.69 ± 0.12	0.71 ± 0.10	0.73 ± 0.09
LDA (Test set)	0.65	0.60	0.70	0.65	0.65
RF (Test set)	0.70	0.70	0.70	0.70	0.70

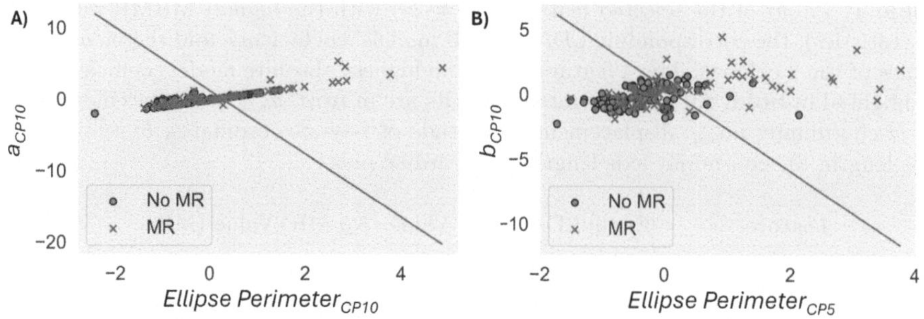

Fig. 2. **(A,B)** Data distribution of 'No MR' class (green circle) and 'MR' class (red cross). The hyperplane projection is calculated from the intercept and coefficients of the LDA model (fitted to the training set and predicted on the test set). The feature pairs are selected based on the LDA coefficients in descending order of absolute values. (Color figure online)

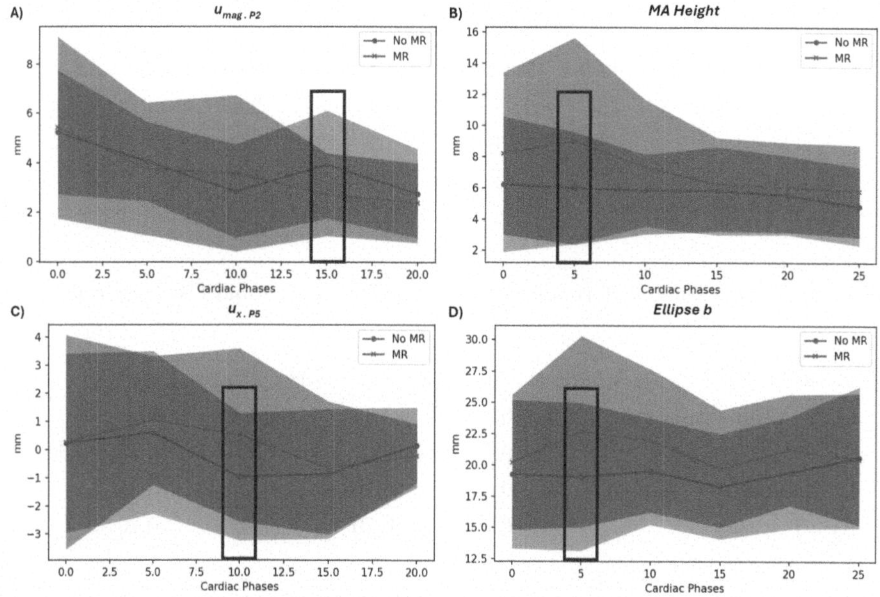

Fig. 3. Mean and standard deviation plots of the 4 most important RF features between the No MR (green) and MR (red) cohort. The specific CP of the selected features are highlighted in the blue box. **(A)** $u_{mag \cdot P_2}$ at CP_{20-15}; **(B)** *MA Height* at CP_5; **(C)** $u_{x \cdot P_5}$ at CP_{15-10}; **(D)** b at CP_5. (Color figure online)

4 Discussion

In this study, we propose a semi-automatic pipeline to accurately identify mitral regurgitation from three clinical long-axis Cine MRI views. We use clinically-interpretable hand-engineered features, which can quantitatively characterise the morphology of the MA, reconstructed from 6 manually annotated MA insertions. We have not included features related to the intensity of the blood in Cine MRI, although flow-related dephasing could in principle contribute to the classification of MR [12]. Flow-induced dephasing of blood signal depends strongly on the imaging parameters, slice positioning and is therefore extremely variable across the patient cohort. We have performed an offline study to evaluate the flow information in classifying MR with the same cohort of patients and dataset using a 1-layer convolutional neural network (CNN). By using the labelled points to extract the flow's pixel intensities across the MV, to create a 2D array to record the intensities across the cardiac cycle. However, the CNN is unable to capture useful features to distinguish between the 2 classes, in which the performance could be hampered by the limitation of Cine MRIs, and phase contrast (i.e. pixel intensity is correlated to flow velocity) could be a better alternative MRI sequence.

Although the MV leaflets cannot be observed in Cine MRI, we show that these images include clinically important MA morphological information that, when optimally combined, can classify MR accurately. In Table 1, the magnitude displacement of the labelled points in Cine 2-ch and 3-ch at mid-diastole (CP_{20-15}) have high relevance with the diagnostic outcomes and high importance in classification, especially in the RF model. This is illustrated in Fig. 3**A**, of mitral septal in Cine 3-ch, which shows the largest class separation at the specific CP (in blue) between the 2 cohorts. The restricted motion of MA is often associated with secondary MR, which the MA becomes dilated and less dynamic [29,30]. The semi-minor length (b) of the ellipse and mitral annulus height during systole also have high relevance and high importance in RF model. From Fig. 3**B** & **D**, both features show good separation between the 2 cohorts at the specific CPs. The higher feature value in b for MR cohort reflects the dilation of the MA as it often seen in patients with MR, especially moderate to severe severity. [31].

In Fig. 2, the 2 sets of features with the highest LDA coefficients show some class separation, even without considering the other 8 features. Also, the plots show the sets of the two classes are contiguous at the LDA decision boundary. The overlapping group at the boundary are largely contributed by mild MR cases, as the MA size can still be within the normal range in primary MR patients [4]. This poses a challenging problem to find a linear combination of features that best separates the two classes. This may suggest that the potential partitions of the feature space into conditional subgroups performed by the RF is crucial, thus explaining the better performance with the RF model.

There are some limitations in the proposed pipeline. First, the landmarks are manually annotated, but we plan to make this step (and thus the entire pipeline) fully automatic. We utilise the labelled landmarks in this study as training data for a CNN that can reliably localise the landmarks from the Cine MRIs. Second, we do not assess the performance of the method separately for primary and secondary MR. We speculate that our method may be better suited for secondary MR, which is expected to bring greater morphological changes to the MA. Third, the limitations of TTE (mentioned in **Introduction**) can lead to inter- and intra-observer disagreements in MR severity stratification, and we therefore do not attempt to classify MR severity.

Our future plans include using the proposed features as one of the auxiliary inputs to a deep NN pipeline that can identify multiple cardiac valvular diseases using clinical Cine MRI. This NN pipeline will potentially not only identify MR, but also classify the MR type (primary/secondary), by bringing in other types of MRI and imaging information. A more detailed automatic analysis of MA features from Cine MRI could enable accurate categorisation of the type of MR (primary/secondary) and locate the possible pathological regions. This additional knowledge is essential for disease management choices (including the types of intervention and/or medical therapy), surpassing the information that can be obtained from standalone echocardiography.

Acknowledgments. This work was supported by: St George's Hospital Charity, the NIHR Imperial Biomedical Research Centre (BRC) and the British Heart Foundation Centre of Research Excellence at Imperial College London (RE/18/4/34215). We acknowledge the computational support provided by the Imperial College Research Computing Service (DOI: 10.14469/hpc/2232).

References

1. Vahanian, A., et al.: 2021 ESC/EACTS guidelines for the management of valvular heart disease: developed by the task force for the management of valvular heart disease of the European Society of Cardiology (ESC) and the European Association for Cardio-Thoracic Surgery (EACTS). Revista espanola de cardiologia (English ed.) **75**(6), 524 (2022)
2. Ennis, D.B., et al.: Changes in mitral annular geometry and dynamics with β-Blockade in patients with degenerative mitral valve disease. Circ.: Cardiovasc. Imag. **3**(6), 687–693 (2010)
3. El Sabbagh, A., Reddy, Y.N., Nishimura, R.A.: Mitral valve regurgitation in the contemporary era: insights into diagnosis, management, and future directions. JACC: Cardiovasc. Imag. **11**(4), 628–643 (2018)
4. Robinson, S., et al.: The assessment of mitral valve disease: a guideline from the British society of echocardiography. Echo Res. Pract. **8**(1), G87–G136 (2021)
5. Enriquez-Sarano, M., Akins, C.W., Vahanian, A.: Mitral regurgitation. Lancet **373**(9672), 1382–1394 (2009)
6. Caruthers, S.D., et al.: Practical value of cardiac magnetic resonance imaging for clinical quantification of aortic valve stenosis: comparison with echocardiography. Circulation **108**(18), 2236–2243 (2003)
7. Marwick, T.H., et al.: Recommendations on the use of echocardiography in adult hypertension: a report from the European Association of Cardiovascular Imaging (EACVI) and the American Society of Echocardiography (ASE). Eur. Heart J. Cardiovasc. Imaging **16**(6), 577–605 (2015)
8. Zoghbi, W.A., et al.: Recommendations for noninvasive evaluation of native valvular regurgitation: a report from the American society of echocardiography developed in collaboration with the society for cardiovascular magnetic resonance. J. Am. Soc. Echocardiogr. **30**(4), 303–371 (2017)

9. Otto, C.M., et al.: 2020 ACC/AHA guideline for the management of patients with valvular heart disease. J. Am. Coll. Cardiol. **77**(4), e25–e197 (2021)
10. Garg, P., et al.: Assessment of mitral valve regurgitation by cardiovascular magnetic resonance imaging. Nat. Rev. Cardiol. **17**(5), 298–312 (2020)
11. Sechtem, U., et al.: Cine MR imaging: potential for the evaluation of cardiovascular function. Am. J. Roentgenol. **148**(2), 239–246 (1987)
12. On, Y., et al.: Automatic aortic valve pathology detection from 3-chamber cine MRI with spatio-temporal attention maps. In: Lecture Notes in Computer Science (including subseries Lecture Notes in Artificial Intelligence and Lecture Notes in Bioinformatics) 13958 LNCS, pp. 648–657 (2023)
13. Kerfoot, E., King, C.E., Ismail, T., Nordsletten, D., Miller, R.: Estimation of cardiac valve annuli motion with deep learning. In: Lecture Notes in Computer Science (including subseries Lecture Notes in Artificial Intelligence and Lecture Notes in Bioinformatics), 12592 LNCS, pp. 146–155 (2021)
14. Xiao, K., Learned-Miller, E., Kalogerakis, E., Priest, J., Fiterau, M.: Machine learning for automated mitral regurgitation detection from cardiac imaging, vol. 14226. LNCS, Springer Nature Switzerland (2023)
15. Vincent, W., Chyou, J.Y., Chung, S., Bhagavatula, S., Axel, L.: Evaluation of diastolic function by three-dimensional volume tracking of the mitral annulus with cardiovascular magnetic resonance: comparison with tissue doppler imaging. J. Cardiovasc. Magn. Reson. **16**(1), 1–14 (2014)
16. Leng, S., et al.: Imaging 4D morphology and dynamics of mitral annulus in humans using cardiac cine MR feature tracking. Sci. Rep. **8**(1), 1–13 (2018)
17. Ding, C., Peng, H.: Minimum redundancy feature selection from microarray gene expression data. In: Proceedings of the 2003 IEEE Bioinformatics Conference, CSB 2003, pp. 523–528 (2003)
18. Peng, H., Long, F., Ding, C.: Feature selection based on mutual information criteria of max-dependency, max-relevance, and min-redundancy. IEEE Trans. Pattern Anal. Mach. Intell. **27**(8), 1226–1238 (2005)
19. Weir, B.S., Cockerham, C.C.: Estimating F-statistics for the analysis of population structure. Evolution **38**(6), 1358 (1984)
20. Kraskov, A., Stögbauer, H., Grassberger, P.: Estimating mutual information. Phys. Rev. E Stat. Phys. Plasmas Fluids Relat. Interdiscip. Top. **69**(6), 16 (2004)
21. Lee Rodgers, J., Nicewander, W.A.: Thirteen ways to look at the correlation coefficient. Am. Stat. **42**(1), 59–66 (1988)
22. Mohammed, M., Khan, M.B., Bashier, E.B.M.: Linear discriminant analysis (2020)
23. Biau, G., Scornet, E.: A random forest guided tour. TEST **25**(2), 197–227 (2016). https://doi.org/10.1007/s11749-016-0481-7
24. Herzog, B.A., Greenwood, J.P., Plein, S.S., Garg, P., Ph, O.H.: Cardiovascular magnetic resonance (2013)
25. Klema, V.C., Laub, A.J.: The singular value decomposition: its computation and some applications. IEEE Trans. Autom. Control **25**(2), 164–176 (1980)
26. Fraiture, L.: A history of the description of the three-dimensional finite rotation. J. Astronaut. Sci. **57**(1–2), 207–232 (2009)
27. Arlot, S., Celisse, A.: A survey of cross-validation procedures for model selection. Statist. Surv. **4**, 40–79 (2010)
28. Hossin, M., Sulaiman, M.N.: A review on evaluation metrics for data classification evaluations. Int. J. Data Min. Knowl. Manage. Process **5**(2), 01–11 (2015)
29. Sadeghpour, A.: Echocardiographic evaluation of mitral geometry in functional mitral regurgitation. J. Cardiothoracic Surg. **3**(1), 1–7 (2008)

30. Stewart, W.J., et al.: Evaluation of mitral leaflet motion by echocardiography and jet direction by doppler color flow mapping to determine the mechanism of mitral regurgitation. J. Am. Coll. Cardiol. **20**(6), 1353–1361 (1992)
31. Silbiger, J.J.: Anatomy, mechanics, and pathophysiology of the mitral annulus. Am. Heart J. **164**(2), 163–176 (2012)

Gaussian Process Emulators for Few-Shot Segmentation in Cardiac MRI

Bruno Viti[1,5(✉)], Franz Thaler[3,4], Kathrin Lisa Kapper[1,5], Martin Urschler[2,5], Martin Holler[1,5], and Elias Karabelas[1,5]

[1] Department of Mathematics and Scientific Computing, University of Graz, Graz, Austria
bruno.viti@uni-graz.at
[2] Institute for Medical Informatics, Statistics and Documentation, Medical University of Graz, Graz, Austria
[3] Institute of Computer Graphics and Vision, Graz University of Technology, Graz, Austria
[4] Gottfried Schatz Research Center: Medical Physics and Biophysics, Medical University of Graz, Graz, Austria
[5] BioTechMed-Graz, Graz, Austria

Abstract. Segmentation of cardiac magnetic resonance images (MRI) is crucial for the analysis and assessment of cardiac function, helping to diagnose and treat various cardiovascular diseases. Most recent techniques rely on deep learning and usually require an extensive amount of labeled data. To overcome this problem, few-shot learning has the capability of reducing data dependency on labeled data. In this work, we introduce a new method that merges few-shot learning with a U-Net architecture and Gaussian Process Emulators (GPEs), enhancing data integration from a support set for improved performance. GPEs are trained to learn the relation between the support images and the corresponding masks in latent space, facilitating the segmentation of unseen query images given only a small labeled support set at inference. We test our model with the M&Ms-2 public dataset to assess its ability to segment the heart in cardiac magnetic resonance imaging from different orientations, and compare it with state-of-the-art unsupervised and few-shot methods. Our architecture shows higher DICE coefficients compared to these methods, especially in the more challenging setups where the size of the support set is considerably small. The code is available on GitLab (https://gitlab.com/bruno_viti/gpe_4_cardiac_fss).

Keywords: Few-Shot Segmentation · Gaussian Process Emulators · Cardiac Segmentation

1 Introduction

Medical imaging techniques like computed tomography (CT) and magnetic resonance imaging (MRI) are established technologies to assess patient health. Automatic medical image segmentation plays an important role by mapping anatomical structures to specific semantic labels which allows more in-depth analysis

O. Camara et al. (Eds.): STACOM 2024, LNCS 15448, pp. 257–268, 2025.
https://doi.org/10.1007/978-3-031-87756-8_26

through follow-up applications [15]. This is especially relevant in cardiac imaging, with cardiovascular diseases being the world's leading cause of death [25]. In recent years, deep learning (DL) models, and in particular Convolutional Neural Networks (CNNs), have been extensively employed to automate and accelerate heart segmentation, obtaining remarkable results in terms of accuracy [5]. CNNs generally require a large amount of labeled data for training, posing a challenge in the medical field with a lack of data and limited manual segmentations. Moreover, CNNs assume that training and test data are drawn from the same distribution, i.e., that they are *i.i.d.* Therefore, in order for the model to perform well on a different data distribution, it would require a large-scale labeled dataset to update the networks' parameters for adaptation. Consequently, several DL approaches have been suggested to face the latter problem. For example, recent works proposed unsupervised domain adaptation or generalization architectures in the cardiac segmentation scenario [4,6,13,18]. These models alleviate the need for extensively labeled data in the target domain by leveraging knowledge from a related source domain where labeled data is more abundant. However, while these techniques can handle perturbations between training and test images from different modalities and sources, they are not designed to adapt to novel image orientation [7]. Another promising technique, which can effectively reduce data dependency, is few-shot segmentation (FSS) [3,23]. In the FSS framework, the model is trained to learn how to segment a novel image, denoted as *query*, having at its disposal only a small set of labeled examples, denoted as *support*. Especially in the medical field, FSS is gaining attention, and several FSS architectures have been proposed [8,14,17,21]. The method's effectiveness highly depends on the mechanism that extracts the information from the support set and integrates this new information with the query image. Most of the existing models rely on the prototype alignment paradigm, which tries to learn a common representation space, where the feature vectors of different object classes can be aligned, e.g. [24]. Differently, Johnander et al. [11] proposed a novel approach based on GPEs, which they exploit to learn a mapping between dense local deep feature vectors and their corresponding mask values. Saha et al. [22] also used FSS in combination with GPEs and applied their method to microscopy images. However, the majority of these approaches [8,14,17,21,24] are limited to predicting binary segmentations. As we also show in this work exemplarily for [14], this results in a non-efficient class-by-class segmentation for multi-label query images [9]. To face this issue, [9] introduced an extension of prototype alignment to perform one-step multi-class segmentation. They employ a self-supervised training approach, facing the challenge of limited labeled image data. However, in [9] they still consider the same type of images during training and testing, namely 2D short-axis cardiac MRI.

In contrast, our goal is to cope with the scarcity of data in the cardiac setting and to make segmentation models more adaptable to different cardiac image orientations. Therefore, we propose a model for multilabel cardiac image segmentation, which combines a U-Net-like architecture [20] with GPEs as extractors of information from the support set [11]. Specifically, we use the contracting

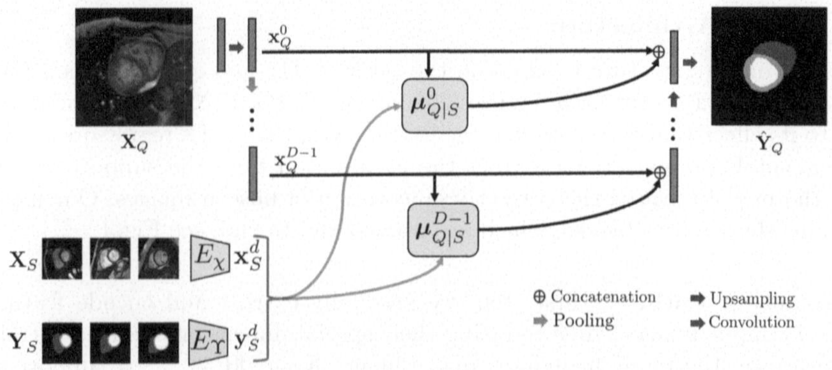

Fig. 1. Overview of a training episode of the proposed method. The encoder $E_{\mathcal{X}}$ encodes the MRIs of the query and the support set, while E_{Υ} encodes the support masks. We train the GPE with the support's features, $\mathbf{x}_S^d$ and $\mathbf{y}_S^d$. Then, given a new point $\mathbf{x}_Q^d$, we infer the mean $\boldsymbol{\mu}_{Q|S}^d$. This additional information is then passed to the decoder D_ζ similar to a skip-connection, and the mask of the query input image is predicted.

branch of the U-Net to bring both the query and the support images into the latent space, in which the GPEs are trained to learn the relationship between the support images and the corresponding masks. This additional information is aggregated into the expanding branch, leading to a more accurate segmentation of the query. We evaluate our model on the public M&Ms2 dataset [2,16], and assess its ability to generalize from 2D short-axis (SA) to long-axis (LA) cardiac MRI.

2 Methods

2.1 Problem Definition

Few-shot segmentation aims to segment novel images given only a tiny set of labeled examples from the same distribution. In a fully supervised setting, training and test data are assumed to follow the same distribution, and further, each training instance is composed of an image and a segmentation mask. In the FSS framework, each instance, denoted as *episode*, includes a small set of labeled images, denoted as the *support* set, and an unlabeled image, denoted as the *query* set, both following the same distribution. More formally, we denote the support set as $\{(\mathbf{X}_{\mathrm{S},k}, \mathbf{Y}_{\mathrm{S},k})\}_{k=0}^{\mathrm{K}-1}$, including K pairs of images, $\mathbf{X}_{\mathrm{S},k} \in \mathbb{R}^{\mathrm{H}_0 \times \mathrm{W}_0}$ and the corresponding segmentation masks, $\mathbf{Y}_{\mathrm{S},k} \in \{0, 1, \ldots, \mathrm{L}-1\}^{\mathrm{H}_0 \times \mathrm{W}_0}$ where L is the number of labels and $\mathrm{H}_0, \mathrm{W}_0$ refer to the height and width of the image, respectively. The task is to predict the segmentation mask $\mathbf{Y}_{\mathrm{Q}}$ of the query image $\mathbf{X}_{\mathrm{Q}}$. During training, each episode includes a support and query set drawn from the domain distribution $\mathcal{D}_{\mathrm{tr}}$. For testing, the model predicts the segmentation mask of a novel image from the distribution $\mathcal{D}_{\mathrm{te}}$, having only a few examples of that distribution at its disposal.

2.2 Model Architecture

Our network architecture is based on the work of [11], and combines their GPE regressor with a U-Net architecture [20]. Although the U-Net has been chosen due to its effectiveness in handling medical images, the GPEs regressors are the fundamental components to extract the information from the support set and lead the model towards the correct segmentation of unseen images. Our model contains three major blocks, which are trained end-to-end, see Fig. 1.

Contracting Path. As first step, we gradually extract and encode features from the input images while reducing their spatial dimensions. For each training instance, the input includes a query image $\mathbf{X}_{\mathrm{Q}} \in \mathrm{I\!R}^{\mathrm{H}_0 \times \mathrm{W}_0}$, a support set of images $\mathbf{X}_{\mathrm{S}} \in \mathrm{I\!R}^{\mathrm{H}_0 \times \mathrm{W}_0 \times \mathrm{K}}$, and the corresponding masks $\mathbf{Y}_{\mathrm{S}} \in \{0, \ldots, \mathrm{L}-1\}^{\mathrm{H}_0 \times \mathrm{W}_0 \times \mathrm{K}}$. The three inputs are moved into the latent space to extract their deep features. We employ the same feature extractor for $\mathbf{X}_{\mathrm{Q}}$ and $\mathbf{X}_{\mathrm{S}}$ as both are MRI. However, we need to train another encoder for the support masks $\mathbf{Y}_{\mathrm{S}}$. At the end of the contracting path, we have

$$\begin{aligned} \mathbf{x}_{\mathrm{Q}} &= E_{\mathcal{X}}(\mathbf{X}_{\mathrm{Q}}) \in \mathrm{I\!R}^{\mathrm{H} \times \mathrm{W} \times \mathrm{F}}, \\ \mathbf{x}_{\mathrm{S}} &= E_{\mathcal{X}}(\mathbf{X}_{\mathrm{S}}) \in \mathrm{I\!R}^{\mathrm{H} \times \mathrm{W} \times \mathrm{KF}}, \\ \mathbf{y}_{\mathrm{S}} &= E_{\mathcal{Y}}(\mathbf{Y}_{\mathrm{S}}) \in \{0, \ldots, \mathrm{L}-1\}^{\mathrm{H} \times \mathrm{W} \times \mathrm{KF}}, \end{aligned}$$

where F is the number of features.

GPEs Few-Shot Learning. GPEs offer a flexible framework for modeling complex functions and are suitable for working with small datasets [19]. In our case, the training points are the support features, which can be reshaped as $\mathbf{x}_{\mathrm{S}} \in \mathrm{I\!R}^{\mathrm{HWK} \times \mathrm{F}}$, and $\mathbf{y}_{\mathrm{S}} \in \{0, \ldots, \mathrm{L}-1\}^{\mathrm{HWK} \times \mathrm{F}}$ is the corresponding output. With this reshaping, we assume that we have HWK training points, and each of them lives in $\mathrm{I\!R}^{\mathrm{F}}$. As GPEs exhibit a computational complexity of $\mathcal{O}(N^3)$ for datasets of size N [19], our GPE will scale as $\mathcal{O}((\mathrm{HWK})^3)$, making it necessary to work in a latent space in which $\mathrm{H} \ll \mathrm{H}_0$ and $\mathrm{W} \ll \mathrm{W}_0$. The GPE aims to predict $\mathbf{y}_{\mathrm{Q}}$ given a new input $\mathbf{x}_{\mathrm{Q}}$. The underlying assumption is that the two outputs $\mathbf{y}_{\mathrm{S}}$ and $\mathbf{y}_{\mathrm{Q}}$ are jointly Gaussian, i.e.

$$\begin{pmatrix} \mathbf{y}_{\mathrm{S}} \\ \mathbf{y}_{\mathrm{Q}} \end{pmatrix} \sim \mathcal{N}\left(\begin{pmatrix} \boldsymbol{\mu}_{\mathrm{S}} \\ \boldsymbol{\mu}_{\mathrm{Q}} \end{pmatrix}, \begin{pmatrix} \boldsymbol{\Sigma}_{\mathrm{SS}} & \boldsymbol{\Sigma}_{\mathrm{SQ}} \\ \boldsymbol{\Sigma}_{\mathrm{QS}} & \boldsymbol{\Sigma}_{\mathrm{QQ}} \end{pmatrix} \right), \tag{1}$$

where $\boldsymbol{\mu}_{\mathrm{S}}$ and $\boldsymbol{\mu}_{\mathrm{Q}}$ are the prior means, which for simplicity are set to $\mathbf{0}$. The covariance matrices $\boldsymbol{\Sigma}$ encode the relationship between different points in the input space and quantify the similarity between them. Given two points $\mathbf{x}_i, \mathbf{x}_j \in \mathrm{I\!R}^{\mathrm{F}}$, the covariance matrix is defined by the kernel function $\boldsymbol{\Sigma}_{i,j} = k(\mathbf{x}_i, \mathbf{x}_j)$, where $k(\cdot, \cdot)\colon \mathrm{I\!R}^{\mathrm{F}} \times \mathrm{I\!R}^{\mathrm{F}} \to \mathrm{I\!R}$. Ideally, for two similar points $\mathbf{x}_1$ and $\mathbf{x}_2$ we would expect similar outputs $\mathbf{y}_1$ and $\mathbf{y}_2$. In our model, we employ the squared exponential kernel function, due to its smoothness and interpretability, defined as

$$k(\mathbf{x}_1, \mathbf{x}_2) = s^2 e^{-\frac{1}{2} \frac{||\mathbf{x}_1 - \mathbf{x}_2||^2}{2l^2}}, \tag{2}$$

where s and l are two additional hyperparameters that control the amplitude of the function values and its smoothness. Using the Gaussian assumption in (1), we can infer the distribution of $\mathbf{y}_\mathsf{Q}$ as

$$\mathbf{y}_\mathsf{Q} | \mathbf{x}_\mathsf{Q}, \mathbf{x}_\mathsf{S}, \mathbf{y}_\mathsf{S} \sim \mathcal{N}(\boldsymbol{\mu}_{\mathsf{Q}|\mathsf{S}}, \boldsymbol{\Sigma}_{\mathsf{Q}|\mathsf{S}}),$$

where

$$\begin{aligned}\boldsymbol{\mu}_{\mathsf{Q}|\mathsf{S}} &= \boldsymbol{\Sigma}_{\mathsf{SQ}}^{\mathsf{T}} \left(\boldsymbol{\Sigma}_{\mathsf{SS}} + \sigma^2 \boldsymbol{I}\right)^{-1} \mathbf{y}_\mathsf{S}, \\ \boldsymbol{\Sigma}_{\mathsf{Q}|\mathsf{S}} &= \boldsymbol{\Sigma}_{\mathsf{QQ}} - \boldsymbol{\Sigma}_{\mathsf{SQ}}^{\mathsf{T}} \left(\boldsymbol{\Sigma}_{\mathsf{SS}} + \sigma^2 \boldsymbol{I}\right)^{-1} \boldsymbol{\Sigma}_{\mathsf{SQ}}.\end{aligned}$$

Here, the term $\sigma^2 \boldsymbol{I}$ is used to bypass numerical issues and to account for noise in the data [1]. The hyperparameter σ represents the variance of noise that affects the observations. This hyperparameter, as well as s and l, are learnable and therefore trained alongside the network weights. Thanks to the Gaussian assumption, we can infer the conditional mean of the output $\mathbf{y}_\mathsf{Q}$, which represents the closest prediction of $\mathbf{y}_\mathsf{Q}$ in the latent space according to the GPE [11].

Expanding Path. We equip our model with an additional mechanism that is able to infer the posterior distribution of the query mask $\mathbf{y}_\mathsf{Q}$, given the image-mask pairs of the support set. At this point, we combine the deep features of the U-Net with the GPE prediction. To do that, we concatenate the posterior mean $\boldsymbol{\mu}_{\mathsf{Q}|\mathsf{S}}$ with the representation of the image in the latent space $\mathbf{x}_\mathsf{Q}$, obtaining $\mathbf{z}_\mathsf{Q} = (\boldsymbol{\mu}_{\mathsf{Q}|\mathsf{S}}, \mathbf{x}_\mathsf{Q})$. The latter is given as input to the decoder D_ς, which predicts the mask of the query image $\hat{\mathbf{Y}}_\mathsf{Q} = D_\varsigma(\mathbf{z}_\mathsf{Q})$. It should be noted that, as proposed in [11], we can successfully leverage the skip-connection structure in the few-shot learning paradigm. In addition to learning the distribution of $\mathbf{y}_\mathsf{Q}$ in the deep latent space, we can do the same after each level of the U-Net encoder. More formally, given our encoders E_χ and E_Υ with depths D, we denote by $\mathbf{x}_\mathsf{S}^d$ and $\mathbf{y}_\mathsf{S}^d$, the output of the encoders after d levels. Following the previous terminology, we have $\mathbf{x}_\mathsf{S}^{\mathrm{D}-1} = \mathbf{x}_\mathsf{S}$, $\mathbf{x}_\mathsf{S}^0 = \mathbf{X}_\mathsf{S}$, $\mathbf{y}_\mathsf{S}^{\mathrm{D}-1} = \mathbf{y}_\mathsf{S}$ and $\mathbf{y}_\mathsf{S}^0 = \mathbf{Y}_\mathsf{S}$. The same idea applies to $\mathbf{x}_\mathsf{Q}$. Standard U-Net skip-connections concatenate the output of level d, $\mathbf{x}_\mathsf{Q}^d$, of the encoder, with the features in the decoder at the same level. We can combine this mechanism with our GPE regressor. For each level d of the encoders, we assume that

$$\mathbf{y}_\mathsf{Q}^d | \mathbf{x}_\mathsf{Q}^d, \mathbf{x}_\mathsf{S}^d, \mathbf{y}_\mathsf{S}^d \sim \mathcal{N}(\boldsymbol{\mu}_{\mathsf{Q}|\mathsf{S}}^d, \boldsymbol{\Sigma}_{\mathsf{Q}|\mathsf{S}}^d).$$

At this point, we construct the corresponding $\mathbf{z}_\mathsf{Q}^d = (\boldsymbol{\mu}_{\mathsf{Q}|\mathsf{S}}^d, \mathbf{x}_\mathsf{Q}^d)$ and pass it to the level d of the decoder D_ς. In the shallower levels, where $d < \mathrm{D}-1$, the resolution of the images and masks is too large; therefore, we down-sample them to the same size as used in the last level to make the GPEs computationally feasible.

2.3 Training Procedure

The construction of meaningful query and support instances is crucial for FSS. In our setup, the model must be trained with different anatomical views to segment a novel view in the test set. To achieve that in the training phase, we treat SA slices from the base to the apex of the heart as different anatomical views based on their size, but with the same image orientation. Specifically, we divide our training dataset into 10 new subsets based on the dimension of the heart, via counting the pixels of the ground truth. In this way, each subset contains cardiac MRIs with similar appearance and GPEs can be trained with different views. During training, each query-support instance is constructed with images from the same subset. For testing, we used a different orientation in the anatomical view, which corresponds to the LA slices.

3 Experiments and Results

We evaluate our method for a scenario where only SA 2D MRI slices are available with ground truth for training, while LA slices of different patients have to be predicted. In this scenario, it is beneficial to have a method that only requires few labeled LA slices in order to optimally segment the whole LA set. We denote our training set as $\{\mathbf{X}_{\text{tr},i}\}_{i=0}^{\text{N}-1} \in \mathcal{D}_{\text{tr}}$ and our testing set as $\{\mathbf{X}_{\text{te},i}\}_{i=0}^{\text{M}-1} \in \mathcal{D}_{\text{te}}$, where $\mathcal{D}_{\text{tr}} = \{\text{SA}\}$, $\mathcal{D}_{\text{te}} = \{\text{LA}\}$ hence $\mathcal{D}_{\text{tr}} \neq \mathcal{D}_{\text{te}}$. We compare our methodology with different baselines. We first measure the ability of a fully supervised method to generalize from SA to LA, having at its disposal only a few LA labeled samples. For this task, we use the widely used nnU-Net due to its good performance on a wide variety of medical image segmentation datasets [10]. Additionally, we compare our model to CAT-Net, a network for medical image FSS [14]. Finally, we also utilize the unsupervised CSDG method [18], a recently published method for domain generalization.

Dataset. For our evaluations, we use the M&Ms-2 dataset [2,16]. The cohort contains 360 patients with different pathologies of the right ventricle and left ventricle, as well as healthy subjects. M&Ms-2 is split into 200 patients for training and 160 for testing. Two pathologies are only present in the test set to evaluate the generalization to unseen pathologies. For each patient, the dataset provides 2 annotated LA MRI slices and about 20 annotated SA MRI slices. As described above, we use all of the SA slices for training and the LA slices for testing.

Implementation Details. The two encoders, $E_\mathcal{X}$ and E_Υ, contain five levels, each consisting of two convolutional blocks followed by ReLU activations, dropout, and batch normalization, ending with max pooling. For decoding, we use bilinear up-sampling at each level, followed by one convolution. The preprocessing involves intensity normalization to zero mean and unit variance, and bilinear resampling to 1.375 mm $\times$ 1.375 mm in-plane resolution. We use the ground truth mask to crop a centered bounding box with a fixed size of 128×128 pixels. We train our model for 1500 epochs, adopting the Adam optimizer [12]. We employ a weighted cross-entropy loss with weight 0.1 for the background and 0.3 for the rest of the labels. The batch size is set to 2. We perform random data augmentation such as rotation and brightness adjustment. To train the network, we utilized an NVIDIA A100-SXM4-40GB. The training process took approximately 24 h to complete.

Comparisons with Other Methods. We compare our method with three state-of-the-art methods, nnU-Net [10], CAT-Net [14], and CSDG [18]. We trained CAT-Net using LA images. CAT-Net is trained through self-supervision with supervoxels, and at inference the true labels are given in the support set. CSDG is trained only with SA, being a fully unsupervised architecture, while nnU-Net is trained with a varying number of labeled LA slices added to the SA training data. For nnU-Net, the number of shots in Tables 1-2 refers to the number of LA images that are given during its training. These images are augmented to have a balanced training set between SA and LA. The results in Table 1 show that nnU-Net fails to generalize from SA to LA having access to only one labeled image. In comparison, CSDG generalizes better to different MRI orientations, obtaining an average DICE of 65.3%. Further, CAT-Net underperforms reaching only an average DICE of 54.2%. This is likely because the LA images are 2D and thus, superpixel-based segmentations can only be generated in 2D which seems to be disadvantageous to the overall performance. Differently to CAT-Net, our method efficiently exploits the information contained in the support image leading to an average score of 74.5% in the 1-Shot scenario, when relying only on a single LA image. In Table 2, it can be observed that increasing the size of the support set up to 4, our method achieves the best result for all of the labels, outperforming nnU-Net on average by 21.4% in the 2-Shot scenario and by 5.1% in the 4-Shot scenario. When the size of the support set further increases to 6 and 8 images, our model continues to obtain the best score for

Table 1. DICE (%) scores for the 0/1-Shot setting.

Method	Shot	DICE			
		LV	RV	MY	Avg.
CSDG [18]	0	79.1	62.5	54.3	65.3
nnU-Net [10]	1	43.8	30.6	36.5	40.0
CAT-Net [14]	1	71.7	58.8	32.5	54.2
Ours	1	**85.9**	**71.9**	**65.8**	**74.5**

Table 2. DICE (%) scores for different Shot settings.

Method	2-Shot			4-Shot			6-Shot			8-Shot		
	LV	RV	MY	LV	RV	MY	LV	RV	MY	LV	RV	MY
nnU-Net [10]	64.1	47.6	53.2	79.0	66.1	68.7	81.8	**76.1**	**73.0**	85.7	**76.9**	**74.6**
Ours	**87.4**	**73.4**	**68.5**	**87.5**	**74.0**	**68.9**	**87.5**	74.8	68.8	**87.6**	75.3	68.9

the LV, while nnU-Net starts to perform better on the other two labels due to the additional LA images that are provided during training nnU-Net. Figures 2 and 4 show that the performance of our model increasingly improves with the number of images in the support set. The largest improvement for all labels is achieved when the size of the support set is increased from 1 to 2 images. While additional improvements beyond 2 support images are comparatively small for the LV and MY, the RV segmentation performance continues to benefit more, achieving an additional improvement of 2% in terms of DICE when increasing the support set from 2 to 10 images. This behaviour is to be expected as the RV undergoes the largest variation from slice to slice and consequently, benefits the most from a diverse set of support images.

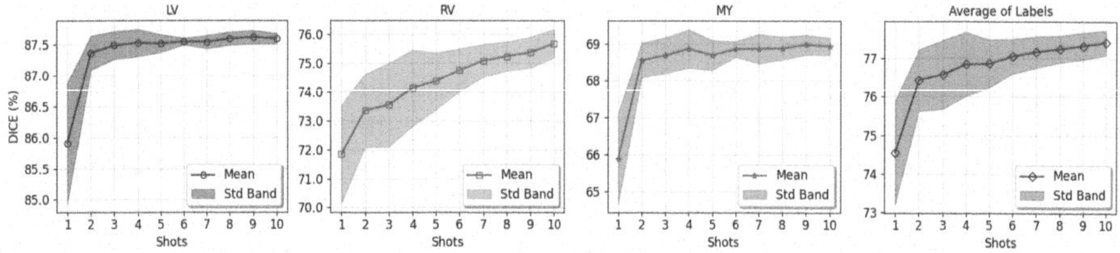

Fig. 2. Average DICE scores (%) ± one standard deviation of our model for an increasing support set from 1 to 10 images.

The impact of different support sets during inference is shown in Table 3, where the standard deviation of a 5-fold cross-validation for the support set is provided. As expected, in the 1-Shot scenario the model retrieves the LA information just from one image leading to a higher variability. Hence, in this scenario, the choice of images in the support set is significant. However, when the number of shots increases, the model is more robust to changes in the support set and the differences in the images affect the accuracy less.

Finally, in Table 4 the fraction of predicted images with a DICE score above a certain threshold is shown. In setting 1 we choose a threshold of 0.9 for the LV and 0.7 for the RV and MY. In setting 2 we fix the thresholds at 0.9, 0.75 and 0.75 for LV, RV and MY, respectively. While in setting 3 we choose as thresholds 0.9, 0.85 and 0.75. As expected, the number of predictions with meaningful segmentation increases with the number of shots, going from 23.8% to nearly 30% in the first scenario. Figure 3 depicts two caveats of the FSS method: the over-segmentation of the atrium, and the tendency of the myocardium to exhibit a ring-shaped pattern rather than an open-shaped configuration. The first problem

Table 3. Standard Deviation (%) scores for different Shot settings.

1-Shot			2-Shot			4-Shot			6-Shot			8-Shot		
LV	RV	MY	LV	RV	MY	LV	RV	MY	LV	RV	MY	LV	RV	MY
1.0	1.7	1.2	0.3	1.3	0.5	0.2	1.3	0.5	0.1	0.7	0.2	0.1	0.5	0.3

Table 4. Fraction (%) of predicted images with DICE above a certain threshold for different Shot settings.

Settings	Fraction %			
	1-Shot	4-Shot	6-Shot	10-Shot
Setting 1	23.75	27.68	28.45	29.53
Setting 2	12.35	12.90	13.54	14.71
Setting 3	5.79	5.61	5.96	6.73

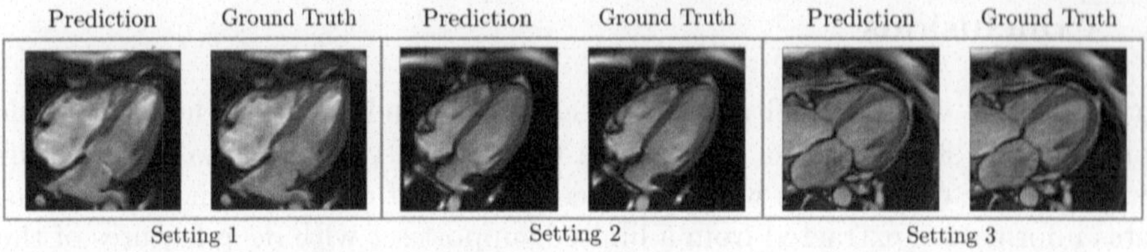

Fig. 3. Qualitative results of our method with different accuracies. On the left, we present a prediction that satisfies setting 1 but not settings 2 and 3, in comparison to the corresponding ground truth. In the center, a prediction meets setting 2 but not setting 3. On the right, the prediction fulfills setting 3.

is due to the absence of a clear gradient between the atrium and the ventricle. However, in most of the cases, the model is able to distinguish the two anatomical structures and segment only the ventricle. The second problem depends on the ring-shaped pattern of the myocardium in the short-axis images used for the training of the model, but, also in this case, extracting the information from the support set is beneficial in order to minimize this error. Moreover, we can argue that even if the percentage of predictions in the third scenario is relatively low, also the predictions in the other settings do not differ too much from the ground truth. However, the number of cases clinically usable is still small, and future works will aim to enhance the results.

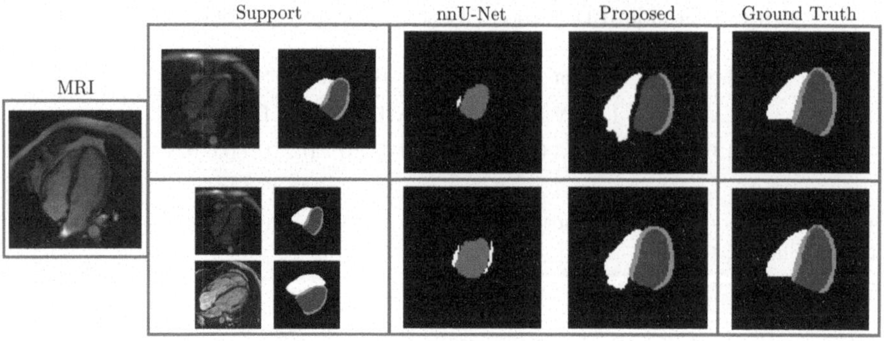

Fig. 4. Qualitative results of our method for an LA slice under 1-Shot (first row) and 2-Shot (second row). From left to right we show: MRI to segment, the support set used in the test phase, nnU-Net prediction, our prediction, ground truth.

4 Conclusions

In this work, we proposed a few-shot cardiac segmentation architecture that combines the strengths of a U-Net with the flexibility of GPEs to segment different cardiac orientations with a limited amount of data. Our model incorporates information extracted from a limited support set with deep features of the CNN, increasing the generalizability of the whole architecture. We evaluated our method for a scenario in which only SA slices are available for training and obtained meaningful heart segmentations for LA orientation. In particular, we obtain a great performance boost using a single support image. As a result, with small effort in terms of annotation, it is possible to work with the differently oriented cardiac images without retraining of the whole network. We plan to further improve our model, in particular, the GPE component by taking into account the conditional variance, and aim to explore the possibility of integrating uncertainty quantification in the predictions.

Acknowledgments and Disclosure of Funding. B. V., K. L. K., and E. K. acknowledge funding from BioTechMed-Graz Young Research Group Grant CICLOPS; F. T. acknowledges funding from the Austrian Science Fund (FWF) CardioTwin grant I6540; M.U. acknowledges funding from FWF grant 10.55776/PAT1748423.

References

1. Andrianakis, I., Challenor, P.G.: The effect of the nugget on Gaussian process emulators of computer models. Comput. Stat. Data Anal. **56**(12), 4215–4228 (2012)
2. Campello, V.M., et al.: Multi-centre, multi-vendor and multi-disease cardiac segmentation: the M&Ms challenge. IEEE Trans. Med. Imaging **40**, 3543–3554 (2021)
3. Chang, Z., Lu, Y., Ran, X., Gao, X., Wang, X.: Few-shot semantic segmentation: a review on recent approaches. Neural Comput. Appl. **35**, 18251–18275 (2023)

4. Chen, C., et al.: improving the generalizability of convolutional neural network-based segmentation on CMR images. Front. Cardiovasc. Med. **7**, 105 (2020)
5. Chen, C., et al.: Deep learning for cardiac image segmentation: a review. Front. Cardiovasc. Med. **7**, 25 (2020)
6. Gao, S., Zhou, H., Gao, Y., Zhuang, X.: BayeSeg: bayesian modeling for medical image segmentation with interpretable generalizability. Med. Image Anal. **89**, 102889 (2023)
7. Guan, H., Liu, M.: Domain adaptation for medical image analysis: a survey. IEEE Trans. Biomed. Eng. **69**(3), 1173–1185 (2021)
8. Hansen, S., Gautam, S., Jenssen, R., Kampffmeyer, M.: Anomaly detection-inspired few-shot medical image segmentation through self-supervision with supervoxels. Med. Image Anal. **78**, 102385 (2022)
9. Hansen, S., Gautam, S., Salahuddin, S.A., Kampffmeyer, M., Jenssen, R.: ADNet++: a few-shot learning framework for multi-class medical image volume segmentation with uncertainty-guided feature refinement. Med. Image Anal. **89**, 102870 (2023)
10. Isensee, F., Jaeger, P.F., Kohl, S., Petersen, J., Maier-Hein, K.H.: nnU-Net: a self-configuring method for deep learning-based biomedical image segmentation. Nat. Methods **18**, 203–211 (2021)
11. Johnander, J., Edstedt, J., Felsberg, M., Khan, F.S., Danelljan, M.: Dense Gaussian processes for few-shot segmentation. In: Computer Vision – ECCV 2022, pp. 217–234. European Conference on Computer Vision (2022)
12. Kingma, D.P., Ba, J.: Adam: a method for stochastic optimization (2017). https://arxiv.org/abs/1412.6980
13. Kumari, S., Singh, P.: Deep learning for unsupervised domain adaptation in medical imaging: recent advancements and future perspectives. Comput. Biol. Med. **170**, 107912 (2023)
14. Lin, Y., Chen, Y., Cheng, K.T., Chen, H.: Few shot medical image segmentation with cross attention transformer. In: International Conference on Medical Image Computing and Computer-Assisted Intervention, pp. 233–243. Springer (2023)
15. Litjens, G., et al.: A survey on deep learning in medical image analysis. Med. Image Anal. **42**, 60–88 (2017)
16. Martín-Isla, C., et al.: Deep learning segmentation of the right ventricle in cardiac MRI: the M&Ms challenge. IEEE J. Biomed. Health Inform. **27**, 3302–3313 (2023)
17. Ouyang, C., Biffi, C., Chen, C., Kart, T., Qiu, H., Rueckert, D.: Self-supervised learning for few-shot medical image segmentation. IEEE Trans. Med. Imaging **41**, 1837–1848 (2022)
18. Ouyang, C., et al.: Causality-inspired single-source domain generalization for medical image segmentation. IEEE Trans. Med. Imaging **42**, 1095–1106 (2022)
19. Rasmussen, C.E., Williams, C.K.I., et al.: Gaussian processes for machine learning, vol. 1. Springer (2006)
20. Ronneberger, O., Fischer, P., Brox, T.: U-net: convolutional networks for biomedical image segmentation. In: Medical Image Computing and Computer-Assisted Intervention–MICCAI 2015: 18th International Conference, Munich, Germany, 5–9 October 2015, proceedings, part III 18, pp. 234–241 (2015)
21. Roy, A.G., Siddiqui, S., Pölsterl, S., Navab, N., Wachinger, C.: 'Squeeze & excite' guided few-shot segmentation of volumetric images. Med. Image Anal. **59**, 101587 (2020)
22. Saha, S., Choi, O., Whitaker, R.: Few-shot segmentation of microscopy images using Gaussian process. In: International Workshop on Medical Optical Imaging and Virtual Microscopy Image Analysis, pp. 94–104 (2022)

23. Shaban, A., Bansal, S., Liu, Z., Essa, I., Boots, B.: One-shot learning for semantic segmentation. arXiv preprint arXiv:1709.03410 (2017)
24. Wang, K., Liew, J.H., Zou, Y., Zhou, D., Feng, J.: Panet: few-shot image semantic segmentation with prototype alignment. In: Proceedings of the IEEE/CVF International Conference on Computer Vision, pp. 9197–9206 (2019)
25. World Health Organization: Cardiovascular diseases. https://www.who.int/health-topics/cardiovascular-diseases. Accessed 28 June 2024

Global Control for Local SO(3)-Equivariant Scale-Invariant Vessel Segmentation

Patryk Rygiel[1(✉)], Dieuwertje Alblas[1], Christoph Brune[1], Kak Khee Yeung[2,3], and Jelmer M. Wolterink[1]

[1] Department of Applied Mathematics, Technical Medical Centre, University of Twente, Enschede, The Netherlands
{p.t.rygiel,d.alblas,c.brune,j.m.wolterink}@utwente.nl
[2] Department of Surgery, Amsterdam UMC location Vrije Universiteit Amsterdam, Amsterdam, The Netherlands
[3] Amsterdam Cardiovascular Sciences, Microcirculation, Amsterdam, The Netherlands

Abstract. Personalized 3D vascular models can aid in a range of diagnostic, prognostic, and treatment-planning tasks relevant to cardiovascular disease management. Deep learning provides a means to obtain such models automatically from image data. Ideally, a user should have control over the included region in the vascular model. Additionally, the model should be watertight and highly accurate. To this end, we propose a combination of a global controller leveraging voxel mask segmentations to provide boundary conditions for vessels of interest to a local, iterative vessel segmentation model. We introduce the preservation of scale- and rotational symmetries in the local segmentation model, leading to generalisation to vessels of unseen sizes and orientations. Combined with the global controller, this enables flexible 3D vascular model building, without additional retraining. We demonstrate the potential of our method on a dataset containing abdominal aortic aneurysms (AAAs). Our method performs on par with a state-of-the-art segmentation model in the segmentation of AAAs, iliac arteries, and renal arteries, while providing a watertight, smooth surface representation. Moreover, we demonstrate that by adapting the global controller, we can easily extend vessel sections in the 3D model. Our code is available on GitHub(https://github.com/MIAGroupUT/SIRE-segmentation.).

Keywords: vessels · segmentation · cardiovascular · geometric deep learning · scale-invariance · rotation-equivariance · data efficiency

1 Introduction

Detailed and topologically correct 3D vascular models can aid in many diagnostic, prognostic, and treatment-planning tasks in patients with cardiovascular

P. Rygiel and D. Alblas—Equal contribution.

O. Camara et al. (Eds.): STACOM 2024, LNCS 15448, pp. 269–279, 2025.
https://doi.org/10.1007/978-3-031-87756-8_27

diseases. Examples include diameter measurements in aneurysms [5], tortuosity measurements [7], fitting stent grafts [14], or hemodynamic analysis with computational fluid dynamics (CFD) [23]. Automatically obtaining 3D models for these purposes has long been a topic of interest [15] that has been primarily targeted with deep learning (DL) recently [6,16].

Most current DL methods for 3D vessel segmentation perform voxel segmentation. These models interpret the *global* context of the image and yield a 3D voxel mask of the vessels, e.g., [10,20]. However, the topological correctness of such voxel masks can only be imposed as a soft constraint [22]. Moreover, the resolution of voxel masks is limited to that of the scanner, which for downstream tasks such as CFD is often insufficient. Additionally, expanding the segmentation's region of interest (ROI) requires adapting annotations and retraining the network, limiting the flexibility of voxel-based segmentation. Alternatively, vessels can be parametrized as generalized cylinders consisting of a manually or automatically [4,8,25] identified centerline and a set of *locally* orthogonal contours [21]. This parametrization has been successfully integrated into machine learning approaches that achieve sub-voxel accuracy [2,17,26]. Performing segmentation locally allows for the incorporation of rotational and scale symmetries present in vessels [2,4] to learn efficiently from few annotations. In contrast to voxel masks, a generalized cylinder is by design a topologically correct representation of the vessel and allows for expanding the segmentation's ROI by elongating the centerline. However, a drawback of local tracking and segmentation is the lack of global context, making it difficult to detect the start and end of the desired vascular centerline. When the user is interested in only a particular segment of the vasculature, stopping criteria based on local heuristics [8,12,25] are unable to constrain a local model without global information.

In this work, we unify global and local segmentation approaches by introducing a global controller for a local segmentation model. Open source multi-organ segmentation models like TotalSegmentator [24] are a useful commodity for the community that we here leverage to provide a set of *global* boundary conditions to constrain *local* tracking and segmentation. This gives us the best of both worlds: we can select only those parts of the vasculature that are relevant to a particular task, and we can exploit local symmetries in vessel segmentation to obtain generalizable models with sub-voxel accuracy. Specifically, we introduce scale-invariant and rotation-equivariant local segmentation, which generalizes between vessels with very different calibres and tortuosities. We reconstruct smooth and watertight surfaces for each artery by fitting a neural field, that represents signed distance function (SDF), to the contours resulting from the local segmentation and blend these surfaces into watertight meshes of the vasculature [3]. We demonstrate the effectiveness of our method on a dataset of patients with abdominal aortic aneurysms (AAAs), where we obtain topologically correct and smooth models of the renal arteries, iliac arteries, and abdominal aorta. Moreover, we show how the user can efficiently adapt the segmentation ROI. Our results show that global and local segmentation are not mutually exclusive, but can be combined in a fruitful synergy to add value for cardiovascular disease management.

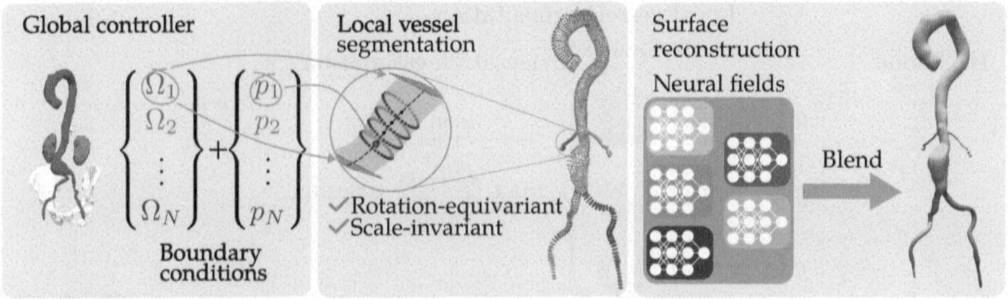

Fig. 1. Proposed framework. A global segmentation method provides boundary conditions $[\Omega_i, p_i]$, such as points or ROIs to control local vessel segmentation. A local joint tracking and segmentation model accurately segments vessels in a rotation-equivariant, scale-invariant manner. Obtained contour sets are reconstructed into a watertight surface using neural fields and blended into a single vascular model [3].

2 Materials and Methods

Figure 1 visualizes our framework, consisting of three elements: (1) extraction of boundary conditions for a global controller, (2) iterative local vessel segmentation subject to these boundary conditions, and (3) reconstruction of a smooth watertight surface i.e., a mesh without self-intersections, nor non-manifold edges or vertices, in which each edge has exactly two incident vertices.

2.1 Global Controller

Iterative local vessel trackers lack a global overview and hence do not know where to start or terminate. In previous works, starting points were sampled from a segmentation mask, and stopping criteria were based on heuristics intrinsic to the tracker [4,25], but this often resulted in early stopping or roaming of the tracker. Instead, we use coarse segmentations of TotalSegmentator [24], an off-the-shelf multi-organ segmentation algorithm, as a global controller for the precise and topologically accurate iterative vessel segmentation module. For each vessel of interest v_i in our 3D vascular model, we use this coarse segmentation to determine the boundary conditions, i.e. a seed point p_i and the set of stopping criteria Ω_i. Seed points are represented as 3D points and stopping criteria are represented as 3D binary masks. The boundaries of these masks define the regions in which segmentation can be performed; hence, crossing these boundaries causes the tracker to stop.

2.2 Local Vessel Tracking and Segmentation

From a local perspective, vessel appearance is subject to change due to varying orientations and diameters of the vessel. For example, the diameters of renal arteries are around 5 mm, iliac arteries range around 10 mm, while AAAs with diameters of 80 mm are not uncommon [1]. Moreover, renal arteries are generally

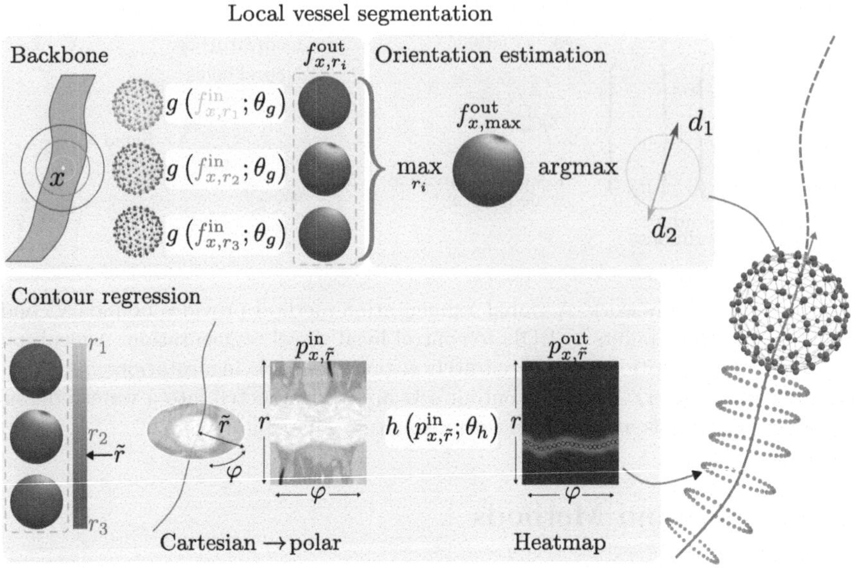

Fig. 2. Overview of our iterative segmentation method. At a point x, local vessel directions d_1, d_2 and optimal scale $\tilde{r}$ are determined based on multi-scale spherical image features. According to these values a scale-adaptive polar image is constructed based on which CNN $h(\cdot;\theta_h)$ predicts lumen radii for each angle in $p^{\text{in}}_{\tilde{r}}$, forming a closed contour in Cartesian space.

more tortuous than the abdominal aorta. This poses a challenge to local artery segmentation methods that use image patches of fixed resolution, size, and orientation,e.g. [2,17,26]. A local segmentation method that is unaffected by vessel diameter and tortuosity is thus desirable.

We propose a scale-invariant, rotation-equivariant local segmentation module that performs iterative joint tracking and segmentation of the vessel, using its cylindrical shape as an anatomical prior. To track vessel v_i, the module is initialized at seed point p_i, provided by the global controller. First, the module estimates up- and downstream vessel directions $\boldsymbol{d}_1$ and $\boldsymbol{d}_2$ at p_i. Subsequently, the module takes a step of size Δ in direction $\boldsymbol{d}_1$. Every η steps, the module delineates the lumen boundary orthogonal to the local vessel orientation. These two steps are shown in Fig. 2. Once one of the boundary conditions in Ω_i is violated, the module terminates. If direction $\boldsymbol{d}_2$ is not yet explored, the module is re-initialized at p_i and traverses the vessel in direction $\boldsymbol{d}_2$ until termination. Both resulting centerline paths are combined into one connected centerline.

Both steps of the iterative segmentation rely on a scale-invariant, rotation-equivariant backbone [4], which extracts features from image data centered around a point $x \in \mathbb{R}^3$ (top-left Fig. 2). This backbone consists of a graph convolutional network (GCN) [11] $g(\cdot;\theta_g)$ operating on the surface of the unit sphere on which local image information within a predefined radius r of x has been projected ($f^{\text{in}}_{x,r}$). The backbone outputs a scalar field $f^{\text{out}}_{x,r}$ on the unit sphere. To adapt to vessels of different calibres, the backbone processes local image information at different scales in parallel. We construct a set of multi-scale inputs $\{f^{\text{in}}_{x,r_j}\}_{j=1,\ldots,m}$ from a pre-defined set of m scales $R = \{r_j\}_{j=1,\ldots,m}$ that we feed through $g(\cdot;\theta_g)$. This results in a set of multi-scale outputs $\{f^{\text{out}}_{x,r_j}\}_{j=1,\ldots,m}$, which are used by the orientation estimation and contour regression modules.

The vessel orientation estimation module first aggregates $\{f^{\text{out}}_{x,r_j}\}_{j=1,\ldots,m}$ into a scale-invariant output $f^{\text{out}}_{x,\max}$ by using a permutation-invariant maximum operation across the scales (top-right Fig. 2) [4]. The local orientations $\boldsymbol{d}_1$ and $\boldsymbol{d}_2$ of the vessel at x are represented by the locations of $f^{\text{out}}_{x,\max}$, at least 90° apart. The vessel orientation estimation module is trained with an MSE loss between $f^{\text{out}}_{x,\max}$ and a target heatmap as in [4].

The contour regression module yields a closed contour delineating the lumen boundary on a 2D plane through x orthogonal to the centerline, while adapting to the vessel calibre (bottom Fig. 2). Closed contours are enforced by estimating the lumen radius for each angle on a polar transformed image $p^{\text{in}}_{x,\tilde{r}} \in \mathbb{R}^{N_r \times N_\varphi}$, centered at x with radius $\tilde{r}$ [2]. Instead of using polar images with fixed $\tilde{r}$ as in [2], we adapt the radius $\tilde{r}$ to the local vessel size, by computing it as the weighted average of the maximum activations on each scale-wise response in $\{f^{\text{out}}_{x,r_j}\}_{j=1,\ldots,m}$. Based on $p^{\text{in}}_{x,\tilde{r}}$, a CNN $h(\cdot;\theta_h)$ predicts a heatmap $p^{\text{out}}_{x,\tilde{r}} \in \mathbb{R}^{N_r \times N_\varphi}$, where for each angle φ_i a corresponding lumen radius is extracted via a differentiable DSNT module [9,19]. These lumen radii are transformed back to Cartesian space to form the predicted lumen contour, hence $h(\cdot;\theta_h)$ can be optimized using a set of ground-truth contours. Network $h(\cdot;\theta_h)$ is translation equivariant and operates on a polar transformed image, implying rotational equivariance in Cartesian space: a rotation of the 2D input plane leads to the same rotation of the output contour. For the loss function we adopt the same setup as in [9,19] consisting of a Euclidean distance term between the predicted and ground truth lumen contours and Jensen-Shannon divergence between the $p^{\text{out}}_{x,\tilde{r}}$ and GT heatmap.

2.3 Surface Reconstruction

Iterative tracking results in a set of contours describing the lumen surface of vessel v_i, but this is not watertight. To reconstruct a watertight surface subject to these contours, we use a neural field (right Fig. 1). This neural field consists of a multi-layer perceptron that is conditioned on spatial coordinates $x \in \mathbb{R}^3$ and is optimized to represent the signed distance function of the lumen surface [18]. As these neural fields are conditioned on continuous coordinates, they have an infinite resolution. Moreover, a smooth transition between connecting vessels is made by taking a smoothed minimum of their signed distance functions [3]. From the neural field, we can obtain a mesh model of any resolution using the marching cubes algorithm.

2.4 Data

We use an in-house dataset consisting of 80 pre-operative contrast-enhanced CT (CTA) scans of AAA patients obtained from Amsterdam UMC, containing at least the thoracic region until the iliac bifurcation. In-plane resolutions and slice thicknesses range between 0.6×0.6 and $1.0\times1.0\,\mathrm{mm}^2$ and 0.5–2.0 mm, respectively. For each patient, we obtained manual annotations consisting of centerlines and locally orthogonal contours of the lumen of the abdominal aorta until the top of the T12 vertebra, the common iliac arteries, and the renal arteries.

3 Experiments and Results

We split the dataset into 14 test cases and 66 train cases and use the same data split for both networks in the local vessel segmentation module. We denote the orientation estimation module as $\mathcal{T} = \{g(\cdot;\theta_g)\}$ and the contour regression module as $\mathcal{S} = \{g(\cdot;\theta_g), h(\cdot;\theta_h)\}$. Note that the orientation estimation and contour regression are trained separately. Both $\mathcal{T}$ and $\mathcal{S}$ were trained with an Adam optimizer, with learning rates of 0.005, 0.001 and batch sizes of 20 and 10 respectively. For both training and inference, we used scales $R = \{5, 10, 15, ..., 80\}$ for the backbone. For tracking we use step sizes $\Delta_{aorta} = 1$ mm, $\Delta_{iliac} = 0.5$ mm, $\Delta_{renal} = 0.3$ mm and delineate contour every $\eta = 5$ steps.

The global controller provided seed points and boundary conditions for tracking the abdominal aorta, iliac arteries and renal arteries. Unless stated otherwise, we used the following heuristics for p_i and Ω_i. p_{aorta} and p_{iliac} are the centers of mass from the aorta and iliac masks provided by TotalSegmentator, respectively. p_{renal} is a bit more challenging to obtain, as TotalSegmentator does not include renal arteries. Instead, we estimate the renal artery by computing the shortest path between the kidneys and the aorta based on image intensities and let p_{renal} be the center of this path. Ω_{aorta} consists of the region above the T12 vertebra and the iliac artery segmentations, Ω_{iliac} contains the aorta and the region below the L5 vertebra, and Ω_{renal} contains the kidneys and the aorta.

3.1 Segmentation Quality

We evaluate the quality of the automatically extracted 3D vascular models on the test set. For each of the 14 test cases, we automatically extract the lumen of the abdominal aorta, renal arteries, and iliac arteries. We compare the quality of these segmentations to an nnU-Net benchmark [13]. We obtained Dice similarity coefficients (DSCs) of 0.84 ± 0.05 and 0.83 ± 0.14 for our method and nnU-Net, respectively, demonstrating that we can fully automatically obtain watertight meshes with sub-voxel accuracy without compromising on overlap metrics.

3.2 Data Efficiency

We train our iterative vessel segmentation method using contour annotations on 2D planes in a 3D images. Therefore, a single 3D scan contains hundreds of training samples. This - in combination with the symmetries that we embed - makes our method very data efficient. We retrained the contour regression model on subsets of 3, 5, 10, and 15 patients randomly selected from the training set to assess this data efficiency. We evaluate the performance of each model using the DSC on the 2D contours in the test set. Figure 3 (left) shows our method achieves a median DSC of at least 0.9, even when only three patients were used for training.

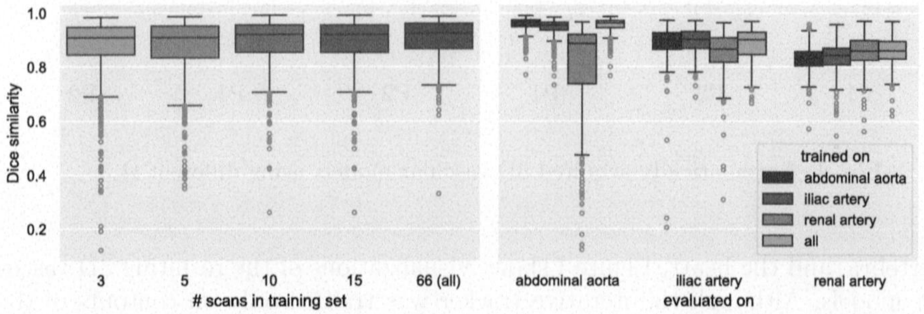

Fig. 3. DSC scores of the contour regression modules on 2D contours in the test set for the data ablation experiments. *Left:* DSC for all contours for data ablation at patient level. *Right:* DSC per vessel for data ablation at the vessel level.

As the contour regression module is scale-invariant, it should generalise to vessels of unseen calibre. To test this property, we perform data ablation at the artery level. We retrain three contour regression modules separately on the lumen contours of the abdominal aorta, iliac arteries, and renal arteries, whose diameters range between 20–55 mm, 10–15 mm, and 5–7 mm, respectively. We assess the performance of the four models using the DSC on the 2D contours of the test set. Figure 3 (right) shows that overall, our model can generalise to arteries of unseen size. For example, the model trained on abdominal aorta contours performs on par with the model trained on iliac arteries when segmenting iliac arteries and vice versa. Moreover, the model trained on all arteries shows performance comparable to the vessel-specific model, indicating that heterogeneity of vessel size in the training data does not deteriorate segmentation performance. However, we observe a performance drop for segmentation of the abdominal aorta for the model trained on renal arteries.

3.3 Modifying Boundary Conditions

The global controller we present provides flexible inclusion of different vessel ROIs in the 3D model. To show this, we extract multiple 3D models while adapting Ω_{aorta}. We include three different regions in Ω_{aorta} to terminate iterative segmentation in the superior direction: until the T10 vertebra, the T12

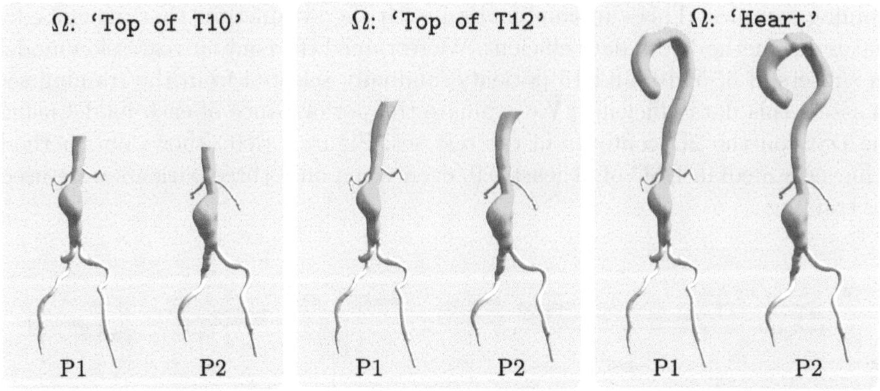

Fig. 4. Automatically acquired 3D vascular models using different Ω_{aorta}.

vertebra, and the heart. Figure 4 shows visualizations of the resulting 3D vascular models. Although the iterative tracker was trained only on contours of the aorta between the T12 vertebra and iliac bifurcation, it still accurately tracks and segments the aortic arch due to its symmetry-preserving properties.

4 Conclusion and Discussion

We have presented a general framework for automatically extracting smooth, topologically correct vascular models, combining strengths of global and local vessel segmentation methods. A global controller provides boundary conditions for a local iterative vessel segmentation model, that exploits local scale- and rotational symmetries to generalise to vessels of unseen sizes and orientations. Our post-processing step provides a neural field of the vascular model that can be reconstructed to a watertight mesh of any desired resolution. We demonstrated the effectiveness of symmetry preservations by automatic extraction of vascular models of abdominal aortic aneurysms, iliac arteries and renal arteries.

By leveraging local scale- and rotational symmetries, our iterative joint tracking and segmentation module generalizes to vessels of sizes (Fig. 3, left) and segments (Fig. 4) unseen during training. Despite this strong generalisation, some symmetries cannot be preserved. For example, the contour regression model trained on renal arteries struggles to segment the abdominal aorta lumen if surrounded by a thrombus, as this was not seen during training. Moreover, generalisation to other image modalities, e.g. MRI or ultrasound requires retraining. However, as the local vessel segmentation step can be trained on contour annotations of a few patients, data collection and annotation costs are low.

The global controller we present in this work gives control over vessel ROIs to incorporate in the final 3D model. In combination with the generalising properties of the local iterative tracker, this allows us to automatically build 3D vascular models in other anatomical regions by providing a different set of heuristics.

We used coarse segmentation masks from TotalSegmentator [24] as input for our global controller, an off-the-shelf tool. In combination with the generalising local segmentation method, it has the potential to automatically segment any vascular region without the need for additional specific annotations. A potential drawback of this approach is that the local segmentation model can only segment those vessel segments for which at least a seed point can be identified. We found, that in some cases, the global controller could not identify renal arteries, thereby affecting overall segmentation performance. Note that we use TotalSegmentator but are not dependent on this specific model. Any existing model that provides binary segmentation masks or seed points could be used, and multiple models could be combined for specialized tasks.

As the boundary conditions for each artery are standardized, the ROIs of the resulting 3D vascular models are consistent across subjects. This enables large-scale studies of the role of hemodynamic and morphological features in the prognosis of cardiovascular diseases and could in future work be extended to quantification of other vessel-specific structures such as intraluminal thrombus and plaques that are important in the analysis of cardiovascular diseases.

In conclusion, we have shown that global constraints can be imposed on local tracking and segmentation models to provide high-quality watertight segmentations of the vasculature.

Acknowledgments. This project has received funding from the European Union's Horizon Europe research and innovation programme under grant agreement No 101080947 (VASCUL-AID). Jelmer M. Wolterink was supported by the NWO domain Applied and Engineering Sciences Veni grant (18192).

References

1. Aggarwal, S., Qamar, A., Sharma, V., Sharma, A.: Abdominal aortic aneurysm: a comprehensive review. Exp. Clin. Cardiol. **16**(1), 11 (2011)
2. Alblas, D., Brune, C., Wolterink, J.M.: Deep learning-based carotid artery vessel wall segmentation in black-blood MRI using anatomical priors (2021)
3. Alblas, D., Brune, C., Yeung, K.K., Wolterink, J.M.: Going off-grid: continuous implicit neural representations for 3D vascular modeling. In: International Workshop on Statistical Atlases and Computational Models of the Heart, pp. 79–90. Springer (2022)
4. Alblas, D., Suk, J., Brune, C., Yeung, K.K., Wolterink, J.M.: SIRE: scale-invariant, rotation-equivariant estimation of artery orientations using graph neural networks (2023)
5. Brewster, D.C., Cronenwett, J.L., Hallett, Jr, J.W., Johnston, K.W., Krupski, W.C., Matsumura, J.S.: Joint Council of the American Association for Vascular Surgery and Society for Vascular Surgery: Guidelines for the treatment of abdominal aortic aneurysms. report of a subcommittee of the joint council of the American association for vascular surgery and society for vascular surgery. J. Vasc. Surg. **37**(5), 1106–1117 (2003)
6. Chen, C., et al.: Deep learning for cardiac image segmentation: a review. Front. Cardiovasc. Med. **7**, 25 (2020)

7. Ciurică, S., et al.: Arterial tortuosity: novel implications for an old phenotype. Hypertension **73**(5), 951–960 (2019)
8. Friman, O., Hindennach, M., Kühnel, C., Peitgen, H.O.: Multiple hypothesis template tracking of small 3D vessel structures. Med. Image Anal. **14**(2), 160–171 (2010)
9. Gajowczyk, M., et al.: Coronary ostia localization using residual u-net with heatmap matching and 3D DSNT. In: Lian, C., Cao, X., Rekik, I., Xu, X., Cui, Z. (eds.) Machine Learning in Medical Imaging, pp. 318–327. Springer Nature Switzerland, Cham (2022)
10. Gharleghi, R., et al.: Automated segmentation of normal and diseased coronary arteries-the ASOCA challenge. Comput. Med. Imaging Graph. **97**, 102049 (2022)
11. de Haan, P., Weiler, M., Cohen, T., Welling, M.: Gauge equivariant mesh CNNs: anisotropic convolutions on geometric graphs. CoRR **abs/2003.05425** (2020). https://arxiv.org/abs/2003.05425
12. van Harten, L.D., de Jonge, C.S., Beek, K.J., Stoker, J., Išgum, I.: Untangling and segmenting the small intestine in 3D cine-MRI using deep learning. Med. Image Anal. **78**, 102386 (2022)
13. Isensee, F., Jaeger, P.F., Kohl, S., Petersen, J., Maier-Hein, K.H.: nnU-Net: a self-configuring method for deep learning-based biomedical image segmentation. Nat. Methods **18**(2), 203–211 (2021)
14. Larrabide, I., Radaelli, A., Frangi, A.: Fast virtual stenting with deformable meshes: application to intracranial aneurysms. In: Metaxas, D., Axel, L., Fichtinger, G., Székely, G. (eds.) MICCAI 2008. LNCS, vol. 5242, pp. 790–797. Springer, Heidelberg (2008). https://doi.org/10.1007/978-3-540-85990-1_95
15. Lesage, D., Angelini, E.D., Bloch, I., Funka-Lea, G.: A review of 3D vessel lumen segmentation techniques: models, features and extraction schemes. Med. Image Anal. **13**(6), 819–845 (2009)
16. Litjens, G., et al.: State-of-the-Art deep learning in cardiovascular image analysis. JACC Cardiovasc. Imaging **12**(8 Pt 1), 1549–1565 (2019)
17. Lugauer, F., Zhang, J., Zheng, Y., Hornegger, J., Kelm, B.M.: Improving accuracy in coronary lumen segmentation via explicit calcium exclusion, learning-based ray detection and surface optimization. In: Medical Imaging 2014: Image Processing, vol. 9034, pp. 993–1002. SPIE (2014)
18. Ma, B., Han, Z., Liu, Y.S., Zwicker, M.: Neural-pull: Learning signed distance function from point clouds by learning to pull space onto surface. In: International Conference on Machine Learning, pp. 7246–7257. PMLR (2021)
19. Nibali, A., He, Z., Morgan, S., Prendergast, L.A.: Numerical coordinate regression with convolutional neural networks. CoRR **abs/1801.07372** (2018). http://arxiv.org/abs/1801.07372
20. van Praagh, G.D., et al.: Automated multiclass segmentation, quantification, and visualization of the diseased aorta on hybrid PET/CT–sequoia. Med. Phys. (2024)
21. Shani, U., Ballard, D.H.: Splines as embeddings for generalized cylinders. Comput. Vis., Graph. Image Process. **27**(2), 129–156 (1984)
22. Shit, S., et al.: clDice - a topology-preserving loss function for tubular structure segmentation. CoRR **abs/2003.07311** (2020). https://arxiv.org/abs/2003.07311
23. Taylor, C.A., et al.: Patient-specific modeling of blood flow in the coronary arteries. Comput. Methods Appl. Mech. Eng. **417**, 116414 (2023). https://doi.org/10.1016/j.cma.2023.116414, https://www.sciencedirect.com/science/article/pii/S0045782523005388. a Special Issue in Honor of the Lifetime Achievements of T. J. R. Hughes

24. Wasserthal, J., et al.: TotalSegmentator: robust segmentation of 104 anatomic structures in CT images. Radiol.: Artif. Intell. **5**(5), e230024 (2023).https://doi.org/10.1148/ryai.230024, https://doi.org/10.1148/ryai.230024
25. Wolterink, J.M., van Hamersvelt, R.W., Viergever, M.A., Leiner, T., Isgum, I.: Coronary artery centerline extraction in cardiac CT angiography using a CNN-based orientation classifier. CoRR **abs/1810.03143** (2018). http://arxiv.org/abs/1810.03143
26. Wolterink, J.M., Leiner, T., Išgum, I.: Graph convolutional networks for coronary artery segmentation in cardiac CT angiography. In: Graph Learning in Medical Imaging: First International Workshop, GLMI 2019, Held in Conjunction with MICCAI 2019, Shenzhen, China, October 17, 2019, Proceedings 1, pp. 62–69. Springer (2019)

A Self-distillation Bi-atrial Segmentation Network for Cardiac MRI

Waqas Anwaar[1,2,3], Manh The Van[1,2,3], Jiongtong Hu[1,2,3], Zhurong Chen[1,2,3], Wufeng Xue[1,2,3(✉)], and Dong Ni[1,2,3]

[1] National-Reginoal Key Technology Engineering Laboratory for Medical Ultrasound, Guangdong Key Laboratory of Biomedical Measurements and Ultrasound Imaging, School of Biomedical Engineering, Shenzhen University Medical School, Shenzhen University, Shenzhen 518060, China
xuewf@szu.edu.cn

[2] Marshall Laboratory of Biomedical Engineering, Shenzhen University, Shenzhen, China

[3] Medical Ultrasound Image Computing (MUSIC) Lab, Shenzhen University, Shenzhen, China

Abstract. Bi-atrial segmentation in late gadolinium-enhanced MRI (LGE-MRI) is an important step for diagnosing and treating atrial fibrillation. However, in real scenarios, this process is hugely impeded by varying appearance of the atrial shape, intensity variations and large inter- and intra-class distribution among volumes. In this paper, we propose a 3D-based self-distillation network for robustly segmenting both the left atrium, right atrium as well as their walls. Specifically, our network employs a 3D segmentation model to better explore the spatial dependencies information in input volumes. The model is equipped with an atrous spatial pyramid pooling module to better deal with variation in atrial shape and thin boundary details at different scales followed by a spatial and depth self-attention module to learn the long-range relationship between relevant information within volume. In addition, we apply a distillation training scheme to improve the robustness of the segmentation model for huge intensity variations among volumes. The experimental results show that our framework can produce a robust performance for bi-atrial segmentation and attain an average Dice of 83.3% and HD95 of 4.51 (pixels).

Keywords: Bi-atrial segmentation · Self-attention · Self-distillation

1 Introduction

Atrial fibrillation (AF) is the most common type of cardiac arrhythmia and is associated with increased mortality and morbidity, particularly heart failure and stroke [5]. Late gadolinium-enhanced MRI (LGE-MRI) is widely used to

W. Anwaar and M. Van—These authors contributed equally to this work.

O. Camara et al. (Eds.): STACOM 2024, LNCS 15448, pp. 280–288, 2025.
https://doi.org/10.1007/978-3-031-87756-8_28

localize scars and provide guidance for ablation therapy. Segmenting the left atrial (LA), right atrial (RA) and their walls in LGE-MRI provides the structural information for diagnosing and treating atrial fibrillation. However, manual bi-atrial segmentation is time-consuming and expertise-dependent.

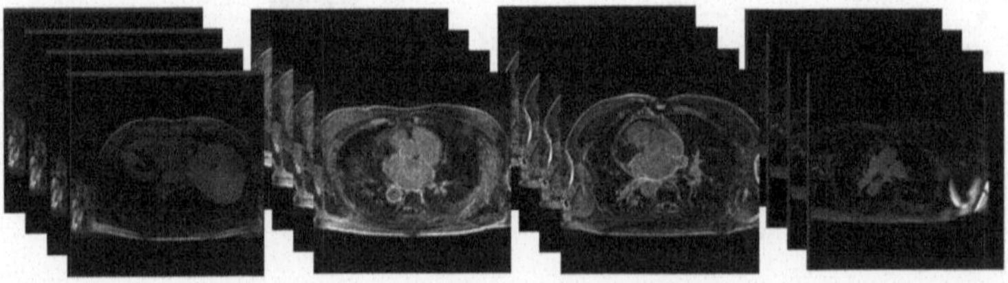

Fig. 1. Automatic bi-atrial segmentation is challenging due to the appearance shift between the volumes and the intra- and inter-class differences. *(for better visualization same intensity level is increased in every volume)*

Recently, deep-learning methods have gained a lot of successes in computer vision tasks such as classification [7], segmentation [11], and detection [2]. Many deep-learning methods have been proposed for the automatic segmenting of left atrial, right atrial, and their walls in cardiac MRI. For example, the STACOM 2018 Left Atrial Segmentation Challenge [15] has collected various CNN methods for only left atrial segmentation. Through the challenge, the UNet [12] was shown to be the most widely used and most easily adaptable architecture for the task. Tan et al. [13] opposed a benchmarking study of deep learning approaches for bi-atrial segmentation on LGE-MRI. Other methods tried to use a cascaded network for bi-atrial segmentation. For example, Xiong et al. [14] proposed a double 2D CNN pipeline in which the first network is for detecting the ROI and the second network is for regional segmentation of the ROI. However, these methods mostly lack considering the varying appearances with respect to intensity and shape and large inter- and intra-class differences among volumes (Fig. 1), which may hinder their performances.

In this paper, we propose an end-to-end 3D-based self-distillation network for robustly segmenting both the left atrium, right atrium as well as their walls. To better explore the spatial dependencies information in input volumes, an end-to-end 3D segmentation network is employed. We crop the ROI and use the model input size accordingly to avoid resizing and improve the capacity of the segmentation model. An atrous spatial pyramid pooling module is applied to better capture the contextual information at different scales especially for intrinsic wall thickness. We use a spatial and depth self-attention module to learn the long-range relationship information. Finally, we apply a self-distillation training scheme to improve the robustness of the segmentation model for different object appearances. The experimental results show that our framework can produce a robust performance for bi-atrial segmentation and attain an average Dice of 83.3% and HD95 of 4.51 pixels.

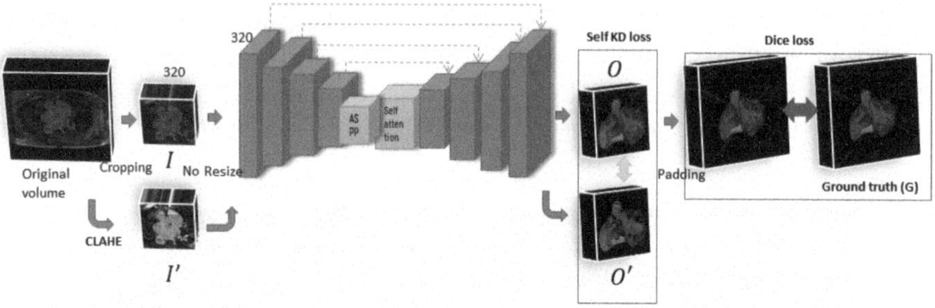

Fig. 2. Illustration of the proposed network for bi-atrial segmentation.

2 Methodology

The schematic illustration of our proposed framework is shown in Fig. 2. The center-cropped and contrast limited adaptive histogram equalization (CLAHE) ROIs serve as the inputs of our customized 3D network. Our network is equipped with atrous spatial pyramid pooling and self-attention modules. The model is trained with Dice loss and self-knowledge distillation loss.

2.1 A End-to-End Resizing-Free 3D Segmentation Network

In MRI segmentation, some methods use 2D UNet [12] to segment each slice of the volume and then aggregate the predictions for the final result. However, because 2D UNet takes a single slice as input, they inherently fail to leverage context from adjacent slices [1]. Voxel information from adjacent slices may be useful for the prediction of segmentation maps. Because of that, we propose to use end-to-end 3D-based work to learn the relationship between the slices. We did not perform any resizing which may cause to reduce the effects of losing structure details. As described in Fig. 2, the original volume is first centre cropped to 320×320 to focus only on the object and get rid of the irrelevant areas. We use the input size of the 3D UNet to 320×320 to avoid resizing the input and make the model bigger capacity. The model output is directly padded to the original size.

This challenge is a multi-class segmentation. Therefore, the sizes of the left atrium, right atrium, and their walls are different. Especially, the atrial walls are very thin. A 3D UNet with that regular convolutional operation can not deal complex shapes specially along the boundaries. To deal with scenario, we apply the atrous spatial pyramid pooling module (ASPP) [4] to aggregate the features at multiple scale.

2.2 Self-attention Module

In cardiac MRI, the relationship information between the slices is important to predict the overall shape of the object. Therefore, we employ the 3D self-

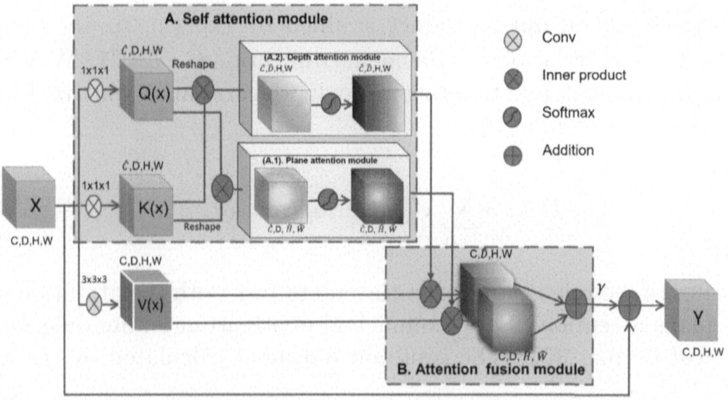

Fig. 3. Illustration of self-attention module.

attention module to force the model to learn the dependencies in different directions utilizing by two submodules named as plane attention module and depth attention module. [10]. Figure 3 shows the structure of the self-attention module which is applied in the bottleneck layer after the ASPP module which outputs the feature maps X having dimension (C,D,H,W), where C denotes channel(No. of feature maps), (D, H and W) denotes the depth, height and width. These feature maps are transformed into two query feature spaces Q(x) and K(x) by applying $1 * 1 * 1$ convolution to reduced channel space $\hat{C}$ which is served as feature encoding and channel pooling for the output of the ASPP module and V(x) as backbone convolutional layer with $3 * 3 * 3$ kernal.

Plane attention module computes the intra-slice dependency by first flattening the H and W of the feature maps into 1 D vector from both the feature spaces Q(x) and K(x). Then the inner product is calculated between these feature spaces along every channel to find out the relationship between every two points with the resultant dimensions of $\hat{H} * \hat{W} =$ (H $*$ W) $*$ (H $*$ W). Which is calculated as:

$$\sigma_{j,i} = Q(x_i)^T K(X_j), \tag{1}$$

Where $\sigma_{j,i}$ indicates the relationship of ith location with j^{th} location on feature map and having the dimension $\hat{C} * \mathrm{D} * \hat{H} * \hat{W}$. Softmax function is then applied to normalize the σ. Finally, to calculate the response of the plane attention, correlation matrix is applied to the convolutional feature maps which can be formulated as:

$$P_{att} = \sum_{i=1}^{N} V(x_i) * Softmax(\sigma_{j,i}) \tag{2}$$

Depth attention module computes the inter-slice dependency by computing the correlation matrix along the depth dimension. The difference with the plane attention module is that we first transpose the Q(x) and K(x) from $\hat{C} * \mathrm{D} * \mathrm{H} * \mathrm{W}$

to $H*W*\hat{C}*D$. Then inner product is allied by Eq. (1). to get the $\sigma_{j,i}$ the result is the depth correlation metrix having dimention $\hat{C}*\hat{D}*H*W$. which then passed through softmax function to get the depth correlation matrix. Finally the depth attention is calculate by:

$$D_{att} = \sum_{i=1}^{N} V(x_i) * Softmax(\sigma_{j,i}) \tag{3}$$

Attention fusion module is applied at the end to fuse both the attention information from plane attention (P_{att}) module and depth attention module(D_{att}) with the input feature map (X). The resultant output is calculated by the following equation:

$$Y = \gamma(P_{att} + D_{att}) + X \tag{4}$$

where γ is the trainable scaling factor which is initialized as 0. With the training process, It gradually learns to apply self-attention features with the network training goes on.

2.3 CLAHE Self-knowledge Distillation

In the challenge, the appearances of objects vary making the model difficult to have robust performance. To deal with this problem, a self-distillation training scheme is proposed. Firstly, a contrast limited adaptive histogram equalization (CLAHE) version of the input is created. Then both the original image and it's CLAHE version are fed onto segmentation network one by one to get two probability predictions. A knowledge distillation loss is used to calculate the mean square error between the two output predictions of the original and CLAHE versions. Minimizing the knowledge distillation loss can help the segmentation model learn the invariant features irrespective of the input differences. The knowledge distillation loss is formulated as follows:

$$L_{SKD} = \|O - O'\|_2, \tag{5}$$

where O and O' are the outputs of the input I and I'. The main Dice loss is calculated as:

$$L_{Dice} = 1 - \frac{1}{C}\frac{2\sum_{i=1}^{C} O_i G_i}{\sum_{i=1}^{C} O_i^2 + \sum_{i=1}^{C} G_i^2}, \tag{6}$$

where the O and G are the output of the input I and its ground truth; C is the number of classes of bi-atrial segmentation.

Finally, the total loss for training the segmentation network can be calculated as:

$$L_{total} = L_{Dice} + \alpha L_{SKD}, \tag{7}$$

where α is the empirical weighting factor for the self-distillation training.

3 Experimental Results

Experiment Materials: The MBAS Challenge offers 200 multi-center 3D LGE-MRIs, the largest one in the field so far. More importantly, the precious data were carefully labeled in consensus with three independent experts for each LGE-MRI scan to obtain one segmentation per scan. The dataset is released as 70 3D LGE-MRI scans for training with their labels, 30 for validation with no label provided, and an additional 100 designated for the final test phase which will be released after the challenge ends. Each volume had a 3 dimensions of 640×640 x 44 or 576×576 x 44, with a spatial resolution of $0.625 \times 0.625 \times 2.5$ mm^3. Our For experimental purpose, we use only labeled data composed of 70 volumes published for the challenge. We split the data into 56 volumes for training and 14 volumes for validation.

Implementation Details: We implemented our method in Pytorch using a single NVIDIA GeForce RTX 3090. The volumes were center-cropped to 320×320. We used the standard augmentation such as random rotation, flipping, and contrast adjustment. The segmentation network was optimized using Adam optimizer with a learning rate of 1e-4 in 200 epochs.

Table 1. Experiment results of different methods for bi-atrial segmentation (Dice is measured in % and HD95 is measured in pixels)

Method	Walls		RA		LA		AVG	
	Dice	HD95	Dice	HD95	Dice	HD95	Dice	HD95
2D UNet [12]	0.618	3.08	0.877	7.42	0.861	2.09	0.785	4.19
3D UNet [6]	0.628	3.32	0.896	4.66	0.898	5.51	0.807	4.49
3D UXNet [9]	0.616	3.44	0.892	4.78	0.882	5.65	0.797	4.61
Swin UNet [3]	0.650	3.24	0.905	4.53	0.902	5.45	0.819	4.47
nnFormer [16]	0.638	3.16	0.899	4.43	0.892	5.62	0.811	4.34
UNetTR [8]	0.562	3.72	0.881	5.33	0.857	5.73	0.767	4.92
3D UNet+ASPP	0.672	3.14	0.911	4.75	0.908	4.93	0.830	4.33
3D UNet+ASPP+self Att	0.678	3.21	0.913	4.41	0.904	4.83	0.832	4.15
3D UNet+ASPP+self Att+self Distil	0.681	3.23	0.911	4.85	0.907	5.43	0.833	4.51

Quantitative Analysis: The quantitative results are shown in Table 1. We performed the experiments with Dice Similarity Coefficient (DSC) and 95% Hausdorff distance (HD95) metrics. We re-implemented different methods including 2D UNet [12], 3D UNet [6], 3D UXNet [9], Swin UNet [3], nnFormer [16], and UNetTR [8]. We also conducted ablation study on the introduced modules such as 3D UNet + ASPP, 3D UNet + ASPP + self-attention, 3D UNet + ASPP +

self-attention + self-distillation. Generally, the 3D-UNet performs better than 2D because it can learn the relationship information between the slices. The Transformer-based UNets can improve the performance of the 3D UNet because they can learn the long-range dependencies. However, the transformer-based UNet is heavy and difficult to optimize. Our end-to-end resizing-free method with self-distillation can help the 3D UNet better without heavy computation. The self-distillation training can help the model work robustly on the three classes of bi-atrial segmentation, especially the atrial walls. As a result, we can see that our proposed method can achieve the highest performance.

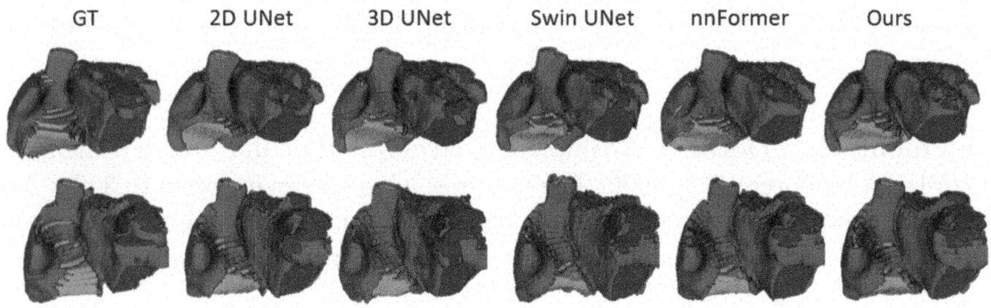

Fig. 4. Visualization of different methods.

Qualitative Analysis: We also visualize the segmentation results of different methods in Fig. 4. We can see that the 2D UNet has low segmentation and poor shapes as it ignores the relationship information between the slices. The 3D UNet and transformer-based can get better visualization results with their 3D convolutional operations. Combining with our introduced modules, our proposed method achieves the best performance with the most plausible results. We can see that the predictions of our methods align with ground truth better for the atrial walls, LA, and RA.

4 Conclusion

In this paper, we present a self-distillation end-to-end 3D segmentation network for cardiac MRI. The network is equipped with a resizing-free, ASPP module and self-attention module to improve the 3D UNet in learning the details of the left atrium, right atrium, and their walls. The self-distillation training scheme can force the model to work robustly toward the appearance shift of the inputs. The quantitative and qualitative experiments show the effectiveness and robustness of the proposed method compared to state-of-the-art methods.

Acknowledgement. The work is partially supported by the Natural Science Foundation of China (62471313), Natural Science Foundation of Guangdong

Province (2024A1515030143), the Shenzhen Science and Technology Program (20220810145705001, SGDX20201103095613036), and CAMS Innovation Fund for Medical Sciences (CIFMS) (2023-I2M-C&T-B-117, 2023-I2M-C&T-B-056).

References

1. Avesta, A., Hossain, S., Lin, M., Aboian, M., Krumholz, H.M., Aneja, S.: Comparing 3D, 2.5D, and 2D approaches to brain image auto-segmentation. Bioengineering **10**(2), 181 (2023)
2. Brunetti, A., Buongiorno, D., Trotta, G.F., Bevilacqua, V.: Computer vision and deep learning techniques for pedestrian detection and tracking: a survey. Neurocomputing **300**, 17–33 (2018)
3. Cao, H., Wang, Y., Chen, J., Jiang, D., Zhang, X., Tian, Q., Wang, M.: Swin-UNet: UNet-like pure transformer for medical image segmentation. In: European Conference on Computer Vision, pp. 205–218. Springer (2022)
4. Chen, L.C., Papandreou, G., Kokkinos, I., Murphy, K., Yuille, A.L.: DeepLab: semantic image segmentation with deep convolutional nets, Atrous convolution, and fully connected CRFs. IEEE Trans. Pattern Anal. Mach. Intell. **40**(4), 834–848 (2017)
5. Chugh, S.S., et al.: Worldwide epidemiology of atrial fibrillation: a global burden of disease 2010 study. Circulation **129**(8), 837–847 (2014)
6. Çiçek, Ö., Abdulkadir, A., Lienkamp, S.S., Brox, T., Ronneberger, O.: 3D U-Net: learning dense volumetric segmentation from sparse annotation. In: Ourselin, S., Joskowicz, L., Sabuncu, M.R., Unal, G., Wells, W. (eds.) MICCAI 2016. LNCS, vol. 9901, pp. 424–432. Springer, Cham (2016). https://doi.org/10.1007/978-3-319-46723-8_49
7. Esteva, A., et al.: Deep learning-enabled medical computer vision. NPJ Digit. Med. **4**(1), 5 (2021)
8. Hatamizadeh, A., et al.: UNETR: transformers for 3D medical image segmentation. In: Proceedings of the IEEE/CVF winter Conference on Applications of Computer Vision, pp. 574–584 (2022)
9. Lee, H.H., Bao, S., Huo, Y., Landman, B.A.: 3D UX-Net: a large kernel volumetric convnet modernizing hierarchical transformer for medical image segmentation. arXiv preprint: arXiv:2209.15076 (2022)
10. Li, M., Hsu, W., Xie, X., Cong, J., Gao, W.: SACNN: self-attention convolutional neural network for low-dose ct denoising with self-supervised perceptual loss network. IEEE Trans. Med. Imaging **39**(7), 2289–2301 (2020)
11. Minaee, S., Boykov, Y., Porikli, F., Plaza, A., Kehtarnavaz, N., Terzopoulos, D.: Image segmentation using deep learning: a survey. IEEE Trans. Pattern Anal. Mach. Intell. **44**(7), 3523–3542 (2021)
12. Ronneberger, O., Fischer, P., Brox, T.: U-Net: convolutional networks for biomedical image segmentation. In: Navab, N., Hornegger, J., Wells, W.M., Frangi, A.F. (eds.) MICCAI 2015. LNCS, vol. 9351, pp. 234–241. Springer, Cham (2015). https://doi.org/10.1007/978-3-319-24574-4_28
13. Tan, Y., Feng, F., Zhao, J.: A benchmarking study of deep learning approaches for bi-atrial segmentation on late gadolinium-enhanced MRIs. In: International Workshop on Statistical Atlases and Computational Models of the Heart, pp. 250–258. Springer (2023)

14. Xiong, Z., Nalar, A., Jamart, K., Stiles, M.K., Fedorov, V.V., Zhao, J.: Fully automatic 3D Bi-atria segmentation from late gadolinium-enhanced MRIs using double convolutional neural networks. In: Statistical Atlases and Computational Models of the Heart. Multi-Sequence CMR Segmentation, CRT-EPiggy and LV Full Quantification Challenges: 10th International Workshop, STACOM 2019, Held in Conjunction with MICCAI 2019, Shenzhen, China, October 13, 2019, Revised Selected Papers 10, pp. 63–71. Springer (2020)
15. Xiong, Z., et al.: A global benchmark of algorithms for segmenting the left atrium from late gadolinium-enhanced cardiac magnetic resonance imaging. Med. Image Anal. **67**, 101832 (2021)
16. Zhou, H.Y., Guo, J., Zhang, Y., Han, X., Yu, L., Wang, L., Yu, Y.: nnFormer: volumetric medical image segmentation via a 3D transformer. IEEE Trans. Image Process. (2023)

Enhanced Deep Unrolled Models Applied to the CMRxRecon2024 Challenge

Bingyu Xin[1(✉)], Meng Ye[1], Leon Axel[2], and Dimitris N. Metaxas[1]

[1] Rutgers University, New Brunswick, USA
{bx64,my389,dnm}@cs.rutgers.edu
[2] NYU Langone Medical Center, New York City, USA

Abstract. This paper presents details of our solution to the CMRxRecon2024 Challenge, where we developed the model, **PromptMR+**, for accelerated cardiac magnetic resonance imaging (CMR) reconstruction by utilizing recent advances in deep unrolled neural networks. Our model, consisting of as many as 32 cascades, was trained on approximately 61K slices, covering 83% of the training dataset provided by the Challenge, which includes raw k-space CMR data with diverse contrasts, anatomical views, and sampling patterns, with acceleration factors ranging from 4 to 24. Submitted under the team `CBIM`, our approach achieved two first places in both tasks of the multi-contrast and random sampling CMR reconstruction, excelling across all evaluation metrics, including SSIM, PSNR, NMSE, and radiology score. The code is available at https://github.com/hellopipu/PromptMR-plus.

Keywords: Accelerated MRI Reconstruction · Deep Unrolled Model · Prompt · Cardiac MRI

1 Introduction

Magnetic resonance imaging (MRI) is an important medical imaging technique for clinical disease diagnosis and treatment planning. However, the MRI acquisition process is time-consuming, due to the Nyquist sampling limit. Accelerated MRI techniques, including compressed sensing (CS), parallel imaging (PI) and partial Fourier, decrease the acquisition time by reducing the fraction of k-space data needed to reconstruct an MR image. With reduced k-space measurements, MRI reconstruction is usually formulated as an inverse problem, which is regularized by image priors and constrained by the acquired measurements. We can estimate the complex MR image $\boldsymbol{x}$ from its undersampled measurements $\boldsymbol{y}$ in k-space as follows:

$$\min_{\boldsymbol{x}} \boldsymbol{f}(\boldsymbol{x}) = \min_{\boldsymbol{x}} \frac{1}{2}||\boldsymbol{A}\boldsymbol{x} - \boldsymbol{y}||_2^2 + \lambda \boldsymbol{R}(\boldsymbol{x}), \tag{1}$$

where $\boldsymbol{A} = \boldsymbol{M}\boldsymbol{\mathcal{F}}\boldsymbol{S}$ is the forward process of acquiring the k-space measurements, which is a combination of coil sensitivity maps $\boldsymbol{S}$, the Fourier transform

O. Camara et al. (Eds.): STACOM 2024, LNCS 15448, pp. 289–300, 2025.
https://doi.org/10.1007/978-3-031-87756-8_29

$\mathcal{F}$ and an undersampling mask $\boldsymbol{M}$. The first term, $\frac{1}{2}||\boldsymbol{A}\boldsymbol{x} - \boldsymbol{y}||_2^2$, represents the data consistency constraint, which enforces data fidelity. In contrast, the second term, $\boldsymbol{R}(\boldsymbol{x})$, serves as the regularization term, encouraging image priors such as sparsity. The weighting parameter λ balances the influence of the two terms. E2E-VarNet [6] proposes to solve the inverse problem by unrolling the conventional truncated iterative gradient descent algorithm while learning the MR image prior from a large-scale fully sampled training dataset:

$$\boldsymbol{x}_{t+1} = \boldsymbol{x}_t - \eta_t \boldsymbol{A}^H(\boldsymbol{A}\boldsymbol{x}_t - \boldsymbol{y}) + \mathbf{CNN}_t(\boldsymbol{x}_t), \tag{2}$$

where $\mathbf{CNN}_t(\boldsymbol{x}_t)$ is a deep neural network. The neural network can provide a data-driven approximation of $-\eta_t \lambda \nabla \boldsymbol{R}(\boldsymbol{x}_t)$, incorporating prior knowledge of fully sampled MR images into the iterative updates. In Eq. (2), $\boldsymbol{A}^H$ is the complex conjugate of $\boldsymbol{A}$, η_t is the scalar learning rate at iteration t, and $t \in \{0, 1, ..., T-1\}$, with T being the total number of truncated iterations.

2 Method

For the solution to this challenge, we adopted the method, **PromptMR+**, proposed in our recent work [10]. PromptMR+ enhances deep unrolled models through: (1) self-adaptive and (2) momentum-accelerated gradient learning, and (3) efficient sensitivity map estimation for multi-coil MRI reconstruction. The architecture of PromptMR+ is shown in Fig. 1. While we provide a brief overview of the method here, more details are available in [10].

2.1 Self-Adaptive Unrolling (SAU)

We improved E2E-VarNet [6] by extending the idea of an adaptive gradient algorithm [2], resulting in an advanced updating formula:

$$\boldsymbol{x}_{t+1} = \boldsymbol{x}_t - \eta_t \boldsymbol{A}^H(\boldsymbol{A}\boldsymbol{x}_t - \boldsymbol{y}) + \mathbf{CNN}_t(\boldsymbol{x}_t, \boldsymbol{A}^H\boldsymbol{A}\boldsymbol{x}_t, \boldsymbol{A}^H\boldsymbol{y}, \boldsymbol{s}_t), \tag{3}$$

where $\boldsymbol{s}_t$ is a learnable auxiliary memory. In essence, $\boldsymbol{s}_t$ serves as a hidden state to track the historical gradient information implicitly. Note that $\boldsymbol{A}^H\boldsymbol{y}$ yields $\boldsymbol{x}_0$, which represents the image transformed from zero-filled k-space.

2.2 Momentum Acceleration (MA)

We also extended the momentum-accelerated algorithm [4] in the multi-level feature space across different unrolling stages, as follows,

$$\text{Momentum}(\boldsymbol{F}_0^m, ..., \boldsymbol{F}_t^m) = \text{CAB}(\text{Conv}_{1\times1}(\text{Concat}(\boldsymbol{F}_0^m, ..., \boldsymbol{F}_t^m))), \tag{4}$$

where $\boldsymbol{F}_t^m$ is the input feature of the m-th level decoder layer at the t-th iteration satge. CAB is the channel attention block, which can fuse the multi-stage information in a self-adaptive manner through learned attention weights.

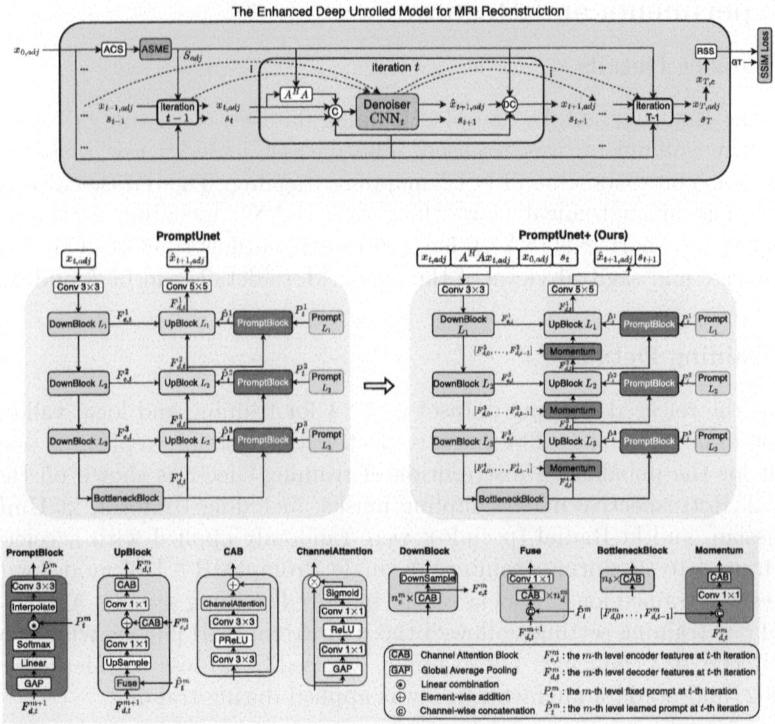

Fig. 1. First row: Overview of the enhanced deep unrolled model for MRI reconstruction. Middle row: Enhanced CNN denoiser incorporating the proposed techniques into PromptUnet [9]. Bottom row: Detailed architecture of each component in the denoiser.

2.3 Adjacent Sensitivity Map Estimation (ASME)

For mutli-coil MRI reconstruction, sensitivity map estimation is an essential step. This process is spatially, temporally, and contrastively correlated, making it beneficial to consider information in adjacent slices simultaneously. Unlike the coil-by-coil estimation used in E2E-VarNet [6], we propose an adjacent SME strategy. Spefically, we estimate the n-th sensitivity map set $\boldsymbol{S}^n_{adj}$, encompassing the n-th coil's sensitivity maps for all $2a+1$ adjacent slices, as following:

$$\boldsymbol{S}^n_{adj} = \mathbf{SME}(\mathbf{ACS}(\boldsymbol{x}^n_{0,adj})), \tag{5}$$

where $\boldsymbol{S}^n_{adj} = [\boldsymbol{S}^n_{c-a}, \ldots, \boldsymbol{S}^n_{c}, \ldots, \boldsymbol{S}^n_{c+a}]$. This sensitivity map set concatenates the sensitivity map of the central slice $\boldsymbol{S}^n_c$ with those of its $2a$ adjacent slices for the n-th coil, where $n \in [0, N-1]$, with N being the total number of coils for each slice. The function $\mathbf{ACS}(\boldsymbol{x}^n_{0,adj})$ represents the Auto-Calibration Signal (ACS) for the n-th coil across all $2a+1$ adjacent slices.

3 Experiments and Results

3.1 Dataset Details

During the challenge, a fully sampled training dataset, consisting of cases from 200 healthy volunteers, was released. The dataset includes raw k-space data with various contrasts: cine, T1/T2 mapping, tagging. The dataset also covers a wide range of anatomical views: long axis (LAX), including 2-, 3-, and 4-chamber views, short axis (SAX), left ventricular outflow tract (LVOT), as well as transverse and sagittal views of the aorta. More details can be found in [8].

3.2 Training Details

We split the released training dataset as 5 : 1 for training and local validation, resulting in 60,564 and 15,230 slices, respectively. We simply replicated slices to account for the unbalanced distribution of training slices, as shown on the left of Fig. 2. Retrospective undersampling masks, including Uniform, kt-Uniform, kt-Gaussian, and kt-Radial (pseudo), were randomly applied with acceleration ratios from 4 to 24 during training. A single PromptMR+ [10] model with 32 cascades was trained for 12 epochs on the balanced training set on 4 A100 GPUs. Most of our training settings followed those of PromptMR [9], the winner of the CMRxRecon2023 challenge. For example, we used SSIM loss with learning rate of 0.0002, and no data augmentation was applied during training.

Fig. 2. Left: Training slice balance; Right: SSIM comparison over each data set on the local validation set under 10× uniform undersampling.

3.3 Results

We made quantitative and qualitative analyses of the reconstruction results on the local validation set of 15,230 images. We compared the reconstruction performance between different methods: zero-filling, GRAPPA [3], PromptMR [9] (12 cascades), and PromptMR+ [10] (12 and 32 cascades).

Quantitative Results. We used SSIM, PSNR, NMSE as quantitative metrics to evaluate the reconstruction accuracy. In Table 1, PromptMR+ (12) consistently outperformed PromptMR (12) across all undersampling mask types and acceleration factors, on all three metrics, in this highly diverse validation set. Increasing the number of cascades from 12 to 32 in PromptMR+ can further enhance the performance. On the right of Fig. 2, we also display the SSIM results of each case in the local validation set.

Table 1. MRI reconstruction performance comparison between PromptMR (PMR) and PromptMR+ (PMR+) under different undersampling masks and acceleration factors on the local validation set.

Task	Mask	Acc	NMSE ($\times 10^{-2}$) (↓)			PSNR (dB) (↑)			SSIM (%) (↑)		
			PMR (12)	PMR+ (12)	PMR+ (32)	PMR (12)	PMR+ (12)	PMR+ (32)	PMR (12)	PMR+ (12)	PMR+ (32)
1	Uniform	4	0.40	0.33	0.33	42.27	42.99	43.28	97.73	98.00	98.09
		8	0.89	0.74	0.70	37.70	38.57	38.94	95.22	95.81	96.05
		10	1.18	0.97	0.91	36.36	37.27	37.68	94.25	94.95	95.24
	Avg		0.82	0.68	0.65	38.78	39.61	39.97	95.73	96.25	96.46
2	kt-Uniform	8	0.75	0.62	0.58	38.63	39.53	39.90	95.90	96.43	96.61
		16	1.44	1.18	1.10	35.46	36.35	36.73	93.49	94.24	94.54
		24	2.05	1.73	1.60	33.75	34.57	34.96	91.80	92.68	93.06
	kt-Gaussian	8	0.93	0.76	0.72	37.85	38.72	39.06	95.51	96.05	96.23
		16	1.45	1.19	1.13	35.63	36.55	36.90	93.70	94.43	94.69
		24	1.85	1.51	1.42	34.46	35.38	35.73	92.53	93.38	93.70
	kt-Radial	8	0.53	0.47	0.44	40.58	41.12	41.33	96.94	97.18	97.27
		16	0.89	0.77	0.73	37.80	38.49	38.75	95.19	95.63	95.79
		24	1.28	1.10	1.02	36.01	36.78	37.11	93.71	94.36	94.61
	Avg		1.24	1.04	0.97	36.69	37.50	37.83	94.31	94.93	95.17

Qualitative Results. The quantitative metrics used above have limitations in their ability to comprehensively evaluate MRI reconstruction results, particularly for cardiac MRI, where accurate recovery of small anatomical structures is crucial for clinical diagnosis. In Fig. 3, we qualitatively compare representative reconstruction results from the local validation set. Among all the methods, PromptMR+ demonstrates superior performance in aliasing artifact removal and in reconstructing fine anatomical details.

3.4 Ablation Study

Number of Cascades. Previous works commonly set the number of cascades in deep unrolled models to 8 or 12. In line with recent trends in model scaling [7], we designed PromptMR+ with 32 cascades. In Table 2, we present comprehensive evaluations of PromptMR and PromptMR+ across 4, 8, 12, and 32 cascades. The results are displayed in Fig. 4. We can observe an approximately linear improvement in both PSNR and SSIM as the number of cascades increases (on a logarithmic scale), indicating consistent performance gains with higher cascade counts. Due to memory limitations, data for PromptMR (32) was unavailable. Notably, thanks to the efficient sensitivity map estimation, the PromptMR+ (32) consumes even less GPU memory than PromptMR (4) during tests.

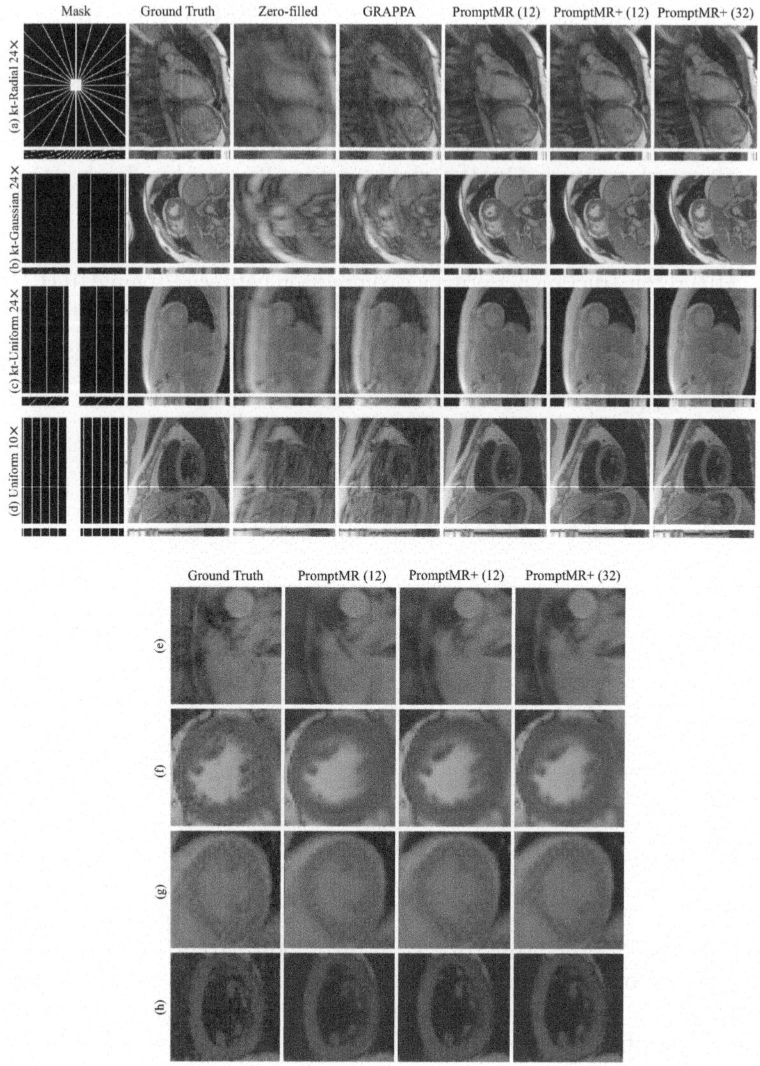

Fig. 3. Qualitative comparison of different reconstruction methods on the local validation set. Yellow arrows indicate the horizontal location for the temporal or contrast profile, while red arrows highlight artifacts in the baseline PromptMR results: (a) Cine LAX-2CH: Streaking artifacts from radial sampling are not fully removed; (b) Cine SAX: Blurring of the papillary muscle adjacent to the myocardial wall results in apparent myocardial wall thickening; (c) Tagging: Tagging lines in the anterior wall are blurred; (d) T1-weighted: The papillary muscle appears thickened. (e)-(g) are zoomed-in views of (a)-(d), respectively. Overall, PromptMR and PromptMR+ perform similarly well across various contrasts, views, and undersampling masks. However, PromptMR+ shows superior performance in aliasing artifact removal, structural sharpness and reconstruction accuracy, which are critical for clinical diagnosis. (Color figure online)

Different Components in PromptMR+. In Table 3, we incrementally added each component described in Sect. 2 to the baseline PromptMR (12), in order to assess their influence. Among these components, self-adaptive unrolling (B) proved most effective in enhancing MRI reconstruction performance, while momentum acceleration (C) provided further gains. Additionally, adjacent sensitivity map estimation (D) significantly reduced both memory consumption and inference time.

Unseen Contrasts. After the challenge, fully sampled phase-contrast (Flow2D) and black-blood imaging data were released. In Table 4, we evaluated the out-of-distribution (OOD) performance on these two validation datasets. Our models showed similar improvements over PromptMR, with the exception of a slight drop in SSIM for black-blood imaging. This is probably because black-blood contrast is rarely seen in the training dataset. Apart from a small portion of T1-weighted images in the T1 mapping dataset, which exhibit black-blood contrast, bright-blood and gray-blood contrasts are more commonly seen in the training dataset. We further note that, since the training of our model is solely supervised on RSS [5] reconstruction, phase information is ignored, resulting in phase image recovery of Flow2D deviating from the ground truth, as shown in Fig. 5.

Table 2. Comparison of PromptMR and PromptMR+ with varying cascade numbers (Cas.). Results were evaluated using 10× uniform undersampling.

Method	Cas.	Training Time	Training Mem.	Test Time	Test Mem.	Param.	PSNR	SSIM
PromptMR	4	3.8 h/epoch	31.4 GB	0.14 s/slice	8.7 GB	28 M	34.88 dB	92.94%
	8	5.1 h/epoch	36.3 GB	0.17 s/slice	9.1 GB	55 M	36.02 dB	93.98%
	12	5.9 h/epoch	40.9 GB	0.19 s/slice	9.4 GB	82 M	36.36 dB	94.25%
PromptMR+	4	1.9 h/epoch	14.0 GB	0.07 s/slice	3.6 GB	31 M	36.18 dB	94.06%
	8	3.1 h/epoch	20.0 GB	0.11 s/slice	4.1 GB	62 M	36.99 dB	94.71%
	12	4.3 h/epoch	26.3 GB	0.14 s/slice	4.5 GB	93 M	37.27 dB	94.95%
	32	11.2 h/epoch	55.5 GB	0.31 s/slice	6.7 GB	245 M	37.68 dB	95.24%

Table 3. Performance improvement of components incrementally added to the baseline PromptMR with 12 cascades. Results were evaluated using 10× uniform undersampling on the local validation set. SAU: self-adaptive unrolling; MA: momentum acceleration; ASME: adjacent sensitivity map estimation.

Method	Component	Training Time	Training Mem.	Test Time	Test Mem.	Param.	PSNR	SSIM
A (PromptMR)	Baseline	5.9 h/epoch	40.9 GB	0.19 s/slice	9.4 GB	81.5 M	36.36 dB	94.25%
B	+ SAU	6.0 h/epoch	41.9 GB	0.19 s/slice	9.5 GB	82.3 M	37.09 dB	94.81%
C	+ MA	6.3 h/epoch	45.4 GB	0.20 s/slice	9.6 GB	92.9 M	37.22 dB	94.93%
D (PromptMR+)	+ ASME	4.3 h/epoch	26.3 GB	0.14 s/slice	4.5 GB	92.9 M	37.27 dB	94.95%

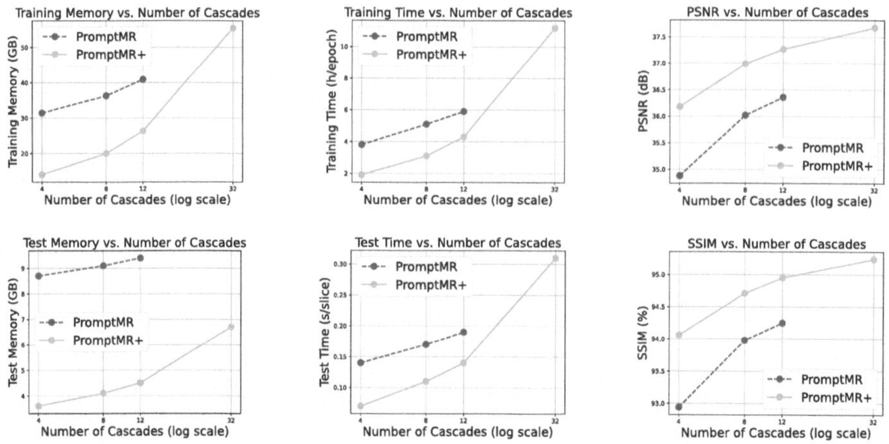

Fig. 4. Comparison of PromptMR and PromptMR+ with different numbers of cascades. Results of PromptMR (32) are unavailable, due to memory limitations.

Table 4. Performance on unseen contrasts evaluated under 10x uniform undersampling on the official validation set.

Method	Flow2D			Black-Blood		
	NMSE	PSNR	SSIM	NMSE	PSNR	SSIM
PromptMR (12)	0.88%	35.59 dB	95.46%	1.88%	33.17 dB	88.73%
PromptMR+ (12)	0.74%	36.30 dB	95.93%	1.78%	33.44 dB	88.60%
PromptMR+ (32)	0.65%	36.90 dB	96.29%	1.69%	33.65 dB	88.58%

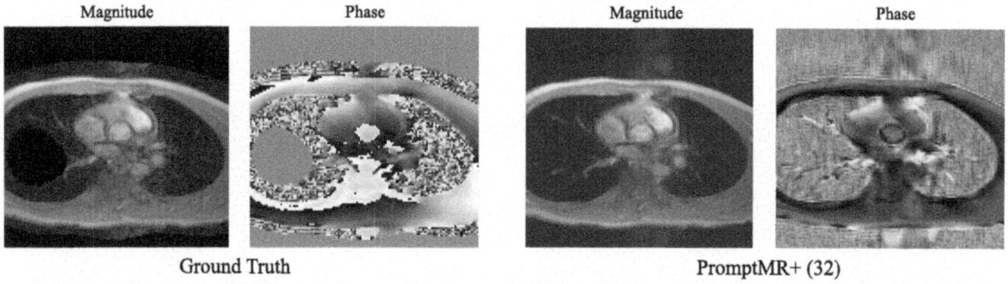

Fig. 5. Magnitude and phase image reconstructions on unseen Flow2D dataset under 10× uniform undersampling. The phase images are shown without phase unwrapping.

Intermediate Visualization. In Fig. 6, we display the intermediate results across cascades for the example in Fig. 3 (a). PromptMR (12) struggles to remove streaking artifacts, while PromptMR+ (12) successfully removes them by the 5th cascade. Interestingly, for both PromptMR and PromptMR+, the reconstruction performance does not consistently improve with increasing cascades (see Fig. 6(f)). In some cascades, the performance significantly degrades, even becoming worse than the zero-filled reconstruction. We believe that this phenomenon warrants further investigation in future work on deep unrolled MRI reconstruction models.

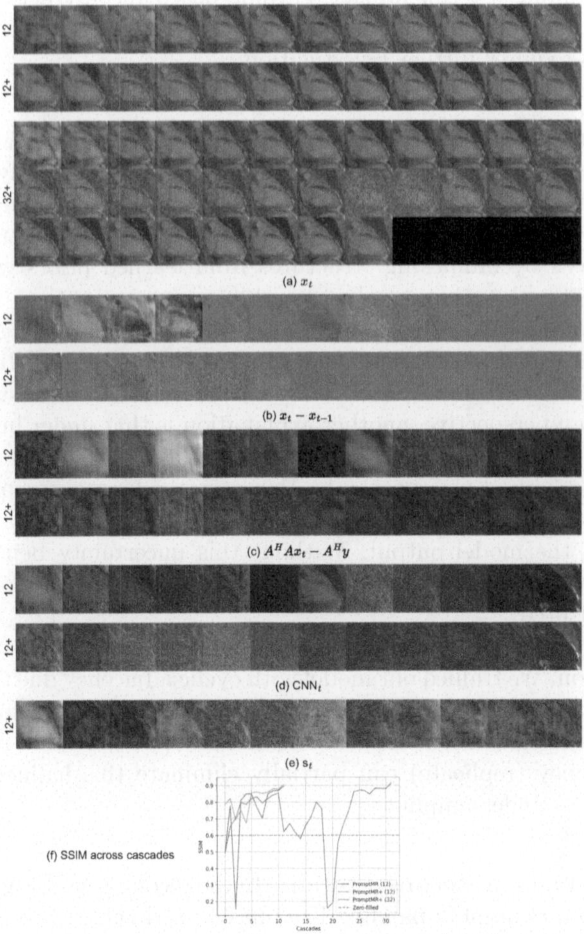

Fig. 6. Display of intermediate results under 24× kt-Radial undersampling. 12: PromptMR (12), 12+: PromptMR+ (12), 32+: PromptMR+ (32).

4 Discussion

Training Dataset Quality. MRI acquisition of dynamic objects is challenging. Complex cardiac and respiratory motion, patients' inability to hold their breath, imperfect gradient shimming, thermal fluctuations, instrumental imperfections and other factors can significantly impact the final image quality. Although quality control was applied during data preparation [8], a few cases with degraded quality were still observed, as shown in Fig. 7. Since MRI reconstruction serves as the foundation for downstream image analysis, understanding how low-quality data in the training set can impact learning-based reconstruction model performance, and developing strategies to mitigate its negative effects, is critically important and requires further investigation.

Hallucinations in Reconstruction. In the first row of Fig. 8, we show that when the acceleration factor increases, PromptMR+ (32) falsely reconstructs the myocardial wall, *i.e.*, the wall becomes more and more thickened. It appears that the learning-based reconstruction method exhibits a tendency to handle challenging areas by mimicking structures from learned priors when possible-even if incorrect-rather than producing a blurry reconstruction. However, this behavior can lead to catastrophic outcomes, as the visually pleasing but incorrect reconstructions would fail to alert clinicians to potential issues. Instead, a visibly blurry result could immediately alert clinicians of the reconstruction issue. From a Bayesian perspective, another explanation is that under high undersampling factors, such reconstructions could represent a plausible posterior sampling result given the limited measurements. However, the large discrepancy between the learning-based reconstruction and ground truth can signal a high level of uncertainty in the model output. Without this uncertainty being exposed or considered, the reliability of the learning-based MRI reconstruction results cannot be fully ensured.

One factor contributing to this hallucination is adjacent reconstruction under high acceleration. We trained our model with cyclic adjacency due to the periodic nature of cardiac motion, which works for cine but not for tagging, due to tag fading in the later phases of a cardiac cycle. In Fig. 9, we show that reconstruction without adjacency (replicate) can partially eliminate this hallucination under high-acceleration undersampling.

Banding Artifacts in Reconstruction. In the second row of Fig. 8, increasing acceleration factors amplify banding-like aliasing artifacts [1], caused by Cartesian undersampling, along the phase-encoding direction in PromptMR+ (32) reconstructions.

Denoising in Reconstruction. As acceleration factors increase, the reconstruction results in Fig. 8 exhibit a stronger denoising effect compared to the ground truth. With fewer input measurements, the model tends to averagely adjust reconstruction uncertainties over different areas in an image, resulting in a smoother output.

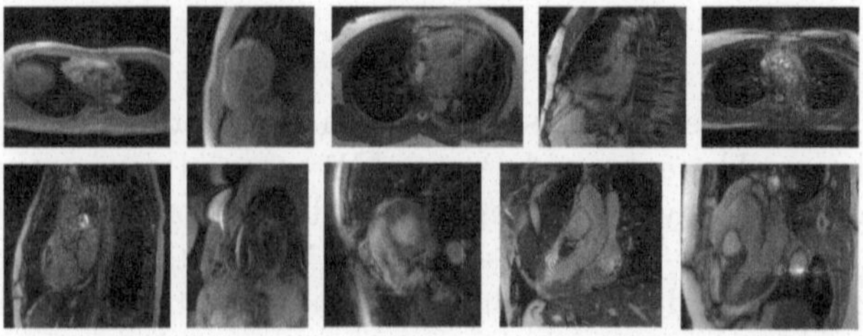

Fig. 7. Examples of poor-quality images in the official training dataset.

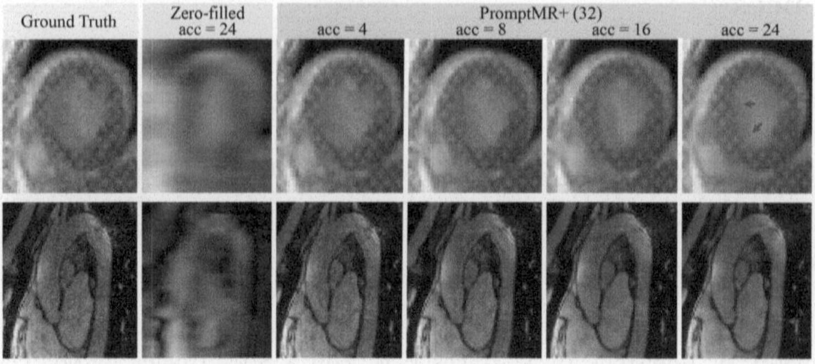

Fig. 8. Learning-based reconstruction results under kt-Uniform undersampling. The first row shows hallucinations of a thickened myocardial wall, with associated hallucinated extra tags, while the second row highlights banding-like aliasing artifacts at high acceleration factors. As acceleration factors increase, the reconstructions in both rows exhibit stronger denoising effect compared to the ground truth.

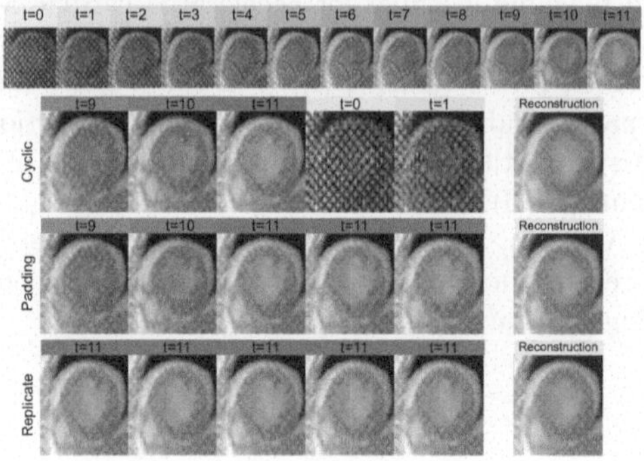

Fig. 9. Reconstruction using different adjacent strategies under 24× kt-Uniform undersampling. The replicate method, without incorporating adjacent information, can partially eliminate the hallucination of the thickened myocardial wall.

5 Conclusion

In this paper, we present details of our winning method in the CMRxRecon2024 Challenge. The enhanced PromptMR+ model demonstrates superior performance and efficiency over PromptMR and other baseline MRI reconstruction methods, on a large and diverse cardiac MRI dataset. Extensive experiments, visualizations, and analyses highlight the advantages and limitations of our method.

References

1. Defazio, A., Murrell, T., Recht, M.: MRI banding removal via adversarial training. Adv. Neural. Inf. Process. Syst. **33**, 7660–7670 (2020)
2. Duchi, J., Hazan, E., Singer, Y.: Adaptive subgradient methods for online learning and stochastic optimization. J. Mach. Learn. Res. **12**(7) (2011)
3. Griswold, M.A., et al.: Generalized autocalibrating partially parallel acquisitions (GRAPPA). Magn. Resonance Med. Official J. Int. Soc. Magn. Resonance Med. **47**(6), 1202–1210 (2002)
4. Polyak, B.T.: Some methods of speeding up the convergence of iteration methods. USSR Comput. Math. Math. Phys. **4**(5), 1–17 (1964)
5. Roemer, P.B., Edelstein, W.A., Hayes, C.E., Souza, S.P., Mueller, O.M.: The NMR phased array. Magn. Reson. Med. **16**(2), 192–225 (1990)
6. Sriram, A., et al.: End-to-end variational networks for accelerated MRI reconstruction. In: Martel, A.L., et al. (eds.) MICCAI 2020. LNCS, vol. 12262, pp. 64–73. Springer, Cham (2020). https://doi.org/10.1007/978-3-030-59713-9_7
7. Sutton, R.: The bitter lesson. Incomplete Ideas (blog) **13**(1), 38 (2019)
8. Wang, Z., et al.: Cmrxrecon2024: a multi-modality, multi-view k-space dataset boosting universal machine learning for accelerated cardiac MRI. arXiv preprint arXiv:2406.19043 (2024)
9. Xin, B., Ye, M., Axel, L., Metaxas, D.N.: Fill the k-space and refine the image: prompting for dynamic and multi-contrast MRI reconstruction. In: International Workshop on Statistical Atlases and Computational Models of the Heart, pp. 261–273. Springer, Cham (2023)
10. Xin, B., Ye, M., Axel, L., Metaxas, D.N.: Rethinking deep unrolled model for accelerated MRI reconstruction. In: European Conference on Computer Vision, pp. 164–181. Springer, Cham (2024)

Reducing the Number of Leads for ECG Imaging with Graph Neural Networks and Meaningful Latent Space

Giacomo Verardo[1(✉)], Daniel F. Perez-Ramirez[1,2], Samuel Bruchfeld[3], Magnus Boman[3,4], Marco Chiesa[1], Sabine Koch[3], Gerald Q. Maguire Jr.[1], and Dejan Kostic[1]

[1] KTH Royal Institute of Technology, Stockholm, Sweden
{verardo,mchiesa,maguire,dmk}@kth.se
[2] RISE Computer Science, Stockholm, Sweden
[3] Karolinska Institute, Stockholm, Sweden
{samuel.bruchfeld,magnus.boman,sabine.koch}@ki.se
[4] MedTechLabs, Stockholm, Sweden

Abstract. ECG Imaging (ECGI) is a technique for cardiac electrophysiology that allows reconstructing the electrical propagation through different parts of the heart using electrodes on the body surface. Although ECGI is non-invasive, it has not become clinically routine due to the large number of leads required to produce a fine-grained estimate of the cardiac activation map. Using fewer leads could make ECGI practical for clinical patient care. We propose to tackle the lead reduction problem by enhancing Neural Network (NN) models with Graph Neural Network (GNN)-enhanced gating. Our approach encodes the leads into a meaningful representation and then gates the latent space with a GNN. In our evaluation with a state-of-the-art dataset, we show that keeping only the most important leads does **not** increase the cardiac reconstruction and onset detection error. Despite dropping almost 140 leads out of 260, our model achieves the same performance as another NN baseline while reducing the number of leads. Our code is available at github.com/giacomoverardo/ecg-imaging.

Keywords: Deep Learning · ECG Imaging · Graph Neural Networks

1 Introduction

ECG Imaging (ECGI) is based on solving the inverse problem of cardiac electrophysiology to reconstruct the electrical propagation through different parts of the heart by using electrodes on the body surface together with cardiac and torso geometry. It offers a non-invasive method to map and analyze cardiac electrical activity with high precision, aiding in the diagnosis and treatment of arrhythmias such as atrial fibrillation and ventricular tachycardia. ECGI has typically required hundreds of leads. This paper presents an approach to reduce the number of leads needed to reconstruct the cardiac activation map from the Body Surface Potential Mapping (BSPM) measurements.

O. Camara et al. (Eds.): STACOM 2024, LNCS 15448, pp. 301–312, 2025.
https://doi.org/10.1007/978-3-031-87756-8_30

Our main contributions are:

1. We present a novel approach for redundant lead removal in ECGI, which relies on a Graph Neural Network (GNN)-gating mechanism in the Frequency Modulated Möbius (FMM) latent space. To the best of our knowledge, we are the first to propose using this combination when solving the inverse problem of ECGI.
2. We evaluate our model on a state-of-the-art dataset[1] and show that we can reduce the number of leads required for ECGI by up to a hundred out of 260, thus potentially making ECGI feasible for clinical use. Moreover, we show that our model outperforms a state-of-the-art model in terms of cardiac reconstruction loss and onset detection even when a large number of leads are removed.

2 Background

This section provides information about Artificial Intelligence (AI) and non-AI-based ECGI (Sect. 2.1), and the sub-networks composing our model: convolutional layers (Sect. 2.2), FMM encoding (Sect. 2.3), and the GNN architecture (Sect. 2.4).

2.1 ECG Imaging

ECGI is a non-invasive cardiophysiology technique that combines information from BSPM and geometry information (usually obtained with a CT scan) to reconstruct the electrical propagation inside the heart. ECGI is a broad field and includes different tasks, such as reconstruction of the cardiac activation map [2,3], detection of the earliest activation sites [13] during a heartbeat, and estimation of the cardiac Transmembrane Potential (TMP) [5]. The forward problem of ECGI refers to the computation of the BSPM (*i.e.*, the potentials measured with electrodes) from a known electrical cardiac source. This relationship can be expressed with a linear model:

$$\mathbf{x}_{bspm}(t) = \mathbf{H}\mathbf{x}_{acti}(t) + \mathbf{n}(t) \tag{1}$$

where the matrix $\mathbf{H}$ can be estimated from geometric information (e.g., torso and heart meshes) and the electrical conductivity of tissues within the torso, $\mathbf{x}_{acti}(t)$ is the source cardiac electrical activity, $\mathbf{x}_{bspm}(t)$ is the signal measured from the electrodes, and $\mathbf{n}(t)$ represents random additive noise.

Although the linearity of the equation may suggest that reconstructing the cardiac source is a straightforward task, the ECGI inverse problem is ill-posed [14]: even though the relation between the cardiac electrical source and the corresponding BSPM measurement is linear, a negligible error in the measurements may lead to considerable errors in the estimated cardiac source:

$$\hat{\mathbf{x}}_{acti}(t) = \mathbf{H}^{-1}\mathbf{x}_{bspm}(t) \tag{2}$$

Tikhonov regularization has been used to include *a priori* knowledge of x_{acti} and reduce the impact of the ill-posed problem. AI has also been proposed to learn the relation between x_{acti} and x_{bspm} or a suitable regularization [1,12].

[1] The number of employed leads in the evaluated dataset is $N = 260$. The cardiac and torso geometry is provided by Computed Tomography (CT) scans.

ECGI with Tikhonov Regularization. Solving the inverse problem for ECGI requires two steps: (i) defining the forward H matrix in Eq. (1) from geometrical and conductivity properties of the torso (for instance, by means of the Boundary Element Method (BEM)) and (ii) estimating the cardiac electrical source while stabilizing the solution. Tikhonov regularization is a standard technique to penalize predicted $\hat{x}_{acti}$ with high norms. The general formulation is:

$$J(\mathbf{x}) = ||\mathbf{H}\mathbf{x}_{acti} - \mathbf{x}_{bspm}||^2 + \lambda^2||\mathbf{R}\mathbf{x}_{acti}||^2 \tag{3}$$

where λ is the regularization parameter and $\mathbf{R}$ the regularization matrix, which is the identity matrix for zero-order Tikhonov regularization. Tikhonov regularization thus involves solving an optimization problem that balances the fit of the forward model to the measured data with a regularization term to stabilize the solution, minimizing $J(x)$.

ECGI via AI-Based Models. Neural Network (NN)-based solutions have been applied to ECGI. They employ Long-Short Term Memory (LSTM) layers, fully connected models, convolutional networks [2,4,5], or GNNs [8] to include the geometry input (e.g.,obtained from the CT scan or Magnetic Resonance Imaging (MRI) [21]) and the BSPM measurements to estimate the cardiac electrical source. This allows learning from available data instead of solving an optimization problem for each heartbeat. Once the NN is trained, inference can be performed to retrieve an estimate in much less time than the optimization time of standard solvers for inverse problems.

Although ECGI is a viable alternative to invasive approaches, it is clinically impractical since it requires hundreds of electrodes to be placed on the patient's body by means of either single electrodes or vests [11]. Multiple non-AI-based methods have been proposed to detect the earliest activation site from a reduced number of electrodes [13] or even from standard 12-lead Electrocardiograms (ECGs) [11] with increasing error as the number of leads is reduced. Instead, we argue that AI-based techniques can be beneficial for ECGI when a limited set of leads is available.

2.2 Convolutional Neural Networks

Convolutional Neural Networks (CNNs) [10] are state-of-the art NNs that detect increasingly complex patterns in the input as their depth increases. Each layer in a CNN performs a convolution operation between the inputs and a set of trainable parameters (*i.e.* the kernel), thus producing an output which is higher where the input is closer to the kernel's representation. CNNs have been extensively used for 1D signals, such as ECGs [2,7], since they can recognize specific patterns in the ECG input, such as the shapes of the PQRST waves, and thus CNNs are employed to solve numerous tasks. However, CNNs have seldom been employed to encode ECGs into explainable, lead-specific latent space [20].

2.3 FMM Formulation

The FMM model [15] is a time-series oriented mathematical formulation that can encode simple waves as a set of interpretable parameters. In particular, they can be used to model the depolarization and repolarization waves in each ECG lead [16], thus compressing them into a lower-dimensional representation that retains the original information. These parameters can be extracted with the following formulas:

$$W(t_i) = \mu(t_i) + e(t_i) = M + A\cos(\varphi(t_i)) + e(t_i) \tag{4}$$
$$\varphi(t) = \beta + 2\arctan(\omega\tan(t-\alpha)) \tag{5}$$
$$\mathbf{e} = (e(t_1),\ldots,e(t_n))^T \sim \mathcal{N}(\mathbf{0},\sigma^2\boldsymbol{I}) \tag{6}$$

where each parameter in the tuple $\theta = (A,\alpha,\beta,\omega,M)$ can be either shared or not between leads and the parameters are: $A \in \mathbb{R}^+$ is the amplitude of the wave; $\alpha \in [0,2\pi)$ is the angular position of the peak (lead-shared); $\beta \in [0,2\pi)$ is the peak direction; $\omega \in [0,1]$ indicates the width of the lobe of the peak (lead-shared); $M \in \mathbb{R}$ is the constant offset of the lead; and $\mathbf{e} \in \mathbb{R}$ is the error between the real heartbeat and one reconstructed from FMM parameters.

Different NN-based models have been proposed to automatically extract the FMM parameters from ECG data [20,23]. Using a model trained with these FMM parameters results in an inference time that is three orders of magnitude less than the time to solve the optimization problem [16]. However, these studies are limited to 12-leads ECG and have never been applied for ECGI.

2.4 Graph Neural Networks

GNNs have become a versatile tool for addressing various inference tasks on graph-structured data [17,22]. Conceptually, a GNN is composed of message-passing layers that recursively update each node's feature vector by considering information from its neighboring nodes [6]. Stacking K GNN layers involves creating node embedding vectors that consider their K-hop neighborhood by utilizing the graph's structure and the relationships between nodes [6,9]. Mathematically, for a graph $G = \langle V,E\rangle$ with nodes $v \in V$ and edges $(v,u) \in E$, at GNN layer i, the hidden feature vector h_v for node v is updated as:

$$h_v^i = g_1\left(\text{AGG}_{u\in\mathcal{N}(v)}\left[g_2\left(h_u^{i-1}\right)\right], h_v^{i-1}\right), \tag{7}$$

where g_1 and g_2 are non-linear transformations, AGG is an order-invariant aggregation function, and $\mathcal{N}(v)$ represents the set of neighboring node feature vectors of node v [6,9,18]. These embedding vectors are generally processed further with linear layers to produce the final output based on the specific task. For instance, node regression can be achieved by feeding each h_v $\forall v \in V$, through a final regression layer. Although GNNs have already been employed for ECGI to map torso to heart geometries [8], we claim that they are also beneficial to detect the most crucial leads to reconstruct the cardiac activation map.

3 Methodology

Our aim is to reconstruct the cardiac activation map from the BSPM measurements and the CT scans. We do so by integrating two novel components: (i) an FMM-based encoded latent space that supports hundreds of leads and (ii) a gating Graph

Attention (GAT)-based GNN that selects and weights the latent space, as depicted in Fig. 1. We argue that such a solution can determine the importance of the different leads in the latent space, while still allowing the retrieval of information from all the leads into the latent space (*e.g.* to compute shared FMM parameters). Section 3.1 describes our pipeline, including inputs, 1D-convolutional encoder, FMM latent space, 3D-convolutional decoder, and employed loss functions. We describe our GNN gating mechanism in Sect. 3.2.

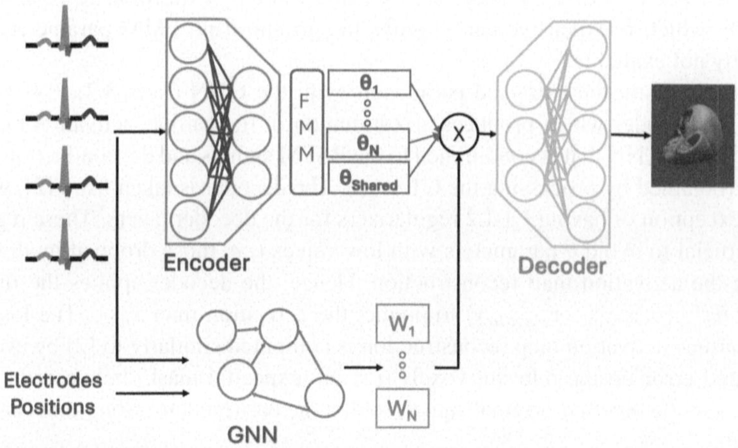

Fig. 1. Model architecture, which includes a 1D convolutional encoder, an FMM latent space for hundreds of leads, a model gating GNN and a 3D convolutional decoder. The FMM latent space includes lead-exclusive and lead-shared θ parameters.

3.1 Encoder-Decoder Pipeline

Our input consists of 3D tensors (the cardiac activation map and conduction velocity map obtained from the CT scan) and the measured BSPM data. Our encoder extracts the estimated FMM parameters from the BSPM sensors by means of stacked convolutional blocks with different kernel sizes and number of kernels per layer, in a similar fashion as [7]. The extracted parameters are $\theta = f_{\phi_{enc}}(x_{bspm})$, which includes both lead-shared and not shared parameters:

$$\begin{aligned} &\alpha_j, \omega_j \in \theta_{shared}, \\ &A_{ij}, \beta_{ij} \in \theta_i, \quad i \in \{1, \ldots, N\}, \quad j \in \{1, \ldots, M\}, \end{aligned} \tag{8}$$

where N is the number of leads, M is the number of ECG waves in the BSPM inputs, θ_{shared} represents the parameters shared across all leads, and θ_i represents the parameters specific to the i-th lead. Inspired by [20], we designed our FMM layer to handle parameters for hundreds of leads. The layer has two outputs: the θ parameters and the reconstructed BSPM signal $\hat{x}_{bspm}$ from θ. We can hence define a loss function in the latent space to constrain θ to be close to the real FMM parameters by minimizing the mean squared error between original and reconstructed BSPM signals of length L:

$$\text{Loss}_{bspm} = \frac{1}{NL} \sum_{i=1}^{N} \sum_{j=1}^{L} (x_{bspm,ij} - \hat{x}_{bspm,ij})^2 \tag{9}$$

Additionally, we constrain the FMM latent space to a meaningful representation by encouraging through a loss term Loss_{α} the α parameters to be close to the detected peaks in the BSPM leads. By doing so, we avoid running a warm-up regression phase as in [20], which is expensive and requires the ground-truth FMM parameters which are usually not available.

After weighting the extracted parameters with the GNN (Sect. 3.2), θ is given as input to the decoder which produces an estimation of the cardiac activation map. The decoder is a 3D CNN that is conditioned to the BSPM signals and the conduction system velocity (obtained by processing the CT scan). The decoder is taken from [2], with the notable exception of having L1-L2 regularizers for the decoder layers. These regularizers are crucial to avoid θ parameters with low values (*i.e.* those dropped by the GNN) affecting the activation map reconstruction. Hence, the decoder applies the function: $\hat{x}_{acti} = h_{\phi_{dec}}(\theta, x_{bspm}, x_{conduct})$ to predict the activation map x_{acti}. The loss value for the cardiac activation map reconstruction is computed similarly to [2] by averaging the squared error on the relevant voxels (*i.e.* on a specific mask) between the reconstruction and the *inverted* original signal, so that the earliest activation points are better identified:

$$\text{Loss}_{acti} = \frac{1}{|\mathcal{M}|} \sum_{(i,j,k) \in \mathcal{M}} ((1 - x_{acti,ijk}) - \hat{x}_{acti,ijk})^2 \tag{10}$$

where $\mathcal{M}$ is the set of indices defined by the 3D mask x_{mask}. We refer the reader to [2] for further implementation details. The full loss function is therefore:

$$\text{Loss} = \gamma_{bspm} \cdot \text{Loss}_{bspm} + \gamma_{\alpha} \cdot \text{Loss}_{\alpha} + \text{Loss}_{acti} \tag{11}$$

where γ_{bspm} and γ_{α} are loss weight ratios that determine the relative importance of the BSPM loss and α loss compared to the activation loss.

3.2 Graph Neural Network-Gating

We use a GNN to gate the θ_i FMM parameters from the latent space. The GNN includes multiple blocks of GAT attention layers with skip connections and layer normalization.

Our GNN produces as output a set of N w_i scalar weighs, one for each lead. The θ parameters can then be weighted as:

$$\theta'_{ij} = \begin{cases} \theta_{ij}, & \text{if } \theta_{ij} \in \theta_{shared} \\ w_i \cdot \theta_{ij}, & \text{otherwise} \end{cases} \tag{12}$$

where $\mathbf{w} = g_{\phi_{gnn}}(\text{Adj}, x_{bspm})$ and Adj is the adjacency matrix for the electrodes, computed as radial basis function of their positions: $k(x, y) = \exp\left(-\frac{\|x-y\|^2}{2\ell^2}\right)$, ℓ being the length scale parameter. The choice of a suitable ℓ allows a sufficient number of neighboring electrodes to attend to the signals of each electrode. Hence, a node with high importance can reduce the weight of its neighboring nodes, thus removing the redundancy of the available leads and therefore reducing the number of necessary leads.

During testing, the number of leads needed can be reduced by dropping those corresponding to the lowest w_i. The corresponding θ_i parameters are therefore zeroed out, as well as excluding the input leads x_{bspm} from the input to the encoder and the decoder. This approach enables training using *all* the leads, thereby enhancing the robustness of learning to reconstruct the FMM parameters since they are encoded from every lead. Moreover, our model can be dynamically evaluated on *any* number of leads, thus providing a more complete and compact analysis of the impact of lead removal for ECGI compared to full training sessions with a reduced set of leads (which would require $\sum_{k=0}^{N} \binom{N}{k} = 2^N$ different models, one for each possible set of leads).

3.3 Metrics

We evaluate the performance of our method based on two metrics: reconstruction error on the activation map (Mean Absolute Error (MAE)) and onset prediction error. The computation of the MAE follows the same formula of Eq. (10), where the absolute value is employed instead of the square function. Hence, the MAE provides a measure of the normalized average temporal distance between ground-truth and predicted cardiac activation map. The onset prediction error is instead trickier to compute, since the identification of an onset depends on the employed policy. One simple policy is the selection of the earliest activation point whose neighbours are activated within a specific time interval. Although such solution is straightforward to implement, it lacks generalizability on multiple onsets (as in the dataset employed in Sect. 4) and robustness with respect to sparsity of the predicted activation map, which is AI-generated.

For these reasons, we propose to employ clustering algorithms to estimate the onsets on both predicted and ground-truth activation maps. We perform such computation in three steps: (i) we extract the 10% percentile of the activation map, which represents the neighborhood of the activated points at the beginning of the onset, (ii) we weight the extracted points according to their activation time and (iii) we perform K-Means clustering, where K is the number of onsets, which is available during testing. In order to assign the predicted onset to the best estimated onset from the initial points of the ground-truth activation map, we solve a Linear assignment problem (LAP) where the cost matrix is the euclidean distance. The impact of few activated outlier points is thus attenuated by the large number of points used as input to the clustering model.

4 Experimental Results

We evaluate our model on the available-on-request dataset from [2,3], which consists of simulated data of source cardiac activation map (generated with the Eikonal model), correspondent measurement on the torso and conduction map for 7 patients (5 for training and 2 for testing). We train our model with multiple learning rates, early stopping, and learning rate decay for a maximum number of 240 epochs. We then compute the average and standard deviation of the test loss for the activation map reconstruction across the test samples. We evaluate the impact of dropping leads by zeroing out the input leads and the extracted non-shared parameters θ_i in a range between 0 and $N = 260$ with a step size of 20, as explained in Sect. 3.2. We compare our model performance with the Variational Autoencoders (VAE) from [2]. Details of the training procedure are available in the code repository [19].

Compared to the VAE baseline, our approach achieves lower training, validation, and test loss, thus achieving a more reliable reconstruction of the activation map. Table 1 shows the test and validation loss for different learning rates: (i) for our model with multiple ratios between cardiac activation loss and signal loss and (ii) for our baseline. The results show that the test loss is reduced by up to 6.9% in terms of mean squared error on the prediction of the cardiac activation map. The accuracy of our approach can be visually estimated by considering one example of cardiac activation map ground-truth Fig. 5a, prediction Fig. 5b, and the absolute difference between them.

Table 1. Comparison of our model with the VAE baseline for multiple hyper-parameters: learning rate, γ_{bspm} and γ_α. The last two are not applicable for the VAE baseline.

Model	LR	$\gamma_{\mathbf{bspm}}$	γ_α	MAE [ms]	MSE [$\times 10^{-3}$]	Onset Error [mm]
Ours	1e−5	0.1	1.0	131.7 ± 14.7	27.8 ± 5.6	29.99 ± 11.17
	1e−5	1.0	1.0	128.8 ± 16.3	26.9 ± 6.3	29.76 ± 11.63
	1e−5	1.0	10.0	131.8 ± 14.7	27.8 ± 5.6	29.81 ± 11.11
	1e−4	0.1	1.0	130.2 ± 19.1	27.5 ± 7.4	30.19 ± 12.74
	1e−4	1.0	1.0	**127.1 ± 16.1**	**26.1 ± 6.3**	**27.47 ± 10.66**
	1e−4	1.0	10.0	128.9 ± 15.9	26.9 ± 6.2	28.96 ± 11.94
VAE	1e−4	n/a	n/a	**133.4 ± 16.3**	**28.0 ± 6.7**	**30.36 ± 9.14**
	1e−5	n/a	n/a	145.0 ± 9.7	32.2 ± 4.1	43.88 ± 10.35

Our model achieves the same test reconstruction loss for the cardiac activation map as our baseline when almost 140 leads are dropped. Figure 2b shows the test activation loss of our model as we progressively remove the leads with the lowest weight w_i from the inputs. Although the impact on the loss is dramatic when a large number of leads are ignored, our model remains robust to lead removal up to almost 50 leads. Moreover, our model achieves a lower test loss than the baseline, even when 134 leads are removed. Hence, we argue that NN-based learning of important leads is a feasible technique to reduce the burden of placing hundreds of electrodes for ECGI.

Our model achieves lower prediction error on the estimated onsets compared to the baseline. Table 1 provides the average and standard deviation on the error on the position of the onsets estimated by clustering on the firstly activated points, as defined in Sect. 3.3. Our results prove that our model is able to better detect the predicted position of 9.5% compared to our baseline. However, the high standard deviation indicates high variability between different test samples. Indeed, as shown in Fig. 3, test samples

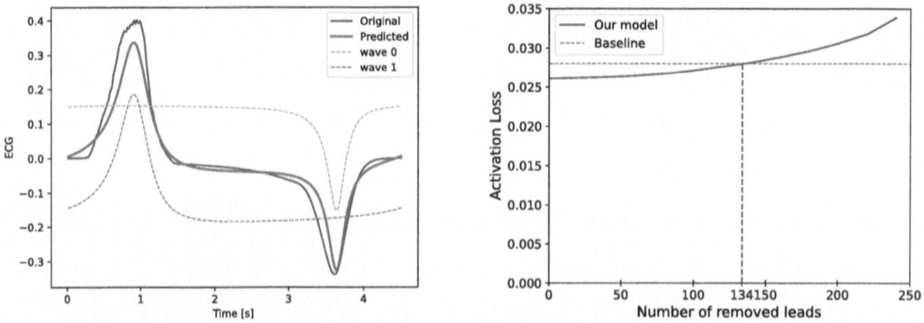

(a) Latent FMM reconstruction (b) Test loss vs. number of removed leads

Fig. 2. Results for our model with learning rate $1e-4$ and loss weight ratio 1.0: (a) shows the FMM reconstruction in the latent space for lead 100, proving that our training procedure can extract meaningful FMM parameters (b) shows the increase in test activation loss when leads are removed, demonstrating that (i) up to 50 leads can be removed with negligible performance degradation (ii) up to 134 leads can be removed while having lower loss than the baseline.

with a single onset demonstrate much lower prediction error (17.9 mm) compared to two (30.9 mm) and three onsets (29.9 mm). Hence, finding multiple onsets is the most challenging task. Figure 3a shows that we can remove up to 150 leads without affecting the estimated onset prediction error for the one onset case. Interestingly, the performance of the model increases when few leads are removed. Such leads have almost zero weight and do not contribute to the activation map reconstruction. The specific sample depicted in Fig. 4 graphically shows that removing 150 leads reduces the onset prediction by only 7.5 mm. Instead, spatial resolution is significantly affected by lead removal in the two- and three-onset cases, which are more challenging due to the superimposition of propagation from multiple pacing sites.

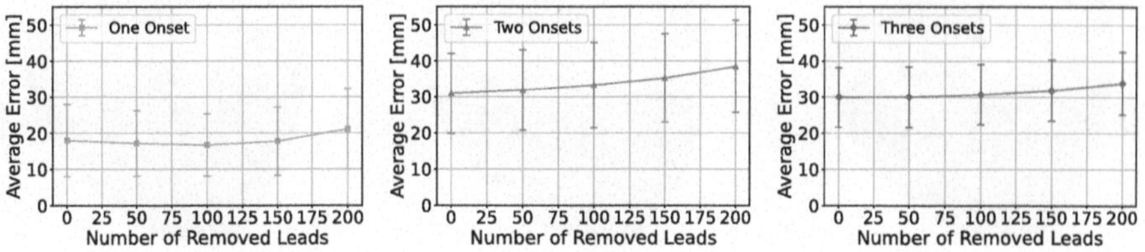

(a) One onset (b) Two onsets (c) Three onsets

Fig. 3. Prediction error for the pacing site when varying the number of removed leads for our model with learning rate $1e-4$, $\gamma_{\text{bspm}} = \gamma_{\alpha} = 1.0$. Figures (a), (b), (c) depicts the average error for test samples with one, two and three onsets respectively.

Our model achieves a meaningful, explainable latent space by means of FMM coefficients estimation. Figure 2a shows the reconstruction obtained from the estimated

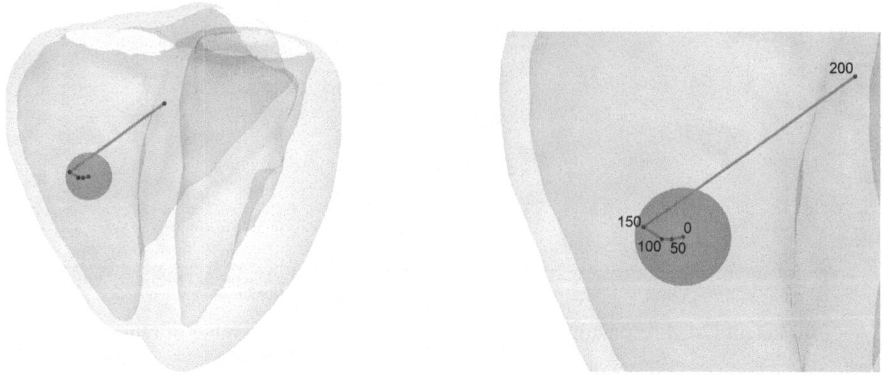

Fig. 4. Results for our model with learning rate $1e-4$ and $\gamma_{\text{bspm}} = \gamma_{\alpha} = 1.0$: (a) shows the full heart with the predicted onsets when multiple leads are removed by means of our GNN-gating mechanism while (b) zooms-in on the predicted onsets. The black numbers above the points indicate the number of zeroed-out leads. For the sample shown, if up to 150 leads are removed, the distance to the predicted pacing site with all the leads is within 7.5 mm (green sphere) (Color figure online)

FMM parameters in the latent space, which closely resembles the real BSPM signals. Hence, we argue that our latent space can indeed be used as a compressed version of the inputs, thus allowing θ_i to be correctly gated. Compared to [20], our model does not need expensive warm-up regression to achieve a meaningful latent space. That is due to the fact that: (i) our dataset only includes the QRS complex and T wave, which can be both represented with a single FMM wave from 8 and (ii) having more leads as input is beneficial to the accuracy of the estimation, as pointed out by [16].

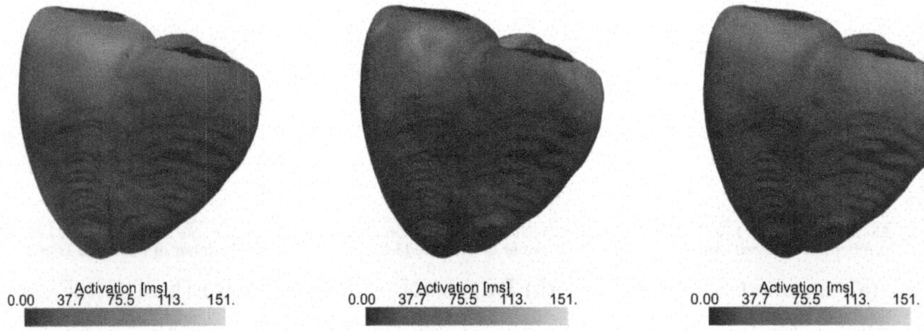

(a) Ground-truth (b) Predicted (Ours) (c) Predicted (VAE)

Fig. 5. Example cardiac activation map: (a) ground-truth, (b) prediction with our approach, (c) prediction with VAE baseline. The reconstruction capability depends on the sample, patient, and training hyper-parameters. Table 1 provides a quantitative analysis.

5 Conclusions

In conclusion, this study has investigated the possible reduction of the number required leads when performing ECGI tasks. Our GNN-based approach to lead-gating on FMM-enabled latent space allows reducing the number of employed leads by 50 with only a slight increase in the error for cardiac activation map reconstruction. Our findings also indicate that it is possible to encode ECG data from BSPM to a FMM representation with a convolutional model without requiring computing the ground-truth FMM coefficients. As future work we will investigate how the different components of our model affect the reconstruction capabilities. Moreover, we will train our models for a finer grid of hyper-parameters to show the robustness of our approach.

Acknowledgements. This work was supported by the Swedish Research Council (project "Scalable Federated Learning", no. 2021-04610) and by the King Abdullah University of Science and Technology (KAUST) Office of Research Administration (ORA) under Award No. ORA-CRG2021-4699. Daniel F. Perez-Ramirez is supported by the Swedish Foundation for Strategic Research (SSF). We acknowledge Euro HPC Joint Undertaking for awarding us access to the Leonardo supercomputer located at the CINECA data center in Bologna, Italy.

References

1. Adler, J., Öktem, O.: Solving ill-posed inverse problems using iterative deep neural networks. Inverse Prob. **33**(12), 124007 (2017). https://doi.org/10.1088/1361-6420/aa9581
2. Bacoyannis, T., Krebs, J., Cedilnik, N., Cochet, H., Sermesant, M.: Deep learning formulation of ECGI for data-driven integration of spatiotemporal correlations and imaging information. In: Coudière, Y., Ozenne, V., Vigmond, E., Zemzemi, N. (eds.) FIMH 2019. LNCS, vol. 11504, pp. 20–28. Springer, Cham (2019). https://doi.org/10.1007/978-3-030-21949-9_3
3. Bacoyannis, T., Ly, B., Cedilnik, N., Cochet, H., Sermesant, M.: Deep learning formulation of electrocardiographic imaging integrating image and signal information with data-driven regularization. EP Europace **23**(Supplement_1), i55–i62 (2021). https://doi.org/10.1093/europace/euaa391
4. Chen, K.W., Bear, L., Lin, C.W.: Solving inverse electrocardiographic mapping using machine learning and deep learning frameworks. Sensors **22**(6) (2022). https://doi.org/10.3390/s22062331
5. Ghimire, S., Dhamala, J., Gyawali, P.K., Sapp, J.L., Horacek, M., Wang, L.: Generative modeling and inverse imaging of cardiac transmembrane potential. In: Frangi, A.F., Schnabel, J.A., Davatzikos, C., Alberola-López, C., Fichtinger, G. (eds.) MICCAI 2018. LNCS, vol. 11071, pp. 508–516. Springer, Cham (2018). https://doi.org/10.1007/978-3-030-00934-2_57
6. Gilmer, J., Schoenholz, S.S., Riley, P.F., Vinyals, O., Dahl, G.E.: Neural message passing for quantum chemistry. In: Proceedings of the 34th International Conference on Machine Learning (ICML), vol. 3, pp. 2053–2070 (2017)
7. Jang, J.H., Kim, T.Y., Lim, H.S., Yoon, D.: Unsupervised feature learning for electrocardiogram data using the convolutional variational autoencoder. PLOS ONE **16**(12), 1–16 (2021). https://doi.org/10.1371/journal.pone.0260612
8. Jiang, X., et al.: Hybrid neural state-space modeling for supervised and unsupervised electrocardiographic imaging. IEEE Trans. Med. Imaging **PP** (2024). https://doi.org/10.1109/TMI.2024.3377094, epub ahead of print

9. Kipf, T.N., Welling, M.: Semi-supervised classification with graph convolutional networks. In: Proceedings of the 5th International Conference on Learning Representations (ICLR). ICLR (2017). https://openreview.net/forum?id=SJU4ayYgl
10. Lecun, Y., Bottou, L., Bengio, Y., Haffner, P.: Gradient-based learning applied to document recognition. Proc. IEEE **86**(11), 2278–2324 (1998). https://doi.org/10.1109/5.726791
11. Misra, S., et al.: Initial validation of a novel ECGI system for localization of premature ventricular contractions and ventricular tachycardia in structurally normal and abnormal hearts. J. Electrocardiol. **51**(5), 801–808 (2018). https://doi.org/10.1016/j.jelectrocard.2018.05.018
12. Ongie, G., Jalal, A., Metzler, C.A., Baraniuk, R.G., Dimakis, A.G., Willett, R.: Deep learning techniques for inverse problems in imaging. IEEE J. Sel. Areas Inf. Theory **1**(1), 39–56 (2020). https://doi.org/10.1109/JSAIT.2020.2991563
13. Parreira, L., et al.: Electrocardiographic imaging (ECGI): what is the minimal number of leads needed to obtain a good spatial resolution? J. Electrocardiol. **62**, 86–93 (2020). https://doi.org/10.1016/j.jelectrocard.2020.07.004
14. Potyagaylo, D., Chmelevsky, M., Kalinin, A.: Single-layer based algorithms for solving the inverse problem of ECG. In: 2019 Computing in Cardiology (CinC), pp. 1–4 (2019). https://doi.org/10.22489/CinC.2019.408
15. Rueda, C., Larriba, Y., Peddada, S.D.: Frequency modulated möbius model accurately predicts rhythmic signals in biological and physical sciences. Sci. Rep. **9**, 18701 (2019). https://doi.org/10.1038/s41598-019-54569-1
16. Rueda, C., Rodríguez-Collado, A., Fernández, I., Canedo, C., Ugarte, M.D., Larriba, Y.: A unique cardiac electrocardiographic 3D model. toward interpretable AI diagnosis. iScience **25**(12), 105617 (2022). https://doi.org/10.1016/j.isci.2022.105617
17. Scarselli, F., Gori, M., Tsoi, A.C., Hagenbuchner, M., Monfardini, G.: The graph neural network model. IEEE Trans. Neural Netw. **20**(1), 61–80 (2009). https://doi.org/10.1109/TNN.2008.2005605
18. Veličković, P., Casanova, A., Liò, P., Cucurull, G., Romero, A., Bengio, Y.: Graph attention networks. In: Proceedings of the 6th International Conference on Learning Representations (ICLR). ICLR (2018). https://doi.org/10.17863/CAM.48429
19. Verardo, G.: ECGI Repository (2024). https://github.com/giacomoverardo/ecg-imaging. Accessed 15 Aug 2024
20. Verardo, G., et al.: FMM-head: enhancing autoencoder-based ECG anomaly detection with prior knowledge (2023). https://doi.org/10.48550/arXiv.2310.05848
21. Webber, M., Joy, G., Bennett, J., Chan, F., Falconer, D., Shiwani, H., et al.: Technical development and feasibility of a reusable vest to integrate cardiovascular magnetic resonance with electrocardiographic imaging. J. Cardiovasc. Magn. Reson. **25**(1), 73 (2023). https://doi.org/10.1186/s12968-023-00980-7
22. Wu, Z., Pan, S., Chen, F., Long, G., Zhang, C., Yu, P.S.: A comprehensive survey on graph neural networks. IEEE Trans. Neural Netw. **32**(1), 4–24 (2021). https://doi.org/10.1109/TNNLS.2020.2978386
23. Yang, Y., Rocher, M., Moceri, P., Sermesant, M.: Explainable electrocardiogram analysis with wave decomposition: application to myocardial infarction detection. In: Statistical Atlases and Computational Models of the Heart. Regular and CMRxMotion Challenge Papers, pp. 221–232. Springer, Cham (2022). https://doi.org/10.1007/978-3-031-23443-9_21

Rotor Core Projection Ablation (RCPA): Novel Computational Approach to Catheter Ablation Therapy for Atrial Fibrillation

George Obada[1](✉), Shaheim Ogbomo-Harmitt[1], Maria Deprez[1], Robin Van Den Abeelen[2], Nele Vandersickel[2], and Oleg Aslanidi[1]

[1] King's College London, London, UK
george-teodor.obada@kcl.ac.uk
[2] Ghent University, Ghent, Belgium

Abstract. Atrial fibrillation (AF) poses significant challenges for healthcare systems, warranting novel approaches for its treatment. This study introduces a computational framework leveraging cardiac magnetic resonance imaging data and image-based modelling for guiding patient-specific catheter ablation (CA) therapy for AF. By employing Direct Graph Mapping (DGM) to identify stable AF drivers (rotors) in the image-based atrial models, we developed a novel, efficient AF ablation technique, Rotor Core Projection Ablation (RCPA). The RCPA method achieved 100% success rate in terminating AF after a single procedure, with 92.3% success rate for recurrent AF cases. Notably, RCPA minimizes the amount of ablated tissue, potentially improving treatment outcomes. The study highlights the importance of personalized CA procedures tailored to individual patients, thereby enhancing efficacy and reducing the risk of AF recurrence.

Keywords: Atrial Fibrillation · Catheter Ablation · Medical Imaging · Cardiac Modelling · Re-entrant Drivers

1 Introduction

1.1 Atrial Fibrillation Prevalence and Management

Atrial Fibrillation (AF) is the most widespread cardiac arrhythmia affecting up to 2.5% of the world's population, which also increases the chances of stroke five-fold [1]. However, the mechanisms perpetuating and sustaining AF are poorly understood, and as a result, the success of its clinical treatments and patient outcomes are suboptimal.

Radiofrequency Catheter Ablation (RFCA) has emerged as a first-line treatment in AF management, to provide a long-term curative effect. This procedure relies on a catheter inserted into the left atrium (LA) that delivers heat to destroy the source of arrhythmogenesis, thereby restoring the sinus rhythm (SR). The cornerstone RFCA procedure is pulmonary vein isolation (PVI), which involves encircling the four PVs. This therapy is particularly effective in paroxysmal AF (PAF) patients, in which arrhythmogenic excitations usually originate from the PVs in the LA [2].

O. Camara et al. (Eds.): STACOM 2024, LNCS 15448, pp. 313–322, 2025.
https://doi.org/10.1007/978-3-031-87756-8_31

1.2 Patient-Specific Treatments and Modelling

However, RFCA is not as effective for persistent AF (PsAF) or long-standing AF [3], with arrhythmia recurring in half of PsAF after 1-year follow-up [4]. This may be explained by structural remodelling of LA in such patients, in which the rotors settle in areas of low conduction velocity (CV) or dense fibrosis [5]. Multiple ablation patterns were tested both in-silico and in-vivo that showed promising long-term outcomes [6], including LA roof ablation (PVI with additional linear lesions) or identifying and targeting abnormal circuits with Complex Fractionated Atrial Electrograms (CFAE). However, even recent advancements in the field highlight the recurrent pattern of AF, with AF-free survival of only 40.2% after a single procedure and 52.2% after repeated applications [7]. Therefore, understanding the mechanisms of AF recurrence and perpetuation is crucial, necessitating novel patient-tailored CA procedures.

Computational simulations play a crucial role in devising such patient-specific treatments [5, 8]. Modelling has leveraged medical imaging data, such as Cardiac Magnetic Resonance (CMR), to study the formation, maintenance, and later termination of AF drivers in image-based models, contributing to treatment planning tailored to a patient.

This study aims to *(i)* create a novel ablation technique, suitable for both PAF and PsAF patients, using a novel Direct Graph Method (DGM) software [9] for rotor-mapping, and *(ii)* assess the likelihood of AF recurrence after successful ablation in-silico.

2 Methods

2.1 Patient Data and Pre-processing

The CMR data used in the study has been acquired in a previous study to investigate AF recurrence post-ablation [10]. 13 CMR-based LA models were selected to capture a wide range of characteristics, including disease status (PAF, PsAF, and long-standing AF), fibrosis burden and distribution, fibre orientation, therefore different LA electrical activation patterns. According to the Utah Scale quantifying the fibrosis burden [11], the study presents patients in Utah II ($n = 1$), Utah III ($n = 7$) and Utah IV ($n = 5$) stages.

For each patient, Image Intensity Ratio (IIR) was calculated by dividing Late Gadolinium Enhancement (LGE) CMR image intensity in each vertex by the mean LA blood pool intensity. Figure 1 describes the process of 3D LA mesh generation: (i) LGE-CMR acquisition and LA segmentation, (ii) IIR normalising and (iii) thresholding of fibrotic tissue. Atrial fibre orientation was mapped onto the mesh using an LA fibre atlas derived from a human ex-vivo Diffusion Tensor MRI dataset, using the universal atrial coordinate system [10, 12]. In step (iii), LA tissue was classified into six levels based on IIR, specifically healthy tissue (level 0) corresponded to an IIR ≤ 1.2, and dense fibrosis (level 5) to IIR ≥ 1.32 [13]. The intermediate values of IIR resulted in 4 discrete levels of fibrosis border zone (FBZ) [14]. The meshes were triangular surfaces and were resampled to achieve an average 340 μm edge length [10].

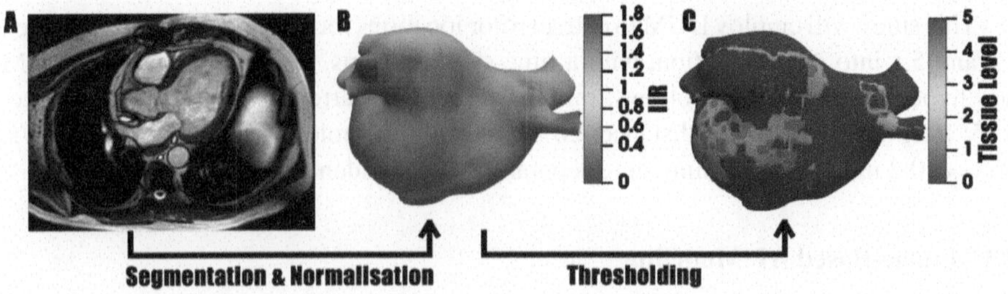

Fig. 1. Image-based atrial modelling workflow. **A**: LGE-CMR acquisition and LA segmentation. **B**: IIR normalization. **C**: Tissue thresholding to delineate fibrotic areas.

2.2 Direct Graph Method (DGM)

Recent approaches for RFCA included rotor ablation for AF termination, such as Focal Impulse and Rotor Modulation (FIRM) [15] based on phase singularity (PS) mapping. However, PSs can identify all types of rotational electrical activity, including transient, unstable re-entrant drivers (RDs), and ablating them may lead to acute outcomes.

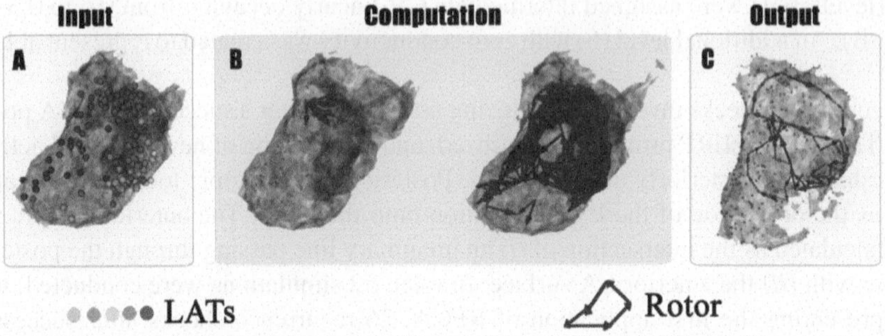

Fig. 2. DGM method of rotor detection in the ventricles [9]. **A:** LATs are converted to nodes for the network. **B:** DGM main computation algorithm. **C:** Rotors identified as output.

A novel approach described by Vandersickel et al. [9] aims to identify stable rotors using graph theory, solely upon the local activations time (LATs), derived for individual vertices in 3D LA meshes for each simulation. The DGM computation pipeline is outlined in Fig. 2. In Fig. 2A, the LATs serve as nodes in the network, which can be acquired either in-vivo (electro-anatomical mapping) or in-silico, from simulations. Figure 2B shows an iterative traversal through DGM's computation algorithm, where neighbouring nodes are considered if they fall within a pre-set distance threshold. Velocity is then calculated based on the distance and LATs, and the initial excitation network is established by iteratively applying these steps to all nodes. A smoother cardiac activation is achieved by interpolating over the computed network. All potential rotors are subsequently identified using an iterative breadth first search approach [9], but only cycles with the lowest variance, indicative of more constant CV are retained. This aligns with the current understanding of rotors, which are typically characterized by a constant CV [16].

This study will employ DGM to extract rotor locations for a period of 4 s, i.e., between 1 s and 5 s into AF simulation, with a time-step of 20 ms. For the simulations, DGM was hyperparametrized as follows: minimal number of arrows in a loop: 5; variance: 0.15; intervariability: 0.15; distance: 20 mm; focal threshold: 25 mm; CV_{min}: 0.1 m/s; CV_{max}: 0.2 m/s; jump: 40 mm; core separator between identifiable loops: 50 mm.

2.3 Image-Based AF Modelling

The openCARP software was employed to simulate AF on the 3D LA meshes, and to assess its termination by CA [17]. Two electrophysiology models were used for different regions of the LA: the AF variant of the Courtemanche model for non-fibrotic "healthy" tissue and FBZ, and a cytokine-related and TGF-β1 remodelling model for dense fibrosis [18, 19]. Pacing at the end of the effective refractory period (PEERP) with five pre-pacing beats at a fixed basic cycle length of 270 ms (f = 3.7 Hz) was used to induce AF [20], which was run for 5 s in each patient (Fig. 3A). The ectopic location leading to AF was in one of the PVs, whereas in SR waves emerged in the LA from the Bachmann's Bundle. The Tissue conductivities were assigned per LA region to match the CV observed in patients: 0.7 m/s in healthy tissue (level 0), and 0.2 m/s in dense fibrosis (level 5) [21]. FBZ (levels 1–4) were assigned intermediate CV linearly decaying from 0.6 to 0.3 m/s (Fig. 3B). An additional level (6) with zero conductivity was created to represent ablated tissue.

Recurrence checks involved re-inducing arrhythmia after a successful RFCA procedure. The same PEERP protocol was utilized, and the location of new electrical activity was tested both posteriorly and anteriorly. Posteriorly, the ectopic location was calculated as the projection of the PV intersection onto the mesh. The anterior ectopic was then calculated as the intersection of *(i)* an imaginary line passing through the posterior ectopic with *(ii)* the anterior LA surface. In total, 52 simulations were conducted, with 13 representing the first application of RFCA, 26 recurrence checks after successful termination, and 13 redo RFCA procedures.

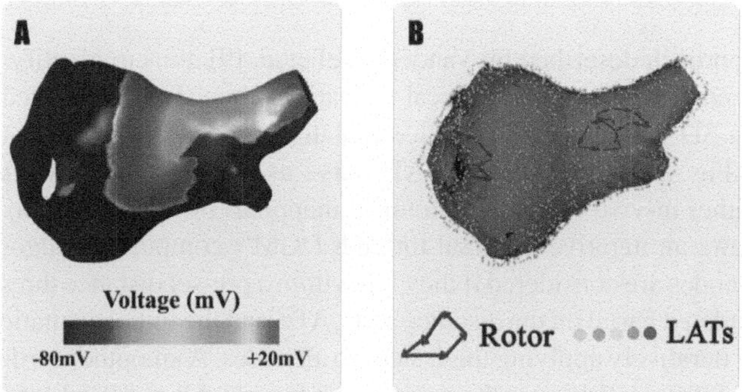

Fig. 3. Application of DGM for identifying re-entrant loops in the atria. **A**: Voltage map showing a propagating rotor in LA. **B**: Corresponding rotor loop identified by DGM.

2.4 Rotor Core Projection Ablation (RCPA)

The new ablation approach, termed Rotor Core Projection Ablation (RCPA), is outlined in Fig. 4. RCPA comprises 3 steps: (i) PVI ablation encircling the 4 PVs, (ii) connecting RCP points via geodesic lines on the LA mesh, and (iii) an additional geodesic line linking steps (i) and (ii). It was found that PVI and RCP alone could create an anatomical barrier around the RCP line, serving as an anchoring site for a rotor. The geodesic lines were computed as the shortest path connecting two points over the mesh, using Dijkstra's algorithm [22].

In addition to RCPA, PVI was also assessed as a standalone ablation technique, serving as a baseline for comparison with the novel technique.

The procedure for creating the RCPA is depicted in Fig. 4. On the right, the PVs are manually segmented for each patient, and the PVI is applied as a set of small circular lesions around the PVs, at the proximal site of the LA. Each circle is 3.5 mm in thickness, corresponding to the diameter of the catheter tip [23].

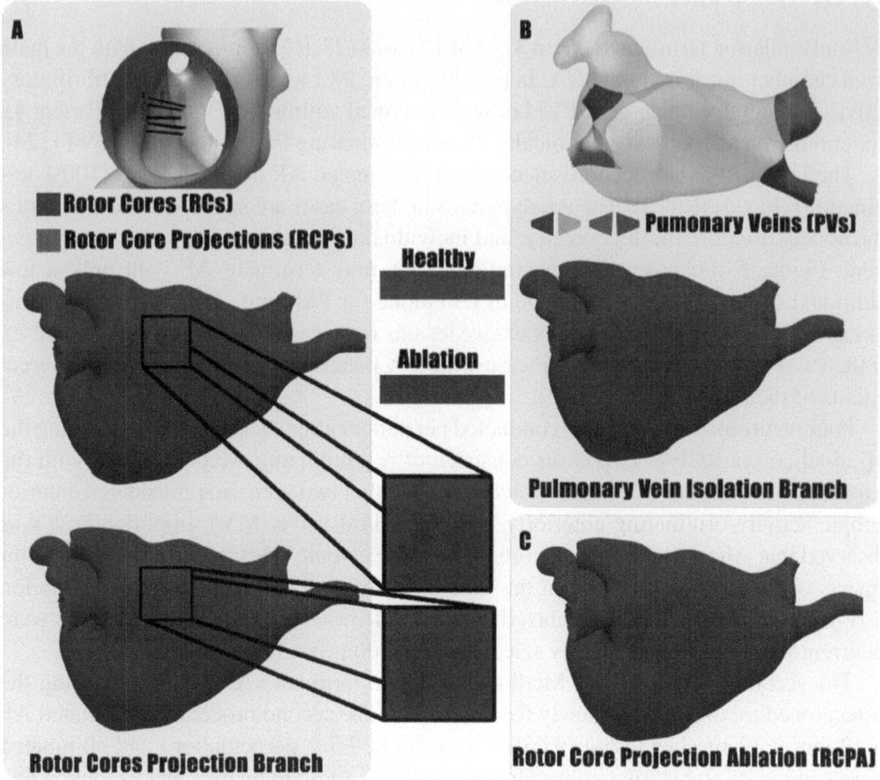

Fig. 4. RCPA workflow. **A**: Rotor Cores Projection Branch, RCPs points are created as the projection of RCs onto the LA mesh; **B**: PVI Branch, PVI is applied using segmentation of the PVs. **C:** The two workflows then merge to produce RCPA.

The projection of RC points may lead to the creation of the same or a smaller number of RCP points, as multiple RC projections can overlap. These RCP points are subsequently sorted by the angle they form with the origin, to create a polygon (Fig. 4A). The polygon is then delineated using geodesic lines, forming what is referred to as the RCP Line, depicted in Fig. 4A.

The calculation of the third 'close-the-loop' step involves determining the line with the minimum number of ablation lesions from the RCP Line to any of the four PVs. If the initial ablation using RCPA successfully terminated the arrhythmia, but recurrence was still possible, the same algorithm of identifying the new rotors, projecting them, and connecting them to the current ablation pattern was applied. In essence, RCPA represents an iterative in-silico process of identifying rotors and terminating arrhythmia through targeted ablation.

3 Results

3.1 AF Termination and Recurrence

PVI-only ablation terminated AF in 3 out of 13 cases (23.1% termination), with the main atrial chamber unexcited post-CA. In patients where PVI was successful, the fibrillatory activity did not disappear post-PVI but was contained within the PVs (Fig. 5, Patient 4), a phenomenon also observed clinically as self-terminating PV Tachycardia (PVT) [24].

The initial RCPA procedure successfully terminated AF in all 13 cases (100% termination) (Fig. 5, Row **3**). The AF dynamics in these cases are specific to each patient's fibrosis distribution, mesh geometry, and individual tissue electrical properties arrangement. Figure 5 illustrates that the initial RCPA may terminate AF with only a few additional ablation lesions compared to PVI alone for PAF (e.g., Patient 4), also terminating the PVT, or result in more extensive lesions for non-PAF patients (e.g., Patient 2). In the PsAF and long-standing AF cases, the CA patterns varied and covered different extents of the atria (e.g., Patient 3).

Four recurrence checks were conducted per patient after successfully terminating the AF in-silico via RCPA. Two cases consisted of re-initializing AF posteriorly, with the ectopic foci located between the PVs whereas the other two scenarios considered anterior ectopic activity originating anteriorly, near the mitral valve (MV). Importantly, it was observed that AF recurrence was more likely if the ectopic originates from the posterior aspects of the LA. At least one of the four cases tested led to arrhythmia perpetuation or re-initialization in all 13 patients, displayed in Row 4, Fig. 5. If multiple cases were recurrent, only one was manually selected for a redo procedure.

The second procedure (see Methods) aimed to terminate AF via RCPA, using the same procedure of targeting newly formed rotors. The second procedure terminated AF in 12 out of 13 meshes tested (92.3% termination, 7.7% perpetuation) and eliminated more sources of arrhythmogenesis. In some patients (e.g., Patients 1 and 2), only a few more ablation lesions were required compared to the initial procedure, whereas in others (e.g., Patients 3 and 4) more extensive lesions were necessary.

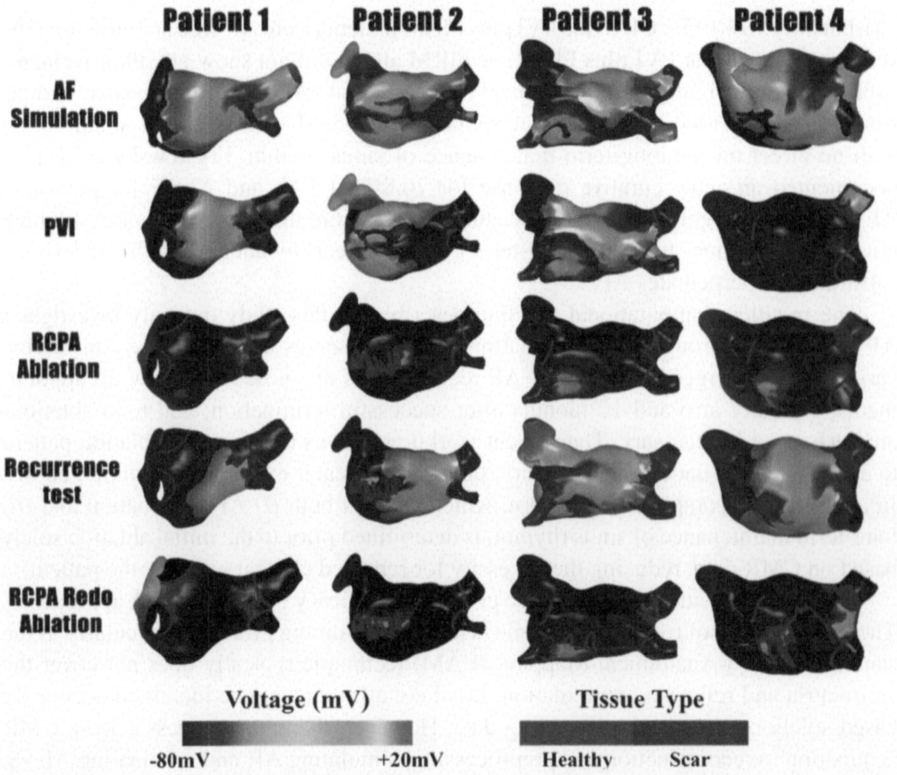

Fig. 5. Ablation outcomes for 4 patients. Row **1**: Initial AF Simulation. Row **2**: Stand-alone Pulmonary Vein Isolation. Row **3**: RCPA, first procedure leading to AF Termination. Row **4**: Failed recurrence test (arrhythmia re-initialization post-ablation). Row **5**: RCPA, redo procedure.

4 Discussion

The main objective of this study was to validate Direct-Graph Mapping (DGM) in identifying rotors and to develop a novel ablation technique that terminates AF in-silico. The PVI technique was implemented first using the PV segmentations. The RCPA method was afterwards simulated by connecting the RCP points with geodesic lines, which were then joined to the PVI pattern with a 'close-the-loop' line. In the dataset presented in this study comprised of 13 segmented LA models with fibrosis, the PVI technique achieved a 23% success rate in terminating the AF. In contrast, the novel RCPA method successfully terminated the arrhythmia in all patients after a single procedure, and further recurrent arrhythmias in 12 out of 13 patients, achieving a 92.3% termination rate.

A meta-analysis performed in 2018 aimed to investigate AF termination by CA in patients undergoing (i) PVI alone and (ii) PVI + FIRM ablation [25]. The study comprised 15 trials: 10 with PVI plus FIRM (n = 511), and 5 with PVI only (n = 295), involving mostly non-paroxysmal cases, with no statistically significant difference between the groups ($P = 0.21$). The analysis found a 58% success rate in PVI alone (95%

CI), but only 50% (95% CI) of the PVI plus FIRM patients were AF-free at follow-up. The study concluded that PVI plus FIRM, or FIRM alone, did not show any improvements over PVI alone, highlighting the clinical importance of PVI [25]. Additionally, various studies have reported spontaneous termination of the AF during FIRM ablation, but with no effect on the long-term maintenance of sinus rhythm. Prystowsky et al. [26] documented an acute curative response for 76.8% of PAF and 27.6% for persistent AF patients undergoing FIRM. These studies underscore the importance of developing novel CA techniques to ensure long-term curative effects by addressing the underlying substrate that perpetuates AF.

The in-silico computational pipeline described in this study not only investigates AF termination through catheter ablation but also assesses the recurrence considering various ectopic triggers. Clinically, AF recurrence is diagnosed at follow-up appointments, typically at 6 and 12 months after successful termination, and redo ablations are performed if necessary. The current workflow allows tailoring the ablation pattern to an individual's underlying structure, targeting a greater extent of arrhythmogenesis. In addition, the complete CA pattern, which ensures both *(i)* AF termination and *(ii)* long-term maintenance of sinus rhythm, is determined prior to the initial ablation solely based on CMR data, reducing the necessity for repeated clinical visits by the patient.

Moreover, consideration should be given to the efficacy of the novel CA approaches. The identification of rotors in the clinic is a time-consuming process, particularly as the standard Electro-Anatomical Mapping (EAM) technique typically does not cover the entire atria and relies on extrapolation. In-silico, all rotors can be identified accurately based solely on the acquired imaging data. However, the entire process – from CMR acquisition to reconstruction and preprocessing, simulating AF, and terminating AF via CA with additional recurrence checks – took, on average, 7 h per individual patient, which is a limitation of this study. However, novel deep learning (DL) algorithms can perform such segmentations and predictions in several minutes [27].

5 Conclusion

In conclusion, this study presents a novel computational approach for investigating AF termination through RFCA and assessing recurrence risks. By leveraging medical imaging data, the study demonstrates the ability to tailor CA procedures to individual patient anatomies, thereby improving efficacy while reducing unnecessary ablation. The study also highlights the potential of the proposed RCPA method to achieve AF termination in all patients after a single procedure (100% success), with promising results of 92.3% termination for recurrent cases. The RCPA method relies on the DGM Software to identify rotor cores based on AF simulations, which are then projected onto the LA mesh and targeted with ablation. Multiple studies reported a short-term success of in-vivo rotor ablation, and using RCPA may improve those outcomes by ablating only stable rotational activity that sustains AF. This study underscores the importance of advancing computational methods to improve AF treatment outcomes and pave the way for personalised therapeutic strategies.

Acknowledgements. This work was supported by funding from the EPSRC Centre for Doctoral Training in Smart Medical Imaging (EP/S022104/1) and by a project grant from the British Heart Foundation (PG/23/11552).

References

1. Wolf, P.A., Abbott, R.D., Kannel, W.B.: Atrial fibrillation as an independent risk factor for stroke: the Framingham Study. Stroke **22**(8), 983–988 (1991). https://doi.org/10.1161/01.STR.22.8.983
2. Marrouche, N.F., et al.: Association of atrial tissue fibrosis identified by delayed enhancement MRI and atrial fibrillation catheter ablation: the DECAAF study. JAMA **311**(5), 498–506 (2014). https://doi.org/10.1001/JAMA.2014.3
3. Marrouche, N.F., et al.: Efficacy of LGE-MRI-guided fibrosis ablation versus conventional catheter ablation of atrial fibrillation: the DECAAF II trial: study design. J. Cardiovasc. Electrophysiol. **32**(4), 916–924 (2021). https://doi.org/10.1111/JCE.14957
4. Dretzke, J., et al.: Predicting recurrent atrial fibrillation after catheter ablation: a systematic review of prognostic models. EP Europace **22**(5), 748–760 (2020). https://doi.org/10.1093/EUROPACE/EUAA041
5. Roy, A., et al.: Identifying locations of re-entrant drivers from patient-specific distribution of fibrosis in the left atrium. PLoS Comput. Biol. **16**(9), e1008086 (2020). https://doi.org/10.1371/JOURNAL.PCBI.1008086
6. Zakeri, R., et al.: The burden of proof: the current state of atrial fibrillation prevention and treatment trials. Heart Rhythm **14**(5), 763–782 (2017). https://doi.org/10.1016/J.HRTHM.2017.01.032
7. Yubing, W., et al.: Long-term outcome of radiofrequency catheter ablation for persistent atrial fibrillation. Medicine **97**(29), e11520 (2018). https://doi.org/10.1097/MD.0000000000011520
8. Sakata, K., et al.: Assessing the arrhythmogenic propensity of fibrotic substrate using digital twins to inform a mechanisms-based atrial fibrillation ablation strategy. Nat. Cardiovasc. Res. **3**, 857–868 (2024). https://doi.org/10.1038/s44161-024-00489-x
9. Van Nieuwenhuyse, E., et al.: DG-mapping: a novel software package for the analysis of any type of reentry and focal activation of simulated, experimental or clinical data of cardiac arrhythmia. Med. Biol. Eng. Comput. **60**(7), 1929–1945 (2022). https://doi.org/10.1007/S11517-022-02550-Y
10. Roney, C.H., et al.: Predicting atrial fibrillation recurrence by combining population data and virtual cohorts of patient-specific left atrial models. Circ. Arrhythm. Electrophysiol. **15**(2), e010253 (2022). https://doi.org/10.1161/CIRCEP.121.010253
11. Mahnkopf, C., et al.: Evaluation of the left atrial substrate in patients with lone atrial fibrillation using delayed-enhanced MRI: implications for disease progression and response to catheter ablation. Heart Rhythm **7**(10), 1475–1481 (2010). https://doi.org/10.1016/J.HRTHM.2010.06.030
12. Roney, C.H., et al.: Constructing a human atrial fibre atlas. Ann. Biomed. Eng. **49**(1), 233–250 (2021). https://doi.org/10.1007/S10439-020-02525-W/FIGURES/8
13. Benito, E.M., et al.: Left atrial fibrosis quantification by late gadolinium-enhanced magnetic resonance: a new method to standardize the thresholds for reproducibility. Europace **19**(8), 1272–1279 (2017). https://doi.org/10.1093/EUROPACE/EUW219
14. Morgan, R., Colman, M.A., Chubb, H., Seemann, G., Aslanidi, O.V.: Slow conduction in the border zones of patchy fibrosis stabilizes the drivers for atrial fibrillation: insights from multi-scale human atrial modeling. Front. Physiol. **7**, 474 (2016). https://doi.org/10.3389/fphys.2016.00474

15. Tilz, R.R., et al.: Focal impulse and rotor modulation ablation vs. pulmonary vein isolation for the treatment of paroxysmal atrial fibrillation: results from the FIRMAP AF study. Europace **23**(5), 722–730 (2021). https://doi.org/10.1093/EUROPACE/EUAA378
16. Waks, J.W., Josephson, M.E.: Mechanisms of atrial fibrillation – reentry rotors and reality. Arrhythm Electrophysiol Rev. **3**(2), 90 (2014). https://doi.org/10.15420/AER.2014.3.2.90
17. Plank, G., et al.: The openCARP simulation environment for cardiac electrophysiology. Comput. Methods Programs Biomed. **208**, 106223 (2021). https://doi.org/10.1016/J.CMPB.2021.106223
18. Loewe, A., Wilhelms, M., Dössel, O., Seemann, G.: Influence of chronic atrial fibrillation induced remodeling in a computational electrophysiological model. Biomedizinische Technik **59**, 929 (2014). https://publikationen.bibliothek.kit.edu/1000064025. Accessed 10 Aug 2023
19. Roney, C.H., et al.: Modelling methodology of atrial fibrosis affects rotor dynamics and electrograms. EP Europace **18**(suppl_4), iv146–iv155 (2016). https://doi.org/10.1093/europace/euw365
20. Azzolin, L., Schuler, S., Dössel, O., Loewe, A.: A reproducible protocol to assess arrhythmia vulnerability in silico: pacing at the end of the effective refractory period. Front. Physiol. **12**, 656411 (2021). https://doi.org/10.3389/fphys.2021.656411
21. Plank, G., Costa, C.M., Prassl, A.J.: Parametrization strategies for matching activation sequences in models of ventricular electrophysiology. In: Proceedings of the Annual International Conference of the IEEE Engineering in Medicine and Biology Society, EMBS, pp. 1534–1537 (2013). https://doi.org/10.1109/EMBC.2013.6609805
22. Dijkstra, E.W.: A note on two problems in connexion with graphs. Numer. Math. (Heidelb.) **1**(1), 269–271 (1959). https://doi.org/10.1007/BF01386390/METRICS
23. Ghzally, Y., Ahmed, I., Gerasimon, G.: Catheter Ablation. StatPearls (2023). https://www.ncbi.nlm.nih.gov/books/NBK470203/. Accessed 11 Aug 2023
24. Oral, H., et al.: Mechanistic significance of intermittent pulmonary vein tachycardia in patients with atrial fibrillation. J. Cardiovasc. Electrophysiol. **13**(7), 645–650 (2002). https://doi.org/10.1046/J.1540-8167.2002.00645.X
25. Mohanty, S., et al.: Long-term outcome of pulmonary vein isolation with and without focal impulse and rotor modulation mapping: insights from a meta-analysis. Circulation: Arrhythmia Electrophysiol. **11**(3), e005789 (2018). https://doi.org/10.1161/CIRCEP.117.005789
26. Henley, P., et al.: Single-center experience of the FIRM technique to ablate paroxysmal and persistent atrial fibrillation. J. Cardiovasc. Electrophysiol. **30**(4), 493–502 (2019). https://doi.org/10.1111/JCE.13832
27. Muffoletto, M., et al.: Toward patient-specific prediction of ablation strategies for atrial fibrillation using deep learning. Front. Physiol. **12**, 674106 (2021). https://doi.org/10.3389/fphys.2021.674106

Automated Pipeline for Regional Epicardial Adipose Tissue Distribution Analysis in the Left Atrium

Èric Lluch[1(✉)], Eduardo Castañeda[1,2], Maximilian E. R. Weiss[1], Anamaria Vizitiu[3], Athira Jacob[4], Jitin Jami[5], Chloé Audigier[6], Felix Meister[1], Viorel Mihalef[4], and Tiziano Passerini[4]

[1] Siemens Healthineers, Digital Technologies and Innovation, Erlangen, Germany
eric.lluch@siemens-healthineers.com

[2] Pattern Recognition Lab, Department of Computer Science, Friedrich-Alexander University Erlangen-Nürnberg, Erlangen, Germany

[3] Siemens SRL, Brasov, Romania

[4] Siemens Healthineers, Digital Technology and Innovation, Princeton, USA

[5] Machine-Learning and Data Analytics, Friedrich-Alexander-Universität Erlangen-Nürnberg, Erlangen, Germany

[6] Advanced Clinical Imaging Technology, Siemens Healthineers AG, Lausanne, Switzerland

Abstract. Atrial fibrillation (AF) is the most common cardiac arrhythmia, affecting approximately 3% of the global population and rising to 11% in individuals over 80 years old. The distribution of epicardial adipose tissue (EAT) around the atria plays a significant role in the pathophysiology of AF, yet its regional analysis remains underexplored. This paper introduces an automated pipeline capable of segmenting 14 atrial regions from patient CT angiograms, a notable advancement over previous attempts to automate the creation of synthetic anatomical or physics-based twins. Our fully automatic pipeline requires no human supervision and provides the regional EAT quantification successfully on 91.9% of the analyzed patients for which the LA and its structures were automatically segmented (1797 out of 1954 patients), including the automatic segmentation of the aorta and pericardium to assess EAT volume around the left atrium. By applying principal component analysis of the regional EAT to analyze a separate dataset of 84 individuals who were clinically labeled, we identified a distinct group that contained all the patients with AF. This finding suggests that there is a significant link between the pattern of fat distribution and AF. This novel approach has the potential to enhance our understanding of AF and improve patient outcomes by enabling large-scale, regional fat analysis in an automated manner on all patients with accurate segmentation of left atrial structures.

Keywords: Epicardial adipose tissue · atrial fibrillation · automatic pipeline

O. Camara et al. (Eds.): STACOM 2024, LNCS 15448, pp. 323–333, 2025.
https://doi.org/10.1007/978-3-031-87756-8_32

1 Introduction

Atrial fibrillation (AF) is the most common sustained cardiac arrhythmia, affecting millions worldwide and increasing the risk of stroke, heart failure, and overall mortality. Recent studies have identified a significant association between atrial fat, particularly epicardial adipose tissue (EAT), and the pathogenesis of AF [1–3]. EAT is a visceral fat depot located between the myocardium and visceral pericardium that has been implicated in the structural and electrical remodeling of the atria predisposing individuals to AF. The study by S. Hatem et al. described the role of EAT in promoting fibrosis and inflammation in atrial tissues, which contribute to AF maintenance and recurrence [3]. The role of EAT volume as a marker of increased AF risk has been investigated using imaging modalities such as computed tomography (CT) and magnetic resonance imaging (MRI) across different patient cohorts. An incremental increase in anterior EAT volume was demonstrated in patients with paroxysmal and persistent AF when compared to control groups to [2, 4–6]. An association between increased EAT and increased risk of AF recurrence post-ablation was also shown [1, 2], highlighting the potential utility of EAT volume quantification for also predicting treatment response.

However, these methods entail several limitations. There is a lack of standardization in the definition of Hounsfield Units (HU) thresholds that correspond to EAT, though $(-190, -30)$ HU is a commonly used range [2, 4]. Moreover, many publications have not analyzed the association between accumulation of EAT in specific atrial regions and increased AF risk, focusing instead on the global amount of EAT [2, 5, 6], which may overlook critical local effects influencing AF pathogenesis, as suggested by [3]. Current EAT analysis methods are primarily image-based, relying on manual or semi-automatic segmentation techniques that do not incorporate detailed geometric information from the left atrium (LA) structures, such as the pulmonary veins (PVs) or the left atrial appendage (LAA). The lack of automation limits the possibility of scaling the analysis to a big population of patients. The missing detailed anatomical consideration limits the accuracy and reproducibility of regional EAT assessments. Additionally, the focus has predominantly been on the overall volume of EAT rather than its regional distribution, which might be more relevant for understanding the relationship between EAT and AF and its recurrence.

To address these limitations, we propose an automated pipeline for regional EAT analysis. This novel approach includes automatic segmentation of EAT in CT images. Furthermore, we subdivide the LA into 14 anatomically relevant regions using automatically detected landmarks and project the EAT onto these regions. We then use unsupervised clustering methods to develop biomarkers based on regional EAT distribution. We further demonstrate the utility of these biomarkers in a separate cohort of 84 patients with clinical labels where we show that these clusters correlate with control versus AF populations. The main objective of this study is to develop a model of regional distribution of EAT in the left atrium. Using this model, we identify common patterns of EAT distribution in subjects with and without AF recurrence. We believe this approach has the potential to enhance the understanding of EAT's role in AF pathogenesis and improve patient stratification for targeted treatment.

2 Methods

2.1 Data

Retrospective multi-center data of cardiac CT angiograms (CTA) of 3620 patients was used for this study. A second set of 84 multi-center data cardiac CTA images with disease labels (AF: 49 vs controls: 35) was used to analyze the influence of EAT in AF classification.

2.2 Epicardial Adipose Tissue (EAT) Segmentation

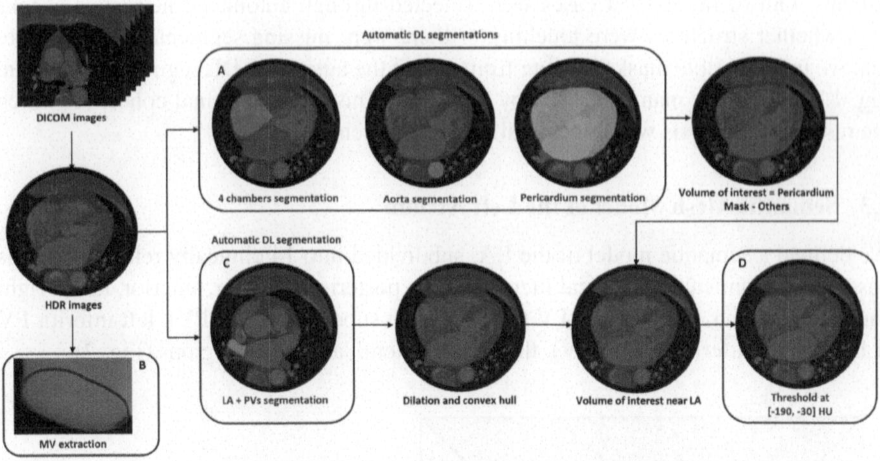

Fig. 1. Image processing pipeline to extract the EAT

The first step of the pipeline involves segmentation of EAT in the whole heart from the CT images. The pericardium, left and right ventricles, left and right atria and aorta were delineated using previously validated deep learning-based segmentation algorithms [7, 12], which are based on a 3D Image-to-Image (I2I) segmentation network and a conditional variational autoencoder (cVAE). The ventricles, atria and aorta masks were subtracted from the pericardium mask to obtain candidate regions for EAT computation. To delineate substructures more precisely in the left atrium, we used a similar DL network, trained to segment the left atrium body (LA), 4 PVs, and LAA. The model was trained on a cohort of 3620 CT volumes with ground truth masks, divided on a patient level into train (3506) and validation (114) sets. The network architecture, and other training details, were adopted as described in [7]. On a separate testing set of 191 volumes, the network achieved average dice scores of 0.91 for LA, 0.81 each for LIPV, LSPV and RIPV respectively, 0.83 for RSPV and 0.82 for LAA. The 3D shape of the mitral valve annulus was estimated with a multi-step process. In the shape initialization step, the expected shape was generated based on a mean shape model. This model was obtained by Procrustes analysis on a large database of annotated left atrium anatomies. The mean shape (represented by a tetrahedral mesh) was then aligned with the imaging data based on 4 automatically detected anatomical landmarks (anterior/inferior, lateral/septal annulus points). An affine transformation was estimated based on the position of the landmarks and the corresponding points in the shape model. Thin plate splines interpolation was used to apply the transformation to the remaining mesh points (Fig. 1, B).

To define an epicardial region of interest in the left atrium, the LA mask was dilated by 2 voxels in each direction and subsequently intersected with the volume of interest (EAT around the heart). Finally, voxels in the candidate region with intensities within [−190, −30] HU were selected to get the final EAT region as shown in Fig. 1, D. The segmentation models were developed prior to the work described in the current study, and the models were run as-is, with no modifications, on a sub-cohort of 2770 patients. Out of these 1954 cases were selected through automated heuristics checks (e.g. whether structures were touching or there were missing segmentation masks) to remove incompatible masks coming from any of the automatic DL segmentations from Fig. 1, followed by manual review by 3 experts. This forms the final cohort on which the rest of the analysis was done, totaling 1954 patients.

2.3 Semantic Mesh Model of the Left Atrium

We defined a semantic model of the LA, subdivided into 14 clinically relevant regions based on anatomical and clinical literature: The posterior, superior, anterior, LAA, right carina, left carina, left superior PV (LSPV), right superior PV (RSPV), left inferior PV (LIPV), right inferior PV (RIPV), floor, roof, lateral, and septal regions (Fig. 2).

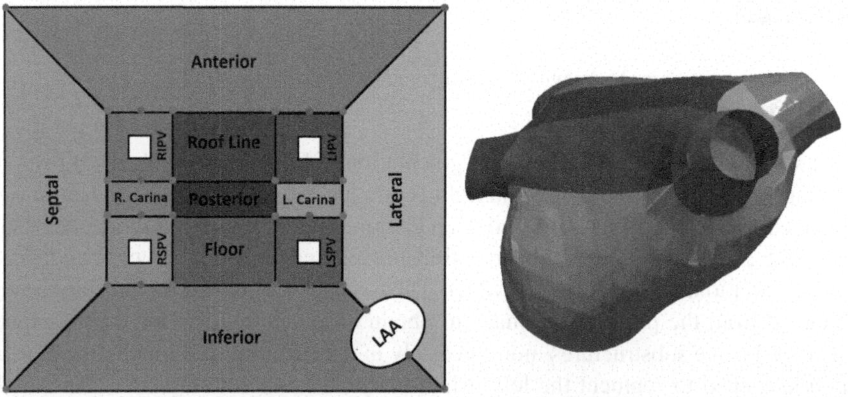

Fig. 2. 2D representation of the LA surface, including the defined regional subdivision. The outside boundary represents the Mitral Valve. Red points represent automatically computed landmarks, while blue points represent landmarks derived based on equidistant subdivisions of boundary edges. Black lines represent geodesic lines drawn between landmark points.

- 3D mesh from imaging (Fig. 3, step 1). An initial triangular mesh representation of the LA and its substructures was obtained by thresholding the segmentation masks.

- PVs clipping (Fig. 3, step 2). Orifices were created in the PVs by clipping the mesh with a plane, 15mm away from the common edge with the LA body.
- Mitral valve (MV) clipping (Fig. 3, step 3). The MV orifice was created based on the MV annulus defined in the previous pipeline (Sec. 2.2). Additionally, the detection of the MV is performed via an AI-based landmark detection algorithm [8]. This one estimates an ellipsoid defined by a set of 58 points with consistent anatomical locations that describe the MV. To improve the algorithm robustness, the ellipsoid can be adjusted to match the position of the PVs. (Fig. 3, step 3).
- Landmarks detection (Fig. 3, step 4). Landmarks defining the boundaries between the PVs and the LA body were detected automatically by drawing a line through the centre of mass of the contiguous PV boundaries (left PVs and right PVs separately, see Fig. 2). The closest points to the line in both the carina and the anterior and inferior regions are selected respectively as the furthest and closest points (red points on the PV contours in Fig. 2). These points split each PV in 2 halves and define 2 derived landmark points as the barycenters of the split PV contours (blue points on in Fig. 2). An equal approach is followed to find landmarks on the distal contour of the PVs. Landmarks in the LAA boundary were estimated by drawing a line between the center of mass of the LAA ostium and the center of mass of the LSPV. The closest point in the LAA boundary to the line and the LSPV, and the closest point to the line and furthest from the LSPV boundary are selected (red points on LAA ostium in Fig. 2).
- Geodesic lines definition (Fig. 3, step 5, 6): the landmarks are used to define geodesic lines that define the 14 desired anatomical regions.

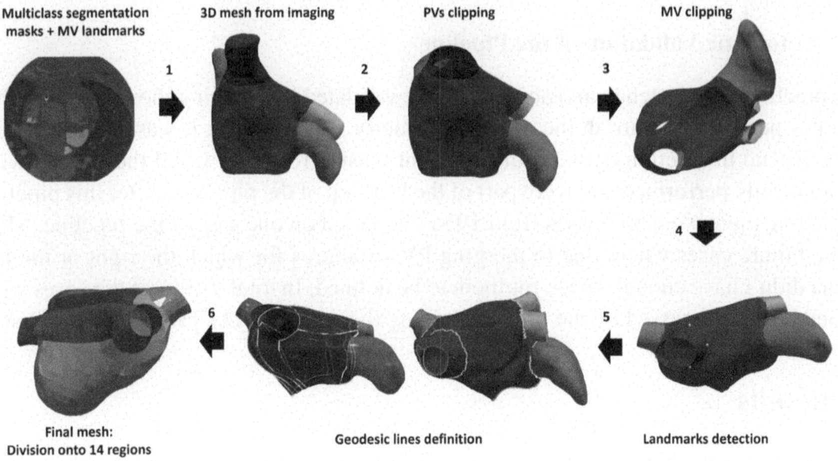

Fig. 3. Automatic pipeline for mesh processing to generate 14 relevant anatomical regions.

2.4 Regional Analysis of the EAT

The EAT segmentation in the epicardial region of interest of the left atrium as described in Sec. 2.2 (Fig. 4, left panel) was attributed to the 14 LA regions using a projection method, computing the intersection of the EAT volume with the surface normal vectors as in the middle of Fig. 4. The projection on the LA surface was computed by adding the total EAT from the sum of all intersections along the normal with the EAT volume.

Fig. 4. Projection of EAT onto each of the regions using the normal projection.

Using this approach, regional EAT distribution for a patient data set is represented with a 14-dimensional vector, each component representing the regional burden of EAT. We performed principal component analysis of this vector representation over the test set of 1797 patient data sets. We considered the first 7 principal components, representing 90% of the variance, and computed the corresponding weights. Finally, we clustered the weights. Moreover, for each principal component we qualitatively analyzed the corresponding regional EAT burden patterns and their mode of variation.

2.5 Automatic Validation of the Pipeline

The pipeline's consistency and accuracy were validated by ensuring the correct number of holes per LA structure in the clipped geometry. Additionally, it was confirmed that there was no interaction between the different masks from Fig. 1. All these steps were automatically performed and were part of the last step of the pipeline. After this pipeline validation, there were 157 cases from 1954 that failed on one step of the pipeline. Most of the failure cases where due to merging PV structures for which the right or the left carina didn't have enough space for them to be defined. In total 91.9% of the cases were automatically processed by the pipeline, giving the regional EAT projection as a result.

3 Results

3.1 Regional Fat Distribution Analysis and Its Correlation with AF

The regional fat distribution analysis was conducted on the segmented structures, focusing on the epicardial adipose tissue (EAT) in 14 anatomically relevant regions. Boxplots were generated to visualize the distribution of EAT/area in each region (Fig. 5). The first component of the PCA analysis (PC1) explained 59% of the variance, revealing the most relevant regions to the variance in EAT distribution.

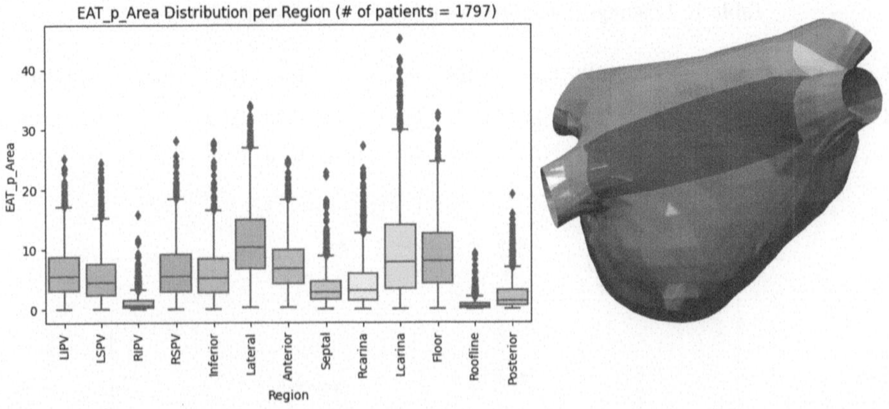

Fig. 5. EAT volume/Region Area distribution (mm^3/mm^2) per region on 1797 patients. Showing the relevance of the regions normalized to the area.

Then, K-means clustering was used to extract 2 clusters from the principal components, providing insights into patterns of EAT distribution among the patient cohort (See Fig. 6).

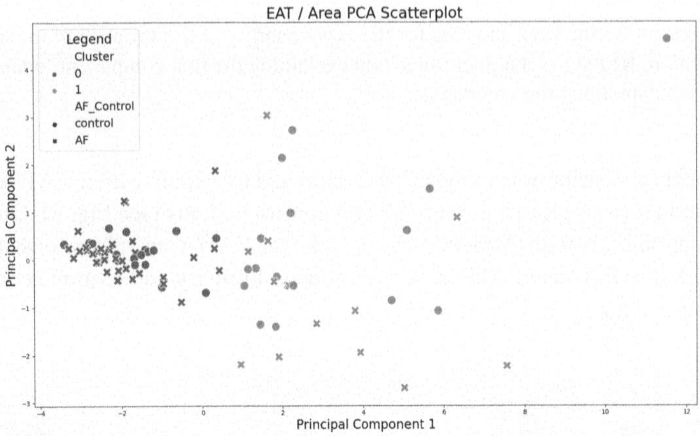

Fig. 6. Automatic clustering of the data based on 2 principal component analysis. Correlation of AF (marked with cross) and control (marked with a circle) labels with the clusters

The regions explaining the most variance have the highest loadings in Table 1 and Fig. 7 (RSPV and Anterior). The regions highlighted by the PCA and clustering analysis were compared with known clinical observations and literature such as the anterior and posterior accumulation of EAT [9–11]. Furthermore, we analyzed 84 patients for whom disease labels were available.

Table 1. Loadings of the first two principal components per region.

Load	RSPV	Ant	Floor	LSPV	Inf	LIPV	Post	Sep	Lat	Rcar	Roof	Lcar	RIPV
PC1	0.88	0.87	0.85	0.84	0.84	0.80	0.79	0.78	0.78	0.69	0.69	0.67	0.53
PC2	0.08	0.16	0.07	0.14	0.30	0.07	0.26	0.30	0.28	0.55	0.27	0.06	0.65

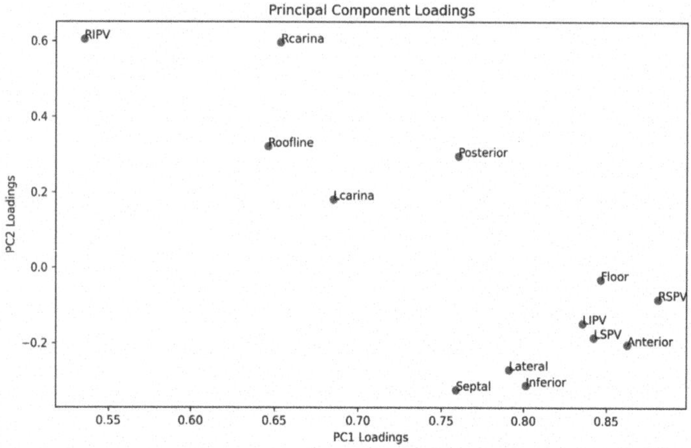

Fig. 7. Importance of the EAT and area for the explainability of the variance of the first components of variation. RSPV has a higher impact on explaining the first component, while RIPV has a higher impact explaining the second one

A regional fat distribution analysis was conducted to visualize their EAT/Area distribution, including examples from both AF and control patients (see Fig. 8). Our findings indicated that patients with AF tended to have lower EAT/Area values, possibly due to the larger LA in AF patients. This observation is preliminary and requires confirmation with additional data.

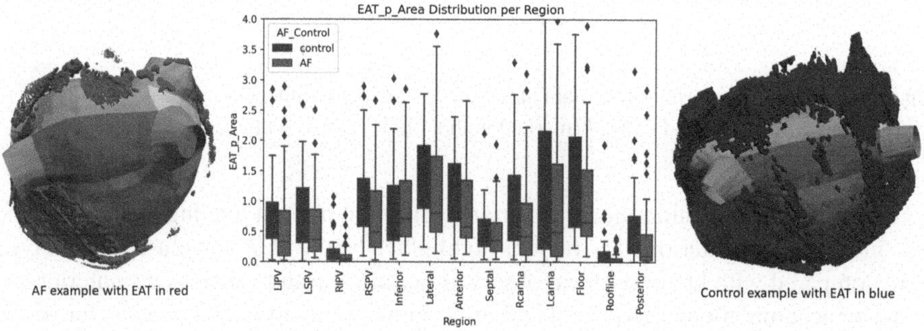

Fig. 8. EAT/Area distribution per region for AF (# 49) and control (# 35) patients.

The comparison of EAT_p_Area between control and AF groups across different regions did not yield statistically significant differences after applying the Benjamini-Hochberg correction for multiple comparisons (all corrected p-values > 0.05; Table 2). A K-means cluster was also performed for the 84 patients and these clusters were visualized together with the disease labels (Fig. S1). This analysis aimed to identify whether the labeled patients aligned with the clusters identified by the PCA, providing a potential link between regional EAT distribution and AF status.

Table 2. Statistical significance of AF vs controls per region together with the corrected p-value from the Benjamini-Hochberg test.

Region	Control mean	AF mean	T-statistic	p-value	Corrected p-value
LIPV	0.80	0.66	0.92	0.36	0.52
LSPV	0.83	0.59	1.55	0.13	0.27
RIPV	0.19	0.10	1.74	0.09	0.23
RSPV	1.16	0.75	2.18	0.03	0.23
Inferior	0.93	0.93	−0.01	1.00	1.00
Lateral	1.43	1.31	0.53	0.60	0.65
Anterior	1.16	0.87	1.73	0.09	0.23
Septal	0.54	0.48	0.71	0.48	0.62
RCarina	1.01	0.64	2.10	0.04	0.23
LCarina	1.31	1.11	0.64	0.52	0.62
Floor	1.41	1.12	1.26	0.21	0.35
Roofline	0.19	0.07	1.97	0.06	0.23
Posterior	0.59	0.40	1.25	0.22	0.35

The runtime of the pipeline after parallelization with 32 CPU cores was 7.1 s per patient on average to perform each step from the segmentations to the creation of the mesh and the EAT projection, allowing for rapid processing of large datasets (see Table 3 for detailed computational times without parallelization).

Table 3. Run times of each processing step of the pipeline in seconds. The pipeline is parallelized with CPU, so the timings are decreased linearly with more clusters.

Runtime per step	Average [s]	Standard dev. [s]
EAT mask generation	2.87	0.04
4 chamber segmentation	7.61	0.11
Aorta segmentation	9.94	0.48

(continued)

Table 3. (*continued*)

Runtime per step	Average [s]	Standard dev. [s]
LA + PV segmentation	6.94	0.06
Mitral valve extraction	4.80	0.11
Pericardium segmentation	11.31	0.50
Mask-to-Mesh conversion	1.71	0.29
PV clipping	1.45	0.18
MV clipping	0.97	0.12
LA Refinement	51.09	20.12
Geodesic line placement	121.43	12.56
Regional projection of EAT	6.61	2.54

4 Discussion

The automated pipeline for segmentation and analysis of EAT around the atria demonstrates advancements over existing methods, which are primarily manual or semi-automatic. Our pipeline incorporates advanced AI techniques to achieve accurate and consistent segmentations. This approach reduces the need for extensive manual intervention, providing a robust and scalable solution for analyzing large datasets. The detailed analysis of regional EAT distribution offers insights into the pathophysiology of AF. Our findings suggest that certain regions, such as the anterior or the RSPV, are particularly significant in explaining the distribution of total EAT around the LA. While the observed differences in EAT/Area between control and AF groups suggest potential trends, these differences were not statistical significance after correction for multiple comparisons. This lack of significance is likely due to the limited sample size in our study. Future studies with larger clinically-labeled sample sizes are needed to validate these preliminary findings and provide more definitive conclusions. The ability to automatically quantify and analyze regional EAT could be a valuable tool for clinicians, aiding in the identification of patients at higher risk. One of the primary limitations of this study is the lack of ground truth labels for a comprehensive evaluation of the segmentation accuracy as well as a lack of labels on disease vs controls.

Future research should aim to address failure cases by ensuring adequate separation between merged PVs. This can be achieved by adding sufficient voxels between the touching veins, thereby creating enough space to accurately define the right or left carina. Future research should aim to correlate the automated regional EAT analysis with clinical labels to validate the relevance of the identified regions. Further investigations could explore the combination of EAT features with shape analysis, leveraging the automated pipeline to prepare meshes and regional areas consistently across populations. Developing a parametric mesh for point correspondences could facilitate detailed morphometric studies and improve our understanding of the relationship between EAT distribution and AF. Ultimately, these advancements could lead to more precise diagnostic tools and targeted therapies, significantly impacting AF research and treatment strategies.

Acknowledgments. The work was supported by the European Union's Horizon 2020 Research and Innovation Program under grant agreement No. 965286 (MAESTRIA).

Disclosure of Interests. The authors have no competing interests to declare that are relevant to the content of this article.

Disclaimer. The concepts and information presented in this presentation are based on research results that are not commercially available. Future commercial availability cannot be guaranteed.

References

1. Anagnostopoulos, I., et al.: Epicardial adipose tissue and atrial fibrillation recurrence following catheter ablation: a systematic review and meta-analysis. J. Clin. Med. **12**(6369) (2023)
2. Takahashi, N., Abe, I., Kira, S., Ishii, Y.: Role of epicardial adipose tissue in human atrial fibrillation. J Arrhythmia **39**, 93–110 (2023)
3. Hatem, S.N., Sanders, P.: Epicardial adipose tissue and atrial fibrillation pathophysiology and clinical implications. Cardiovasc. Res. **120**(3), 567–577 (2023)
4. Nagashima, K., et al.: Association between epicardial adipose tissue volumes on 3-dimensional reconstructed CT images and recurrence of atrial fibrillation after catheter ablation. Circ J **75**(11), 2559–2565 (2011)
5. Oba, K., et al.: Effect of the epicardial adipose tissue volume on the prevalence of paroxysmal and persistent atrial fibrillation. Circ. J. **82**(7),1778–1787 (2018)
6. Nakamori, S., Nezafat, M., Ngo, L.H., Manning, W.J., Nezafat, R.: Left atrial epicardial fat volume is associated with atrial fibrillation: a prospective cardiovascular magnetic resonance 3D dixon study. J Am Heart Assoc. **7**(6) (2018)
7. Aquino, G.J., et al.: Diagnostic accuracy and performance of artificial intelligence in measuring left atrial volumes and function on multiphasic CT in patients with atrial fibrillation. Eur. Radiol. **32**(8), 5256–5264 (2022)
8. Grbic, S., Voigt, I., Mansi, T., Georgescu, B., Ionasec, R., Comaniciu, D.: Aortic and mitral valve modeling from multi-modal image data. In: Medical Image Recognition, Segmentation and Parsing, pp. 363–382, Academic Press, New York (2016)
9. Tsao, H.M., et al.: Quantitative analysis of quantity and distribution of epicardial adipose tissue surrounding the left atrium in patients with atrial fibrillation and effect of recurrence after ablation. Am. J. Cardiol. **107**(10), 1498–1503 (2011)
10. Nakatani, Y., Sakamoto, T., Yamaguchi, Y., Tsujino, Y., Kinugawa, K.: Epicardial adipose tissue affects the efficacy of left atrial posterior wall isolation for persistent atrial fibrillation. J Arrhythmia. **36**, 652–659 (2020)
11. Ohguchi, S., et al.: Regional left atrial conduction velocity in the anterior wall is associated with clinical recurrence of atrial fibrillation after catheter ablation: efficacy in combination with the ipsilateral low voltage area. BMC Cardiovasc. Disord. **22**(1), 457 (2022)
12. Zheng, Y., Yang, D., John, M., Comaniciu, D.: Multi-part modeling and segmentation of left atrium in C-arm CT for image-guided ablation of atrial fibrillation. IEEE Trans. Med. Imaging **33**(2), 318–331 (2013)

Low-Rank Conjugate Gradient-Net for Accelerated Cardiac MR Imaging

Jaykumar H. Patel[1,2](✉), Brenden T. Kadota[1,2], Calder D. Sheagren[1,2], Mark Chiew[1,2], and Graham A. Wright[1,2]

[1] Department of Medical Biophysics, University of Toronto, Toronto, ON, Canada
jaykumar.patel@mail.utoronto.ca

[2] Physical Sciences Platform, Sunnybrook Research Institute, Toronto, ON, Canada

Abstract. Cardiovascular diseases (CVDs) remain the leading cause of mortality and morbidity worldwide. Both diagnosis and prognosis of these diseases benefit from high-quality imaging, which cardiac magnetic resonance imaging provides. CMR imaging requires lengthy acquisition times and multiple breath-holds for a complete exam, which can lead to patient discomfort and frequently results in image artifacts. In this work, we present a Low-rank tensor U-Net method (LowRank-CGNet) that rapidly reconstructs highly undersampled data with a variety of anatomy, contrast, and undersampling artifacts. The model uses conjugate gradient data consistency to solve for the spatial and temporal bases and employs a U-Net to further regularize the basis vectors. Currently, model performance is superior to a standard U-Net, but inferior to conventional compressed sensing methods. In the future, we aim to further improve model performance by increasing the U-Net size, extending the training duration, and dynamically updating the tensor rank for different anatomies.

Keywords: Multi-contrast Imaging · k-t imaging · Deep subspace learning · Low-Rank Tensor Model · U-Net and Unrolled Model

1 Introduction

Cardiovascular diseases (CVDs) remain the leading cause of mortality and morbidity worldwide, necessitating early diagnosis, disease monitoring, and prognosis assessment post-treatment, all involving imaging [16]. Cardiac magnetic resonance (CMR) imaging, a non-invasive modality devoid of ionizing radiation, is often employed for diagnosing and managing CVDs. CMR imaging utilizes various tissue contrast mechanisms to provide detailed and comprehensive assessments of the heart's structure and function for a range of cardiac pathologies and morphologies.

CMR is a powerful tool that offers dynamic imaging at high spatiotemporal resolution and varying image contrast for accurate diagnosis. T1 and T2 mapping are crucial in CMR imaging due to their ability to provide quantitative

O. Camara et al. (Eds.): STACOM 2024, LNCS 15448, pp. 334–344, 2025.
https://doi.org/10.1007/978-3-031-87756-8_33

assessments of myocardial tissue properties. T1 mapping evaluates extracellular volume and diffuse myocardial fibrosis for detecting cardiomyopathies [6,9], while T2 mapping identifies edematous regions within myocardial tissue. Cine imaging, with its high temporal resolution, captures cardiac motion, allowing the measurement of ventricular volumes and ejection fraction. Cine imaging is the gold standard for assessing ventricular function and plays a pivotal role in determining cardiac health [8]. Tagging techniques also provide functional information regarding myocardial strain and deformation [1]. Additionally, imaging of the aorta using CMR is vital for assessing aortic pathologies such as aneurysms and dissections [14]. These advanced CMR techniques enhance diagnostic accuracy by providing detailed spatiotemporal information about cardiac and vascular structures and functions.

However, a significant drawback of these techniques is the extended acquisition time, which can lead to patient discomfort and frequently results in image artifacts from cardiac, respiratory, and patient motion, compromising spatial resolution and signal-to-noise ratio (SNR) [12]. This necessitates re-scans, the use of motion compensation or correction methods, and the use of triggering techniques to mitigate artifacts and/or accelerated undersampled scans to reduce the acquisition duration. These limitations diminish patient throughput and accessibility. To shorten image acquisition times, parallel imaging techniques such as Sensitivity Encoding (SENSE) and Generalized Autocalibrating Partial Parallel Acquisition (GRAPPA) are employed to exploit multi-channel spatial redundancy and accelerate scans by acquiring a fraction of the fully-sampled data [5,10]. Dynamic undersampled datasets can be reconstructed using methods that exploit spatiotemporal (k-t) redundancy, including compressed sensing and low-rank approaches. While compressed sensing techniques are effective and can leverage sparsity in the k-t domain, they often require extensive empirical parameter tuning tailored to each patient and imaging contrast, which can reduce generalizability.

In contrast, low-rank tensor formulations and subspace regularization methods provide a more flexible model, where correlations between global and local features are exploited across spatiotemporal domains. This flexibility enables low-rank methods to deliver superior reconstruction quality and consistency, making them a more robust choice across diverse imaging contrasts [15]. These approaches improve the overall image quality of reconstructions by addressing varied undersampled spatial patterns that arise from cardiac motion. For instance, spatiotemporal images for T1/T2 mapping, cine imaging, and aorta imaging, which are composed of spatial images acquired over different temporal domains (inversion times, echo times, cardiac phases, or respiratory phases), benefit significantly from these techniques [7,19].

Low-rank tensor formulations can be employed to set singular vectors with lower-magnitude singular values - corresponding to noise and artifacts - to zero [2]. This process enhances image quality by filtering out undesirable elements. Additionally, subspace regularization can further refine dynamic image quality. Low-rank algorithms are particularly advantageous due to their ability to compactly represent a wide range of signal profiles, making them highly suitable for generalizing across different imaging contrasts [13].

To generalize over the cardiac dynamic contrast subspace, separate spatial and temporal regularizers can also be added to improve image quality and preserve dynamic spatial information across generalized image contrasts. In this paper, we propose a temporally dynamic generalized MRI reconstruction method using a subspace low-rank model, which aims to recover spatiotemporal bases with the U-Net backbone of the generalized accelerated dynamic cardiac images.

2 Methods

2.1 Dataset

Briefly, this method was trained on the CMRxRecon2024 dataset [18]. The dataset consists of 2D Cartesian-sampled MRI images of the heart and aorta. A prospective acceleration factor of 2 was used to acquire the data, with GRAPPA reconstruction being used to create a fully sampled k-space. Subsequently, data was undersampled using various *k-t* methods, which include uniform undersampling, Gaussian-density undersampling, and pseudo-radial undersampling. Each pattern preserved a fully sampled central k-space region for coil sensitivity calculation.

2.2 Algorithm

Define X to be the image we want to reconstruct, with $X_q, q = 1, \ldots, Q$ being images from individual coils. We define our forward model as

$$A : \mathbb{C}^{N_x \times N_y} \to \mathbb{C}^{N_x \times N_y \times N_c}; \quad X \mapsto (PFS_1X, PFS_2X, \ldots, PFS_QX). \tag{1}$$

Here, S_q represents multiplication by the sensitivity maps for coil q; F represents the Fourier Transform; and P represents the projection onto the sampled points. Similarly, we can define $\tilde{P}$ to be the projection onto the points not sampled by P and define a corresponding operator

$$\tilde{A} : \mathbb{C}^{N_x \times N_y} \to \mathbb{C}^{N_x \times N_y \times N_c}; \quad X \mapsto (\tilde{P}FS_1X, \tilde{P}FS_2X, \ldots, \tilde{P}FS_QX). \tag{2}$$

We denote the k-space data by $Y \in \mathbb{C}^{N_x \times N_y \times N_z}$, with the relation $AX = Y$ (Fig. 1).

The reconstruction method is shown in Figure 1, we decompose $X = C \otimes T$ to be a tensor product of a spatial basis C and a temporal basis T. We aim to jointly solve for data consistency of X and spatial regularization of C, denoted as $R_s(C)$,

$$\hat{C}, \hat{T} = \arg\min_{C,T} \left\{ \frac{1}{2} \|Y - A(C \otimes T)\|_2^2 + \lambda R_s(C) \right\}. \tag{3}$$

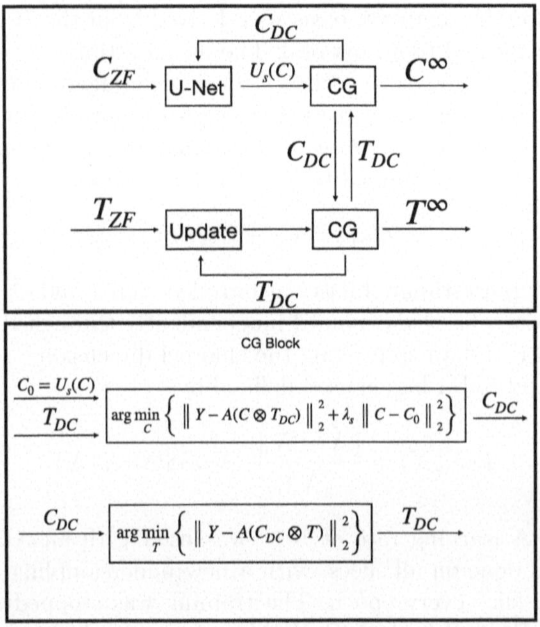

Fig. 1. Visual overview of algorithm. Initial spatial and temporal bases C_{ZF} and T_{ZF} are fed into the model. Spatial bases are fed through a U-Net, before a conjugate gradient-based update step is used to generate data-consistent spatial bases. Subsequently, a separate conjugate gradient update step is used to generate data-consistent temporal bases. These steps are iterated until convergence.

We aim to solve this problem via an alternating minimization of conjugate gradient data consistency on C and T separately, with a trained model U_s that provides regularization of spatial bases at each step of the reconstruction:

$$\hat{C}^k = \arg\min_C \left\{ \left\| Y - A(C \otimes T^{k-1}) \right\|_2^2 + \lambda \left\| C - U_s(C^{k-1}) \right\|_2^2 \right\} \tag{4}$$

$$\hat{T}^k = \arg\min_T \left\{ \left\| Y - A(C^k \otimes T) \right\|_2^2 \right\}. \tag{5}$$

As a final step, to ensure consistency with acquired data points, we set

$$X = A^H (Y + \tilde{A}(C^\infty \otimes T^\infty)). \tag{6}$$

2.3 Model Training

We train our Low-rank tensor U-Net model (LowRank-CGNet) in an end-to-end manner with a U-Net spatial denoiser and conjugate gradient data consistency steps. The U-Net [11] has an initial filter size of 32 with 4 down-sampling layers. We make a modification to the original U-Net by adding instance norm after each convolution which has been shown to improve texture recovery [17]. The initial estimate for the temporal basis was derived from the 16×16 centre box of k-space and truncated to a rank of 3. The initial estimate of the spatial basis was determined by solving Eq. 4 with $\lambda = 0$ using conjugate gradient for 4 steps. The initial estimates were then passed through the unrolled architecture with alternating steps of conjugate gradient data consistency and spatial denoising. The U-Net denoisers were trained to map to the residuals

$$\hat{C}_{k+1} = C_k + U_s(C_k) \tag{7}$$

where C_k is the input estimated bases at unrolled step k and $\hat{C}_{k+1}$, are the final denoised bases. Each basis was passed independently through the network with the complex and real data split along the channel dimension.

A mixed ℓ^1 and SSIM loss is used defined by

$$L(\hat{X}, X) = \frac{\|\hat{X} - X\|_1}{\|\hat{X}\|_1} + \beta \text{SSIM}(\hat{X}, X) \tag{8}$$

with $\beta = 10^{-1}$. A learning rate of 10^{-3} was used with an ADAM optimizer. The training was done on all slices, with a new undersampling mask randomly selected for each slice every epoch. The training was stopped after 20 epochs on 4× NVIDIA P100 Pascal (16G HBM2 memory) which took around 48 h. All models were implemented in PyTorch and trained using PyTorch Lightning. Additionally, we trained a comparison U-Net model directly on aliased zero-filled images with the same training parameters as above.

K-space data was pre-processed by normalizing the maximum k-space point to 1 and zero padding of k_x and k_y to [512, 256]. A batch size of 4 was used for both models and the internal validation was done on 10% of the training dataset.

2.4 Experiments

Model performance in undersampled images was compared to the fully sampled reference images, within the validation set. Peak signal-to-noise ratio (PSNR), structural similarity index (SSIM), and normalized mean squared error (NMSE) metrics were used. The results for our new method were benchmarked against conventional parallel imaging reconstruction methods, including CG-SENSE, ℓ^1-ESPIRiT, and U-Net, all as scored by the Synapse validation portal.

3 Results

LowRank-CGNet was successfully implemented, and a model was trained to convergence. The model's performance on the internal validation set during the training was SSIM: 0.91, PSNR: 35.5 and NMSE: 0.066. Comparing quantitative metrics on the CMRxRecon validation set, LowRank-CGNet slightly outperforms the conventional U-Net, but does not perform at the level of ℓ^1-ESPIRiT (Table 1). This may be due to the limited generalization capacity of the model

Fig. 2. Comparison of reconstructed Short-Axis CINE MRI images at varying acceleration rates using Uniform, Gaussian, and Radial sampling patterns. Each row displays images at different acceleration factors (R = 4, 8, 12, 16, 20, 24), showing the reconstructed images, corresponding error maps, and sampling masks for each technique. The first column presents the Ground Truth for reference. As the acceleration rate increases, image quality degradation is observed across all sampling patterns, with Radial sampling showing relatively less error at higher acceleration rates compared to Uniform and Gaussian sampling.

Table 1. CMRxRecon Validation dataset results for LowRank-CGNet and reference reconstruction methods, as scored by the online Synapse portal.

Metric	Zero-Filled	CG-SENSE	ℓ^1-ESPIRiT	U-Net	Proposed
SSIM [1]	0.46	0.56	**0.67**	0.56	0.66
PSNR [dB]	19.42	21.92	25.40	23.25	**30.06**
NMSE [1]	0.29	0.21	**0.15**	0.22	0.31

when compared to compressed sensing, which requires several tunable parameters. Further, ℓ^1-ESPIRiT is an iterative algorithm that requires more computational time than LowRank-CGNet, which may be infeasible for the time constraints of real-world clinical applications. Our model performed similarly on the CMRxRecon testing set, with PSNR = 30.6dB, SSIM = 0.69, NMSE = 0.31.

Visually, LowRank-CGNet images appear natural and well spatially encoded. Figure 3 contains sample 20-fold accelerated radial cine long-axis images from multiple reconstruction techniques. Zero-filled images have blurring and streaking artifacts, which are marginally corrected by CG-SENSE. ℓ^1-ESPIRiT images are blurrier, with some unnatural block-artifacts arising from the compression. U-Net images appear reasonably sharper compared to ℓ^1-ESPIRiT, but have significant artifacts in the blood pool signal. LowRank-CGNet images look visually similar to ℓ^1-ESPIRiT, with more complex blood-pool signal variations than ℓ^1-ESPIRiT. When comparing our model performance across sampling patterns and acceleration factors, we observe increasing blur artifacts as the acceleration factor increases, which is most prominent on the uniform and Gaussian sampling pattern (Fig. 2).

Figure 4 shows sample T1-weighted and T2-weighted images from the mapping sequence, acquired with an 8-fold accelerated uniform undersampling pattern. Magnetization recovery dynamics are present, but spatial blurring and aliasing artifacts remain. Figure 5 shows sample tagging images from two representative slices of a 20-fold accelerated radial acquisition. Images appear visually free of spatial aliasing, and the tag lines appear as expected, encoding myocardial motion. Figure 6 shows sample transversal and sagittal aorta images, acquired with a 24-fold Gaussian-density undersampling pattern. Transversal images have good image quality, free of aliasing artifacts, whereas the sagittal images are affected by flow artifacts which compromises the reconstruction quality. Regions of high signal such as the right ventricle appear visually accurate, but regions of low and intermediate signal such as the ascending aortic arch have reduced image quality.

4 Discussion

Overall, we have developed and implemented a Low-rank tensor U-Net model, LowRank-CGNet, that rapidly reconstructs highly undersampled data with a

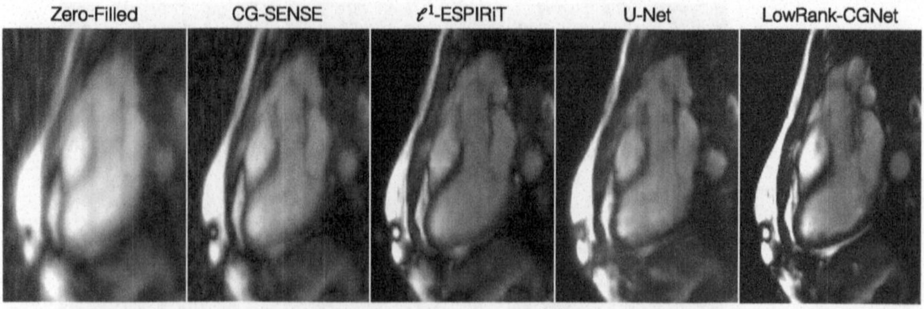

Fig. 3. Comparison of reconstruction methods on a sample cine long-axis dataset, acquired with a radial undersampling pattern and 20-fold acceleration. Columns represent different reconstruction techniques.

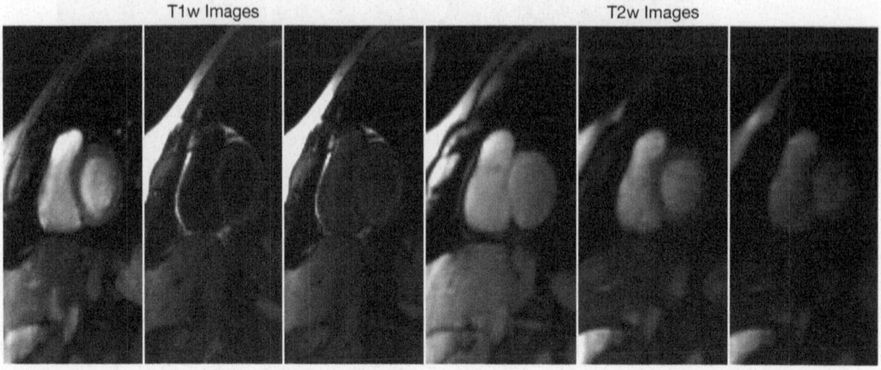

Fig. 4. Sample LowRank-CGNet reconstructions for T1-weighted (T1w) and T2-weighted (T2w) images from the mapping dataset, acquired with an eight-fold uniform undersampling pattern. Left three columns: T1w images. Right three columns: T2w images.

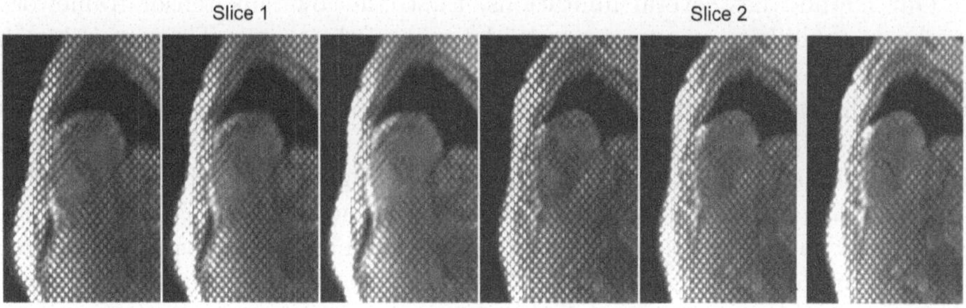

Fig. 5. Sample tagging images for three time-frames in two representative slices acquired with a 20-fold radial undersampling pattern. Images are generally free from aliasing artifacts, and tag lines are visible, as expected.

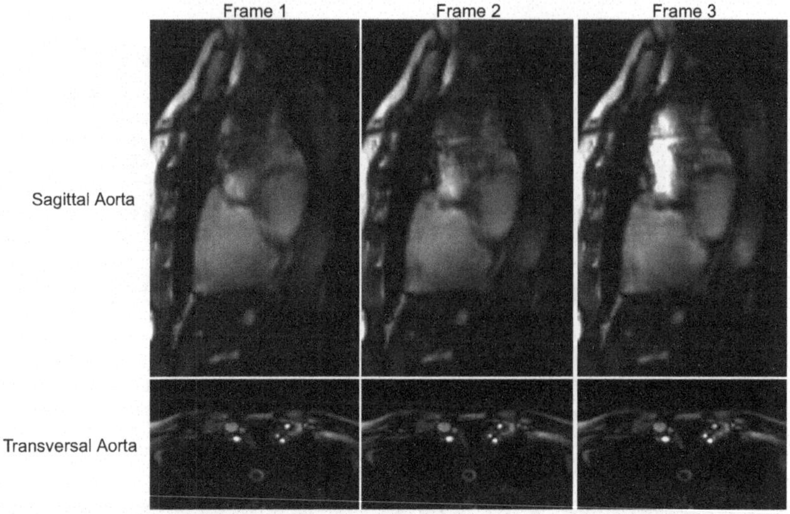

Fig. 6. Sample time-frames for one representative slice of transversal and sagittal aorta images, acquired with a 24-fold Gaussian undersampling pattern. Transversal images have reasonable image quality, whereas sagittal images are corrupted by flow artifacts. Note the changes in blood flow in the sagittal images, where the region of hypointensity travels up the ascending aorta from the leftmost image to the rightmost image, and a hyperintensity appears just past the aortic valve.

variety of anatomy, contrast, and undersampling artifacts. Image quality is generally adequate, with variations in the degree of undersampling artifacts across anatomies and acceleration factors. Generally, LowRank-CGNet performed the worst on uniformly accelerated data and better on Gaussian-density and radial data. This agrees with prior literature that temporal incoherence enables highly accelerated reconstruction with minimal artifacts [3]. Further, residual artifacts in radial undersampling are more benign than the one-dimensional Cartesian undersampling, due to the oversampling of low-order k-space.

Our method has several limitations. First, the low-rank tensor framework requires a tunable parameter, namely the rank of the tensor. The optimal rank may vary across modalities, since mapping images have three to nine temporal frames, whereas cine images can have 10 to 15 or more temporal frames. As such, we will investigate changing the tensor rank based on the anatomy type. Second, residual aliasing artifacts in data such as eight-fold accelerated mapping indicate that the model may not be sufficiently regularizing the data. As such, we will look to increase the number of model parameters to generalize better to unknown anatomy. While LowRank-CGNet images appear sharper with fewer spatial artifacts compared to ℓ^1-ESPIRiT, inadequate data scaling may reduce the quantitative scores, affecting model performance.

A final limitation of this work is that the reference data was prospectively undersampled by a factor of 2, with parallel imaging used to estimate the missing k-space. As shown by Frenklak et al., increasing prospective acceleration factors causes overly optimistic model performance within the given dataset, which leads to biased downstream perception of algorithm performance [4]. A potential extension of the challenge could be recovering the two-fold undersampled k-space data directly from subsampled measurements of the acquired k-space lines.

5 Conclusion

In conclusion, we have developed and implemented a Low-rank tensor U-Net method, LowRank-CGNet, that rapidly reconstructs highly undersampled data with a variety of anatomy, contrast, and undersampling artifacts. In future, we hope to further improve the model performance by increasing the U-Net size and dynamically updating the tensor rank for different anatomies.

Acknowledgments. This work utilized the computing resources of the Digital Research Alliance of Canada. JP, CS, and GW received funding from the Canadian Institutes of Health Research (PJT178299), while BK and MC received support from the Canada Research Chairs Program and the NSERC Discovery Grant (RGPIN/2023-03410).

References

1. Axel, L., Dougherty, L.: MR imaging of motion with spatial modulation of magnetization. Radiology **171**(3), 841–845 (1989)
2. Cao, T., et al.: Alternating low-rank tensor reconstruction for improved multiparametric mapping with cardiovascular MR multitasking. Magn. Resonance Med. (2024)
3. Feng, L., et al.: Golden-angle radial sparse parallel MRI: combination of compressed sensing, parallel imaging, and golden-angle radial sampling for fast and flexible dynamic volumetric MRI. Magn. Reson. Med. **72**(3), 707–717 (2014). https://doi.org/10.1002/mrm.24980
4. Frenklak, E., Arefeen, Y., Tamir, J.I.: Parallel imaging reconstruction in public datasets biases downstream analysis in retrospective sampling studies. In: Proceedings of the 2024 International Society of Magnetic Resonance in Medicine Meeting (2024)
5. Griswold, M.A., et al.: Generalized autocalibrating partially parallel acquisitions (GRAPPA). Magn. Reson. Med. **47**(6), 1202–1210 (2002)
6. Henningsson, M., Malik, S., Botnar, R., Castellanos, D., Hussain, T., Leiner, T.: Black-blood contrast in cardiovascular MRI. J. Magn. Reson. Imaging **55**(1), 61–80 (2022). https://doi.org/10.1002/jmri.27399
7. Ma, S., Du, H., Mei, W.: Dynamic MR image reconstruction from highly undersampled (k, t)-space data exploiting low tensor train rank and sparse prior. IEEE Access **8**, 28690–28703 (2020)
8. Menchón-Lara, R.M., Simmross-Wattenberg, F., Casaseca-de-la Higuera, P., Martín-Fernández, M., Alberola-López, C.: Reconstruction techniques for cardiac cine MRI. Insights Imaging **10**, 1–16 (2019)

9. Perea, R.J., et al.: T1 mapping: characterisation of myocardial interstitial space. Insights Imaging **6**, 189–202 (2015)
10. Pruessmann, K.P., Weiger, M., Scheidegger, M.B., Boesiger, P.: SENSE: sensitivity encoding for fast MRI. Magn. Reson. Med. **42**(5), 952–962 (1999)
11. Ronneberger, O., Fischer, P., Brox, T.: U-net: convolutional networks for biomedical image segmentation. arXiv arXiv:1505.04597
12. Sheagren, C.D., et al.: Motion-compensated T1 mapping in cardiovascular magnetic resonance imaging: a technical review. Front. Cardiovasc. Med. **10** (2023). https://www.frontiersin.org/articles/10.3389/fcvm.2023.1160183
13. Sheagren, C.D., Kadota, B.T., Patel, J.H., Chiew, M., Wright, G.A.: Accelerated cardiac parametric mapping using deep learning-refined subspace models. In: International Workshop on Statistical Atlases and Computational Models of the Heart, pp. 369–379. Springer, Cham (2023)
14. Stankovic, Z., Allen, B.D., Garcia, J., Jarvis, K.B., Markl, M.: 4D flow imaging with MRI. Cardiovasc. Diagn. Ther. **4**(2), 173 (2014)
15. Tanner, J., Vary, S.: Compressed sensing of low-rank plus sparse matrices. Appl. Comput. Harmon. Anal. **64**, 254–293 (2023)
16. Tseng, W., Su, M., Tseng, Y.: Introduction to cardiovascular magnetic resonance: technical principles and clinical applications. Acta Cardiologica Sinica **32**(2), 129 (2016)
17. Ulyanov, D., Vedaldi, A., Lempitsky, V.: Improved texture networks: maximizing quality and diversity in feed-forward stylization and texture synthesis. http://arxiv.org/abs/1701.02096
18. Wang, Z., et al.: CMRxRecon2024: A Multi-Modality, Multi-View K-Space Dataset Boosting Universal Machine Learning for Accelerated Cardiac MRI
19. Yaman, B., Weingärtner, S., Kargas, N., Sidiropoulos, N.D., Akçakaya, M.: Low-rank tensor models for improved multidimensional MRI: application to dynamic cardiac T1 mapping. IEEE Trans. Comput. Imaging **6**, 194–207 (2019)

SegBAW-Net: Segmentation of Bi-Atria and Wall Network Offering Valuable Insights into Challenge Data

Arthur L. Lefebvre[2](✉), Ishan Vatsaraj[2], Carolyna A. P. Yamamoto[1,2], Kensuke Sakata[1], Brock Tice[2], Natalia A. Trayanova[1,2], and Eugene G. Kholmovski[1,2]

[1] Alliance for Cardiovascular Diagnostic and Treatment Innovation (ADVANCE), Johns Hopkins University, Baltimore, MD, USA

[2] Department of Biomedical Engineering, Johns Hopkins University School of Medicine, Baltimore, MD, USA

alefebv4@jh.edu

Abstract. Accurate segmentation of bi-atrial structures and their walls in patients with atrial fibrillation is essential for detailed anatomical analysis and patient-specific treatment planning. Determining the thickness of the atrial wall is particularly challenging due to significant regional variability and low image resolution, which existing approaches often overlook by applying a fixed thickness. Accurate segmentation of both atria and their walls can substantially improve outcome of catheter ablation by providing patient-specific treatment strategies.

To address challenges in atrial segmentation, we developed and validated Segmentation of Bi-Atria and Wall Network (SegBAW-Net), a multistage deep neural network designed to automatically segment the left and right atria along with their walls.

The performance of SegBAW-Net was evaluated using the Dice Similarity Coefficient and the 95% Hausdorff Distance metrics. Data triaging was applied to ensure robust training and validation. The network was trained using 3D late gadolinium-enhanced magnetic resonance images provided by the MICCAI MBAS 2024 Challenge, which included 70 scans for training, 30 for validation, and 100 for testing. A key contribution of this paper is the in-depth analysis of both the data and ground truth by two experts.

Keywords: Segmentation · Late gadolinium enhancement · Deep learning · Atrial fibrillation · Bi-Atria

1 Introduction

Atrial fibrillation (AF) is the most prevalent sustained cardiac arrhythmia, associated with significant morbidity, mortality, and healthcare costs [1]. While

N. A. Trayanova, E. G. Kholmovski—Indicates equal contribution by senior authors.

O. Camara et al. (Eds.): STACOM 2024, LNCS 15448, pp. 345–356, 2025.
https://doi.org/10.1007/978-3-031-87756-8_34

catheter ablation has emerged as a common treatment modality, its efficacy is often limited by structural remodeling of the atria, including fibrosis and fat accumulation. Pre-ablation late gadolinium-enhanced magnetic resonance imaging (LGE-MRI) is crucial for visualizing this remodeling to guide the treatment decisions [2]. A critical challenge is the precise segmentation of both the left (LA) and right atria (RA), along with their walls. Standard segmentation techniques often assume a fixed wall thickness since the atrial walls are thin, while their thickness varies significantly across different regions. This fails to capture the true variability, leading to inaccurate anatomical reconstructions [3]. Bi-atrial segmentation is difficult due to the distinct but interconnected anatomy of the LA and RA and the complications arising due to variations in size, shape, and position of the pulmonary veins and atrial appendages [4]. Segmenting the walls is challenging since the thin atrial walls and their proximity to extracardiac structures also contribute to partial volume effects in LGE-MRI. Additionally, inconsistent image quality and inter-patient variability in atrial anatomy amplify these challenges. While manual segmentation is possible, it is time-consuming and subject to significant inter-operator variability, even among experts. This underscores the need for a robust, fully automated method for atrial segmentation. We developed SegBAW-Net: Segmentation of Bi-Atria and Wall Network, a multistage deep neural network designed to automatically segment both the left and right atria, along with their walls. SegBAW-Net refines segmentation accuracy through a progressive, multi-network process that enhances precision, particularly in regions with abrupt anatomical changes and varying wall thickness. By focusing on both atrial and wall segmentation, SegBAW-Net enables more detailed and patient-specific anatomical reconstructions, required for patient-specific treatment planning in AF ablation.

2 Methods

The clinical data used in this work was provided by MICCAI MBAS 2024 - Multi-class Bi-Atrial Segmentation Challenge [5]. The data were collected across multiple centers using consistent MRI scan. All training and validation images have dimensions of either $44 \times 640 \times 640$ or $44 \times 576 \times 576$, with voxel size of $2.5 \times 0.625 \times 0.625$ mm.

The challenge aimed to perform automatic medical semantic 3-class bi-atrial segmentation. The dataset includes 70 3D LGE-MRI scans for training, with corresponding ground truth (GT) segmentations for the LA, RA, and atrial walls. Thirty MRIs without GT were used for model validation through online submission of the predictions. An additional 100 scans were designated for the final test phase.

2.1 Data Quality Assessment

The performance of neural networks (NN) in segmentation tasks is greatly influenced by the quality of input image data and the precision of segmentation GT annotations [6,7]. Our team included a cardiac MRI specialist and an atrial segmentation expert to inspect the data and identify any lower-quality image datasets that could negatively impact the training of SegBAW-Net.

Figure 1 shows examples of LGE-MRI scans provided for bi-atrial segmentation, evaluated by the experts in our team for the scan quality. Poor quality training data represented 22% of the dataset according to the cardiac MRI specialist and 24% according to the atrial segmentation expert. We evaluated the performance of SegBAW-Net with and without the low-quality training data in Sect. 3.

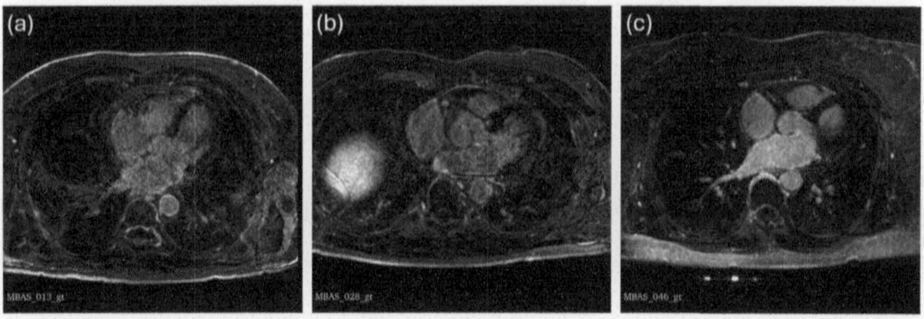

Fig. 1. Representative examples of scans of various quality: (a) poor quality (very noisy, blurred cardiac anatomy, and ghosting artifacts); (b) fair quality (noisy and blurred cardiac anatomy); (c) good quality (good signal to noise ratio and well defined cardiac anatomy).

The results of quality analysis of training and validation datasets are presented in Fig. 2. Figure 2(a) provides the quality scores assigned to the dataset according to the cardiac MRI specialist in our team. The data is color-coded and categorized into three main quality levels: good quality in green, fair quality in yellow, and poor quality in red. Each quality category is then subdivided according to specific quality attributes.

The radial layout of this figure allows a quick visual assessment of the dataset's quality distribution, with each segment highlighting specific issues affecting groups of scans. Figure 2(b) shows a similar categorization by the atrial segmentation expert in our team as percentage of each category remains consistent.

In the same manner, we also analyzed the quality of the ground truth segmentation provided by challenge organizers, associated with the scans. One figure is constructed for each regional segmentation and they are provided in the Appendix: Fig. 5 for the LA, Fig. 6 for the RA, Fig. 7 for the LA wall, and Fig. 8 for the RA wall. Common issues that were found in the segmentations analysis include fixed wall thickness, anatomical overlaps, incorrect wall locations, and under- or over-segmentation, especially in regions such as the RA appendage, LA appendage, atrial floors, and roofs. Specific faults, like RA wall over-segmentation, LA wall thickness inconsistencies, incomplete floor regions,

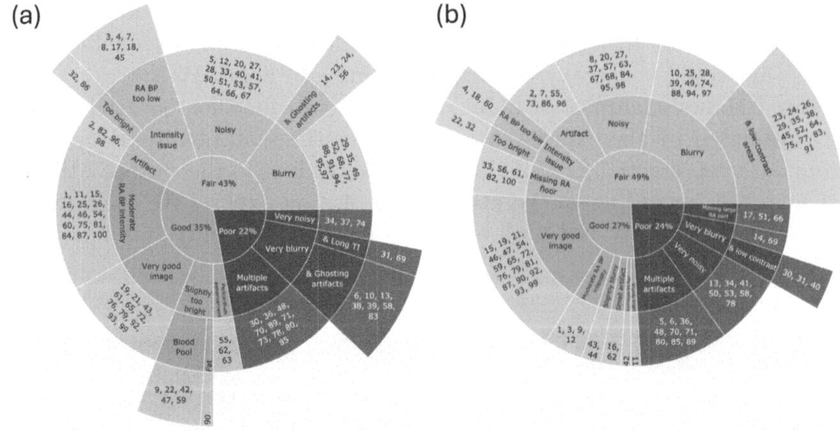

Fig. 2. Quality analysis of training and validation datasets by: (a) cardiac MRI specialist; (b) atrial segmentation expert. The numbering corresponds to sample IDs provided by the challenge organizers.

and overlaps between atrial regions, highlight challenges in defining the complex atrial boundaries.

2.2 The Neural Network Architecture

The global architecture of SegBAW-Net is presented in Fig. 3. It follows a sequential multi-network approach, consisting of three NN models operating in two stages to segment the LA, RA, and atrial wall.

In the first stage, the initial segmentation network was selected from the top performing state-of-the-art networks on an internal dataset with characteristics similar to the challenge dataset. This internal data set includes a holdout test set of 30 scans, which includes both pre- and post-ablation scans. The scans have image dimensions of either $44 \times 640 \times 640$ or $44 \times 576 \times 576$ voxels, with voxel size of $2.5 \times 0.625 \times 0.625$ mm.

The segmentation performance of various state-of-the-art architectures on our internal dataset, shown in Table 1, demonstrates their effectiveness in segmenting the LA, RA, and atrial wall. We primarily assessed 3 types of network architectures: CNN based, transformer-based, and UMamba architectures. First, we implemented nnU-Net models [8], including MedNeXt [9], with different configurations (3D full resolution, low resolution and ensemble). The nnU-Net ensemble combines predictions from multiple models and configurations to improve segmentation accuracy and robustness. Furthermore, we trained region-specific [10] and non-region-specific segmentation models to assess their impact on performance. Region-specific training focuses on segmenting anatomical substructures, using specialized data and labels to improve the accuracy of the targeted areas.

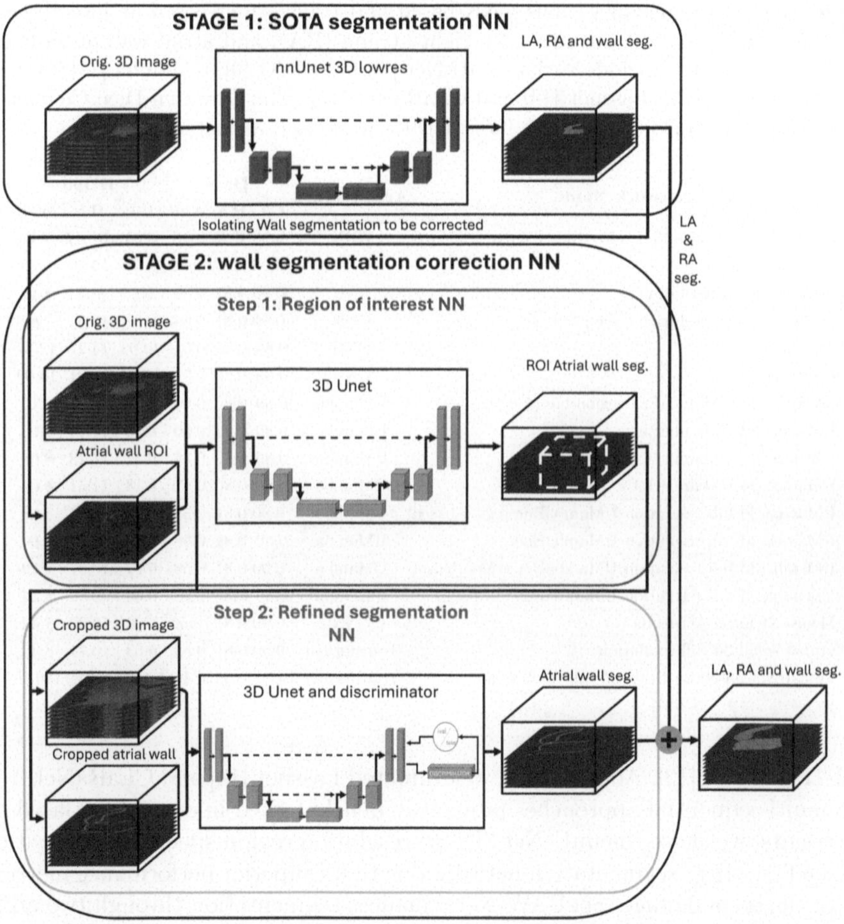

Fig. 3. SegBAW-Net architecture. Stage 1: LGE-MRI images are passed through the first NN (nnUNet 3D Lowres) to obtain predictions for the left atrium (LA), right atrium (RA), and atrial wall. **Stage 2:** *Step 1:* LGE-MRI images and the original LA wall segmentation masks are passed into the region of interest (ROI) network, which identifies the region containing the atrial walls. The image volume and the LA wall ROI are then cropped to the identified ROI and resampled. *Step 2:* The cropped LGE-MRI images and original LA wall ROI segmentation masks are input into the refined segmentation network to refine the LA wall segmentation. Step 2 includes a discriminator for adversarial training. Finally, the LA and RA segmentations are combined with the new refined wall segmentation to produce the final 3-class label segmentation.

We further explored the UMamba models, including UMambaBot and UMambaEnc, in 2D and 3D full resolution configurations [11]. UMambaEnc (UMamba Encoder) captures features through convolutional layers, while UMambaDec (UMamba Decoder) reconstructs spatial details for accurate segmentation. We also integrated transformer-based networks such as vtUNet [12]

Table 1. Performance of various CNN-based and transformer-based architectures in segmentation of the left atrium (LA), right atrium (RA), and atrial wall on in-house dataset. The models include various nnU-Net [8], MedNeXt [9], UMamba [11], vtUNet [12], TransUNet [13], Lassnet [14] and CLaRaNet [15]. The Sørensen-Dice Coefficient (Dice) and the Hausdorff distance 95% (HD95) were used as metrics

Network Name	Architecture	Dice			HD95		
		LA	RA	Wall	LA	RA	Wall
nnUnet 3d low resolution region-based	CNN	0.90	0.82	0.58	6.04	10.47	7.76
nnUnet ensemble	CNN	0.90	0.82	0.59	6.04	10.59	8.22
nnUnet 3d full resolution	CNN	0.90	0.81	0.58	5.89	10.54	8.11
nnUnet 3d full resolution region-based	CNN	0.90	0.81	0.58	6.01	10.54	7.82
CLaRaNet	CNN	0.85	0.76	0.51	10.04	14.21	12.87
LASSNet	CNN	0.85	0.76	0.51	10.04	14.21	13.09
UMamba 2d UMambaEnc region-based	UMamba	0.89	0.80	0.56	6.32	11.40	8.15
UMamba 2d UMambaEnc	UMamba	0.89	0.80	0.56	6.56	11.75	8.00
UMamba 2d UMambaBot region-based	UMamba	0.89	0.80	0.56	6.57	11.13	8.08
UMamba 2d UMambaBot	UMamba	0.89	0.80	0.57	6.18	11.51	8.03
UMamba 3d full resolution UMambaEnc region-based	UMamba	0.90	0.81	0.58	6.10	10.85	8.08
UMamba 3d full resolution UMambaEnc	UMamba	0.89	0.81	0.58	6.51	11.09	8.60
UMamba 3d full resolution UMambaBot region-based	UMamba	0.90	0.81	0.58	6.07	10.87	8.29
UMamba 3d full resolution UMambaBot	UMamba	0.89	0.81	0.58	6.05	10.81	8.24
MedNeXt Large kernel 3x3x3	CNN	0.90	0.81	0.59	6.02	10.82	8.42
vtunet small 3d full resolution	Transformer	0.90	0.81	0.58	6.33	10.71	8.40
TransUNet based on Brats 3d full resolution	Transformer	0.70	0.35	0.46	130.04	205.09	136.16

and TransUNet [13]. Additionally, we evaluated Lassnet [14] and CLaRaNet [15], two multi-sequential approaches previously benchmarked in our work. Based on the results, we chose the nnU-Net 3D low-resolution region-specific configuration for the first-stage segmentation network due to its superior performance metrics.

In the second stage, SegBAW-Net enhances segmentation through two additional steps. This approach has proven effective for other heart structures, such as the left ventricle, as demonstrated in the study by Popescu et al. [16]. First, a Region of Interest (ROI) network detects the anatomical location of the atria. Then, the segmentation network refines this initial segmentation to improve the accuracy of the left atrium (LA), right atrium (RA), and atrial wall. This multistage approach enables precise identification and segmentation of anatomical regions, ultimately enhancing the quality of the final segmentation output.

2.3 Implementation

In total, 70 LGE-MRI images, each annotated with LA and RA cavity and atrial wall GT, were used to train both stages of the SegBAW-Net architecture. To prevent overfitting, early stopping was applied. The NNs for Stage 1 were trained for 1000 epochs (approximately 18 h of computational time), while the NNs in Stage 2 were trained for 162 epochs in the first step (approximately 6 h) and 135 epochs in the second step (approximately 6 h). All networks used the Adam optimizer with an adaptive learning rate (LR) initialized at 10^{-3}. The models were implemented in PyTorch [17] and TensorFlow [18] and trained on either an Nvidia A100 or RTX A6000 GPU.

The loss function was based on the Sørensen-Dice Coefficient (Dice), which quantifies the overlap between two areas (in 2D) or volumes (in 3D). Dice values range from 0 to 1, where a Dice score of 0 indicates no overlap and 1 indicates a perfect match. Given the ground truth and predicted 3D volumes P and $\widehat{P}$, respectively, the Dice-based loss function is defined as follows:

$$l(P, \widehat{P}) = 1 - Dice(P, \widehat{P}) = 1 - \frac{2 \cdot (P \cap \widehat{P})}{|P| + |\widehat{P}|} \tag{1}$$

where P and $\widehat{P}$ represent the ground truth and predicted 3D image volumes.

3 Experimental Results

The MBAS 2024 Challenge was divided into three different phases: the training phase, the validation phase, and the test phase. The results obtained from the validation phase are presented in the following section. To evaluate the performances of SegBAW-Net, the Dice and the Hausdorff distance 95% (HD95) were used as metrics for each image volume.

Table 2. Validation Phase Results. "SegBAW-Net - Stage 1 only" are the segmentation results after the first training stage, "SegBAW-Net - Both stages on full dataset" are the segmentation results after the two training stages, "SegBAW-Net - Both stages on high-quality dataset" are the segmentation results after the two training stages but trained on high-quality data only.

Network or Team Name	Dice			HD95		
	LA	RA	Wall	LA	RA	Wall
SegBAW-Net - Stage 1 only	0.91	0.87	0.64	4.69	7.79	5.26
SegBAW-Net - Both stages on full dataset	0.91	0.87	0.62	4.69	7.79	6.32
SegBAW-Net - Both stages on high-quality dataset	0.91	0.87	0.61	4.69	7.79	7.48

For the validation, three different sets of results were submitted: segmentation results after Stage 1 only, segmentation results after Stage 2 trained on the full dataset, and segmentation results after Stage 2 trained solely on high-quality data. In this case, "high-quality data" refers to high- and medium-quality scans selected by both experts for the training set.

The segmentation performance metrics presented in Table 2 show high performance for LA and RA cavities (blood pools) segmentations and adequate performance for atrial wall segmentation. The results show that adding the second stage did not enhance segmentation performance, with the metrics generally being lower compared to using the first stage alone. This suggests that the added complexity of the second stage, especially when trained on the entire dataset, may not generalize well to all cases. Potential reasons for this include data inconsistencies or overfitting to specific anatomical features.

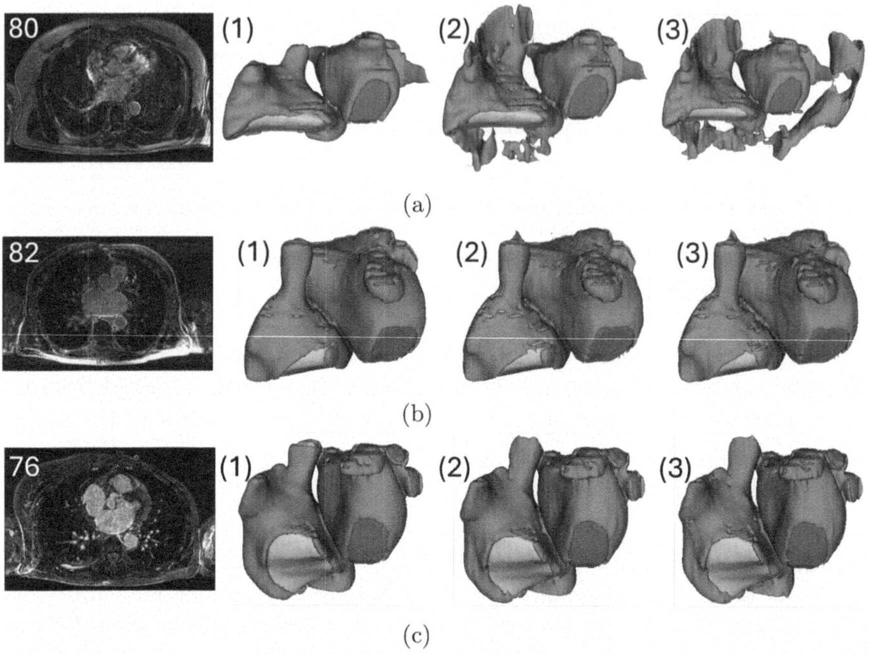

Fig. 4. Segmentation results from the validation phase tested across three scan quality categories: (a) Poor, (b) Fair, and (c) Good. Scans were selected based on consensus of both experts. Results in (1) show segmentation results after Stage 1, (2) show segmentation results after Stage 2 with training on the full dataset, and (3) show segmentation results after Stage 2 with training on the high-quality dataset only.

Figure 4 reveal some notable findings which were observed for SegBAW-Net performance on datasets with different image qualities. Models trained with both stages perform substantially worse on low-quality data. For fair-quality data, the segmentations are comparable to or slightly less accurate than those from Stage 1 alone. However, on high-quality data, models trained with both stages achieve similar performance to Stage 1-only predictions and, in some cases, provide more refined segmentation around the valve, appendage, and vein regions.

These results emphasize the importance of data quality assessment and management. While the additional complexity introduced by the second stage did not universally improve segmentation accuracy, particularly on lower-quality data, the refinement in segmentation around specific regions (e.g., the valve, appendage, and vein areas) on high-quality data suggests that our model has potential for improvement with high quality data. The findings also highlight the challenge of generalizing complex models to heterogeneous datasets, where overfitting or data inconsistencies can hinder performance. Moving forward, we plan to focus on improving data preprocessing techniques, enhancing the robustness of our model to variation in image quality, and exploring alternative architectures that could better handle lower-quality scans. Additionally, these results provide insight into the trade-off between model complexity and generalization ability, which will be crucial for developing more effective segmentation models for clinical applications.

4 Conclusion

In this study, we presented SegBAW-Net, a multistage deep neural network designed for segmenting the left and right atria and their respective walls from late gadolinium enhancement magnetic resonance imaging (LGE-MRI). Using a sequential, multi-network approach, SegBAW-Net achieved effective segmentation of atrial structures in the MICCAI MBAS 2024 dataset.

Our performance evaluation demonstrated SegBAW-Net's robustness, particularly on high-quality data, with strong results in both Dice and 95% Hausdorff distance metrics. However, challenges with segmentation of lower-quality scans highlighted the impact of image quality on segmentation accuracy, emphasizing the need for effective data quality assessment and management. Additionally, our analysis of the dataset quality provides a foundation for improved handling of diverse data quality in future developments.

A Appendix

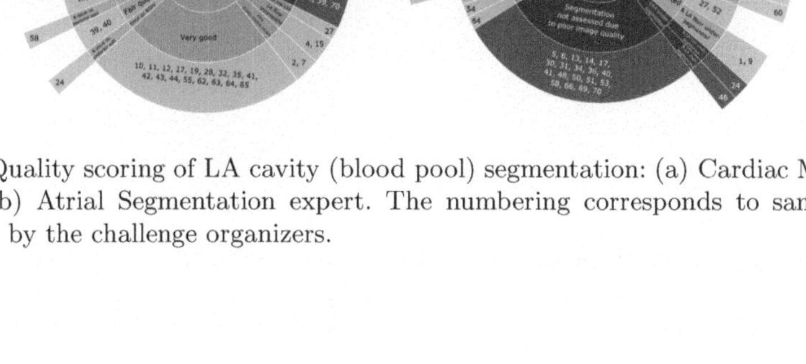

Fig. 5. Quality scoring of LA cavity (blood pool) segmentation: (a) Cardiac MRI specialist; (b) Atrial Segmentation expert. The numbering corresponds to sample IDs provided by the challenge organizers.

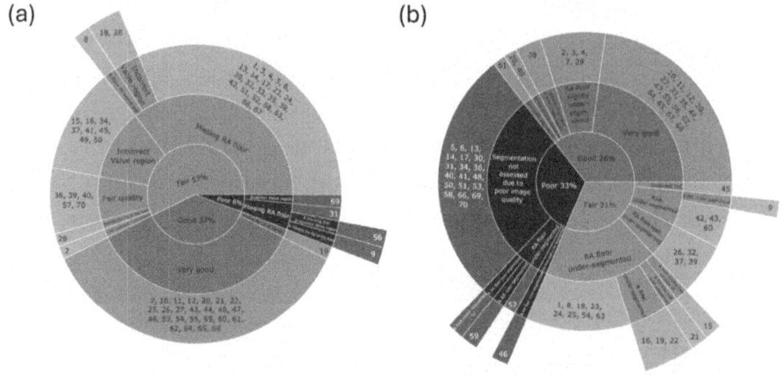

Fig. 6. Quality scoring of RA cavity (blood pool) segmentation: (a) Cardiac MRI specialist; (b) Atrial Segmentation expert. The numbering corresponds to sample IDs provided by the challenge organizers.

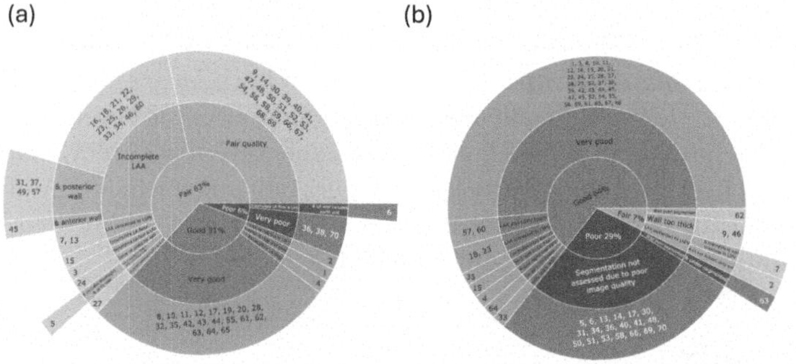

Fig. 7. Quality scoring of LA wall segmentation: (a) Cardiac MRI specialist; (b) Atrial Segmentation expert. The numbering corresponds to sample IDs provided by the challenge organizers.

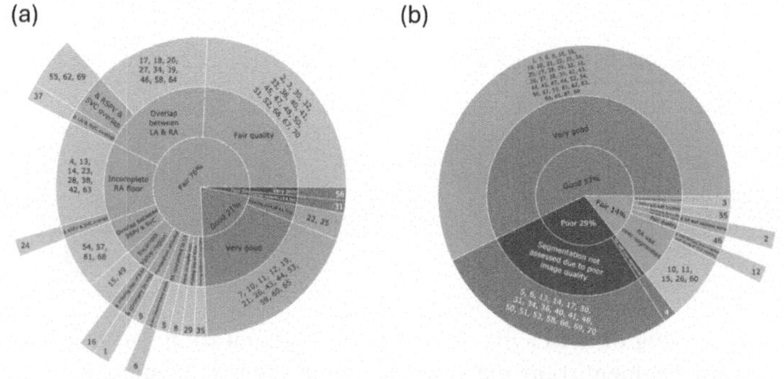

Fig. 8. Quality scoring of RA wall segmentation: (a) Cardiac MRI specialist; (b) Atrial Segmentation expert. The numbering corresponds to sample IDs provided by the challenge organizers.

References

1. Wolf, P.A., Mitchell, J.B., Baker, C.S., Kannel, W.B., D'Agostino, R.B.: Impact of atrial fibrillation on mortality, stroke, and medical costs. Arch. Intern. Med. **158**(3), 229–234 (1998)
2. Mont, L., Roca-Luque, I., Althoff, T.F.: Ablation lesion assessment with MRI. Arrhythmia Electrophysiol. Rev. **11**, e02 (2022). https://doi.org/10.15420/aer.2021.63
3. Li, L., Zimmer, V.A., Schnabel, J.A., Zhuang, X.: Medical image analysis on left atrial LGE MRI for atrial fibrillation studies: a review. Med. Image Anal. **77**, 102360 (2022). https://doi.org/10.1016/j.media.2022.102360
4. Garre, P., Vazquez-Calvo, S., Ferro, E., Althoff, T., Roca-Luque, I.: Impact of LGE-MRI in arrhythmia ablation. Appl. Sci. **13**(6), 3862 (2023)
5. Xiong, Z., et al.: A global benchmark of algorithms for segmenting the left atrium from late gadolinium-enhanced cardiac magnetic resonance imaging. Med. Image Anal. **67**, 101832 (2021)
6. Taran, V., Gordienko, Y., Rokovyi, A., Alienin, O., Stirenko, S.: Impact of ground truth annotation quality on performance of semantic image segmentation of traffic conditions. In: Advances in Computer Science for Engineering and Education II, pp. 183–193. Springer International Publishing, Cham (2020)
7. Güneş, A.M., van Rooij, W., Gulshad, S., Slotman, B., Dahele, M., Verbakel, W.: Impact of imperfection in medical imaging data on deep learning-based segmentation performance: an experimental study using synthesized data. Med. Phys. **50**(10), 6421–6432 (2023)
8. Isensee, F., Jaeger, P.F., Kohl, S.A., Petersen, J., Maier-Hein, K.H.: nnU-Net: a self-configuring method for deep learning-based biomedical image segmentation. Nat. Methods **18**(2), 203–211 (2021)
9. Roy, S., et al.: Mednext: transformer-driven scaling of convnets for medical image segmentation. In: International Conference on Medical Image Computing and Computer-Assisted Intervention, pp.405–415. (MICCAI) (2023)
10. Isensee, F., Jäger, P.F., Full, P.M., Vollmuth, P., Maier-Hein, K.H.: nnU-Net for brain tumor segmentation. In: Brainlesion: Glioma, Multiple Sclerosis, Stroke and Traumatic Brain Injuries: 6th International Workshop, BrainLes 2020, Held in Conjunction with MICCAI 2020, Lima, Peru, 4 October 2020, Revised Selected Papers, Part II 6, pp. 118–132. Springer International Publishing, Cham (2021)
11. Ma, J., Li, F., Wang, B.: U-mamba: enhancing long-range dependency for biomedical image segmentation. arXiv preprint (2024). arXiv:2401.04722
12. Peiris, H., Hayat, M., Chen, Z., Egan, G., Harandi, M.: A robust volumetric transformer for accurate 3D tumor segmentation. In: International Conference on Medical Image Computing and Computer-Assisted Intervention, pp. 162–172. Springer Nature Switzerland, Cham (2022)
13. Chen, J., et al.: Transunet: transformers make strong encoders for medical image segmentation. arXiv preprint (2021). arXiv:2102.04306
14. Lefebvre, A.L., et al.: LASSNet: a four steps deep neural network for left atrial segmentation and scar quantification. In: Challenge on Left Atrial and Scar Quantification and Segmentation, pp. 1–15. Springer Nature Switzerland, Cham (2022)
15. Lefebvre, A., et al.: PO-05-009 combined left and right atrial network (CLARANET): a deep learning (DL) model for reproducible LGE-MRI automatic segmentation of left and right atria (La/Ra) in atrial fibrillation (AF) treatment planning. Heart Rhythm **21**(5), S553–S554 (2024)

16. Popescu, D.M., et al.: Anatomically informed deep learning on contrast-enhanced cardiac magnetic resonance imaging for scar segmentation and clinical feature extraction. Cardiovasc. Digit. Health J. **3**(1), 2–13 (2022)
17. Paszke, A., et al.: Pytorch: an imperative style, high-performance deep learning library. Adv. Neural. Inf. Process. Syst. **32**, 8024–8035 (2019)
18. Martin, A.: TensorFlow: large-scale machine learning on heterogeneous systems. Software available from tensorflow.org (2015). https://www.tensorflow.org/

ResNet-Based Convolutional Framework for Segmenting Left Atrial Scars and Cavities

Malitha Gunawardhana(✉), Fangqiang Xu, Yun Gu, and Jichao Zhao

Auckland Bioengineering Institute, University of Auckland, Auckland, New Zealand
malitha.gunawardhana@auckland.ac.nz

Abstract. Atrial fibrillation (AF) poses a significant health concern due to its irregular and rapid heart rhythm. Accurate segmentation of left atrial (LA) scars and cavities from Late Gadolinium-Enhanced Magnetic Resonance Imaging (LGE-MRI) is essential for effective AF ablation procedures. In this study, we presented a novel approach utilizing a neural network architecture comprising three nnU-Nets. Departing from the original backbones, we leveraged the ResNet backbone to enhance segmentation performance. Our methodology was evaluated on the LAScarQS2022 dataset on two distinct tasks: Task-1 involved segmentation of both LA cavity and scar regions, while Task-2 focused solely on cavity segmentation. For Task-1 scar segmentation, our proposed model achieved promising results with a Dice score of 0.601. Cavity segmentation under Task-1 demonstrated even higher performance, yielding a Dice score of 0.946 and a Hausdorff distance (HD) of 10.486. In Task-2, our model achieved competitive performance metrics, with a Dice score of 0.941 and HD of 13.362. Our proposed method demonstrated superior performance, indicating its robustness and generalization capability. Our findings underscored the effectiveness of leveraging ResNet backbones and nnU-Net architecture for accurate segmentation of scars and cavities from LGE-MRI in the context of AF treatments, offering promising avenues for improved clinical outcomes in AF management.

Keywords: LA segmentation · Atrial Scar · LGE-MRI · Deep Learning · nnU-Net

1 Introduction

Atrial fibrillation (AF) stands as the most prevalent cardiac arrhythmia worldwide, affecting millions of individuals [2]. It is characterized by an irregular and often rapid heart rhythm, which can lead to various complications such as stroke, heart failure, and other cardiovascular issues. The global incidence of AF has surged from 33.5 million to 59 million people between 2010 and 2019, with

Supplementary Information The online version contains supplementary material available at https://doi.org/10.1007/978-3-031-87756-8_35.

O. Camara et al. (Eds.): STACOM 2024, LNCS 15448, pp. 357–367, 2025.
https://doi.org/10.1007/978-3-031-87756-8_35

projections indicating further increases to 15.9 million in America by 2050 and 17.9 million in Europe by 2060 [14]. The precise identification and quantification of atrial scars are pivotal for effective AF management, profoundly influencing patient selection and treatment planning. Late gadolinium-enhanced MRI (LGE-MRI), facilitated by advancements in MRI technology and contrast agents such as Gadolinium, emerges as a critical imaging modality for detecting atrial scar tissue. Studies from the University of Utah highlighted the association between left atrial (LA) wall fibrosis and outcomes in AF ablation procedures [1].

However, manual segmentation of the LA scar and cavity is challenging due to its labour-intensive and time-consuming nature. Deep learning methods also face significant difficulties because of the thinness of the atrial wall and the similarity in intensities of anatomical structures within the images. Additionally, there is a substantial class imbalance in the images, as illustrated in Fig. 1. Typically, the LA scar accounts for only about 0.04% of the pixels and the cavity for about 0.46%, with the remaining pixels corresponding to the background. This severe class imbalance further complicates the task of accurate segmentation. Even if the region of interest appears at the centre of the image, cropping may not effectively balance the classes, as the distribution of classes within the cropped area remains imbalanced.

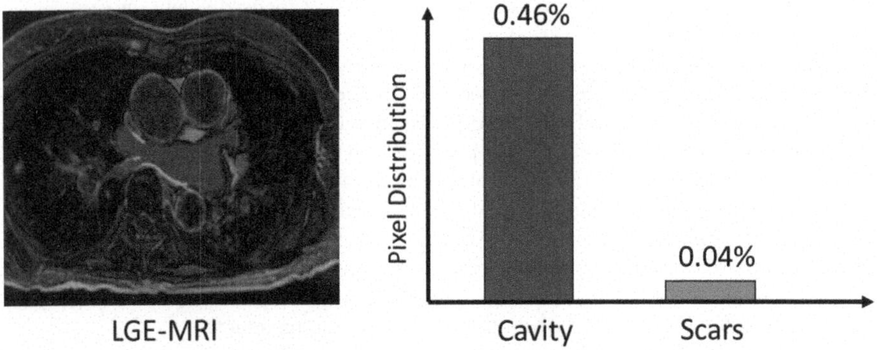

Fig. 1. Pixel distribution of LA cavity (red) and scars (green). LGE-MRI: Late gadolinium-enhanced magnetic resonance image; LA, left atrium. (Color figure online)

The growing interest in LA scar and cavity segmentation was underscored by previous challenges [24,26]. Proposed solutions encompassed various approaches, including uncertainty-based methods [26], curriculum learning [7], and semi-supervised learning [18]. Notably, most proposed methods were based on deep learning, underscoring the progress in this field.

This study presents a novel deep learning-based framework comprising three nnU-Nets [5] with ResNet blocks [4] to segment both LA scars and cavities effectively. Residual connections, initially proposed for image classification, have become integral to many deep learning architectures and have recently been incorporated into U-Net for medical image segmentation [8,11]. However, to the best of our knowledge, our work is the first to utilize ResNet blocks within the nnU-Net specifically for LA cavity and scar segmentation. Notably, our framework achieves higher performance even with lesser data compared to other models. Our paper is organized as follows. The next section elaborates on the methodology, detailing the model architecture. Then, we provide insights into the experimental setup, including information about the dataset, evaluation methods, and implementation details. Following the presentation of results, we summarise our findings and outline avenues for future research.

2 Methodology

The proposed architecture, illustrated in Fig. 2, where we enhanced the nnU-Net architecture by integrating ResNet blocks [4] within the encoder, which demonstrated superior performance compared to the nnU-Net's original configuration [6].

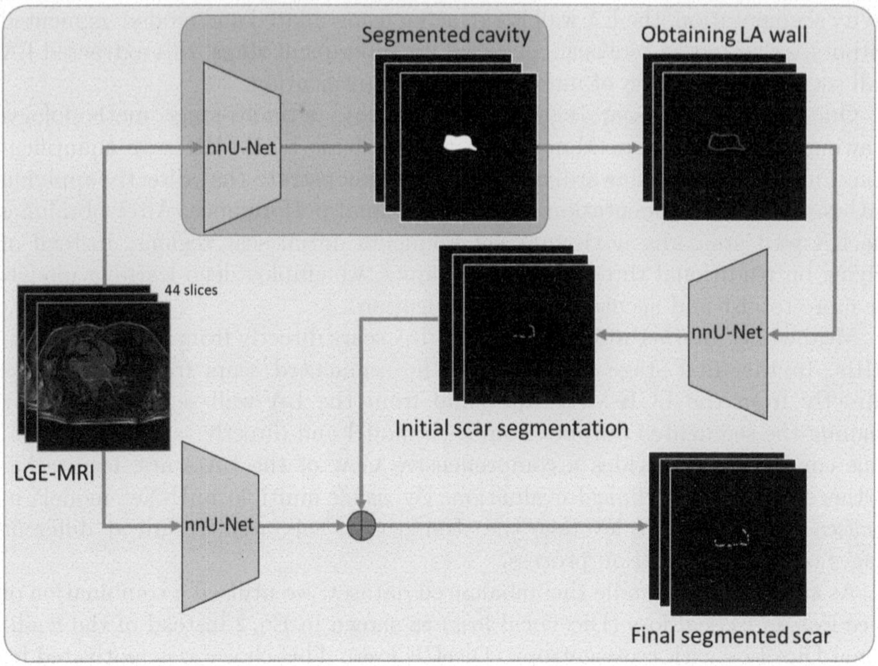

Fig. 2. The proposed architecture for left atrial (LA) scar and cavity segmentation begins with inputting Late Gadolinium Enhancement Magnetic Resonance Imaging (LGE-MRI) scans consisting of 44 slices. Initially, the nnU-Net model segments the LA cavity from the LGE-MRIs, producing binary masks of the segmented cavity, which is highlighted by a purple box. These masks are then used to extract the LA wall. Following this, another nnU-Net model performs initial scar segmentation from the LA wall. Meanwhile, a separate nnU-Net segments LA scars directly from the input LGE-MRIs. In the final stage, the segmented scars from both paths are combined. (Color figure online)

Residual connections allow for deeper network architectures without performance degradation, helping the model learn more discriminative features and preventing the vanishing gradient issue. In our implementation, residual connections are incorporated at each depth level of the nnU-Net. These connections operate over two convolutional blocks within each depth level, mathematically represented as:

$$\text{Output}_d = \mathcal{F}(\text{Input}_d) + \text{Input}_d \tag{1}$$

where $\mathcal{F}$ denotes the function learned by the two convolutional blocks within the depth level.

The first stage of our pipeline involves segmenting the LA cavity from the input LGE-MRI scans using an initial nnU-Net model, which produces binary masks representing the segmented cavity. The purple box in Fig. 2 highlights the cavity segmentation. Only this part is used for cavity segmentation, and subsequent processes are used for scar segmentation. The performance of this step is crucial as it forms the foundation for subsequent processes. Following the cavity segmentation, the LA wall is extracted using dilated and eroded segmented outputs, as described by Karim et al. [9]. In subsequent steps, this extracted LA wall serves as the region of interest for scar identification.

Our approach to scar segmentation employs a multi-stage methodology, drawing inspiration from Marrouche et al. [17] for LGE-MRI scar quantification. Furthermore, Gunawardhana et al. [3] demonstrate that directly applying nnU-Net for scar segmentation yields suboptimal performance. After obtaining the LA wall area, the next nnU-Net segments initial scar regions. Instead of relying on traditional thresholding techniques, we employ deep learning models for more robust and accurate scar identification.

Meanwhile, another nnU-Net segment LA scars directly from the input LGE-MRIs. In the final stage, we combine the segmented scars from both paths (directly from the LGE-MRI input and from the LA wall segmentation) by aligning the segmented outputs from each model and directly taking the union. This combination provides a comprehensive view of the LA scars, facilitating further analysis and clinical evaluation. By using multiple nnU-Net models in a cascaded fashion, we leverage the strengths of this architecture at different stages of the segmentation process.

As suggested, to handle the imbalanced dataset, we utilized a combination of Dice loss and Focal loss (DiceFocal loss) as shown in Eq. 2 instead of the traditional Dice loss with cross-entropy (DiceCE loss). This choice was motivated by the superior performance of DiceFocal loss in handling class imbalanced datasets, as demonstrated in previous studies [16].

$$DiceFocalLoss = \alpha \times DiceLoss + \beta \times FocalLoss \tag{2}$$

where L_{Dice} and L_{Focal} are defined as follows.

$$DiceLoss = 1 - \frac{2\sum_{i=1}^{N} p_i g_i}{\sum_{i=1}^{N} p_i + \sum_{i=1}^{N} g_i} \tag{3}$$

$$FocalLoss = -\frac{\sum_{i=1}^{N}(1-p_i)^{\gamma}g_i log(p_i)}{N} \quad (4)$$

where N is the total number of pixels, p_i is the predicted probability for the i-th pixel, g_i is the ground truth label for the i-th pixel.

3 Experimental Setup

3.1 Dataset

This study utilized the LAScarQS2022 challenge dataset, comprising 194 Late Gadolinium Enhancement Magnetic Resonance Imaging (LGE-MRI) images sourced from three international centres, each with unique spatial resolutions: University of Utah ($0.625 \times 0.625 \times 2.5$ mm), Beth Israel Deaconess Medical Center ($1.4 \times 1.4 \times 1.4$ mm), and King's College London ($1.3 \times 1.3 \times 4.0$ mm). Each image includes 44 slices.

The challenge is divided into two tasks: Task-1, which involves the segmentation of the left atrium (LA) scar and cavity, and Task-2, focusing on LA cavity segmentation. For Task-1, the dataset provides 60 LGE-MRI images with corresponding labels for training and 10 unlabeled images for validation. Task-2 includes 130 labeled and 20 unlabeled LGE-MRI images. The validation labels and test set were not publicly released by the challenge organizers.

3.2 Evaluation Method

For evaluating Task-1, we use five different metrics: 1) Dice score, 2) Jaccard score, 3) accuracy, 4) sensitivity, and 5) specificity. For Task-2, the evaluation criteria include 1) Dice score, 2) Jaccard Score, 3) Hausdorff distance (HD), and 4) Mean/Average surface distance (MSD/ASD). Due to the presence of gaps and breaks in segmentation, which are less common in cavity segmentation, HD and AHD are not applied to scar evaluation. Although [27] presents the performance of various models in the LAScarQS 2022 challenge, these results are not directly comparable to ours due to our access being limited to only the training data.

3.3 Implementation Details

For Task-1, which involves both cavity and scar segmentation, we utilized a dataset comprising 60 labelled training images. From this dataset, we randomly selected 50 images for training and 10 images for testing, as the validation images do not have public labels. For Task-2, which focuses solely on cavity segmentation, we used a dataset of 130 labelled images, with 115 images designated for training and 15 for testing, also selected randomly. Both tasks were performed using five-fold cross-validation. In Task-1, scar and cavity segmentation were handled independently, whereas, for Task-2, only the cavity segmentation nnU-Net (highlighted in the purple box in Fig. 2) was employed.

All methods were implemented using PyTorch 2.0.1. The models were trained individually on a 32 GB Tesla V100 GPU, with a learning rate of 0.01, a batch size of 2, and over 1000 epochs. We employed the SGD optimizer, with a weight decay of 3×10^{-5} and a momentum of 0.99. Each nnU-Net was trained using 3D full-resolution configurations, utilizing nnU-Net V2 independently. For the values of the parameters α, β, and γ, we selected 1 for both α and β, and 2 for γ as recommended by the original paper [16]. DiceFocal loss is used for every nnU-Nets in our approach.

Table 1. The performance of our proposed model in both Task-1 and Task-2 is evaluated across seven matrices.

Metrics	Task-1		Task-2
	Scar	Cavity	Cavity
Dice Score	0.601	0.946	0.941
Jaccard Score	0.430	0.898	0.881
HD	-	10.486	13.362
ASD	-	0.572	0.629
Accuracy	0.999847	-	-
Sensitivity	0.530131	-	-
Specificity	0.999964	-	-

4 Results

Table 1 showcases the performance metrics of our model on the test dataset for both tasks, and in Table 2, we compare the performance between our approach with other methods, assessing their efficacy across scar and cavity segmentation tasks.

In our study, our model demonstrated strong performance across various evaluation metrics for both Task-1 and Task-2 cavity segmentation. For Task-1, the model achieved a Dice score of 0.946, a Jaccard score of 0.898, an HD of 10.486, and an ASD of 0.572. For Task-2, the model attained a Dice score of 0.937, a Jaccard score of 0.881, an HD of 13.362, and an ASD of 0.629. Furthermore, our model exhibited notable performance in scar segmentation with a Dice score of 0.601 and a Jaccard score of 0.430.

able 2 presents a comparative analysis of our model's performance against other methodologies using Dice scores, which are the primary evaluation metric used by the challenge organizers. Consequently, we adopt the same metric for our evaluation. The first part of the table displays test results from the LAScarQS challenge[1]. The second section compares results from other studies on LA cavity

[1] https://zmiclab.github.io/projects/lascarqs22/result.html.

Table 2. Comparison of Dice scores for scar and cavity segmentation across various methods. The table begins with LAScarQS challenge test results, followed by independent LA cavity segmentation studies. The final section assesses the performance of the nnU-Net and our approach with and without ResNet blocks.

Method	Task-1		Task-2
	Scar	Cavity	Cavity
Deep U-Net [7]	0.595	**0.947**	0.831
UGformer [15]	0.582	-	-
nnU-Net [19]	0.561	0.908	-
LASSNet [12]	0.549	0.922	0.792
Semi-supervised [18]	0.478	0.828	-
Cross-Domain [13]	0.188	-	-
MSWS [10]	-	0.940	0.824
ViTUNet [23]	-	-	0.803
MADRU-Net [21]	-	-	0.919
ARW-Net [20]	-	-	0.929
Ours with original blocks	0.594	0.939	0.937
Ours with ResNet blocks	**0.601**	0.946	**0.941**

segmentation, which are not part of the challenge but use the dataset solely for evaluating their methods. The final section contrasts the results of our approach with those of the original nnU-Net and ResNet block implementations.

Our approach, which incorporates ResNet blocks, demonstrates significant performance improvements over existing methods. For Task 1, our method outperforms the original nnU-Net by more than 0.7% in cavity segmentation. Similarly, for Task 2, we observe a 0.4% improvement in cavity segmentation compared to the original nnU-Net. Furthermore, our approach surpasses the performance of other two-stage networks, such as the work presented by Lefebvre et al. [12]. Although [19] also employs nnU-Net for Task 1, their approach utilizes multi-class segmentation. Additionally, the test set and challenge test set used in their work differ from ours, as theirs is not publicly available

The qualitative segmentation results for the LA cavity and scar are presented in Fig. 3 and Fig. 4. A comparative analysis of the outcomes for Task 1 and Task 2 is shown in Fig. 3, where the cavity is depicted in blue and scars in red. Additionally, Fig. 4 illustrates the segmentation results across three anatomical orientations: axial, sagittal, and coronal. Our method demonstrates high accuracy in predicting cavity segmentation for both Task 1 and Task 2. The visualizations were created using Amira 3D software [22] for Fig. 3 and ITK-SNAP software [25] for Fig. 4.

5 Discussion

According to the results, our approach with ResNet blocks remains competitive even with a smaller training set compared to the challenge dataset. The integration of ResNet blocks consistently improves performance across all tasks. This is evident when comparing our method with and without ResNet blocks and the nnU-Net results. Even without integrating ResNet blocks, our approach achieves higher performance in cavity segmentation in Task-2 compared to other state-of-the-art methods such as MADRU-Net and ARW-Net.

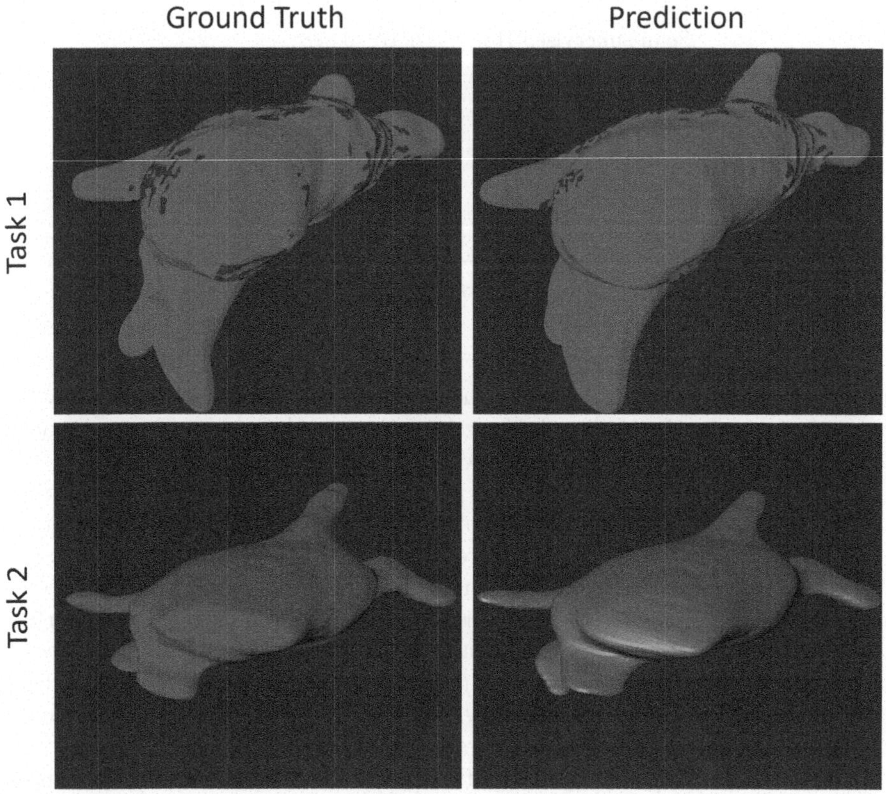

Fig. 3. Comparison of ground truth and predicted segmentation for left atrial (LA) cavity (blue) and LA scar (red). The top row displays results for Task-1, and the bottom row for Task-2, all visualized using Amira 3D software. (Color figure online)

T As illustrated in Table 1, our approach achieves near-perfect accuracy and specificity, reflecting the impact of class imbalance in the dataset between foreground and background. One of the main observations is that Task-1 cavity segmentation results are higher than the Task-2 cavity segmentation results, even with the larger number of LGE-MRIs in Task-2. This can be attributed

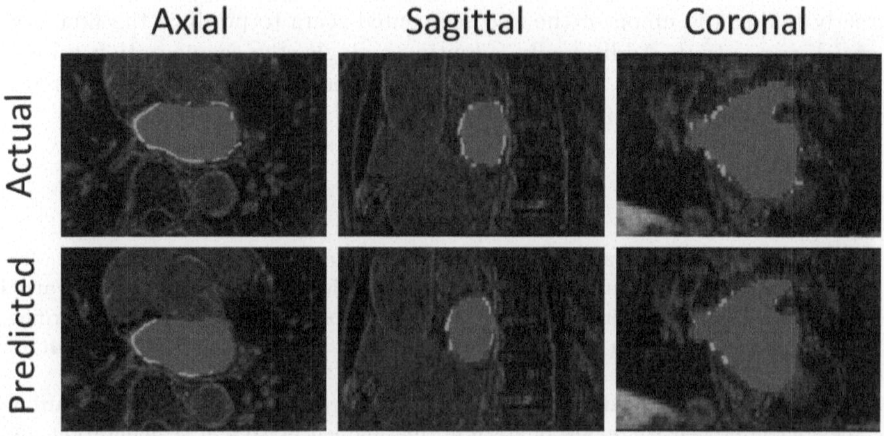

Fig. 4. Comparison of ground truth and predicted segmentation masks for the left atrial (LA) cavity and scar in Task-1, in three anatomical orientations. The cavity is highlighted in red, and the scar is in green. Visualized using ITK-SNAP software. (Color figure online)

to the location of the datasets; since they contain data from three locations, Task-1, likely derived from a single location, provides a more homogeneous data source, making it easier for the model to learn and generalize. In contrast, Task-2 includes data from all three different locations, introducing variability that poses a greater challenge for the segmentation task. Despite this, our method still achieves high performance, demonstrating its robustness and adaptability to varying data conditions.

6 Conclusion

This study introduces a novel methodology for segmenting LA scars and cavities, utilizing three nnU-Nets enhanced with a ResNet backbone, diverging from the original nnU-Net architecture. This approach not only advances current techniques but also sets the stage for future research aimed at refining these methods. The incorporation of ResNet blocks within the nnU-Net architecture yields significant improvements in segmentation performance.

However, it is important to note the limitations of our study. Due to the constraints of the dataset the organisers provided, we could not evaluate our method's performance from a clinical biomarker perspective. Additionally, the absence of comprehensive data from other methodologies precluded a thorough statistical analysis to establish the significance of our approach compared to existing methods.

Looking ahead, the exploration of additional architectures such as DenseNet, Inception, and attention mechanisms within the nnU-Net framework may offer further enhancements. Also, leveraging more advanced and sophisticated methods, such as deep learning-based approaches, would be more beneficial than directly taking the union of the two segmented scars to produce the final segmented scar output. Additionally, adopting a single-step segmentation process could streamline training, reduce computational demands, and minimize errors, thereby potentially boosting model performance.

References

1. Akoum, N., Fernandez, G., Wilson, B., Mcgann, C., Kholmovski, E., Marrouche, N.: Association of atrial fibrosis quantified using LGE-MRI with atrial appendage thrombus and spontaneous contrast on transesophageal echocardiography in patients with atrial fibrillation. J. Cardiovasc. Electrophysiol. **24**(10), 1104–1109 (2013)
2. Gunawardhana, M., Kulathilaka, A., Zhao, J.: Integrating deep learning in cardiology: a comprehensive review of atrial fibrillation, left atrial scar segmentation, and the frontiers of state-of-the-art techniques. arXiv preprint (2024). arXiv:2407.09561
3. Gunawardhana, M., Xu, F., Zhao, J.: How good nnU-Net for segmenting cardiac MRI: a comprehensive evaluation. Res. Square (2024)
4. He, K., Zhang, X., Ren, S., Sun, J.: Deep residual learning for image recognition. In: Proceedings of the IEEE Conference on Computer Vision and Pattern Recognition, pp. 770–778 (2016)
5. Isensee, F., Jaeger, P.F., Kohl, S.A., Petersen, J., Maier-Hein, K.H.: nnU-Net: a self-configuring method for deep learning-based biomedical image segmentation. Nat. Methods **18**(2), 203–211 (2021)
6. Isensee, F., et al.: nnU-Net revisited: a call for rigorous validation in 3D medical image segmentation. arXiv preprint (2024). arXiv:2404.09556
7. Jiang, L., Li, Y., Wang, Y., Cui, H., Xia, Y., Zhang, Y.: Deep u-net architecture with curriculum learning for left atrial segmentation. In: Challenge on Left Atrial and Scar Quantification and Segmentation, pp. 115–123. Springer (2022)
8. Jin, D., Xu, Z., Harrison, A.P., Mollura, D.J.: White matter hyperintensity segmentation from T1 and FLAIR images using fully convolutional neural networks enhanced with residual connections. In: 2018 IEEE 15th International Symposium on Biomedical Imaging (ISBI 2018), pp. 1060–1064. IEEE (2018)
9. Karim, R., et al.: Evaluation of current algorithms for segmentation of scar tissue from late gadolinium enhancement cardiovascular magnetic resonance of the left atrium: an open-access grand challenge. J. Cardiovasc. Magn. Reson. **15**(1), 105 (2013)
10. Khan, A., Alwazzan, O., Benning, M., Slabaugh, G.: Sequential segmentation of the left atrium and atrial scars using a multi-scale weight sharing network and boundary-based processing. In: Challenge on Left Atrial and Scar Quantification and Segmentation, pp. 69–82. Springer, Cham (2022)
11. Khanna, A., Londhe, N.D., Gupta, S., Semwal, A.: A deep residual U-net convolutional neural network for automated lung segmentation in computed tomography images. Biocybern. Biomed. Eng. **40**(3), 1314–1327 (2020)
12. Lefebvre, A.L., et al.: LASSNet: a four steps deep neural network for left atrial segmentation and scar quantification. In: Challenge on Left Atrial and Scar Quantification and Segmentation, pp. 1–15 (2022)
13. Li, F., Li, W.: Cross-domain segmentation of left atrium based on multi-scale decision level fusion. In: Challenge on Left Atrial and Scar Quantification and Segmentation, pp. 124–132. Springer, Cham (2022)

14. Linz, D., et al.: Atrial fibrillation: epidemiology, screening and digital health. Lancet Reg. Health–Eur. **37** (2024)
15. Liu, T., Hou, S., Zhu, J., Zhao, Z., Jiang, H.: Ugformer for robust left atrium and scar segmentation across scanners. In: Challenge on Left Atrial and Scar Quantification and Segmentation, pp. 36–48. Springer, Cham (2022)
16. Ma, J., et al.: Loss odyssey in medical image segmentation. Med. Image Anal. **71**, 102035 (2021)
17. Marrouche, N.F., et al.: Association of atrial tissue fibrosis identified by delayed enhancement MRI and atrial fibrillation catheter ablation: the DECAAF study. JAMA **311**(5), 498–506 (2014)
18. Mazher, M., Qayyum, A., Abdel-Nasser, M., Puig, D.: Automatic semi-supervised left atrial segmentation using deep-supervision 3dresunet with pseudo labeling approach for lascarqs 2022 challenge. In: Challenge on Left Atrial and Scar Quantification and Segmentation, pp. 153–161. Springer, Cham (2022)
19. Punithakumar, K., Noga, M.: Automated segmentation of the left atrium and scar using deep convolutional neural networks. In: Challenge on Left Atrial and Scar Quantification and Segmentation, pp. 145–152. Springer, Cham (2022)
20. Singh, K.R., Sharma, A., Singh, G.K.: Attention-guided residual W-net for supervised cardiac magnetic resonance imaging segmentation. Biomed. Signal Process. Control **86**, 105177 (2023)
21. Singh, K.R., Sharma, A., Singh, G.K.: MADRU-Net: multi-scale attention-based cardiac MRI segmentation using deep residual U-net. IEEE Trans. Instrum. Measur. **73**, 1–13 (2023)
22. Stalling, D., Westerhoff, M., Hege, H.C.: Amira: a highly interactive system for visual data analysis. Vis. Handb. **38**, 749–767 (2005)
23. Tu, C., et al.: Self pre-training with single-scale adapter for left atrial segmentation. In: Challenge on Left Atrial and Scar Quantification and Segmentation, pp. 24–35. Springer, Cham (2022)
24. Xiong, Z., et al.: A global benchmark of algorithms for segmenting the left atrium from late gadolinium-enhanced cardiac magnetic resonance imaging. Med. Image Anal. **67**, 101832 (2021)
25. Yushkevich, P.A., Gao, Y., Gerig, G.: ITK-SNAP: an interactive tool for semi-automatic segmentation of multi-modality biomedical images. In: 2016 38th Annual International Conference of the IEEE Engineering in Medicine and Biology Society (EMBC), pp. 3342–3345. IEEE (2016)
26. Zhang, Y., Meng, Y., Zheng, Y.: Automatically segment the left atrium and scars from LGE-MRIs using a boundary-focused nnU-Net. In: Challenge on Left Atrial and Scar Quantification and Segmentation, pp. 49–59. Springer, Cham (2022)
27. Zhuang, X.: Left atrial and scar quantification and segmentation: first challenge. In: LAScarQS 2022, Held in Conjunction with MICCAI 2022, Singapore, September 18, 2022, Proceedings. Springer Nature (2023)

EAT-Mamba: Epicardial Adipose Tissue Segmentation from Multi-modal Dixon MRI

Fan Feng[1(✉)], Abdallah Hasaballa[1], Fangqiang Xu[1], Jiayuan Yang[1], Yun Gu[1], Grace Wen[1], Carl-Johan Carlhäll[2,3,4], and Jichao Zhao[1]

[1] Auckland Bioengineering Institute, University of Auckland, Auckland, New Zealand
ffen908@aucklanduni.ac.nz

[2] Division of Diagnostics and Specialist Medicine, Department of Health, Medicine and Caring Sciences, Linköping University, Linköping, Sweden

[3] Center for Medical Image Science and Visualization (CMIV), Linköping University, Linköping, Sweden

[4] Department of Clinical Physiology in Linköping, Department of Health, Medicine and Caring Sciences, Linköping University, Linköping, Sweden

Abstract. Patients with type 2 diabetes (T2D) face an increased risk of cardiovascular complications, highlighting the need for improved diagnostic methods. Epicardial adipose tissue (EAT) is recognized as an important biomarker for cardiovascular diseases (CVDs). Dixon MRI is widely used to depict adipose tissue by deriving fat and water signals, providing complementary information through multi-modality images: in-phase, out-of-phase, fat phase and water phase. This study introduces EAT-Mamba, a novel multi-modal deep learning-based approach for automated segmentation and quantification of EAT from Dixon MRI. The model architecture integrates a U-shape structure with a multi-modal fusion layer and Res-Mamba blocks to capture features across various modalities and scales. The Res-Mamba block is designed with a Mamba block, based on state space sequence models (SSMs) for long-range context learning, and a convolutional neural network (CNN)-based residual block for local feature extraction. Comprehensive evaluations were conducted on a Dixon MRI dataset involving 90 subjects, comparing EAT-Mamba against Res-UNet and UNet++ across different modality combinations. EAT-Mamba achieved a Dice similarity coefficient of 0.873, a Hausdorff distance 95 of 3.34 mm, and an average symmetric surface distance of 0.591 mm with four modalities, consistently outperforming the other methods. Notably, the use of out-of-phase and fat phase modalities can enhance EAT segmentation performance. This study provides valuable insights into the use of SSMs and multi-modal images in EAT segmentation, paving the way toward improved cardiovascular risk assessment and treatment planning in clinical practice.

Keywords: Epicardial Adipose Tissue · Dixon MRI · multi-modal · medical image segmentation · Mamba

O. Camara et al. (Eds.): STACOM 2024, LNCS 15448, pp. 368–377, 2025.
https://doi.org/10.1007/978-3-031-87756-8_36

1 Introduction

Patients with type 2 diabetes (T2D) are at a significantly higher risk of cardiovascular complications and heart failure compared to non-diabetic individuals [1]. There is an urgent need to better understand the underlying pathophysiological mechanisms and to identify novel risk factors that can facilitate the early detection of high-risk patients, leading to more aggressive and effective treatments [2]. Studies have demonstrated that epicardial adipose tissue (EAT), due to its close anatomical and functional association with the myocardium and coronary arteries, plays a crucial role in cardiovascular diseases (CVDs). Consequently, EAT could serve as an imaging biomarker for cardiovascular risk assessment and a potential therapeutic target [3, 4].

EAT can be visualized and quantified by computed tomography (CT) and magnetic resonance imaging (MRI) [5, 6]. However, CT is associated with a significant radiation exposure, posing risks, especially in longitudinal studies requiring multiple scans and younger individuals. In contrast, MRI, particularly utilizing the Dixon water-fat-separated technique, is currently regarded as the gold standard for assessing subcutaneous and visceral adipose tissue, especially hepatic adipose tissue, by providing complementary information: in-phase (IP) images, out-of-phase (OP) images, fat phase (FP) images and water phase (WP) images [7]. Dixon MRI provides excellent contrast resolution and tissue characterization, making it ideal for EAT segmentation. Despite these advantages, the identification and quantification of EAT remain challenging in clinical practice due to the manual process being time-consuming, labor-intensive, and subject to inter- and intra-observer variability [8]. To address these challenges, deep learning-based methods have emerged as the mainstream framework for automatic medical image segmentation, offering significant improvements in accuracy, efficiency, and reproducibility [9].

Several studies have applied deep learning to achieve automatic EAT segmentation, demonstrating significant potential in this area [10]. Convolutional neural networks (CNNs) have been particularly effective, leveraging the powerful feature extraction capabilities to delineate EAT from images with high accuracy and efficiency. However, most existing works have focused on CT-based segmentation [11–18], with relatively little exploration of segmenting EAT from MRI [19, 20]. Notably, none of these studies focus on Dixon MRI, which provides rich multi-modality images suited for EAT identification and distribution analysis around four heart chambers.

In a previous work, a CNN-based pipeline was designed to segment EAT from Dixon MRI, using fat maps as inputs by fusing multi-information from three phases (FP, WP, OP) [21]. However, this approach overlooked the information from IP images, which could provide additional valuable contextual information to improve the segmentation performance. Moreover, CNN-based methods generally exhibit poor sensitivity to long-range contextual relations, which are often crucial in medical image segmentation tasks. An emerging alternative to capture long-range context is state space models (SSMs) [22], and Mamba further improves SSMs with a selective mechanism, reporting promising results in medical image segmentation [23, 24].

In this study, we introduce EAT-Mamba, a novel multi-modal deep learning-based approach for EAT segmentation from Dixon MRI. The model architecture combines the U-shape structure with Res-Mamba blocks to capture localized fine-grained features and long-range features at various scales. The Res-Mamba block comprises a Mamba block to learn long-range context across the spatial dimension of features, followed by a channel fusion layer to capture relations between channels and leverage multi-modal features, and a Residual block to learn short-range contextual information. Comprehensive experiments are conducted to demonstrate the effectiveness of this method. Our contributions can be summarized as follows:

- We introduce the first multi-modal SSM-based model for multi-level feature integration that effectively segments EAT from Dixon MRI.
- Comprehensive evaluations of all possible combinations of the four modalities showcase the accuracy of EAT-Mamba, providing valuable insights for future investigations into Mamba's potential in EAT segmentation from multi-modal Dixon MRI.

2 Methods

2.1 Dixon MRI Dataset

Experiments were conducted on a multi-modal Dixon MRI dataset from a sub-study (HEALTH) of the Swedish CArdioPulmonarybiolmage Study (SCAPIS) [25], involving 90 subjects: 45 with T2D and 45 controls matched according to age, sex and smoking. The

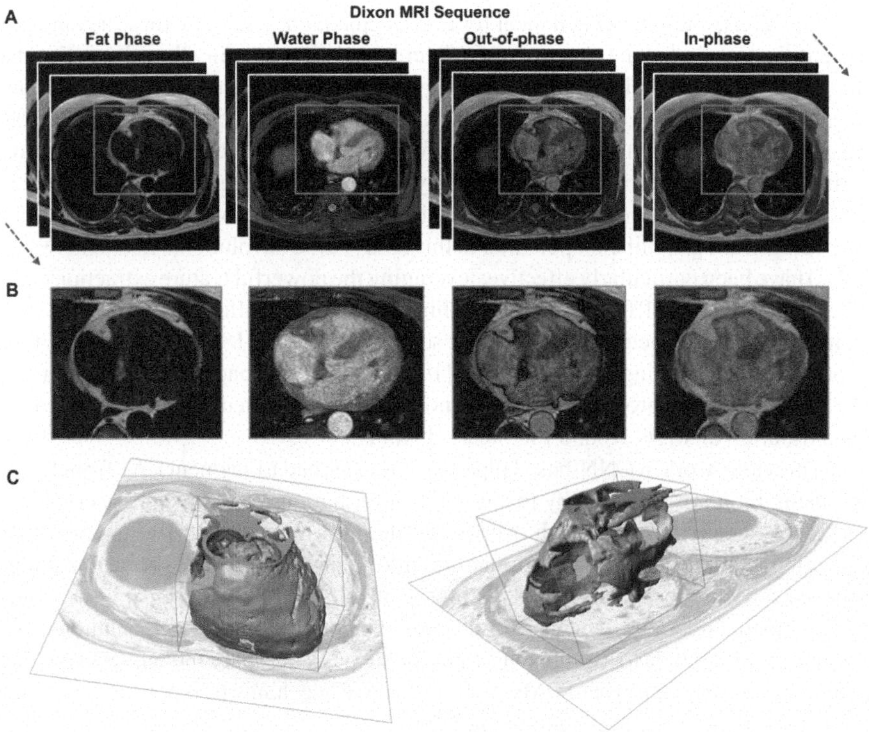

Fig. 1. The 3D multi-modal Dixon MRI sequence. **A,** The Dixon MRI sequence consists of four phases: fat phase (FP), water phase (WP), out-of-phase (OP) and in-phase (IP) images. Each phase contains the same number of slices. **B,** Epicardial adipose tissue (EAT) is identified as interior pixels highlighted in orange. The different phases show varying contrasts that contain both structural and contextual information. **C,** 3D reconstruction of EAT segmentation from multi-modal Dixon MRI. (Color figure online)

study complied with the Declaration of Helsinki and was approved by the Linköping Ethical Review Board (Dnr 2023–02089-02). Images were reconstructed at a spatial resolution of 0.67 mm × 0.67 mm × 1.5 mm. Each three-dimensional (3D) Dixon MRI dataset comprises 50 to 82 slices along the z-axis, with an in-plane dimension of 448 × 448 pixels (Fig. 1).

2.2 EAT-Mamba

The proposed EAT-Mamba employs a U-shape architecture to effectively capture multi-scale image features effectively. The model uses an input-level fusion strategy to preserve the rich feature information from different modalities [26, 27]. Central to its design is the Res-Mamba block, which enhances segmentation performance by acquiring and leveraging long-range contextual information alongside local features. Each block comprises an SSM-based Mamba block for long-range context learning, followed by a channel fusion layer and a Residual block for local feature extraction. Figure 2 provides an overview of the proposed EAT-Mamba.

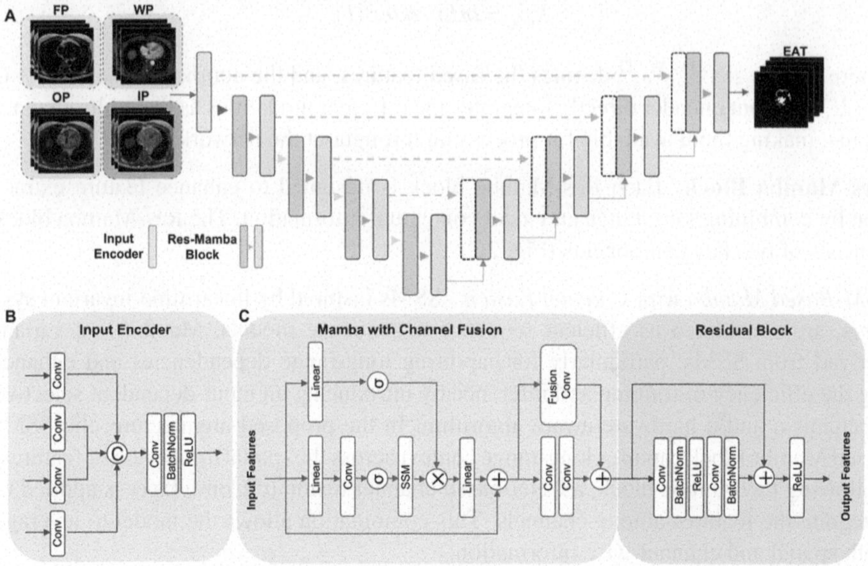

Fig. 2. Overview of the EAT-Mamba and architectures of different blocks. **A,** EAT-Mamba follows the U-shape framework with Res-Mamba blocks in both the encoder and decoder paths. **B,** The input encoder utilizes an input-level fusion strategy to combine multi-modal images from Dixon MRI. **C,** Res-Mamba block is designed to enhance feature extraction by combining long-range and local contextual information. It contains an SSM-based Mamba block for capturing long-range contexts, a channel fusion layer, and a Residual block for short-range contextual features related to low-level image attributes.

EAT-Mamba Architecture. The EAT-Mamba model is based on the widely used U-Net architecture, which is particularly effective for medical image segmentation due to its encoder-decoder structure and skip connections. These skip connections between corresponding layers of the encoder and decoder allow the model to combine low-level and high-level features effectively. By integrating multi-scale image features, the EAT-Mamba architecture enhances segmentation accuracy and performance (Fig. 2A).

Input Encoder with Multi-modal Fusion. To fully exploit the rich information available in multi-modal Dixon MRI images, an input-level fusion strategy is employed. This involves concatenating the different phase images (FP, WP, OP, IP) along the channel dimension, and features from different modalities are preserved, providing a comprehensive representation of EAT. These features can be fully exploited in all layers. As shown in Fig. 2B, the input encoder takes images from different modalities *m*:

$$I = [I_{FP}; I_{WP}; I_{OP}; I_{IP}] \tag{1}$$

where $I_m \in \mathbb{R}^{H \times W \times D}$ denotes the source image scan for the modality *m* and $m \in \{FP, WP, OP, IP\}$. The input tensor is then projected through the input encoder block (InEncoder):

$$X_{enc} = InEncoder\,(I) \tag{2}$$

where $X_{enc} \in \mathbb{R}^{C \times H \times W \times D}$ denotes the fused features, and the number of channels is $C = 32$. The input encoder directly integrates the different modalities as the multi-channel inputs, making them available for processing throughout the network.

Res-Mamba Block. Each Res-Mamba block is designed to enhance feature extraction by combining long-range and local contextual information. The Res-Mamba block consists of two key components (Fig. 2C):

SSM-Based Mamba with Channel Fusion. SSMs inspired by linear time-invariant systems, are considered as efficient sequence-to-sequence models. Mamba is a variant derived from SSMs, particularly for capturing long-range dependencies and enhancing the efficiency of training and inference by introducing an input-dependent selective mechanism and a hardware-aware algorithm. In the proposed architecture, the SSM-based Mamba block captures long-range context across the spatial dimension of features. Following the Mamba block, a CNN-based channel fusion (cfConv) layer is applied to integrate the features among channels. This combination allows the model to leverage both spatial and channel-wise information:

$$M_{ssm} = Mamba(X) \tag{3}$$

$$F_{cf} = cfConv\,(M_{ssm}) \tag{4}$$

where M_{ssm} represents the long-range spatial context obtained from the Mamba block, which is then passed onto the channel fusion layer. F_{cf} is the resulting output from the channel fusion layer.

Residual Block. The Residual Block follows the channel fusion layer to extract local features. It helps in preserving spatial hierarchies and refines the features by learning short-range contextual information. The residual connections allow the network to learn incremental changes rather than complete transformations, enhancing the overall feature extraction process. This approach not only boosts the learning capacity but also aids in maintaining the integrity of both high-level and low-level features throughout the network.

2.3 Training and Evaluation

The model is implemented in Pytorch 2.3.1 + cu118, Mamba-SSM 2.0.3, and Monai 1.3.1, running on an NVIDIA Tesla V100 GPU. The input size for the network is set to 320×320 pixels, with a batch size of 4. Various data augmentation techniques are employed, including random rotation, cropping, elastic deformation, and intensity shifts, to enhance the robustness and generalization of the model. The training process utilizes the AdamW optimizer with an initial learning rate of 0.0001, running for 200 epochs. The loss function is the unweighted sum of Dice loss and cross-entropy loss, which is known to be robust across medical image segmentation tasks. The dataset is divided into three subsets: 60% for training, 10% for validation, and 30% for testing. To assess the segmentation performance of the EAT-Mamba, Dice similarity coefficient (DSC), 95% Hausdorff distance (HD95), and Average symmetric surface distance (ASSD) are utilized to indicate segmentation accuracy and provide insights into the sensitive boundary agreement. Our method is compared against Res-UNet and UNet++ across all the combinations of available modalities. For a fair comparison, all methods are evaluated on the same dataset and data splitting.

3 Results and Discussion

Different phase images of Dixon MRI exhibit varying image contrasts, enabling a comprehensive representation of cardiac structures from multiple perspectives. Although FP and WP are derived from IP and OP, they possess distinct statistical properties, typically considered as multi-modal images [28]. Therefore, we thoroughly evaluated and compared the segmentation performance of the model across different phase combinations, allowing us to extract features from various views and leverage complementary information. Table 1 summarizes the comparison among the proposed EAT-Mamba against Res-UNet and UNet++ on all the 15 possible combinations of available modalities.

In summary, EAT-Mamba outperforms the other two methods across all modality combinations, demonstrating its effectiveness. The proposed model achieved a DSC of 0.873, an HD95 of 3.34 mm, and an ASSD of 0.591 mm with four modalities, with improvements of 1.6% in DSC, 26.0% in HD95, and 35.7% in ASSD compared to UNet++. Although the DSC improved by 0.6% compared to Res-UNet, the HD95 improved by 3.5% and the ASSD improved by 13.0%, indicating significant advancements in boundary accuracy and overall segmentation precision. Generally, the inclusion of more modalities enhances the model's performance, with multi-modal fusion networks outperforming those using single modalities. Incorporating the OP and FP modalities can improve segmentation accuracy: the FP modality highlights fat signals, while the OP provides clear image boundaries. Although fat appears with low intensity in the WP,

Table 1. Comparison of the proposed method EAT-Mamba, and Res-UNet [29], UNet++ [30]. Missing and available modalities are denoted by ○ and •, respectively. Dice similarity coefficient (DSC), 95% Hausdorff distance (HD95), and Average symmetric surface distance (ASSD) are employed for evaluation with every combination setting of modalities.

Modalities				EAT-Mamba			Res-UNet			UNet++		
FP	**WP**	**OP**	**IP**	DSC	HD95	ASSD	DSC	HD95	ASSD	DSC	HD95	ASSD
•	•	•	•	0.873	**3.340**	**0.591**	0.868	3.461	0.679	0.859	4.515	0.919
○	•	•	•	**0.874**	3.438	0.630	0.864	3.420	0.621	0.866	4.067	0.795
•	•	○	•	0.858	3.693	0.638	0.850	3.684	0.671	0.846	4.446	0.819
•	○	•	•	0.867	4.102	0.794	0.866	3.250	0.640	0.856	4.261	0.772
•	•	•	○	0.872	3.488	0.687	0.868	3.366	0.606	0.862	4.075	0.710
•	•	○	○	0.852	3.811	0.718	0.850	3.514	0.673	0.843	4.314	0.739
•	○	○	•	0.847	3.791	0.696	0.843	3.320	0.710	0.838	4.343	0.804
○	○	•	•	0.862	3.965	0.637	0.854	3.979	0.707	0.851	4.457	0.771
•	○	•	○	0.869	3.356	0.595	0.868	3.074	0.617	0.859	3.977	0.729
○	•	•	○	0.868	3.569	0.710	0.861	3.476	0.652	0.858	3.711	0.744
○	•	○	•	0.838	4.592	0.787	0.836	3.873	0.680	0.830	4.594	0.757
•	○	○	○	0.828	4.286	0.816	0.826	3.927	0.809	0.820	4.483	0.809
○	○	•	○	0.858	4.018	0.689	0.851	3.845	0.677	0.848	4.420	0.714
○	•	○	○	0.748	6.302	1.171	0.744	6.307	1.181	0.737	6.469	1.327
○	○	○	•	0.706	6.039	1.360	0.708	6.233	1.188	0.699	6.395	1.412

this phase can clearly depict the structure of the four heart chambers, which is crucial for subsequent research on fat distribution and multi-class image segmentation. The IP provides strong imaging signals for both fat and chambers and includes all structural information. However, the boundaries in the IP images are less clear, resulting in the poorest segmentation performance when used as a single modality input. Figure 3 shows that EAT-Mamba can achieve high-quality segmentation for EAT with four modalities available.

EAT-Mamba employs input-level fusion, which is simple and preserves the intrinsic image feature. However, it does not fully exploit the relationships among different modalities. In contrast, layer-level fusion, with its dense connections among layers, could better capture the complex interactions between modalities. This strategy could potentially help the network learn more valuable information and achieve even better performance. Future work could explore the benefits of layer-level fusion with Mamba to further enhance the EAT segmentation.

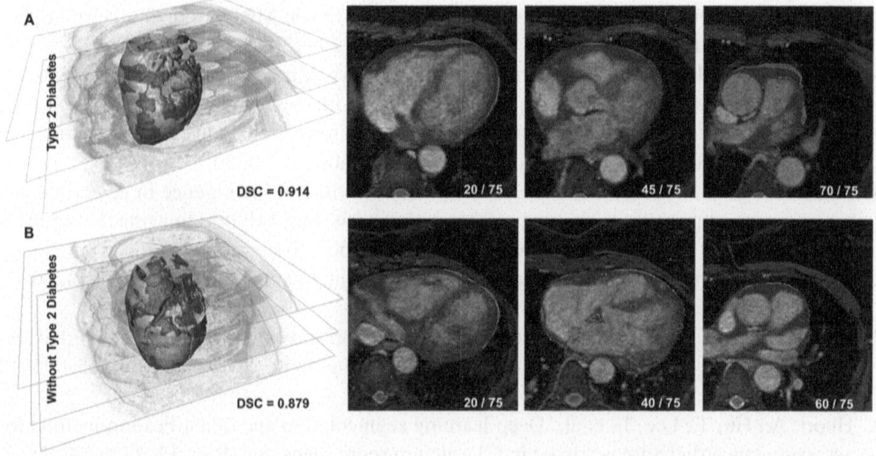

Fig. 3. Visualization of segmentation results in 3D and 2D slices. (**A, B**) The 3D reconstruction of the hearts from T2D (top) and control (bottom) groups, with referenced EAT volumes, are 167.14 cm^3 and 109.77 cm^3, respectively. The ground truth of EAT is indicated in orange, while predicted EAT is in violet. 2D images correspond to the cranial, middle and caudal parts of hearts. DSC, Dice similarity coefficient; T2D, type 2 diabetes. (Color figure online)

4 Conclusion

This paper presents EAT-Mamba, a novel multi-modal deep learning-based pipeline for automated segmentation and quantification of EAT from Dixon MRI. EAT-Mamba combines a U-shape structure with Res-Mamba blocks to capture localized fine-grained features and long-range features at various scales, achieving higher segmentation accuracy. This study provides valuable insights into the potential of SSMs and multi-modal images in EAT segmentation.

References

1. Christensen, R.H., Von Scholten, B.J., Hansen, C.S., et al.: Epicardial adipose tissue predicts incident cardiovascular disease and mortality in patients with type 2 diabetes. Cardiovasc. Diabetol. **18**, 114 (2019)
2. Yamazaki, H., Tauchi, S., Machann, J., et al.: Fat Distribution patterns and future type 2 diabetes. Diabetes **71**, 1937–1945 (2022)
3. Song, D.K., Hong, Y.S., Lee, H., et al.: Increased epicardial adipose tissue thickness in type 2 diabetes mellitus and obesity. Diabetes Metab. J. **39**, 405 (2015)
4. Christensen, R.H., Von Scholten, B.J., Lehrskov, L.L., et al.: Epicardial adipose tissue: an emerging biomarker of cardiovascular complications in type 2 diabetes? Ther. Adv. Endocrinol. **11**, 204201882092882 (2020)
5. Roset-Altadill, A., Domenech-Ximenos, B., Cañete, N., et al.: Epicardial space: comprehensive anatomy and spectrum of disease. Radiographics **44**, e230160 (2024)
6. Nakamori, S., Amyar, A., Fahmy, A.S., et al.: Cardiovascular magnetic resonance radiomics to identify components of the extracellular matrix in dilated cardiomyopathy. Circulation. CIRCULATIONAHA.123.067107 (2024)

7. Homsi, R., Meier-Schroers, M., Gieseke, J., et al.: 3D-Dixon MRI based volumetry of peri- and epicardial fat. Int. J. Cardiovasc. Imaging **32**, 291–299 (2016)
8. Klingensmith, J.D., Elliott, A.L., Fernandez-del-Valle, M., Mitra, S.: Automated segmentation of cardiac adipose tissue in Dixon magnetic resonance images. JBGC. **8**, 1 (2017)
9. Xu, F., Tu, W., Feng, F., et al.: Dynamic Position Transformation and Boundary Refinement Network for Left Atrial Segmentation http://arxiv.org/abs/2407.05505 (2024)
10. Zhang, L., Sun, J., Jiang, B., et al.: Development of artificial intelligence in epicardial and pericoronary adipose tissue imaging: a systematic review. Eur. J Hybrid Imaging. **5**, 14 (2021)
11. Hu, Y., Jiang, S., Yu, X., et al.: Automatic epicardial adipose tissue segmentation in pulmonary computed tomography venography using nnU-Net. Quant. Imaging Med. Surg. **13**, 6482–6492 (2023)
12. He, X., Guo, B.J., Lei, Y., et al.: Automatic segmentation and quantification of epicardial adipose tissue from coronary computed tomography angiography. Phys. Med. Biol. **65**, 095012 (2020)
13. Hoori, A., Hu, T., Lee, J., et al.: Deep learning segmentation and quantification method for assessing epicardial adipose tissue in CT calcium score scans. Sci. Rep. **12**, 2276 (2022)
14. Zhang, Q., Zhou, J., Zhang, B., et al.: Automatic epicardial fat segmentation and quantification of CT scans using dual U-Nets with a morphological processing layer. IEEE Access. **8**, 128032–128041 (2020)
15. He, X., Guo, B., Lei, Y., et al.: Automatic epicardial fat segmentation in cardiac CT imaging using 3D deep attention U-Net. In: Landman, B.A. and Išgum, I. (eds.) Medical Imaging 2020: Image Processing, p. 84. SPIE, Houston, United States (2020)
16. Siriapisith, T., Kusakunniran, W., Haddawy, P.: A 3D deep learning approach to epicardial fat segmentation in non-contrast and post-contrast cardiac CT images. PeerJ Comput. Sci. **7**, e806 (2021)
17. West, H.W., Siddique, M., Williams, M.C., et al.: Deep-learning for epicardial adipose tissue assessment with computed tomography. JACC: Cardiovas. Imaging. **16**, 800–816 (2023)
18. Li, X., Sun, Y., Xu, L., et al.: Automatic quantification of epicardial adipose tissue volume. Med. Phys. **48**, 4279–4290 (2021)
19. Daudé, P., Ancel, P., Confort Gouny, S., et al.: Deep-learning segmentation of epicardial adipose tissue using four-chamber cardiac magnetic resonance imaging. Diagnostics **12**, 126 (2022)
20. Bard, A., Raisi-Estabragh, Z., Ardissino, M., et al.: Automated quality-controlled cardiovascular magnetic resonance pericardial fat quantification using a convolutional neural network in the UK biobank. Front. Cardiovasc. Med. **8**, 677574 (2021)
21. Feng, F., Carlhäll, C.-J., Tan, Y., et al.: FM-Net: a fully automatic deep learning pipeline for epicardial adipose tissue segmentation. In: Camara, O., Puyol-Antón, E., Sermesant, M., Suinesiaputra, A., Tao, Q., Wang, C., and Young, A. (eds.) Statistical Atlases and Computational Models of the Heart. Regular and CMRxRecon Challenge Papers, pp. 88–97. Springer, Cham (2024)
22. Gu, A., Dao, T.: Mamba: Linear-Time Sequence Modeling with Selective State Spaces
23. Ma, J., Li, F., Wang, B.: U-Mamba: Enhancing Long-range Dependency for Biomedical Image Segmentation http://arxiv.org/abs/2401.04722 (2024)
24. Xing, Z., Ye, T., Yang, Y., et al.: SegMamba: Long-range Sequential Modeling Mamba For 3D Medical Image Segmentation http://arxiv.org/abs/2401.13560 (2024)
25. Edin, C., Ekstedt, M., Scheffel, T., et al.: Ectopic fat is associated with cardiac remodeling—A comprehensive assessment of regional fat depots in type 2 diabetes using multi-parametric MRI. Front. Cardiovasc. Med. **9**, 813427 (2022)
26. Zhou, T., Ruan, S., Canu, S.: A review: Deep learning for medical image segmentation using multi-modality fusion. Array. **3–4**, 100004 (2019)

27. Liu, J., Liu, Z., Wu, G., et al.: Multi-interactive feature learning and a full-time multi-modality benchmark for image fusion and segmentation. In: 2023 IEEE/CVF International Conference on Computer Vision (ICCV), pp. 8081–8090. IEEE, Paris, France (2023)
28. Li, X., Dou, Q., Chen, H., et al.: 3D multi-scale FCN with random modality voxel dropout learning for intervertebral disc localization and segmentation from multi-modality MR images. Med. Image Anal. **45**, 41–54 (2018)
29. Alom, Z., Taha, T.M., Asari, V.K.: Recurrent Residual Convolutional Neural Network based on U-Net (R2U-Net) for Medical Image Segmentation. 12
30. Zhou, Z., Siddiquee, M.M.R., Tajbakhsh, N., Liang, J.: UNet++: a nested U-Net architecture for medical image segmentation http://arxiv.org/abs/1807.10165 (2018)

Neural Fields for Continuous Periodic Motion Estimation in 4D Cardiovascular Imaging

Simone Garzia[1,2(✉)], Patryk Rygiel[3], Sven Dummer[3], Filippo Cademartiri[4], Simona Celi[1], and Jelmer M. Wolterink[3]

[1] BioCardioLab, Bioengineering Unit, Fondazione Monasterio, 54100 Massa, Italy
s.garzia95@gmail.com
[2] Department of Information Engineering, University of Pisa, 56126 Pisa, Italy
[3] Department of Applied Mathematics and Technical Medical Centre, University of Twente, 7500 AE Enschede, The Netherlands
[4] Clinical Imaging Department, Fondazione Monasterio, 54100 Massa, Italy

Abstract. Time-resolved three-dimensional flow MRI (4D flow MRI) provides a unique non-invasive solution to visualize and quantify hemodynamics in blood vessels such as the aortic arch. However, most current analysis methods for arterial 4D flow MRI use static artery walls because of the difficulty in obtaining a full cycle segmentation. To overcome this limitation, we propose a neural fields-based method that directly estimates continuous periodic wall deformations throughout the cardiac cycle. For a 3D + time imaging dataset, we optimize an implicit neural representation (INR) that represents a time-dependent velocity vector field (VVF). An ODE solver is used to integrate the VVF into a deformation vector field (DVF), that can deform images, segmentation masks, or meshes over time, thereby visualizing and quantifying local wall motion patterns. To properly reflect the periodic nature of 3D + time cardiovascular data, we impose periodicity in two ways. First, by periodically encoding the time input to the INR, and hence VVF. Second, by regularizing the DVF. We demonstrate the effectiveness of this approach on synthetic data with different periodic patterns, ECG-gated CT, and 4D flow MRI data. The obtained method could be used to improve 4D flow MRI analysis.

Keywords: neural fields · deformation vector fields · 4D flow MRI · diffeomorphism · periodic motion estimation

1 Introduction

Time-resolved three-dimensional flow MRI (4D flow MRI) provides unique insights into patient-specific cardiovascular conditions [5,10]. Visualization, quantification, and interpretation of 4D flow MRI require post-processing, which is typically performed under the assumption that the artery walls are

O. Camara et al. (Eds.): STACOM 2024, LNCS 15448, pp. 378–389, 2025.
https://doi.org/10.1007/978-3-031-87756-8_37

static [5,12,13,23]. However, for reliable determination of hemodynamic parameters such as wall shear stress, the wall deformation over the cardiac cycle should be considered [7–9]. Accurate tracking of the artery wall position remains a major challenge in 4D flow MRI imaging. Previous works have aimed to obtain time-dependent deformation fields based on automatic segmentation in different phases of the cardiac cycle, e.g., [6,18]. However, these techniques are limited by the resolution of the spatial grid and thus struggle to provide deformation fields that reflect the continuous nature of artery wall motion. As an alternative, deep learning-based registration approaches have been proposed, in which a neural network is trained to provide a deformation field, given cardiac imaging frames [1,2,22]. However, such methods rely on the availability of large quantities of training data. In this work, we propose a one-shot image registration approach for motion estimation that does not require training data. Following recent developments in image registration [24,26], we employ implicit neural representations (INRs) [20], also known as neural fields or coordinate networks. INRs represent a signal as a continuous function by associating to the input coordinates the value of an estimated function [19,25]. Considering the target motion as a nonlinear, continuous, time-dependent periodic process, we represent a *time-dependent and periodic* velocity vector field (VVF) with a trainable INR [14,15,21]. The obtained VVF is embedded in an ordinary differential equation (ODE) to estimate the final deformation vector field (DVF), which is trained with a single 4D image and can be applied to images and meshes representing anatomical structures such as the aortic wall. To reflect the periodic nature of cardiovascular data and to ensure periodicity of the time-dependent velocity field, we periodically embed the time input to the INR., Moreover, to impose periodicity on the obtained deformation fields, we propose a novel approach for periodic regularization. We assess the model's effectiveness using synthetic and real data sets and find that the proposed method provides continuous and periodic motion estimation.

2 Methods

2.1 Temporal Diffeomorphic Image Registration

To perform motion estimation, we jointly register multiple time frames within a period T. Instead of estimating the displacement field $\phi_{0\to1}$ between a moving image (I_0) and a fixed one (I_1), we estimate the displacement fields $\phi_{t\to T}$ for a series of images I_t over N time points $\{t_i\}_{i=0}^{N-1}$ with $t_0 = 0$ and $t_{N-1} = T$:

$$\min_{\{\phi_{t_i\to T}\}_{i=0}^{N-2}} \sum_{i=0}^{N-2} \left[\|I_{t_i} \circ \phi_{t_i\to T} - I_T\|_2^2\right], \tag{1}$$

We parameterize $\phi_{t_i\to T}$ via a general ϕ_t. One approach for representing ϕ_t is the large deformation framework, such as the LDDMM method [4]. This approach breaks down the whole deformation process into multiple small steps by parameterizing a diffeomorphic deformation using an ODE:

$$\frac{\partial \phi_t(\hat{P})}{\partial t} = v(\phi_t(\hat{P}), t),\ \phi_0(\hat{P}) = \hat{P},\ \phi_{t\to T} := \phi_T \circ \phi_t^{-1} \tag{2}$$

This iterative approach allows the transformations to be invertible and continuous, ensuring a global one-to-one mapping of each point in the two volumes [3,14]. As the ϕ_t perfectly fits with the temporal nature of our data and because its invertibility fits the inductive bias that all the aortic data is diffeomorphic, we implemented a diffeomorphic registration framework using an ODE and a neural network parameterized VVF v.

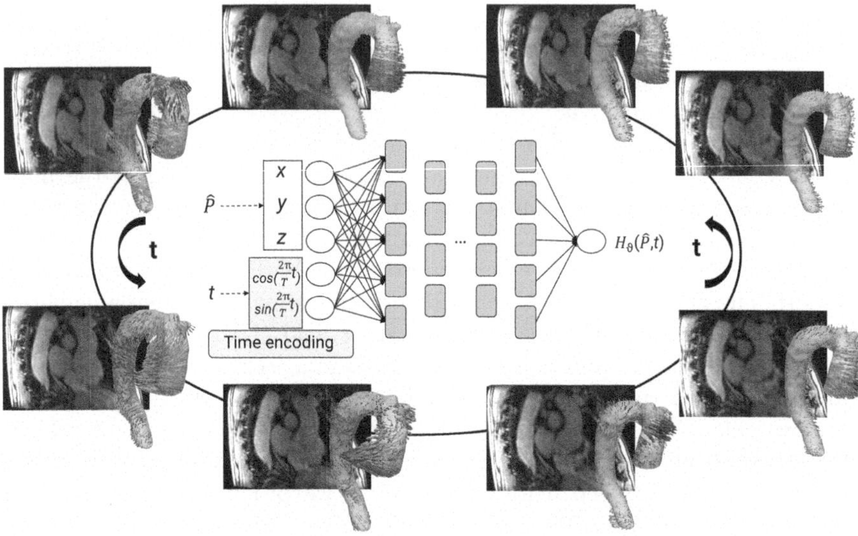

Fig. 1. Overview of the proposed method. We train an implicit neural representation $H_\theta \colon \Omega \times [0, T] \to \mathbb{R}^3$ with spatial domain $\Omega \subset \mathbb{R}^3$, time horizon T, and weights θ to optimize. For every spatial point $\hat{P} = (x, y, z) \in \mathbb{R}^3$ and time frame t, $H_\theta(\hat{P}, t)$ represents the time-dependent velocity field describing the periodic motion at any point in space and time. We obtain a DVF by embedding the VVF H_θ in an ODE. Applying the DVF to the first time frame mesh allows for a full-cycle deformation.

2.2 Implicit Neural Representations

The implicit neural representation that we use to estimate the time-dependent VVF is a neural network that maps a spatiotemporal coordinate to a vector. In particular, an INR representing a time-dependent VVF is a neural network $H_\theta \colon \Omega \times [0, T] \to \mathbb{R}^3$ with spatial domain $\Omega \subset \mathbb{R}^3$, time horizon T, and weights θ. Hence, given a spatial position $\hat{P}$ and time frame t_i, $H_\theta(\hat{P}, t_i)$ is the velocity at $\hat{P}$ at time t_i. By parameterizing the velocity field with a function, we utilize spatial and temporal continuous modeling over multiple timeframes to accurately estimate the continuous DVF in Eq. (2).

2.3 Periodic Velocity Fields

While the standard INR parameterization of the VVF is time-dependent, it does not necessarily consider the periodicity in our cardiovascular imaging data. To ensure our network accurately represents specific time-dependent periodic VVFs, we use an INR $\tilde{H}_\theta \colon \Omega \times \mathbb{R}^2 \to \mathbb{R}$ and let $H_\theta(\hat{P}, t) := \tilde{H}_\theta(\hat{P}, f(t))$ where $f(t)$ enhances the model by encoding the time variable on the unit circle, representing it with two components:

$$f(t) := \left(\cos\left(\frac{2\pi}{T}t\right), \sin\left(\frac{2\pi}{T}t\right)\right)$$

By explicitly encoding the periodicity as an inductive bias, we ensure that the resulting motion inherently satisfies the constraints of periodic behavior. As shown in the numerical experiments, this approach helps to accurately model periodic phenomena by embedding domain-specific knowledge directly into the learning process. Specifically, the approach effectively addresses challenges in representing complex temporal patterns, such as the shrinkage/growing movement of a toy sphere or the expansion and contraction of the aortic wall, as shown in Fig. 1.

2.4 Loss Function

Using H_θ as parameterization of the VVF in Eq. (2), we can in principle optimize the neural network parameters θ via the loss function presented in Eq. (1). However, our data has the inductive bias that it is periodic in time. Although our INR parameterization of the velocity vector field is periodic in time, this does not mean the whole motion is periodic. Consequently, to ensure our network approximately represents periodic motions, such as cardiac vessel movements that repeat each cardiac cycle, we propose a periodic regularisation term, namely R_{cycle}. Specifically, the cycle consistency loss is defined as follows:

$$R_{\text{cycle}} = \left(\frac{1}{n}\sum_{i=1}^{n} \|\mathrm{P}_{0,i} - \phi_T(\mathrm{P}_{0,i})\|_2^2\right)$$

where n is the number of sampled points and $\mathrm{P}_{0,i}$ are the point coordinates at the starting position. In summary, the INR is trained to optimize the following objective function:

$$\begin{aligned} &\min_\theta \quad \left[\sum_{i=0}^{N-2} \left[\|I_{t_i} \circ \phi_{t_i \to T} - I_T\|_2^2\right]\right] + \lambda R_{cycle}, \\ &\text{s.t.} \quad \frac{\partial \phi_t(\hat{P})}{\partial t} = H_\theta(\phi_t(\hat{P}), t),\ \phi_0(\hat{P}) = \hat{P},\ \phi_{t\to T} := \phi_T \circ \phi_t^{-1} \end{aligned} \tag{3}$$

where $\lambda \in \mathbb{R}$ is a regularization constant.

2.5 Data

The model performance was evaluated using three datasets. First, three synthetic datasets were created. Each dataset consisted of a series of 25 3D segmentation images of a sphere that was deformed according to a known pattern: linear growth, exponential growth, or periodic (sinusoidal) growth and shrinkage (Fig. 2a). For each image, a corresponding triangular mesh was provided. It's worth noting that the datasets with linear and exponential growth were used at an early stage to evaluate the operation of diffeomorphic registration. In addition, since they are not periodic, the regularization term R_{cycle} was not applied. Then, five 4D flow MRI datasets were included of three male (15–55y) and two female (77–79y) patients. These datasets were acquired using a 1.5T Optim MR450w scanner (GE Medical Systems). For all datasets, manual segmentation of the end-diastolic and peak-systolic frame was performed using 3D Slicer software [16] and a corresponding triangular mesh was obtained. Reconstructed images had an in-plane resolution of 1.4×1.4 mm^2, and a slice thickness of 2.8 mm. Similar to the synthetic data, each cardiac cycle consisted of 25 images I_t. Finally, three ECG-gated CT data sets including two male (43–80y) and one female (80y) were included to assess the method's performance in alternative modalities. These datasets were acquired with a SOMATOM Force scanner (Siemens Healthineers) with an in-plane resolution of 0.26–0.38$\times$ 0.26–0.38 mm^2, and a slice thickness of 0.69–1 mm. For this specific data, each cardiac cycle consisted of 20 frames.

2.6 Evaluation Metrics

To assess the network performances, temporal image registration was performed on different types of data. To assess deformation quality, we used the Hausdorff Distance (HSD) to compute the distance between meshes and Peak Signal-to-Noise Ratio (PSNR) to assess image similarity. A qualitative evaluation was also carried out comparing different point trajectories in different experiments (Fig. 3).

3 Experiments and Results

3.1 Implementation Details

All models were implemented in PyTorch. The network model $\tilde{H}_\theta$ was an MLP with sinusoidal activation functions [20], accounting for 3 layers with 256 neurons each. The ω hyperparameter, which controls the period of the sinusoidal functions, was set to 6 for the synthetic dataset and 30 for the 4D flow MRI and CT data. To solve the resulting ODE appearing in optimization problem (3), we use the Euler method with $N-1$ timesteps, as implemented in the Torchdiffeq library [11]. We used Adam [17] with a learning rate of 3e-5. Model optimization was conducted for 1000 epochs using 5000 (synthetic dataset) and 10000 (real dataset) randomly sampled points in the spatial domain. For each sequence of images $\{I_{t_i}\}_{i=0,\ldots,N-1}$, the sampled spatial coordinates $\hat{P}$ were normalized to the $[-1,1]^3$ range. Additionally, the time points $t_0, t_1, \ldots, t_{N-1}$ were equally spaced and normalized to the $[0,1]$ interval. Finally, in the numerical experiments below, we get a sequence of spheres or aortic walls by deforming an initial mesh $\mathcal{M}_{t_0}$ via $\phi_t^{-1}(\mathcal{M}_{t_0})$ with ϕ_t the DVF in Eq. (3). The time required for optimization was between 12 min for 5000 points and 25 min for 10000 points for each patient.

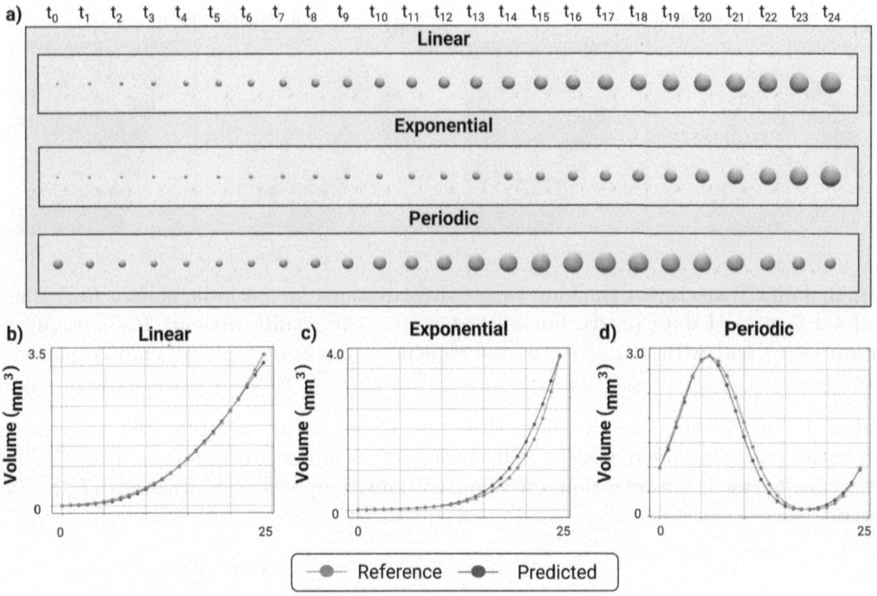

Fig. 2. Synthetic dataset of linear, exponential, and periodic growth patterns (a). Volume comparison of the sphere datasets for linear growth (b), exponential growth (c), and periodic growth (d) between predicted (orange) and reference (blue) lines. (Color figure online)

3.2 Synthetic Data

For each of the three synthetic datasets, an INR was optimized across all time points. For each deformed sphere surface, the volume is calculated and then compared to the ground-truth one. The three plots in Fig. 2(b, c, d) show a comparison of the volumes of the ground truth and extracted spheres for the benchmark experiments across the linear, exponential, and periodic growth patterns respectively. Table 1 lists the average HSD value between predicted and reference meshes over the full cardiac cycle. We qualitatively assessed the effect of periodic regularization in an ablation study. Figure 3 shows the sphere in combination with local point trajectories through the cardiac cycle. From this, it can be seen that regularization enforces periodic motion patterns.

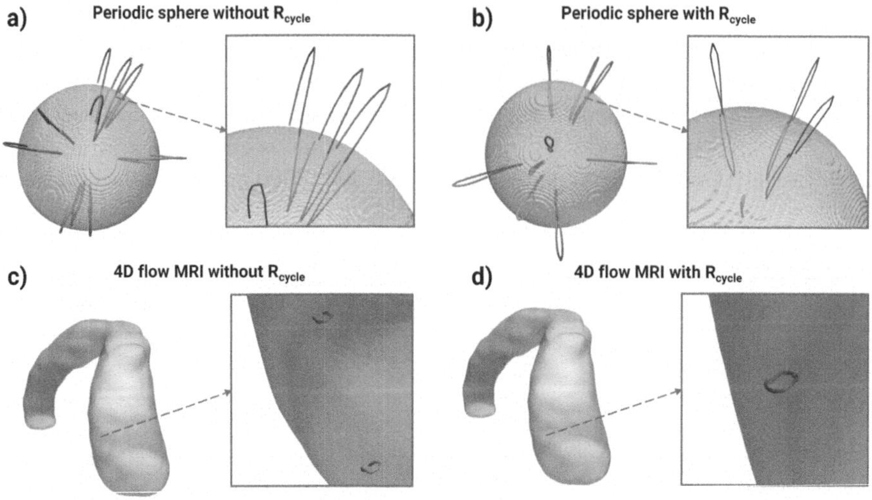

Fig. 3. Point trajectories tracking target deformations for periodic sphere (a, b) and real 4D flow MRI data (c, d). For both datasets, the results without R_{cycle} regularisation (a, c) and with R_{cycle} (b, d) are shown. R_{cycle} is required to enforce periodic motion.

Table 1. Quantitative results. The HSD and PSNR values are reported in columns for both synthetic and real data. The best metric values are reported in bold. The difference between optimization with and without temporal regularization (R_{cycle}) is listed.

		↓ HSD (mm)		↑ PSNR (dB)	
		noR_{cycle}	R_{cycle}	noR_{cycle}	R_{cycle}
Sphere	Linear	8.41	no periodicity	12.78	no periodicity
	Exponential	5.62	no periodicity	11.58	no periodicity
	Periodic	6.23	**4.59**	11.25	10.91
4D flow MRI	Sample 1	7.44	**7.34**	29.65	27.46
	Sample 2	7.34	**6.61**	27.04	26.92
	Sample 3	5.78	**4.98**	25.53	25.34
	Sample 4	5.89	**5.55**	26.32	26.03
	Sample 5	4.27	**3.96**	25.86	25.73
CT data	Sample 1	4.98	**3.62**	18.65	18.45
	Sample 2	5.49	**3.49**	17.37	17.29
	Sample 3	4.33	**3.51**	18.48	18.32

3.3 4D Flow MRI Data

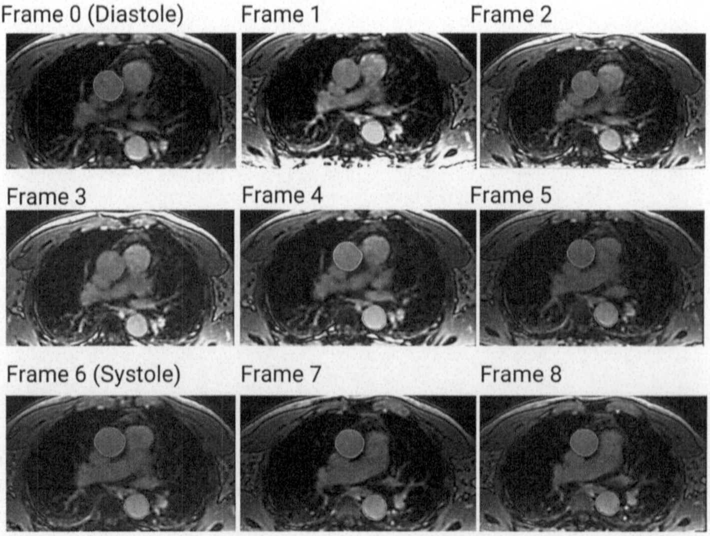

Fig. 4. First 9 frames of 4D flow MRI axial slices ranging from end-diastole to peak-systole. On the same slice, the overlapping of the real target vessel in the magnitude data and the deformed first frame mesh through the estimated DVF is shown over the cardiac cycle.

In the 4D flow MRI experiments, the INR was trained using magnitude images. Differently from the synthetic dataset, the 4D flow MRI datasets account for just two frame segmentations (end-diastole and peak-systole), lacking full-cycle segmentations. For this reason, the HSD was only evaluated in the frames with available meshes, with obtained results shown in Table 1. The table again shows the benefit of the R_{cycle} on the reconstruction performance. Additionally, Fig. 3(c, d) shows that only with R_{cycle} we obtain periodic motions.

Then, a qualitative evaluation was carried out. Figure 4 shows the results for a 4D flow MRI patient. Specifically, the deformed meshes, derived from the DVF application on the first time frame mesh, were overlapped with the first 9 corresponding ground truth time frame magnitude images. The deformed mesh overlaps well with the correct structure inside the magnitude images.

3.4 CT Data

For each frame in the CT data, segmentations were available. The HSD was evaluated in all frames, and the corresponding volume was evaluated and then compared to the ground truth volume. Figure 5 shows a comparison of the volumes of the ground truth and the predicted deformed shapes. Table 1 lists the

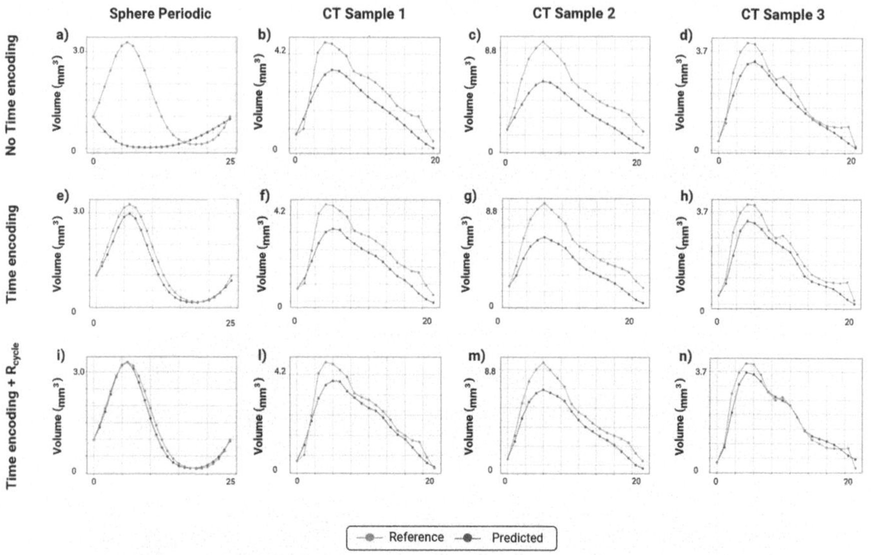

Fig. 5. Volume comparison of the periodic sphere (a,e,i) and CT datasets (b-d,f-h,l-n) with no time encoding (a-d), time encoding (e-h), and time encoding coupled with R_{cycle} (i-n) applied.

average HSD value between predicted and reference meshes over the full cardiac cycle. In this case also, the benefit of time encoding and R_{cycle} on reconstruction performance can be appreciated from both Fig. 5f–h, l–m and Table 1.

4 Discussion and Conclusion

We have presented a time-dependent and periodic VVF modeling approach, continuous in both spatial and temporal domains, based on INRs to be applied to 4D cardiac data. A diffeomorphism was obtained by integrating an ODE solver into our model, and semi-supervised training avoided requiring large datasets. We qualitatively and quantitatively assessed the accuracy of estimated DVFs on synthetic periodic data, and different real data.

Regarding the contribution of R_{cycle}, the point trajectories tracked in both synthetic (Fig. 3b) and real (Fig. 3d) periodic data with regularisation applied showed the model's ability to represent time-dependent and periodic motion even more accurately. This effect can also be noticed in Fig. 5l-n and Table 1. Regarding the effect of time encoding, Fig. 5 shows its effect on different data types. Specifically, on synthetical data with rapid periodic growth patterns (Fig. 5a, e), the network without time encoding was unable to correctly represent the deformation, both in time and space. Regarding periodic cardiac data with minor abrupt deformation (Fig. 5b, c, d, f, g, h), the adoption of time encoding helps to obtain better accuracy, following local volume changes better than the case without time encoding. Indeed, we can notice that the volume evolution is more linear and less realistic in cases without time encoding and more locally refined in the ones that include it. Since the main objective was to build a network with a broad application of cardiac data, time encoding is a fundamental assumption for cases with different temporal deformation patterns. Therefore, the ability of the proposed model to generalize results on different periodic data ensuring continuous representation in space and time but also time-dependency and periodicity, makes it a promising solution for clinical applications.

Even if accurate results are obtained for synthetic full-time sequences, some limitations arise with more complex real data. Specifically, preliminary experiments not shown in the paper demonstrate that, although the volume progression of the aorta during the cardiac cycle follows the correct systole-diastole pattern, a progressive reduction in VVF accuracy arises when the number of time frames shown to the networks increases. In particular, the VVF loses strength and the target wall progressively deforms less and less accurately in relation to more time frames. In future work, this might be addressed by adding additional loss terms that quantify, e.g., similarities between neighboring frames, instead of only the difference with respect to the first frame, as we have now done. Additionally, future work would include an evaluation of the proposed method on different cardiac data employing different quantitative metrics.

Our method requires the optimization of a neural network per imaging sequence. The execution time for this primarily depends on the number of points sampled per iteration and settings of the ODE solver, e.g., the use of the adjoint to limit memory usage. Moreover, we found that the complexity of the data affects the time required for optimization. Alternative approaches such as segmentation-based motion estimation methods [6] or non-neural LDDMM implementations [4] might offer a different trade-off between speed and accuracy and will be compared against in future work.

In summary, we have effectively addressed the limitations of current 4D flow MRI hemodynamic analysis methods, which traditionally assume static artery walls, by leveraging neural fields to estimate continuous time-dependent periodic wall deformations throughout the cardiac cycle. Our method allows us to model time-dependent velocity fields facilitating the deformation of segmentation masks or meshes over time in a Lagrangian framework, enabling detailed visualization and quantification of local wall motion patterns. The periodic regularization R_{cycle} ensures that the periodic nature of cardiovascular data is appropriately captured, enhancing the point trajectories over time. This strategy also helps the network in increasing the DVF estimation accuracy on full-cycle cardiac sequences (Fig. 5). Overall, our neural fields-based method improves the tracking analysis of dynamic vascular structures, paving the way for more accurate and comprehensive hemodynamic assessments in cardiovascular research and clinical practice.

References

1. Ahn, S.S., Ta, K., Lu, A., Stendahl, J.C., Sinusas, A.J., Duncan, J.S.: Unsupervised motion tracking of left ventricle in echocardiography. In: Medical imaging 2020: Ultrasonic Imaging and Tomography, vol. 11319, pp. 196–202. SPIE (2020)
2. Ahn, S.S., et al.: Co-attention spatial transformer network for unsupervised motion tracking and cardiac strain analysis in 3D echocardiography. Med. Image Anal. **84**, 102711 (2023)
3. Ashburner, J.: A fast diffeomorphic image registration algorithm. Neuroimage **38**(1), 95–113 (2007)
4. Beg, M.F., Miller, M.I., Trouvé, A., Younes, L.: Computing large deformation metric mappings via geodesic flows of diffeomorphisms. Int. J. Comput. Vis. **61**, 139–157 (2005)
5. Berhane, H., et al.: Fully automated 3D aortic segmentation of 4D flow MRI for hemodynamic analysis using deep learning. Magn. Reson. Med. **84**(4), 2204–2218 (2020)
6. Bustamante, M., Viola, F., Engvall, J., Carlhäll, C.J., Ebbers, T.: Automatic time-resolved cardiovascular segmentation of 4D flow MRI using deep learning. J. Magn. Reson. Imaging **57**(1), 191–203 (2023)
7. Calò, K., et al.: Impact of wall displacements on the large-scale flow coherence in ascending aorta. J. Biomech. **154**, 111620 (2023)
8. Capellini, K., et al.: A novel formulation for the study of the ascending aortic fluid dynamics with in vivo data. Med. Eng. Phys. **91**, 68–78 (2021)
9. Capellini, K., et al.: Computational fluid dynamic study for aTAA hemodynamics: an integrated image-based and radial basis functions mesh morphing approach. J. Biomech. Eng. **140**(11), 111007 (2018)
10. Celi, S., Martini, N., Emilio Pastormerlo, L., Positano, V., Berti, S.: Multimodality imaging for interventional cardiology. Curr. Pharm. Des. **23**(22), 3285–3300 (2017)
11. Chen, R.T.Q.: Torchdiffeq (2018). https://github.com/rtqichen/torchdiffeq
12. Frydrychowicz, A., et al.: Aortic hemodynamics in patients with and without repair of aortic coarctation: in vivo analysis by 4D flow-sensitive magnetic resonance imaging. Invest. Radiol. **46**(5), 317–325 (2011)
13. Garzia, S., et al.: Coupling synthetic and real-world data for a deep learning-based segmentation process of 4D flow MRI. Comput. Methods Programs Biomed. **242**, 107790 (2023)
14. Han, K., et al.: Diffeomorphic image registration with neural velocity field. In: Proceedings of the IEEE/CVF Winter Conference on Applications of Computer Vision, pp. 1869–1879 (2023)
15. van Harten, L., Van Herten, R.L.M., Stoker, J., Isgum, I.: Deformable image registration with geometry-informed implicit neural representations. In: Medical Imaging with Deep Learning, pp. 730–742. PMLR (2024)
16. Kikinis, R., Pieper, S.D., Vosburgh, K.G.: 3D slicer: a platform for subject-specific image analysis, visualization, and clinical support. In: Intraoperative Imaging and Image-guided Therapy, pp. 277–289. Springer (2013)
17. Kingma, D.P., Ba, J.: Adam: A method for stochastic optimization. arXiv preprint arXiv:1412.6980 (2014)
18. Lian, Z., Lu, Q., Lin, B., Chen, L., Peng, P., Feng, Y.: MRI deep learning-based automatic segmentation of interventricular septum for black-blood myocardial t2* measurement in thalassemia. J. Magn. Reson. Imaging (2023)

19. Mildenhall, B., Srinivasan, P.P., Tancik, M., Barron, J.T., Ramamoorthi, R., Ng, R.: NeRF: representing scenes as neural radiance fields for view synthesis. Commun. ACM **65**(1), 99–106 (2021)
20. Sitzmann, V., Martel, J., Bergman, A., Lindell, D., Wetzstein, G.: Implicit neural representations with periodic activation functions. Adv. Neural. Inf. Process. Syst. **33**, 7462–7473 (2020)
21. Sun, S., et al.: Medical image registration via neural fields. arXiv preprint arXiv:2206.03111 (2022)
22. Ta, K., Ahn, S.S., Stendahl, J.C., Sinusas, A.J., Duncan, J.S.: A semi-supervised joint network for simultaneous left ventricular motion tracking and segmentation in 4d echocardiography. In: Medical Image Computing and Computer Assisted Intervention–MICCAI 2020: 23rd International Conference, Lima, Peru, 4–8 Oct. 2020, Proceedings, Part VI 23, pp. 468–477. Springer (2020)
23. Vali, A., et al.: Semi-automated analysis of 4D flow MRI to assess the hemodynamic impact of intracranial atherosclerotic disease. Magn. Reson. Med. **82**(2), 749–762 (2019)
24. Van Harten, L.D., Stoker, J., Išgum, I.: Robust deformable image registration using cycle-consistent implicit representations. IEEE Trans. Med. Imaging (2023)
25. Wink, N.M., Panknin, C., Solberg, T.D.: Phase versus amplitude sorting of 4D-CT data. J. Appl. Clin. Med. Phys. **7**(1), 77–85 (2006)
26. Wolterink, J.M., Zwienenberg, J.C., Brune, C.: Implicit neural representations for deformable image registration. In: International Conference on Medical Imaging with Deep Learning, pp. 1349–1359. PMLR (2022)

Exploring CNN and Transformer Architectures for Multi-class Bi-Atrial Segmentation from Late Gadolinium-Enhanced MRI

Yichi Zhang[1,2](✉), Yuan Cheng[1,2], and Yuan Qi[1,2]

[1] Artificial Intelligence Innovation and Incubation Institute, Fudan University, Shanghai, China
zhangyichi23@m.fudan.edu.cn
[2] Shanghai Academy of Artificial Intelligence for Science, Shanghai, China

Abstract. Accurate segmentation of atrial anatomical structures from late gadolinium-enhanced magnetic resonance imaging (LGE-MRI) has emerged as a vital tool for mapping fibrosis and guiding targeted therapies, thus improving patient outcomes. In this paper, we investigate the application of CNN-based and transformer-based architectures in the MICCAI 2024 MBAS Challenge, building upon the self-configuration and implementation of nnUNet. The five-fold cross-validation results on the training data indicate that CNN-based segmentation models consistently outperform transformer-based models. In line with the conclusion, the official evaluations show that CNN-based models deliver better performance compared with incorporating transformer-based models, with an average Dice score of 0.6640, 0.8806, and 0.9099, and an average HD95 of 4.5258, 7.6845, and 4.9136 for the atrium wall, left atrium cavity, and right atrium cavity, respectively.

Keywords: Magnetic Resonance Imaging · Image Segmentation · Convolutional Neural Network · Transformer

1 Introduction

As the most prevalent cardiac arrhythmia, atrial fibrillation (AF) is linked to significant health issues and fatalities. Existing clinical treatments perform poorly due to insufficient comprehension of the atrial anatomy that sustains atrial fibrillation within the human heart. In recent years, Late Gadolinium-Enhanced Magnetic Resonance Imaging (LGE-MRI) has become a prevalent method for investigating fibrosis and scarring [2]. Studies involving LGE-MRI on patients with atrial fibrillation have demonstrated that the degree and pattern of atrial fibrosis can be predictive of the success of ablation procedures. Consequently, a direct examination of the atrial structure in individuals with atrial fibrillation

O. Camara et al. (Eds.): STACOM 2024, LNCS 15448, pp. 390–397, 2025.
https://doi.org/10.1007/978-3-031-87756-8_38

is essential for enhancing our understanding and facilitating personalized treatment approaches for the condition [9]. Accurate segmentation of atrial anatomical structure is crucial for mapping the precise areas of fibrosis and guiding targeted therapies, thereby improving patient outcomes.

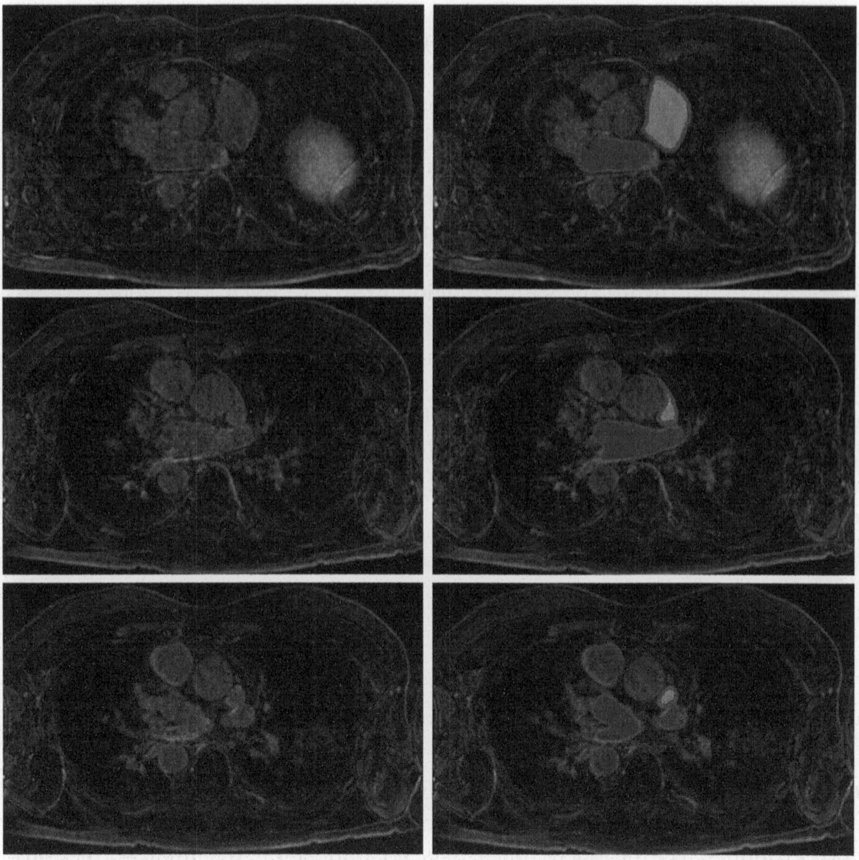

Fig. 1. An example image with corresponding segmentation label from MBAS challenge dataset. The atrium wall, left atrium cavity and right atrium cavity are shown in red, blue and green, respectively. (Color figure online)

Fueled by the availability of annotated data for training, deep learning methods have shown strong capability and achieved state-of-the-art performance in several cardiac MRI segmentation challenges [3,8,13]. The introduction of nnU-Net [7] highlights that careful implementation and configuration of the simple yet efficient U-Net architecture [11] based on Convolutional Neural Network (CNN) is more crucial than modifying the architecture itself and demonstrate cutting-edge performance in several tasks [1,10]. Despite this, there remains a continued interest in incorporating innovative transformer [12] designs to exploit long-range dependencies for medical image segmentation tasks like [4,6], which claims of superior performance compared with CNN-based methods. Given the advancements of each architecture, we aim to evaluate and compare their effectiveness in this challenge task for comparative analysis.

2 Challenge and Evaluation

In this paper, we aim to explore CNN and transformer architectures for MICCAI 2024 Multi-class Bi-Atrial Segmentation (MBAS) Challenge[1], which aims to segment both left and right atria as well as their walls to enhance ablation for atrial fibrillation patients. The challenge dataset consists 200 multi-center 3D LGE-MRIs, which is the largest dataset in the field, with each scan meticulously labeled by three experts. The dataset is split into 70 images for training, 30 images for validation, and 100 images for testing.

The 100 multi-centre LGE-MRIs provided for the training and validation stage are obtained using the same imaging protocol. Each MR image has a 3D dimension of $640 \times 640 \times 44$ or $576 \times 576 \times 44$, with a spatial resolution of $0.625 \times 0.625 \times 2.5\ \text{mm}^3$. While the remaining 100 images are hold by the organizer for testing. An example of the dataset is shown in Fig. 1. The challenge involves the segmentation of multi-class cardiac structures including the Atrium Wall (in red), the Left Atrium Cavity (in blue), and the Right Atrium Cavity (in green). To estimate the performance of segmentation results, two evaluation metrics is used including the Dice Similarity Coefficient (DSC) and 95% Hausdorff distance (HD95).

3 Methods

3.1 Data Preprocessing

For data preprocessing, we used the following three steps in our approach.

Cropping. The process of image cropping aims to remove the boundary region of the 3D medical image that is equal to zero. Compared to the original image before cropping, the cropped image has no impact on the final segmentation result while reducing the image size and improving the computation efficiency.

Resampling. To ensure that each voxel in the image represents a consistent actual spatial spacing, we resampled the original image to a resolution of $0.625 \times 0.625 \times 2.5\ \text{mm}^3$, thereby maintaining uniformity in the spatial representation across the dataset.

Normalization. Normalization is essential for ensuring that the input data of a model has a consistent scale, which helps the model to converge more quickly during training. In our approach, we perform image normalization using the mean and standard deviation of the entire training set.

[1] https://codalab.lisn.upsaclay.fr/competitions/18516.

3.2 Network

For CNN-based architecture, we first develop a 3D U-Net model [5] comprised of encoder-decoder structure with skip connections. The base number of features is set to 32, with the maximum number of features in the deepest layer reaching 320. The network is designed with a series of convolutional layers, each followed by a ReLU activation function, except for the last layer which utilizes a softmax activation for segmentation purposes. The convolutional layers are composed of varying kernel sizes, predominantly $3 \times 3 \times 3$, to capture anisotropic features within the medical images.

For transformer-based architecture, we then develop a UNETR model [6] comprised of an encoder-decoder framework enriched with skip connections. The foundational feature set of UNETR is initialized at 16, scaling up to a substantial 768 features in the profound layers. The configuration accommodates a multi-head self-attention mechanism with 12 attention heads, which serves to capture the intricate interdependencies within the volumetric data.

3.3 Data Split

Among the given 100 MR images, there are totally 70 images with available segmentation labels to train our network, while the other 30 images remained for validation. We make random five-fold cross validation by randomly shuffling the sequence of cases and splitting the training dataset into 5 fixed folds with 14 images in each fold, using 4 folds for training and the other one for cross-validation. Based on the cross validation, we can make a more comprehensive evaluation of our method among different data distributions.

3.4 Implementation Details

All the training procedure of our network is performed using an NVIDIA A100 GPU. During the training phase, we utilized a patch size of $20 \times 320 \times 256$ for training and inference, which allowed for the efficient extraction of relevant features from the images. We use the summation of cross-entropy (CE) loss and Dice loss as the loss function for the training of our network as follows:

$$L_{CE} = -\frac{1}{N}\sum_{i=1}^{N}\sum_{k=1}^{K} y_{ik}\log(\hat{y}_{ik})$$

$$L_{Dice} = 1 - \frac{2}{K}\sum_{k=1}^{K}\frac{\sum_{i=1}^{N} y_{ik}\hat{y}_{ik}}{\sum_{i=1}^{N} y_{ik}^2 + \sum_{i=1}^{N} \hat{y}_{ik}^2}$$

where y_{ik} and $\hat{y}_{ik}$ indicate whether sample i belongs to class k in the ground truth y or segmentation output $\hat{y}$, while N and K indicate the number of samples and classes in the task.

Table 1. The quantitative five-fold cross-validation results of model development procedure of different segmentation methods on the MBAS training set.

Method	Targets	Metrics	fold 0	fold 1	fold 2	fold 3	fold 4
CNN-based (3D nnUNet)	Atrium Wall	DSC	0.7208	0.7128	0.7133	0.7195	0.6962
		HD95	2.57	2.75	2.53	2.46	2.84
	Right Atrium	DSC	0.9210	0.9060	0.9251	0.9166	0.9176
		HD95	2.90	13.85	2.69	16.57	3.15
	Left Atrium	DSC	0.9287	0.9294	0.9278	0.9375	0.9131
		HD95	2.91	2.93	3.11	13.61	4.65
Transformer-based (UNETR)	Atrium Wall	DSC	0.5962	0.5952	0.5763	0.5769	0.5841
		HD95	7.33	6.77	7.05	11.50	23.52
	Right Atrium	DSC	0.7716	0.7929	0.7565	0.7059	0.7643
		HD95	24.77	7.51	14.01	28.75	11.70
	Left Atrium	DSC	0.8963	0.8745	0.8570	0.8921	0.8581
		HD95	5.86	20.29	7.30	17.26	23.04

4 Experiments

4.1 Cross-Validation Results

As introduced, we conduct five-fold cross validation instead of directly use all the training data to achieve more robust generalization ability on unseen data. From the cross-validation results presented in Table 1, we can observe that the CNN-based models demonstrate consistent performance in segmenting the right and left atrium, with DSC scores consistently above 90% across all folds. In contrast, transformer-based models exhibite lower DSC scores for all targets, indicating a less accurate segmentation compared to CNN-based models. The right atrium shows the most significant performance drop, with DSC scores ranging from 70.59% to 79.29%.

Compared to DSC, the HD95 values shows some variability across different folds, indicating that while the segmentation models perform well on average, there are cases where the segmentation accuracy significantly deviates from the ideal. This variability could be due to the inherent complexity of the cardiac structures and the presence of noise in the MR images.

As observed in Fig. 2, we visualize some segmentation results of different methods with comparison to the ground truth in the validation set of our cross-validation split. We can observe that transformer-based method fails to segment atrium wall and produces extremely irregular boundaries. One potential reason for the suboptimal performance of transformer-based methods lies in their foundational approach, which involves partitioning the 3D input volume into a sequence of uniform, non-overlapping small patches. These patches are subsequently processed as input sequences for the transformer model, which might lead to mis-segmentation in regions that exhibit features akin to those of the target cardiac structures (see the second column).

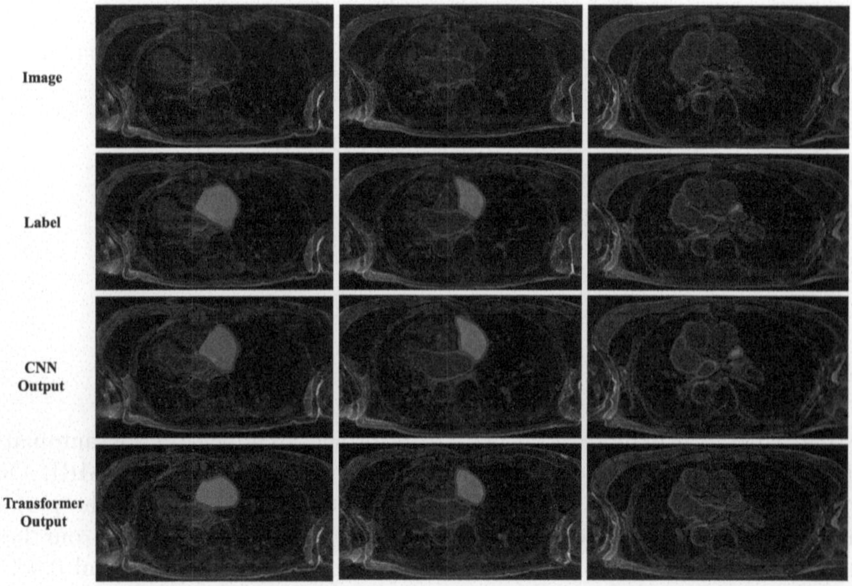

Fig. 2. Visualization of segmentation results of CNN-based and transformer-based methods with comparison to the original images and labels. The atrium wall, left atrium cavity and right atrium cavity are shown in red, blue and green, respectively. (Color figure online)

4.2 Official Validation Results

After the model development procedure, to validate the effectiveness of our segmentation method, we submit our mask predictions of the official 30 validation images for the competition. The inference details are shown as follows.

Validation-1. In our first submission, we use the five CNN-based nnUNet models from the five-fold cross-validation as an ensemble to predict the validation cases. Predictions were consolidated by averaging softmax outputs.

Validation-2. In our second submission, five CNN-based nnUNet models and five transformer-based UNETR models are used as an ensemble to predict the validation cases, where we aim to ensemble these two types of frameworks together to exploiting local and long-range dependency.

The official validation performance is shown in Table 2. In line with the cross-validation results, CNN-based segmentation models deliver better performance compared with incorporating transformer-based models (i.e. the ensemble of both). As a result, we select out the CNN-based segmentation models as our solution for the final testing.

Table 2. The evaluation results of our method on the MBAS validation set.

Targets	Atrium Wall		Right Atrium		Left Atrium	
Metrics	DSC	HD95	DSC	HD95	DSC	HD95
Validation-1	0.6640	4.53	0.8806	7.68	0.9099	4.91
Validation-2	0.6529	4.70	0.8764	7.69	0.9092	5.05

5 Conclusion

In this paper, we explore CNN and transformer architectures for automatic multi-class bi-atrial segmentation from late gadolinium-enhanced MRI. Our experiments find out that transformer-based models fail to match the performance of CNN-based models in the task. For the validation stage, our best performance has achieved an average Dice score of 0.6640, 0.8806 and 0.9099, an average HD95 of 4.5258, 7.6845 and 4.9136 for atrium wall, left atrium cavity and right atrium cavity, respectively.

Acknowledgement. The authors of this paper declare that the segmentation method implemented for the challenge is fully automatic without any manual intervention, and has not used any pre-trained models or additional datasets other than those provided by the organizers. The computations in this research were performed using the CFFF platform of Fudan University.

References

1. Antonelli, M., et al.: The medical segmentation decathlon. Nat. Commun. **13**(1), 4128 (2022)
2. Axel, L.: Modeling of factors affecting late gadolinium enhancement kinetics in MRI of cardiac amyloid. J. Cardiovasc. Magn. Reson. **25**(1), 46 (2023)
3. Bernard, O., et al.: Deep learning techniques for automatic MRI cardiac multi-structures segmentation and diagnosis: is the problem solved? IEEE Trans. Med. Imaging **37**(11), 2514–2525 (2018)
4. Cao, H., et al.: Swin-unet: Unet-like pure transformer for medical image segmentation. In: European Conference on Computer Vision, pp. 205–218. Springer (2022)
5. Çiçek, Ö., Abdulkadir, A., Lienkamp, S.S., Brox, T., Ronneberger, O.: 3d u-net: learning dense volumetric segmentation from sparse annotation. In: International Conference on Medical Image Computing and Computer-Assisted Intervention, pp. 424–432. Springer (2016)

6. Hatamizadeh, A., et al.: Unetr: transformers for 3d medical image segmentation. In: Proceedings of the IEEE/CVF Winter Conference on Applications of Computer Vision, pp. 574–584 (2022)
7. Isensee, F., Jaeger, P.F., Kohl, S.A., Petersen, J., Maier-Hein, K.H.: nnU-Net: a self-configuring method for deep learning-based biomedical image segmentation. Nat. Methods **18**(2), 203–211 (2021)
8. Lalande, A., et al.: Deep learning methods for automatic evaluation of delayed enhancement-MRI. the results of the EMIDEC challenge. Med. Image Anal. **79**, 102428 (2022)
9. Li, L., Zimmer, V.A., Schnabel, J.A., Zhuang, X.: Medical image analysis on left atrial LGE MRI for atrial fibrillation studies: a review. Med. Image Anal. **77**, 102360 (2022)
10. Ma, J., et al.: Abdomenct-1k: is abdominal organ segmentation a solved problem? IEEE Trans. Pattern Anal. Mach. Intell. **44**(10), 6695–6714 (2022)
11. Ronneberger, O., Fischer, P., Brox, T.: U-net: convolutional networks for biomedical image segmentation. In: International Conference on Medical Image Computing and Computer-Assisted Intervention, pp. 234–241. Springer (2015)
12. Vaswani, A., et al.: Attention is all you need. In: Advances in Neural Information Processing Systems, vol. 30 (2017)
13. Xiong, Z., et al.: A global benchmark of algorithms for segmenting the left atrium from late gadolinium-enhanced cardiac magnetic resonance imaging. Med. Image Anal. **67**, 101832 (2021)

EigenBoundaries for Temporally Regularized Segmentation of Echocardiographic Images

Georgios Tziritas(✉)

Computer Science Department, University of Crete, Heraklion, Greece
tziritas@csd.uoc.gr

Abstract. Automatic segmentation of echocardiography videos is critical for assessing various cardiac functions and improving the diagnosis of cardiac diseases. Convolutional Neural Networks (CNNs) have recently demonstrated their ability to segment 2D cardiac ultrasound images, but they suffer from temporal inconsistency. The main approaches for more reliable spatio-temporal analysis include 3D deep learning methods, recurrent segmentation algorithms, and post-processing of 2D segmentations. This paper presents a new efficient method for temporally regularized segmentation of cardiac ultrasound images. To address this problem, we propose a post-processing procedure based on the description of left ventricular, endocardial and epicardial, boundaries by their central distance signatures, which emerges as a powerful representation with excellent temporal coherence. Principal component analysis of the boundary signature is shown to provide a concise model for its representation. In particular, the coefficient of the first component has an interesting functional interpretation. Temporal smoothing of the obtained parametric representation enforces the temporal consistency of the segmentation. Results are given on the TED dataset [11] to illustrate the regularized segmentation and to measure some anatomical cardiac features.

Keywords: echocardiography · segmentation · shape signature · principal component analysis · regularization · anatomical cardiac features

1 Introduction

Echocardiographic image analysis is an essential diagnostic tool for assessing cardiac function [8,13]. Correctly performing automated segmentation of the heart's regions of interest also allows measurement of important parameters, such as left ventricle (LV) ejection fraction, LV area, LV length and myocardial area. Deep learning in medical image analysis, including echocardiographic image segmentation, has been the focus of the scientific community in recent years, especially when incorporating temporal along with spatial information. An evaluation of

O. Camara et al. (Eds.): STACOM 2024, LNCS 15448, pp. 398–410, 2025.
https://doi.org/10.1007/978-3-031-87756-8_39

the clinical applicability of deep learning methods for dynamic cardiac imaging using spatiotemporal image information was presented in [3].

Following the exploration of 2D deep learning approaches, several methods have been developed in recent years to take into account the temporal aspect. In [7] different methods are compared and it is shown that a generic 3D nnUnet CNN for the entire sequence gives better results on a private data set. In [14], a co-learning mechanism is proposed to explore the mutual benefits of cardiac segmentation and myocardium tracking at the appearance and shape levels in a semi-supervised framework. A proxy- and kernel-based video segmentation method for left ventricle localization has also been proposed in a semi-supervised context [15]. In the context of self-supervised segmentation, multi-level semantic adaptation was proposed for few-shot segmentation on cardiac image sequences [2]. A post-processing framework for learning the 2D+time apical long-axis shape of the heart to improve the accuracy and consistency of segmented echocardiographic sequences has also been proposed [11].

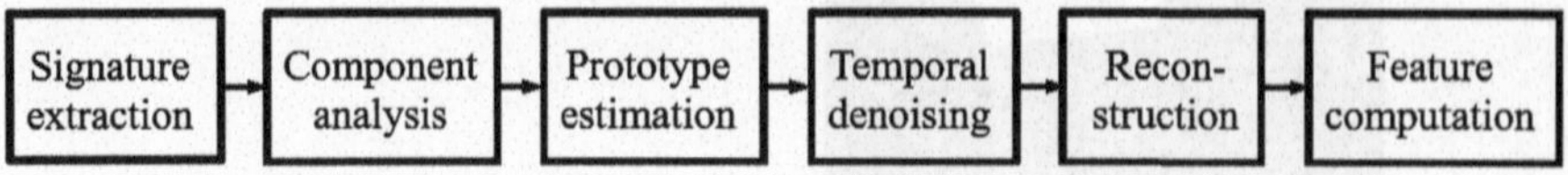

Fig. 1. Flow chart of the proposed post-processing method.

This article presents a new post-processing method for temporal regularization of cardiac ultrasound images. To address this issue, we propose to use the left ventricular boundaries and model their central distance signature using Principal Component Analysis (PCA), assuming that the ultrasound images are aligned. The flow chart of the proposed method is shown in Fig. 1. The PCA is based on the ground truth segmentation maps and is presented in Sect. 2. Since the first component is dominant, a prototype for its temporal evolution is estimated (Sect. 3). The temporal inconsistency appears clearly on the first component time sequence. Therefore, denoising the sequence of the first component results in a temporally consistent segmentation (Sect. 4). After reconstructing temporally smooth segmentation maps, anatomical features can be estimated (Sect. 5).

We evaluated our approach by testing it on the predictions of some deep CNN architectures [5,6,9,10,12], resulting in improved left ventricular endocardial and epicardial segmentation. Finally our method provides a simple and efficient approach to achieve an accurate and temporally consistent segmentation of ultrasound images of the heart. Results are shown on the TED dataset [11][1].

[1] https://humanheart-project.creatis.insa-lyon.fr/databases.html.

2 Representation Basis Computation

The shape of any region can be described by its boundary. A one-dimensional periodic signal can then be used to represent the resulting contour. Among the various approaches used to represent shapes, the central distance signature has interesting properties for nearly convex closed curves. In particular, it is covariant with respect to scale and rotation transformations and invariant with respect to translation. Therefore, the central distance signature is suitable for modeling the shape of the left ventricular, endocardial and epicardial boundaries, as the main transformations during the cardiac cycle are translation, rotation and scale variation.

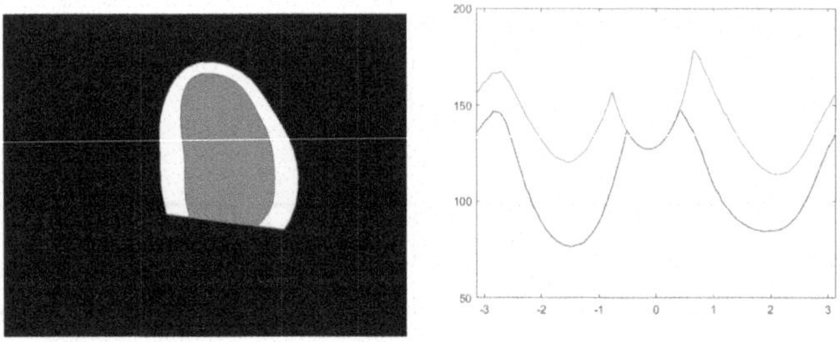

(a) (b)

Fig. 2. Endocardium and epicardium boundaries (a) and their signatures (b). (Color figure online)

Let $(x(\theta), y(\theta))$ be the boundary coordinates, with an appropriate location parameter θ. The central distance is defined as

$$\rho(\theta) = \sqrt{(x(\theta) - x_c)^2 + (y(\theta) - y_c)^2}, \tag{1}$$

where (x_c, y_c) is a central shape point. Since we are considering a representation basis for an entire set of boundaries, the position parameter θ is defined by the angular position, which is determined in exactly the same way over the entire set. Therefore, the entire set of signatures is computed at the same angular positions with a fixed angular interval step. Furthermore, for both the endocardial and epicardial boundaries of the left ventricle, the center is extracted from the epicardial boundary. Then, adaptive representations can be computed separately for endocardium and epicardium.

The basis of the representation is determined from the ground truth segmentation to be the most expressive and accurate. An example of a ground truth segmentation and corresponding signatures of endocardium ('red') and epicardium ('green') is shown in Fig. 2. The angular interval step is set to $1/2°$, so we have $K = 720$ angular positions, from $-\pi$ to π, to calculate the distance of the boundary points from the origin (x_c, y_c). Distance is given in pixels in Fig. 2(b), and the pixel size is $[0.308, 0.308]$.

Let Y be the $N \times K$ matrix containing the set of N known signatures from the ground truth segmentations. The basis of the representation is determined by the eigenvectors of the matrix $C = Y^tY$. We will call these eigenvectors the *eigenboundaries*. Since the number of significant eigenvalues is limited, the basis can be restricted to the eigenvectors corresponding to the most significant eigenvalues. Exactly, the cumulated sum of the first seven eigenvalues are

$$\begin{matrix} 0.988 & 0.992 & 0.996 & 0.998 & 0.998 & 0.999 & 0.999 \\ 0.996 & 0.998 & 0.999 & 0.999 & 1.000 & 1.000 & 1.000 \end{matrix}$$

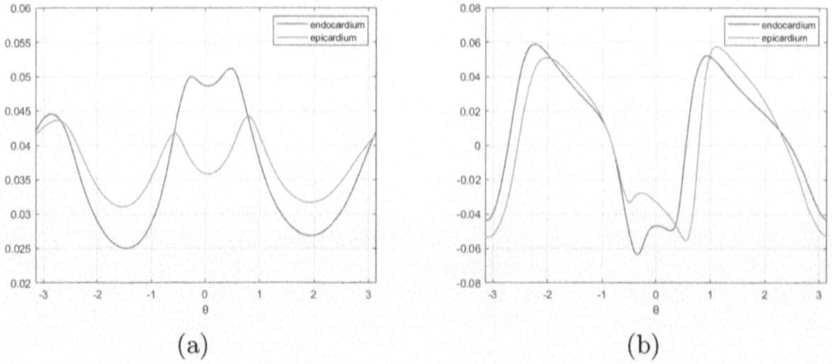

Fig. 3. The two most important eigenboundaries for the representation of the endocardium and epicardium boundaries. (Color figure online)

The first row shows the coefficients for the endocardium and the second row shows the coefficients for the epicardium. Note that only the first component explains approximately 0.99 of the total. Based on these observations, our experiments retain 36 components for endocardium and epicardium. This number is more than sufficient for the ground truth, and at the same time limiting the number of components to 5% results in noise reduction at the boundaries of real machine segmentations.

The two main eigenboundaries for the endocardium ('red') are shown in Fig. 3. Note that the two peaks near the angle origin in Fig. 3(a), where the first eigenboundary is shown, correspond to the two endpoints of the valve, while the peak near the angle $-\pi$ gives the approximate position of the apex. In the same figure are shown in 'green' the two first eigenboundaries for the epicardium, with the same observations regarding the characteristic points.

Comparing the typical LV signatures shown in Fig. 2 with the first eigenboundaries, it can be concluded that they are quite similar, the similarity being measured by the first eigenvalue mentioned above. On the ground truth dataset, the standardised signature is always close to the first eigenboundary.

Having the basis of the representation, we can represent each boundary with the corresponding set of expansion coefficients. Then, for each patient and for the corresponding boundary sequence, we can have the sequence of coefficients of the representation. Figure 4 shows, for two patients, the sequences of coefficients for the principal eigenboundary, both for the endocardium ('red') and for the epicardium ('green'), for one cardiac cycle. It can be seen that the shape of the cardiac function follows a systolic phase with a decreasing coefficient and a diastolic phase with an increasing coefficient. In fact, the minimum position of the sequence of coefficient values gives the approximate position of the end of the systolic phase. An exception is patient number 69, in whom the ejection fraction was extremely low, equal to 13, while the ejection fraction of patient 14 is equal to 75.

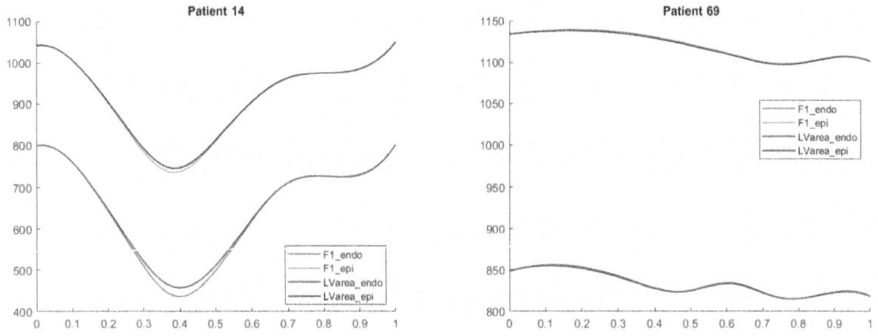

(a) (b)

Fig. 4. The coefficient of the most important eigenvector for the endocardium and the epicardium during one cardiac cycle versus the square root of the corresponding area. (Color figure online)

The same figure shows the corresponding sequences of the square root of the left ventricular area and the area defined by the epicardium. Note the extremely high correlation between the most important coefficient and the area of the respective regions for the ground truth. Indeed, the first coefficient is

$$\sum_{k=1}^{K} \rho(\theta_k) x(\theta_k) \approx \sqrt{\sum_{k=1}^{K} \rho^2(\theta_k)} \approx \sqrt{\frac{KA}{\pi}}, \tag{2}$$

where $\{x(\theta_k), 1 \leq k \leq K\}$ is the first eigenboundary and A is the area of the corresponding region. The distance between the epicardial coefficient and the endocardial coefficient is an approximation of the thickness of the myocardium.

3 Model Construction for the Eigenvector Coefficients

Since the most important eigenvector explains about 99% of the energy of the boundary representation, it is interesting to try to construct a model of the sequence of coefficients of this eigenvector, as this will help to make the segmentation sequence temporally consistent. To construct a generic model, the average coefficient sequence over the entire ground truth set is calculated and used to estimate an initial model for each individual sequence. The average sequence for endocardium is shown in Fig. 5(a). Corresponding average epicardial sequence is similar. We observe that the systolic phase lasts on average about 45% of the time of the cardiac cycle and the diastolic phase about 55%.

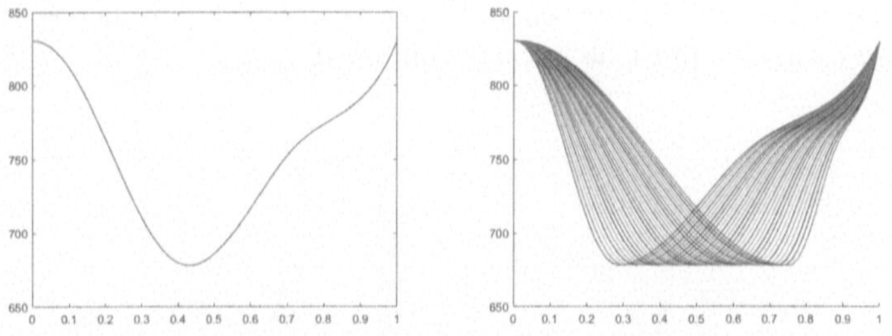

(a) (b)

Fig. 5. (a) Average coefficient of the first eigenboundary for the endocardium and (b) adaptation to the individual duration of the two phases of the cardiac cycle.

However, for individual patients the variation in the division of the total cycle time into systolic and diastolic phases is very large. In order to create a model suitable for each patient, both the position of the end of the systolic phase and the amplitude of the variation for the two phases should be adjusted. If we consider different time division positions for the two phases, we obtain appropriate models, as shown in Fig. 5(b) for the endocardium. The duration of the systolic phase ranges from 30% to 75% of the total cardiac cycle. Similar results are obtained for the epicardium.

For each individual sequence, it is necessary to adjust the separation of the two phases and the amplitude of the change from end diastole to end systole. Three main parameters determine the pattern of the individual sequence: the time division of the two cardiac phases, the mean size of the ventricular area, and the ratio of the area between the end of the diastolic phase and the end of the systolic phase. Therefore, a linear model can be assumed, given the division into systolic and diastolic phases

$$c(t) = b_1 + b_2 f_k(t), \quad (3)$$

where b_1 corresponds to the mean size of the ventricular area, b_2 corresponds to the total change in area, and k indicates the relative duration of the two phases in the heart cycle. $f_k(t)$ is the prototype model shown in Fig. 5(b).

The linear model for the sequence $c(t)$ should be estimated in a robust way, since the data are noisy and have abrupt changes. Robust iteratively reweighted least-squares regression is used with the Cauchy fitting weight function [4]. With respect to the time division of the two phases of the heart cycle, the search is exhaustive. The model is checked for all possible values of k and the total absolute error $e(k)$ is calculated over the sequence,

$$e(k) = \sum_t |c(t) - \hat{b}_1 - \hat{b}_2 f_k(t)|. \tag{4}$$

The time position k that minimizes the error $e(k)$ is chosen. In some cases, it is preferable to estimate the parameters b_1 and b_2 separately in the two phases. The calculation method remains the same, but the number of parameters is doubled and the two estimates must be linearly combined.

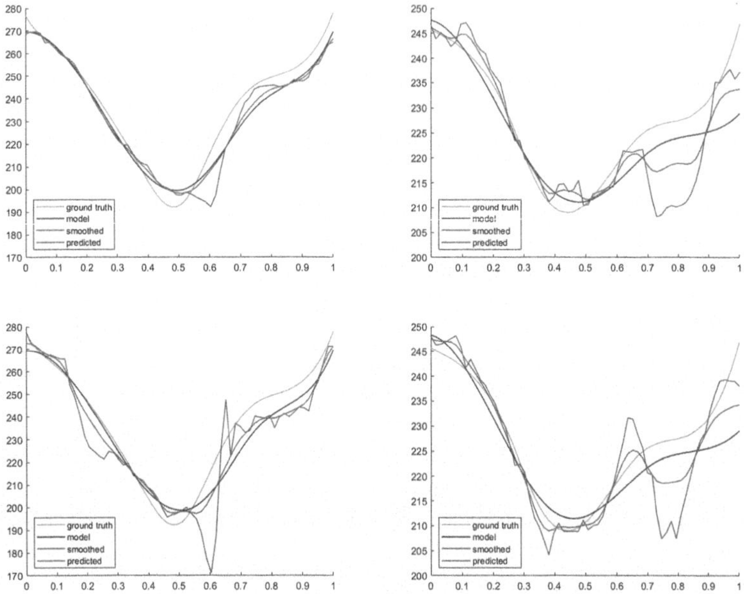

Fig. 6. Coefficient sequence for the segmentation ('red'), the ground truth ('green'), the model ('blue') and after smoothing ('magenta') for the endocardium of two patients. (Color figure online)

Results for the endocardium for two patients are shown in Fig. 6, where the GUDU segmentation method [12] was used for the plots in the first row, the ENet (resp. U-Net) predictor was used for the second row on the left (resp. right). The coefficient is given in mm. The sequence computed from the initial segmentation is shown in ('red'). For the patient on the left side, the segmentation was noisy, while on the right side there is noise and unexpected discontinuities. The prototype of the first coefficient could therefore be used for the detection of temporal inconsistencies in the segmentation results.

A consequence of the model analysis is the finding of a significant correlation between the parameter b_2 and the ejection fraction, specifically the correlation coefficient is 0.7 for the set of experimental data. The values of the estimated parameter $\hat{b}_2$, for the GUDU segmentation method, are shown in Fig. 7(a) for both the endocardium ('red' line) and the epicardium ('green' line). We see that it takes a maximum value for the patient with the maximum ejection fraction (75) and a minimum value for the sequence corresponding to the minimum ejection fraction (13). For the last sequence, where the parameter $\hat{b}_2$ takes a negative value, the above model is not used, but only a smoothed sequence is computed in the next algorithmic step, since it is clear that this sequence does not fit the generic model.

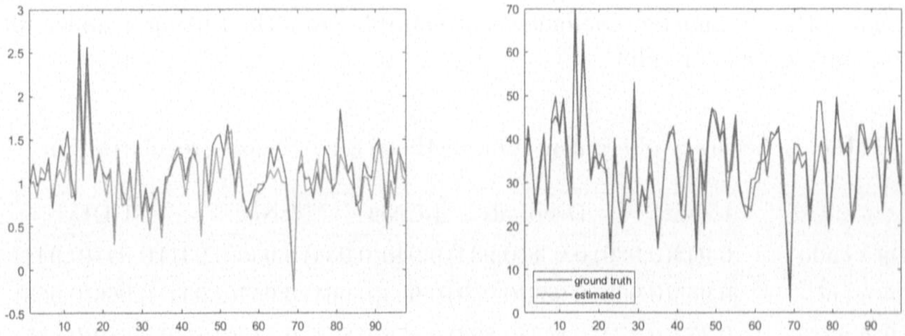

(a) (b)

Fig. 7. (a) Estimation of parameter b_2. (b) Ejection fraction after temporal smoothing. (Color figure online)

4 Temporal Regularization of the Segmented Images

The representation on the basis of eigenvectors allows to obtain temporally smoothed segments, since it is sufficient to smooth the values of the coefficients of the representation. Therefore, smoothing the latent representation yields the regularized segmentation.

Let $c(t)$ be the first coefficient data sequence and $s(t)$ the desired smoothed sequence. The following criterion is proposed to be minimized

$$\Omega(\mathbf{s}) = \|W(\mathbf{c} - \mathbf{s})\|^2 + \mu\|D\mathbf{s}\|^2, \tag{5}$$

where $\mathbf{c}$ is the vector of all values of the data sequence and $\mathbf{s}$ is the vector of the smoothed sequence. The matrix W is diagonal and gives weights for each time point to reduce the weight of outliers [1]. The matrix D models the second derivative to obtain a degree of smoothing μ and gives the filter d of the recurrence relation given below. The direct solution to the above optimization problem is

$$\widehat{\mathbf{s}} = (W'W + \mu D'D)^{-1}W'W\mathbf{c}. \tag{6}$$

It is proposed to implement the smoothing process by an iterative method for the coefficient corresponding to the maximum eigenvalue. The matrix W is re-estimated after each computation, while the initial estimate $f(t)$ is the model described in the previous section. The iterations, starting from $n = 1$, are formulated as follows, $*$ being the symbol of convolution,

$$s^{(n)}(t) = s^{(n-1)}(t) - \beta\left(\epsilon^{(n-1)}(t) + \mu(d * s^{(n-1)}(t))\right), \tag{7}$$

where $\epsilon^{(n-1)}(t) = w^{(n-1)}(t)(s^{(n-1)}(t) - c(t))$ and with $s^{(0)}(t) = f(t)$. The factor $w^{(n-1)}(t)$ is calculated iteratively in order to limit the consequences of the outliers. The other expansion coefficients are simply smoothed using a linear filter, without any *a priori* model.

Table 1. Accuracy of segmentation methods after (before) regularization.

	U-Net	DeepLab	LUNet	ENet	GUDU
Dice endo	0.943(0.939)	0.938(0.934)	0.940(0.934)	0.936(0.931)	0.947(0.944)
Dice epi	0.965(0.963)	0.962(0.959)	0.963(0.961)	0.957(0.954)	0.966(0.965)
Dice myo	0.877(0.870)	0.871(0.864)	0.876(0.871)	0.858(0.852)	0.889(0.885)
Hausdorff endo	3.87(4.13)	4.21(4.62)	4.07(4.38)	4.37(4.77)	3.70(3.97)
Hausdorff epi	3.85(4.16)	4.21(4.79)	4.06(4.58)	4.70(5.18)	3.73(3.90)
Hausdorff myo	4.32(4.82)	4.77(5.57)	4.65(5.32)	5.18(6.05)	4.20(4.61)
EF MAE	2.65(2.74)	2.65(2.85)	2.85(3.08)	2.87(3.23)	2.65(2.71)
EF correlation	0.943(0.934)	0.946(0.938)	0.937(0.932)	0.942(0.926)	0.939(0.940)

Five deep learning segmentation methods are evaluated using statistical and clinical indices on the TED dataset: U-Net [5], LUnet [6], DeepLab [9], ENet [10] and GUDU [12]. The results after regularization are given in Table 1 in comparison to the results before regularization, which are given in parentheses. The Hausdorff distance is measured in mm. In addition, Fig. 6 shows the smoothed coefficient sequence for two patients for qualitative evaluation, compared to the initial sequence, the prototype and the ground truth. Comparing the regularized result in the first row with that in the second row, we see that quite different segmentation results provided by different predictors yield very similar temporally consistent segmentations.

Although the original results showed significant differences between the algorithms, after smoothing the methods have similar performances. Among the indices considered, an improvement in the Hausdorff distance is obtained for all methods. The number of iterations in Eq. (7) was equal to 30 in almost all cases, with $\beta = 0.05$. A higher number of iterations results in a better Dice coefficient and Hausdorff distance, but the EF measurement is worse.

The 2D ejection fraction (EF) is included in the results as it can be calculated after constructing the smoothed segmentations given the end-systolic and diastolic positions. Mean Absolute Error (MAE) and correlation to ground truth are measured for 2D EF. A slight improvement is observed after temporal regularization. Figure 7(b) compares the GUDU segmentation calculated values ('red') with those of the ground truth ('blue').

5 Anatomical Feature Computation

Anatomical features can be calculated from the reconstructed boundary signatures and segmentation maps after smoothing. For comparison with the true values, it is also necessary to compute them from the ground truth segmentation. The measured anatomical features are those considered in [11]: left ventricular (LV) area, myocardial (MYO) area, long axis length, valve width, left ventricular orientation and epicardial (EPI) center position. For the latter, the global displacement due to cardiac function is of interest.

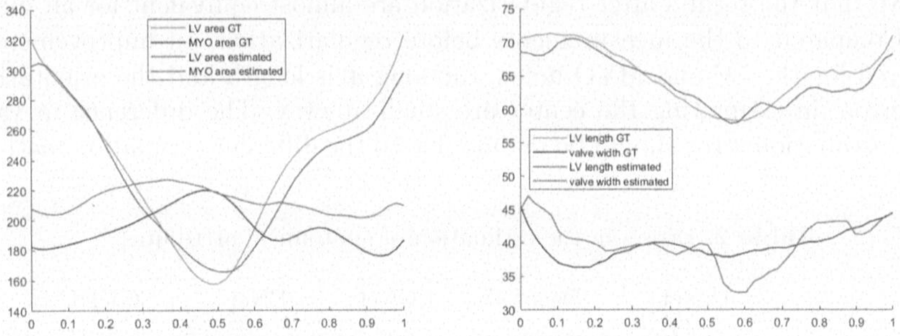

(a) (b)

Fig. 8. Anatomical results for one patient: (a) LV and MYO area. (b) LV length and valve width.

The two areas, the orientation and the centroid are computed in MATLAB® using the function *regionprops*. The entire region of the left ventricle, including the myocardium, is used to calculate the orientation and the centroid. The apex of the left ventricle is localized as the top point of the corresponding segment, since the left ventricle is oriented vertically, as shown in Fig. 2(a). Therefore, to locate the apex, the endocardial border is used. Long axis length is measured as the distance from the center of the base to the apex. The extreme points of the valve are determined directly from the segmentation map for the ground truth and original segmentations.

For the regularized segmentations the extremities of the base are determined by the two vertices of the signature located around the angle origin, as shown in Fig. 2(b) for a ground truth segment, to measure the valve width. The fact that the boundaries of the endocardial and epicardial regions coincide at the valve position is also taken into account for the predicted segment. Therefore, the valve is localized differently for the original segmentation and the reconstructed one after smoothing. The comparison of the corresponding measures may be slightly affected.

Figure 8 shows the results for the GUDU segmentation method for one patient. In (a), the estimated LV and MYO areas are plotted in square mm compared to the ground truth. In (b), the estimated LV length and valve width in mm are shown, also compared to the ground truth.

Table 2 shows the average error in the calculation of anatomical features for the five segmentation algorithms. The absolute mean error is computed for orientation, in degrees. The relative error is measured for all other features

$$\frac{\sum_t |\phi_s(t) - \phi_g(t)|}{\sum_t \phi_g(t)}, \tag{8}$$

where $\phi_s(t)$ is the segmentation-estimated feature value after smoothing and $\phi_g(t)$ is the feature value as computed from the ground truth segments. We observe that the results after regularization are almost equivalent for all methods. Compared to the measurements before regularization, an improvement is observed for the LV and MYO areas, the long axis length and the orientation. The errors in estimating the center are small anyway. The difference in valve width estimation error should be mainly due to the different estimation method.

Table 2. Errors in the estimation of anatomical attributes.

	U-Net	DeepLab	LUNet	ENet	GUDU
LV area	0.051(0.057)	0.060(0.071)	0.067(0.078)	0.056(0.065)	0.054(0.059)
Valve width	0.075(0.070)	0.085(0.083)	0.083(0.081)	0.081(0.083)	0.072(0.063)
Long axis length	0.025(0.028)	0.027(0.032)	0.024(0.027)	0.030(0.035)	0.026(0.029)
Orientation	2.443(2.571)	2.647(2.800)	2.252(2.392)	2.366(2.583)	2.218(2.320)
MYO area	0.057(0.064)	0.077(0.082)	0.100(0.106)	0.091(0.098)	0.086(0.086)
EPI hor center	0.009(0.009)	0.010(0.010)	0.010(0.010)	0.012(0.012)	0.008(0.008)
EPI ver center	0.015(0.014)	0.018(0.018)	0.018(0.018)	0.018(0.018)	0.014(0.014)

6 Discussion and Conclusion

All the algorithmic modules are implemented in MATLAB® and tested on an HP Pavilion laptop (Intel® i7 CPU at 2.60 GHz). The computation time per patient was about 14 s. The first modules of signature computation, coefficient extraction and prototype fitting, using *robustfit* function, take about 10.5 s due to the exhaustive search for the model. The last modules of regularization, reconstruction, feature estimation and evaluation take about 3.5 s.

We have introduced a generic model for representing the left ventricular boundaries, endocardium and epicardium, in long-axis apical echocardiographic images. Based on principal component analysis of the boundary signatures, we have shown that a concise set of parameters describes the boundaries. In particular, the coefficient of the first eigenboundary has strong correlation with the left ventricular area and its temporal variation is related to the cardiac cycle. We proposed an iterative least squares method to smooth this coefficient over time. The reconstructed segmentation maps are then nearly time consistent and are used to estimate anatomical features. The main improvements obtained on five different segmentation results on the TED dataset are in a global temporal consistency and in the measurement of the Hausdorff distance and the anatomical features.

Acknowledgments. The author would like to thank N. Painchaud *et al.* [11] for providing the predictions of their U-Net, DeepLab, LUNet and ENet methods on the full cycle ultrasound sequences to evaluate the proposed approach, and C. Sfakianakis and G. Simantiris for the predictions of the GUDU method. The segmentations used in this work are the result of the method proposed in [12] with a richer set of data augmentations.

References

1. Barron, J.: A general and adaptive robust loss function. In: IEEE/CVF Conference on Computer Vision and Pattern Recognition (2019)
2. Guo, S., et al.: Multi-level semantic adaptation for few-shot segmentation on cardiac image sequences. Med. Image Anal. **73**, 102170 (2021)
3. Hernandez, K., Rienmüller, T., Baumgartner, D., Baumgartner, C.: Deep learning in spatiotemporal cardiac imaging: a review of methodologies and clinical usability. Comput. Biol. Med. **130**, 104200 (2021)
4. Holland, P.W., Welsch, R.E.: Robust regression using iteratively reweighted least-squares. Commun. Stat.-Theory Methods **6**(9), 813–827 (1977)
5. Leclerc, S., Smistad, E., Pedrosa, J., et al.: Deep learning for segmentation using an open large-scale dataset in 2D echocardiography. IEEE Trans. Med. Imaging **38**, 2198–2210 (2019)
6. Leclerc, S., Smistad, E., et al.: LU-Net: a multistage attention network to improve the robustness of segmentation of left ventricular structures in 2-D echocardiography. IEEE Trans. Ultrason. Ferroelectr. Freq. Control **67**, 2519–2530 (2020)
7. Ling, H.J., et al.: Extraction of volumetric indices from echocardiography: which deep learning solution for clinical use? In: Functional Imaging and Modeling of the Heart (2023)
8. McDonagh, T.A., et al.: 2021 ESC guidelines for the diagnosis and treatment of acute and chronic heart failure. Eur. Heart J. **42**, 3599–3726 (2021)
9. Ouyang, D., He, B., Ghorbani, A., et al.: Video-based AI for beat-to-beat assessment of cardiac function. Nature **580**, 252–256 (2020)
10. Painchaud, N., Skandarani, Y., Judge, T., et al.: Cardiac segmentation with strong anatomical guarantees. IEEE Trans. Med. Imaging **39**, 3703–3713 (2020)

11. Painchaud, N., Duchateau, N., Bernard, O., Jodoin, P.-M.: Echocardiography segmentation with enforced temporal consistency. IEEE Trans. Med. Imaging **41**, 2867–2878 (2022)
12. Sfakianakis, C., Simantiris, G., Tziritas, G.: GUDU: geometrically-constrained ultrasound Data augmentation in U-Net for echocardiography semantic segmentation. Biomed. Signal Process. Control **82**, 104557 (2023)
13. Smiseth, O., Torp, H., Opdahl, A., Haugaa, K., Urheim, S.: Myocardial strain imaging: How useful is it in clinical decision making. Eur. Heart J. **37**, 1196–1207 (2016)
14. Wei, H., et al.: Co-learning of appearance and shape for precise ejection fraction estimation from echocardiographic sequences. Med. Image Anal. **84**, 102686 (2023)
15. Wu, H., Lin, J., Xie, W., Qin, J.: Super-efficient echocardiography video segmentation via proxy- and kernel-based semi-supervised learning. In: 37th AAAI Conference on Artificial Intelligence (2023)

Dynamic Cardiac MRI Reconstruction via Separate Optimization of K-Space and Hybrid-Domian Spatial-Temporal Feature Fusion

Xinrui Guo(✉), Liwen Li, and Fumin Guo

Wuhan National Laboratory for Optoelectronics, Huazhong University of Science and Technology, Wuhan, China
xguo5966@gmail.com

Abstract. Cardiac magnetic resonance imaging (CMRI), with excellent soft-tissue contrast and non-invasive nature, is a pivotal clinical imaging technique for the diagnosis and treatment of heart diseases. Deep learning reconstruction methods developed to accelerate CMRI are typically trained and deployed for specific scenarios, which struggles to adapt to the diverse range of clinical situations encountered in practice. These challenges arise due to the variability in multiple contrasts, imaging views, vendors, undersampling patterns, and acceleration rates. To address these issues, we developed an unrolled framework that applies separated optimization on center and periphery of k-space, i.e., the low-frequency and high-frequency components of image, to achieve robust reconstruction of diverse undersampled cardiac data. The proposed approach leverages the information of adjacent frames to complement the k-space center, and uses the designed spatio-temporal reconstruction blocks to mine the prior information about the image details contained in the periphery k-space. Overall image optimization is also incorporated to further improve the reconstruction quality.

Keywords: Dynamic reconstruction · Separate optimization · Unrolled network · Spatial-temporal

1 Introduction

Cardiac magnetic resonance imaging (CMRI) stands as the most precise and dependable non-invasive method for accessing the heart's anatomy, function, and pathology, which has high medical value in the clinical diagnosis and treatment of various cardiovascular diseases (e.g., coronary artery disease, heart failure, atrial fibrillation, etc.) as well as postoperative rehabilitation [8,19]. However, a notable limitation of MRI is its slow imaging speed, which causes patient discomfort and introduces motion artifacts into the images [10]. To avoid the adverse effects of physiological movements such as breathing and heartbeat as

O. Camara et al. (Eds.): STACOM 2024, LNCS 15448, pp. 411–420, 2025.
https://doi.org/10.1007/978-3-031-87756-8_40

much as possible, data acquisition was confined to shorter acquisition windows (e.g., end-diastole of the heart), leading to a high undersampling of k-space data, which increases the difficulty of reconstruction. As we know, CMRI can provide the multi-contrast information, including different modalities, such as cine, mapping, tagging, phase-contrast, and dark-blood imaging. It also involves imaging from various anatomical perspectives, such as long-axis (2-chamber, 3-chamber, and 4-chamber), short-axis, outflow tract, and aortic (cross-sectional and sagittal views) [14]. Furthermore, undersampling patterns for accelerated imaging are diverse, including uniform undersampling and variable-density sampling. This diversity significantly complicates the reconstruction of cardiac MR images.

For reconstruction of undersampled image, the previous approaches mainly include algorithms based on parallel imaging and compressed sensing [2,4,11]. While these classical algorithms are theoretically justified and offer excellent explainability, they are limited by their sensitivity to hyperparameter selection, relatively low feasible acceleration rates, and long reconstruction time for iterative calculations [17]. As an alternative, deep learning (DL) techniques can be used to reconstruct MR images from highly undersampled k-space data within a clinically acceptable reconstruction timeframe [15,16]. Among these, the unrolled network is a prominent and effective reconstruction architecture. It integrates imaging models into the network and learns image priors from extensive data, providing good interpretability [5]. Unfortunately, most conventional DL-based reconstruction models necessitate training and deployment for each specific imaging scenario (e.g., imaging sequence, view, and device vendor), thereby limiting their clinical applicability in real-world settings. Consequently, there is a pressing need to explore effective and generalizable deep learning methods.

In this paper, we proposed an approach based on the advanced unrolled structure, which applied separated optimization on low-frequency and high-frequency components of image according to thier different characterizations. The method learns and combines diferent information from k-space, image domain, and temporal series to enhance the robustness of the network to diverse data.

2 Methods

2.1 Unrolled Reconstruction via Separate Optimization

Traditionally, undersampled MR image acquisition can be modeled as:

$$y = Ax + \varepsilon, \tag{1}$$

where x denotes the 2D images to be reconstructed, y represents the acquired undersampled k-space data, ε is the complex noise in data acquisition. A is an encoding operator that transforms the reconstructed image x to k-space. For multi-channel Cartesian k-space data, the operator A and its adjoint operator A^T can be written as:

$$A = M \circ F \circ \mathcal{E}, \tag{2}$$

$$A^T = \mathcal{R} \circ F^{-1} \circ M, \tag{3}$$

where F represents the Fast Fourier Transform (FFT), and F^{-1} represents the inverse FFT (iFFT). M is the undersampling mask, containing the locations of sampling points. $\mathcal{E}$ and $\mathcal{R}$ denote coil expansion and reduction, respectively.

Prior information about the image x, such as the wavelet sparsity, finite difference sparsity, or low-rank constraints [7], can be incorporated to regularize the solution to the ill-posed optimization problem in Eq. (1) as follows:

$$\hat{x} = \arg\min_{x} \left\{ \frac{1}{2} \|Ax - y\|_2^2 + \lambda R(x) \right\}. \tag{4}$$

Here, $||\cdot||_2$ represents the ℓ^2 norm, and λ denotes the relative weight of the regularization term $R(x)$. While the reconstruction problem in Eq. (4) can be solved numerically via iterative gradient descent, here we performed this procedure using an end-to-end unrolled network that models the regularization operation $R(x)$ using convolutional neural network (CNN).

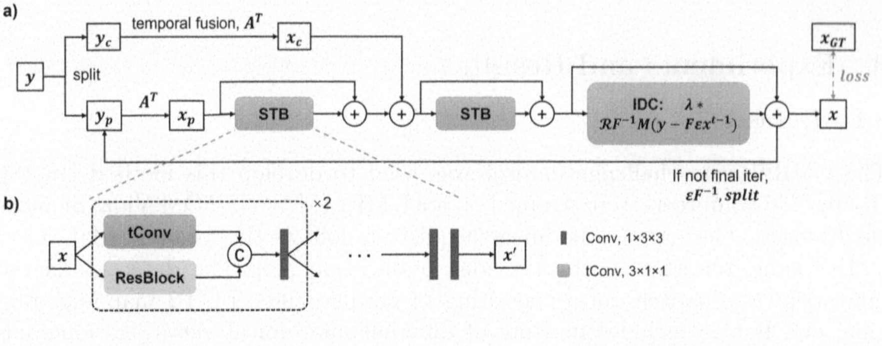

Fig. 1. (a) Schematic of the unrolled reconstruction process. (b) Detailed structure of proposed spatial-temporal block (STB) in (a).

The overall framework of our proposed approach is shown in Fig. 1 (a). We performed the reconstruction for multi-coil 2D image sequence of size (N_c, N_t, N_x, N_y). Unlike conventional unrolled models, we partition the undersampled k-space data y into the center y_c and the periphery y_p, and optimize them separately. This is due to the fact that the similarity in image contrast and the differences in details were found in image sequence obtained from dynamic imaging of moving organs (e.g., the heart), and general CNN may not perform well at high-frequency information [1]. So we think that low-frequency image and high-frequency image need different optimization directions. In the acquired k-space data, the center (i.e., low-frequency components) primarily represents image contrast, while the periphery (i.e., high-frequency components) encodes detail information. Therefore, for a given center size (which was selected by expanding the autocalibration region appropriately based on the aspect ratio of the image, we take as [64, 48] in this work), we obtain a near-fully-sampled k-space center for reconstructing the overall contrast of images by solving for the maximum projection along the temporal dimension. For the peripheral k-space data, we employ a spatial-temporal block (STB) to capture detailed information in image domain and along the temporal dimension. To avoid that the difference between temporal frames may be weakened due to fusion of k-space center, we refine the overall image again using the STB to ensure the feasibility of further optimization after separate optimization and combination. Finally, we incorporate data consistency in image domain (IDC) to ensure alignment between the reconstruction results and the original data. The learnable parameter λ was used to control the step size of IDC. x_c and x_p are images that correspond to y_c and y_p. Through $N = 2$ iterations of reconstruction, the final reconstructed image x was output and used to calculate training loss with the fully sampled ground truth (GT) image x_{GT}. Figure 1 (b) shows the details of STB, which contains parallel temporal convolution (tConv) and residual module [3] (ResBlock) for extracting spatial information in each subblock. The tConv module contains a convolutional layer with a convolutional kernel size of (3,1,1), whereas the convolutional layers in ResBlock all use a convolutional kernel of size (1,3,3).

3 Experiments and Results

3.1 Dataset

The CMRxRecon challenge dataset was used to develop this method [14,18]. Briefly, 330 volunteers were scanned on a 3T MRI (MAGNETOM Vida, Siemens Healthineers, Germany). The image acquisition follows the recommendations of CMR exams reported in the previous publication [13]. The dataset includes multi-contrast k-space data, consisting of cardiac cine, T1/T2 mapping, tagging, etc. It also includes imaging of different anatomical views like long-axis (LAX, including 2-chamber, 3-chamber, and 4-chamber), short-axis (SAX), left ventricul aroutflow tract (LVOT), and aortic (Aorta, transversal and sagittal views). *TrueFISP* sequence was used for cine, phase-constrast, and tagging, and *FLASH* sequence was used for T1/T2 mapping. For T1/T2 mapping, signals are collected at the end of the diastole with ECG triggering. Typically, 5 15 slices are acquired for each contrast. The cardiac cycle is segmented into 12 25 phases with a temporal resolution of around 50 ms. Typical geometrical parameters include: spatial resolution $1.5 \times 1.5\,\text{mm}^2$, slice thickness 8.0 mm, and slice gap 4.0 mm. The different undersampling scenarios include different k-space trajectories: uniform, Guassian, and pseudo radial undersampling with temporal/parametric interleaving; and different acceleration factors: 4×, 8×, 12×, 16×, 20×, 24×, autocalibration signals (ACS) not included for calculations. The central 16 lines (N_y, in ktUniform and ktGaussian) or central 16*16 regions (N_x*N_y, in ktRadial) are always fully sampled to be used as ACS. The coil sensitivity maps for multi-coil data were estimated based on ACS using the Berkeley Advanced Reconstruction Toolbox [12].

To facilitate training, we did the following processing on the raw data: under-sampled k-space data was obtained by multipling the fully sampled k-space data with mask; the k-space data was padded along edges with 0 to $N_x = 528$, $N_y = 256$; along the time dimension N_t, the k-space data was padded using the first and last frame images to $N_t = 12$; then iFFT was applied to get the fully sampled and under-sampled image; the resulting images were normalized using the maximum value of the under-sampled image; finally, the FFT was applied to obtain the corresponding normalized k-space data. The 3D volume data sequence was sliced into 2D data sequence. The reconstructed image is reduced to its original size in k-space by cropping the excess edges.

3.2 Experiments

Implementation Details. The loss for training the model is the combination of l_1 loss and SSIM loss between reconstructed image and GT. The weights of the l_1 loss and the SSIM loss were set as 10 and 1, respectively.

All DL models were implemented using the PyTorch deep learning library. Due to limitations of time and memory, we combine training sets of Aorta, Cine, Mapping and Tagging and randomly chose 0.2% volume data (269 2D sequences in total) from it for training our models. The resulting data was split into internal training set, internal validation set and internal test set in a ratio of 8:1:1. To show quantitative and qualitative results, we then randomly selected another 0.2% of the remaining data that included all modalities, sampling patterns, and acceleration factors for testing. The Adam optimizer was used with default parameters and a learning rate of 10^{-3}, which was adjusted according to the $CosAWR$ scheduler. Batch size was set as one. A single model for use on all image contrasts and acceleration factors was trained for 50 epochs (All models can converge with it) on an NVIDIA Corporation Device 2684 GPU with 24 GB of memory, which took approximately 18 h.

Evaluation Methods. Algorithm reconstruction performance was evaluated using normalized mean square error (NMSE), peak signal-to-noise ratio (PSNR), and structural similarity index (SSIM) between the reconstructed volume image sequence and GT. Lower NMSE and higher PSNR and SSIM indicate greater image quality.

Ablation Studies. A series of ablation experiments were performed to test the efficiency of our model architecture and spatial-temporal block. We used both the classical Unrolled-Net and the proposed SO-Unrolled-Net to generate four models for ablation studies depending on whether or not the tConv module was added to the CNN. The experimental settings are consistent with the $ImplementationDetails$.

Comparative Methods. For the model evaluation, we compare our unrolled model based on separate optimization (SO-Unrolled-Net) with Zero-filling, U-Net [9], and an unrolled model (Unrolled-Net) [6] for reconstruction of 3D cardiac navigators. The Unrolled-Net performed an unrolled optimization on the overall image, using *ResBlock* as the regularization network.

3.3 Results

The reconstructed cine images of different methods are shown in Fig. 2. We chose one slice and one time frame image from the SAX view of one volume cine image to show their reconstruction in the plane and along the time dimension and error maps compared with GT. These qualitative comparisons show that our method can recover more finer and sharper details on small anatomical structures.

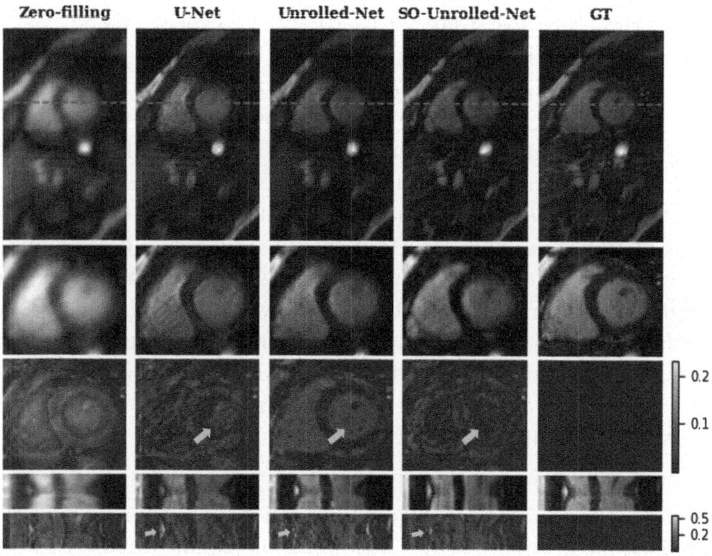

Fig. 2. Reconstruction and absolute error maps of different methods for SAX cine images under acceleration factor of 12×. The top three rows show that the reconstructed images and cropped cardiac regions. The bottom two rows show the reconstruction along the time dimension corresponding to the red dotted line in the images of the first row. Yellow arrows show the difference in anatomical recovery between different reconstruction methods. (Color figure online)

Table 1 shows the quantitative evaluation of different models on diverse cardiac data including Aorta, Cine, Mapping and Tagging. Notably, our proposed method outperforms other comparison methods on all quantitative metrics. We also show in Table 2 the results of quantitative evaluation of multiple contrast images respectively produced by different methods. Our proposed method

Table 1. Quantitative evaluation (NMSE, PSNR, SSIM, mean ± SD) of reconstruction performance of Zero-filling, U-Net, Unrolled-Net and SO-Unrolled-Net.

	NMSE ↓	PSNR (dB) ↑	SSIM ↑
Zero-filling	0.192 ± 0.093	29.661 ± 2.783	0.798 ± 0.061
U-Net	0.169 ± 0.094	30.012 ± 2.452	0.853 ± 0.049
Unrolled-Net	0.156 ± 0.070	30.311 ± 2.121	0.840 ± 0.045
SO-Unrolled-Net	**0.108 ± 0.074**	**32.769 ± 3.556**	**0.873 ± 0.055**

yielded the bset evaluation metrics for all contrast images, which proves the effectiveness and generalization of our approach. And the reconstructed images produced by different methods with multi-contrast are shown in Fig. 3, where we show the zoomed areas of Aorta (sagittal view), Cine (long-axis view), T2 Mapping and Tagging.

Table 2. Quantitative evaluation (average of NMSE, PSNR and SSIM) of reconstruction performance of Zero-filling (ZF), U-Net, Unrolled-Net (Unroll) and SO-Unrolled-Net (SO-Un) on Aorta (Ar), Cine (Cn), Mapping (Mp), Tagging (Tg), respectively.

	NMSE ↓				PSNR (dB) ↑				SSIM ↑			
	Ar	Cn	Mp	Tg	Ar	Cn	Mp	Tg	Ar	Cn	Mp	Tg
ZF	0.254	0.178	0.135	0.132	28.299	30.059	31.787	29.337	0.763	0.826	0.835	0.790
U-Net	0.191	0.156	0.162	0.112	29.503	30.296	30.775	29.914	0.828	0.873	0.867	0.872
Unroll	0.184	0.161	0.149	0.082	29.691	30.240	30.831	31.355	0.809	0.863	0.859	0.863
SO-Un	**0.150**	**0.101**	**0.083**	**0.041**	**31.013**	**32.943**	**34.552**	**34.515**	**0.840**	**0.888**	**0.888**	**0.920**

Table 3. Quantitative evaluation (NMSE, PSNR, SSIM, mean ± SD) of ablation experiments conducted on the Unrolled-Net and SO-Unrolled-Net.

	NMSE ↓	PSNR (dB) ↑	SSIM ↑
Unrolled-Net	0.156 ± 0.070	30.311 ± 2.121	0.840 ± 0.045
Unrolled-Net w tConv	0.129 ± 0.064	31.293 ± 2.241	0.858 ± 0.046
SO-Unrolled-Net w/o tConv	0.128 ± 0.076	31.781 ± 3.414	0.852 ± 0.053
SO-Unrolled-Net	**0.108 ± 0.074**	**32.769 ± 3.556**	**0.873 ± 0.055**

Table 3 demonstrates the results of the ablation studies. According to the results, the effectiveness of our proposed separated optimization unrolling framework and temporal convolution module is confirmed. Both the separate optimization strategy and the STB with temporal convolution contribute to improved reconstruction results. Our approach combined them, yielding the optimal reconstruction outcomes.

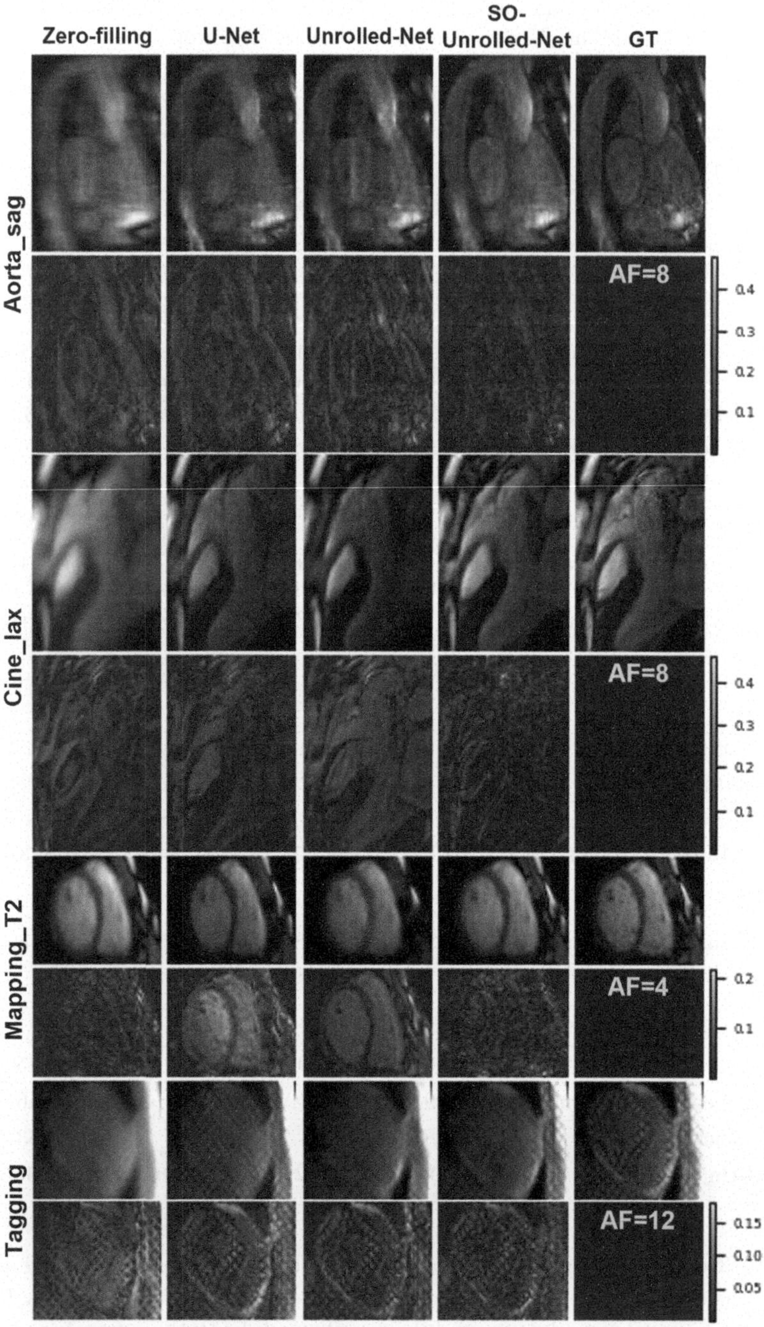

Fig. 3. Reconstruction and error maps of sagittal view of Aorta, LAX of Cine, T2 Mapping and Tagging, with the acceleration factors (AF) of 8, 8, 4, 12 respectively. The cropped cardiac regions were shown.

4 Discussion and Conclusion

In this paper, we introduced SO-Unrolled-Net, a separately optimized unrolling framework for faithful dynamic MRI reconstruction. By leveraging a strategy that optimizes the low-frequency and high-frequency components of the image separately, combining different priori information from k-space, image domain, and temporal series, SO-Unrolled-Net improves the quality and accuracy of dynamic MRI reconstruction, while enhancing robustness to diverse data. Experimental results demonstrated its effectiveness in achieving high-quality reconstructions for multi-contrast images and corresponding various views or weightings such as LAX, SAX, LVOT of Cine and T1, T2 Mapping.

Our method can still be improved. First of all, although our model has achieved good reconstruction results on small datasets, it is necessary to validate our model on larger datasets. Secondly, our framework is flexible and for separate reconstruction of low and high frequencies of image, more advanced modules can be selected for optimization. In addition, the proper selection of k-space center size is worthy of further study. Last but not least, unrolled models take a lot of time and memory to train and test, especially for time series and 3D images, which limits its further development. It's essential to develop a more concise and effective reconstruction framework to improve the reconstruction speed while maintaining sufficient reconstruction quality.

Acknowledgments. This study was funded by the National Natural Science Foundation of China. For their encouragement, support, and research assistance, we would like to thank the following individuals who have contributed to the completion of this work: Graham A. Wright, Calder D. Sheagren and Jaykumar H. Patel.

References

1. Fang, F., Hu, L., Liu, J., Yi, Q., Zeng, T., Zhang, G.: Hfgn: High-frequency residual feature guided network for fast mri reconstruction. Pattern Recogn. **156**, 110801 (2024)
2. Fischer, A., Seiberlich, N., Blaimer, M., Jakob, P., Breuer, F., Griswold, M.: A combination of nonconvex compressed sensing and grappa (cs-grappa). In: Proceedings of the 17th Scientific Meeting of ISMRM, Honolulu, p. 2813 (2009)
3. He, K., Zhang, X., Ren, S., Sun, J.: Deep residual learning for image recognition. In: Proceedings of the IEEE Conference on Computer Vision and Pattern Recognition, pp. 770–778 (2016)
4. Liang, D., Liu, B., Wang, J., Ying, L.: Accelerating sense using compressed sensing. Magn. Reson. Med. Off. J. Int. Soc. Magn. Reson. Med. **62**(6), 1574–1584 (2009)
5. Lyu, J., et al.: The state-of-the-art in cardiac mri reconstruction: results of the cmrxrecon challenge in miccai 2023. arXiv preprint arXiv:2404.01082 (2024)
6. Malavé, M.O., et al.: Reconstruction of undersampled 3d non-cartesian image-based navigators for coronary mra using an unrolled deep learning model. Magn. Reson. Med. **84**(2), 800–812 (2020)
7. Oscanoa, J.A., et al.: Deep learning-based reconstruction for cardiac mri: a review. Bioengineering **10**(3), 334 (2023)

8. Rajiah, P.S., François, C.J., Leiner, T.: Cardiac mri: state of the art. Radiology **307**(3), e223008 (2023)
9. Ronneberger, O., Fischer, P., Brox, T.: U-Net: convolutional networks for biomedical image segmentation. In: Navab, N., Hornegger, J., Wells, W.M., Frangi, A.F. (eds.) MICCAI 2015. LNCS, vol. 9351, pp. 234–241. Springer, Cham (2015). https://doi.org/10.1007/978-3-319-24574-4_28
10. Spieker, V., et al.: Deep learning for retrospective motion correction in mri: a comprehensive review. IEEE Trans. Med. Imaging (2023)
11. Uecker, M., et al.: Espirit-an eigenvalue approach to autocalibrating parallel mri: where sense meets grappa. Magn. Reson. Med. **71**(3), 990–1001 (2014)
12. Uecker, M., et al.: Berkeley advanced reconstruction toolbox. Proc. Intl. Soc. Mag. Reson. Med. **23** (2015)
13. Wang, C., et al.: Recommendation for cardiac magnetic resonance imaging-based phenotypic study: imaging part. Phenomics **1**, 151–170 (2021)
14. Wang, C., et al.: Cmrxrecon: a publicly available k-space dataset and benchmark to advance deep learning for cardiac mri. Sci. Data **11**(1), 687 (2024)
15. Wang, S., et al.: Accelerating magnetic resonance imaging via deep learning. In: 2016 IEEE 13th International Symposium on Biomedical Imaging (ISBI), pp. 514–517. IEEE (2016)
16. Wang, S., et al.: Knowledge-driven deep learning for fast mr imaging: undersampled mr image reconstruction from supervised to un-supervised learning. Magn. Reson. Med. **92**(2), 496–518 (2024)
17. Wang, S., Xiao, T., Liu, Q., Zheng, H.: Deep learning for fast mr imaging: a review for learning reconstruction from incomplete k-space data. Biomed. Signal Process. Control **68**, 102579 (2021)
18. Wang, Z., et al.: Cmrxrecon2024: a multi-modality, multi-view k-space dataset boosting universal machine learning for accelerated cardiac mri. arXiv preprint arXiv:2406.19043 (2024)
19. Xin, B., Ye, M., Axel, L., Metaxas, D.N.: Fill the k-space and refine the image: prompting for dynamic and multi-contrast mri reconstruction. In: International Workshop on Statistical Atlases and Computational Models of the Heart, pp. 261–273. Springer, Heidelberg (2023)

Can Interpretable Learning of Risk Explain Ventricular Arrhythmia Mechanism?

Buntheng Ly[1(✉)], Sonny Finsterbach[2], Marta Nuñez-Garcia[1], Hubert Cochet[1,2], and Maxime Sermesant[1,3]

[1] IHU Liryc, Université de Bordeaux, Bordeaux, France
{buntheng.ly,maxime.sermesant}@inria.fr
[2] CHU Bordeaux, Université de Bordeaux, Bordeaux, France
[3] Inria, Université Côte d'Azur, Sophia Antipolis, Biot, France

Abstract. Ventricular arrhythmia (VA) is the phenomenon most observed leading to the sudden cardiac arrest, which is one of the leading causes of cardiovascular deaths. Although the post-infarction scar on the myocardium was recognised as the main substrate of the arrhythmic event, there is, yet, to be a standard consensus for the characterisation of the arrhythmogenic scars. In this work, we wish to study whether the AI classification model trained on a imaging dataset was able to learn the underlying electrical mechanism of VA through the interpretability method and can be used to pinpoint the arrhythmogenic region. We proposed the edge-weighted graph convolutional network for VA classification, which based its prediction based on the LV wall thickness (WT) extracted from the CT imaging. The model was trained and evaluated on a large retrospective dataset of 600 CT images, and achieved the evaluation score of 83.2%, 80.4% and 84.2% of accuracy, sensitivity and specificity, respectively. Finally, we performed the evaluation of the model interpretability (LIME and integrated gradients) using the data from 10 additional patients for which the electro-anatomical mapping and inducibility VA data were available. Quantitative and visual assessment were done to compare the interpretability output with the reentry channel, manually segmented from the recorded activation map. The quantitative assessment based on the AHA 17-segment model showed the best interpretability method (LIME with quickshift segmentation) could achieve up to 84.5%, 90.0%, 78.9% of balanced accuracy, sensitivity and specificity, respectively. The visual assessment showed a strong correlation between the high coefficient regions of the interpretability outputs and the reentry channel among 7 out of 10 cases.

Keywords: Ventricular Arrhythmia · Interpretable Learning · Graph Convolutional Network · CT Imaging

O. Camara et al. (Eds.): STACOM 2024, LNCS 15448, pp. 421–432, 2025.
https://doi.org/10.1007/978-3-031-87756-8_41

1 Introduction

Ventricular Arrhythmia (VA) is an umbrella term, which refers to the rapid, cyclical and abnormal heart beat and usually covers both ventricular tachycardia (VT) and fibrillation (VF). VA is the leading cause of sudden cardiac arrest (SCA) and death (SCD), which accounted for more than 70% of the recorded SCD cases [13]. SCA and SCD are a major health concern worldwide, which accounted for 50% of the global cardiovascular deaths [25].

In this work, we focus on the post-infarct VA, where the myocardial scar, which resulted from infarction, plays an important role as the anatomical substrate in the reentry mechanism of the arrhythmia. In the reentry mechanism, the healthy myocardium is re-depolarised by the trigger from the action potential exited from the slow conduction zone of the scar region [11]. This creates a self-sustained action potential of the cardiac arrhythmia. The current gold standard for myocardial scar imaging is the late gadolinium enhanced (LGE) MR imaging [9]. However, CT imaging was also proposed, as an alternative, where the scar region could be observed as the wall thinning of the left ventricular (LV) [12]. Radio Frequency Ablation (RFA) could be done as a curative treatment to prevent future on set of VA. Nonetheless, the RFA remains a risky and error-prone procedure, as it requires precise and exhaustive detection of ablation targets.

As the artificial intelligence (AI) found its ways into the medical field, the necessity for an explainable AI (XAI) has become indispensable. While the term explainable and interpretable are usually used interchangeably under the scope of XAI, there is a globally accepted distinction between the two methods, as defined by [7]. The proposed distinctions of the two methods were based on the term "why-", attributed to the explainable method, and "how-", attributed to the interpretable method. While the explainability method, for instance the Grad-Cam [17], would output a saliency map highlighting class defining regions, the concept of coefficient ranking was not as explicitly integrated in the computation as the interpretability method, such as LIME [16]. On top of the explaining the model prediction, XAI methods have also been successfully used as a weakly supervised method for tumour detection [6], as well as, for extracting complex disease markers, for instance Alzheimer disease [1].

Leveraging on a large retrospective dataset, we trained a VA classification model based on the LV (WT) extracted from CT images. In our recent work [14], we used a graph convolutional network (GCN) [10] model for the VA classification task, where the model could be fitted directly on the 3D surface mesh of the LV midwall with the WT feature. The graph model was shown to outperform conventional image-based CNN classification model in classification accuracy. For each VA+ classification, the interpretability maps were computed using LIME and integrated gradients (IG).

In this work, we wish to investigate whether the AI classification model could learn the underlying arrhythmogenic mechanism, through the interpretability methods. Building on the graph model in the previous work, we proposed the edge-weighted GCN-Pooling model, where the edge lengths of the input mesh were used to compute the weights of the adjacency matrix, to explicitly integrate the geometry information in the model optimisation. The edge-aware graph models usually employ more complex methods for edge feature integration including trainable feature extraction model [18], attention mechanism [23] or geometry computation such as Laplacien [5] or diffusion [8]. Nonetheless, we employed a simple and direct method for edge feature integration, as we wish to build a model that emphasis on the influence of the WT heterogeneity. Finally, we evaluated the interpretability outputs on its ability to detect arrhythmogenic region by comparing the high coefficient regions with the VA reentry channel, manually annotated based on the electro-anatomical mappings recorded during the RFA procedures.

2 Method

The Fig. 1 summarises the complete pipeline from the CT image to the interpretability output.

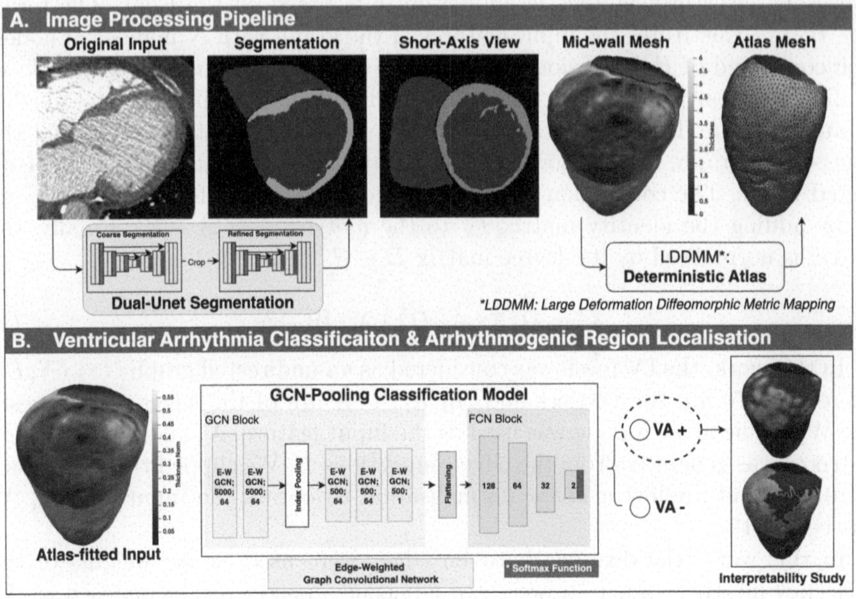

Fig. 1. Automatic pipeline for VA classification and arrhythmogenic region localisation from 3D CT imaging.

2.1 VA Classification Pipeline

Image Processing. First the ventricular masks were segmented from the CT image using a pre-trained Dual-Unet segmentation model [3]. Using the geometry relation between the two ventricles, automatic short-axis orientation was done based on the method proposed in [15]. This step ensured the uniform rotation of all inputs. On the SAX mask, we computed the LV WT and wall depth using the partial differential equation method [24]. The "midwall" surface mesh of the LV was generated at 50% wall depth using the marching cube algorithm and ACVD uniform remeshing [21]. Finally, to ensure the point-to-point correspondent between the meshes, we used the deterministic atlas method proposed by Deformetrica software [2], where a template mesh was registered to all available meshes, after which the WT values were interpolated on to the registered template mesh.

Edge-Weighted GCN-Pooling Model. As shown in Fig. 1.B, the architecture of the VA classification model is consisted of the GCN block and the fully-connected block. We built the model using the GCN layer as the backbone layer. The layer propagated the model outputs based on the Eq. 1. The terms $X \in R^{N \times F}$ constitutes the input features of the graph with N number of nodes, each comprised of F dimension features. The equation produces $X' \in R^{N \times F'}$ of F' dimension output features, according to the number of filters in the weights W and biases b. The term $A \in R^{N \times N}$ is the adjacency matrix, built from the edges of the graph, which allows the information to propagate through the connected nodes. The connection of each node to itself, the self-loop, is accounted for by adding the identity matrix I_N to the matrix, $\hat{A} = A + I_N$. Finally, the matrix is normalised by its degree matrix $D \in R^{N \times N}$.

$$X' = \hat{D}^{-\frac{1}{2}} \cdot \hat{A} \cdot \hat{D}^{\frac{1}{2}} \cdot X \cdot W + b \tag{1}$$

In this work, the LV mesh was considered as an undirected graph $G = (N, E)$, where $n_i \in N$ are the vertices and $(n_i, n_j) \in E$ are of the edges of the mesh. The WT value at each node was used as the input feature, X, and the adjacency matrix A was generated from the edge connections E. We clipped the WT values at 10 mm and applied a linear normalisation to obtain the input feature, X, between $[0, 1]$.

In this work, the distance between edges were used as the weights of the adjacency matrix to generate weighted adjacency matrix A_w, as opposed to the binary weights originally used. The adjacency weights, w_{ij}, for the neighbours $j \in \{1, ..., J\}$ of the node i was calculated as the reverse normalised of the distance to the closest neighbour, where d_{ij} is the euclidean distance between node i and j.

$$w_{ij} = \min_{j \in J}(d_{ij}) \times \frac{1}{d_{ij}} \tag{2}$$

The equation would assign a higher weight to the closer neighbour. From there, the identity matrix was added to obtain the self-loop adjacency matrix $\hat{A}_w =$

$A_w + I_N$, allowing the same weight at the self-loop as the closest neighbour. The $\hat{A}_w$ was then re-injected to the Eq. 1 to obtain the edge-weighted propagation of the output features in the Eq. 3.

$$X' = \hat{D}^{-\frac{1}{2}} \cdot \hat{A}_w \cdot \hat{D}^{\frac{1}{2}} \cdot X \cdot W + b \tag{3}$$

The index pooling layer was integrated to reduce the dimension of the propagated output features. First, a downsampled template was generated using ACVD uniform remeshing [21]. The pooling indices were then determined using the kd-tree nearest neighbour search method between the nodes of the downsampled and original mesh to determine the pooling neighbourhoods. From there, conventional pooling methods can used to pool the output features of the original to the downsampled graph. In this work, the average pooling method was used. During model forwarding at training and inference, we used a workaround, as shown in the Fig. 2 to avoids the computationally expensive uniform remeshing and preserve the point-to-point correspondent of the downsampled mesh for each input.

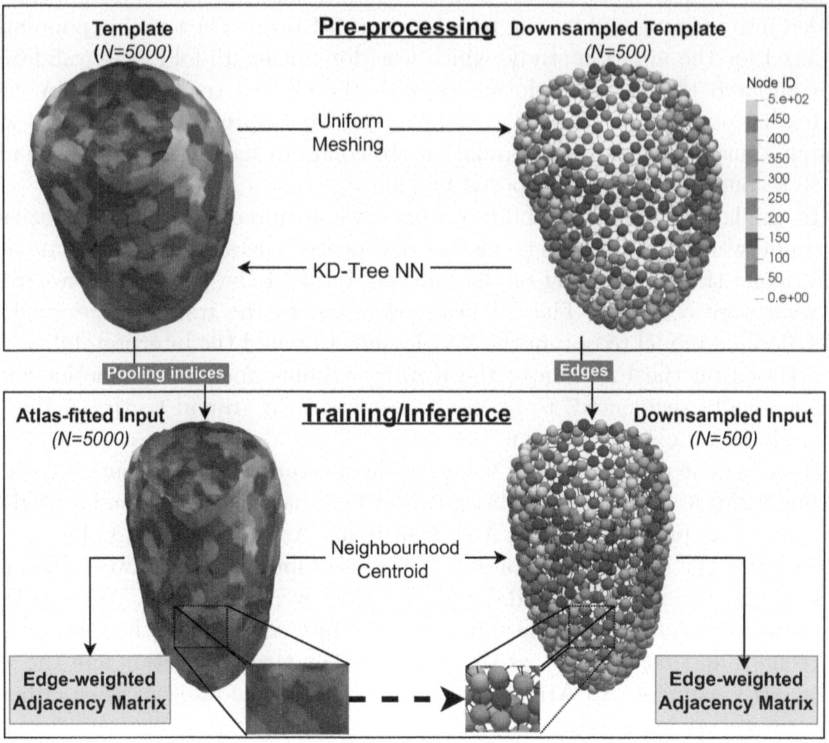

Fig. 2. Computation of the index pooling indices and the edge-weighted adjacency matrices.

2.2 Interpretability Study

We studied the interpretability of the classification model using two interpretability methods: local interpretable model-agnostic explanation (LIME) [16] and integrated gradients [19]. In this work, we substituted the baseline values of the two methods with the normalised healthy WT of **1**. For the LIME method, two methods of input segmentation were tested: quickshift [22] and grid segmentation. The quickshift segmentation, as proposed by the author of LIME, based the segmentation on the node position and feature (WT values), while the grid segmentation, which was done using kd-tree nearest neighbour, only based on the node position. For the IG method, the number of linear steps was set to 50.

3 Experimental Setup

In this study, we used a retrospective dataset collected at the university hospital of Bordeaux between 2010 and 2022. The dataset of 600 patients (27.5% VA+) included the CT image and clinical descriptions of patients with a history of myocardial infarction of at least one month before the scan date. The dataset was divided into training and testing population at 3:1 ratio. The training population was used for the ablation study, which was done using 10-fold cross-validation. We compared the model performance with the clinical correlates of VA using the testing population. The VA correlates and their optimal cut-off values were computed using the training population. To compute the cut-off values, we used the ROC analysis method proposed by [20].

To validate the interpretability outputs, we acquired data from 10 patients, who underwent RFA following a history of repeated sustained VT. The late activation time (LAT) mapping of the induced VT and the pre-operation cardiac CT scan were collected. The CT scans were fed to the trained edge-weighted GCN-Pooling model to obtain the VA classification and the interpretability outputs. Based on the LAT maps, the reentry isthmus and the conduction block were manually segmented, to be used as comparison ground truth, on the CT LV mesh, by an expert clinician.

To define the optimal cut-off values of the interpretability outputs, we set up the quantitative analysis as a classification task based on the standardised LV 17-segment, as proposed by the American Heart Association (AHA) [4].

To define the classification target, the areas of intersection between the manual segmented reentry mask and the AHA regions were calculated. We considered two ratios: *self-ratio*, as the ratio between the intersection and the total area of the reentry, and *region-ratio*, as the ratio between the intersection and the area of the AHA segment. An AHA segment was classified as positive when either of the ratio was at least 10%.

To define the prediction masks of the interpretability outputs, we used the top-k ($k \in [1, 10]$) threshold to define the predictions of the LIME methods and the p percentile ($p \in [90^{th}, 99^{th}]$) threshold to define the prediction of the IG method. The prediction masks from WT, thresholded at $[1, 5]$ mm were also used as the baseline comparison. The AHA predictions of each masks were obtained using the same intersection ratios as the classification target.

4 Results

In Table 1, the ablation study showed that the additional edge-weighted adjacency matrix improved the overall classification accuracy, as well as the sensitivity and specificity of the GCN-Pooling model.

Table 1. Cross-validation results ($mean \pm std$) on the training population.

Model	Acc.	Sens.	Spec.
Edge-Weighted GCN-Pooling	0.805 (±0.044)	0.732 (±0.052)	0.833 (±0.047)
GCN-Pooling	0.785 (±0.052)	0.713 (±0.049)	0.812 (±0.065)

The evaluation results, in Table 2, showed that the edge-weighted model outperformed the baseline models with an accuracy of 83.2% with 80.4% sensitivity and 84.2% specificity.

The results also showed that the current clinical gold standard (LVEF $<$ 35%) only achieved a sensitivity score of 53.7%, which further highlights the necessity for a more robust risk stratification method for VA.

Table 2. Evaluation results on the testing population.

Model	Acc.	Bal. Acc.	Sens.	Spec.
Edge-Weighted GCN-Pooling	0.832	0.823	0.804	0.842
GCN-Pooling	0.818	0.807	0.780	0.833
Scar Size ($> 65.49cm^2$)	0.725	0.712	0.683	0.741
Scar Age ($> 132mnts$)	0.698	0.739	0.839	0.640
LVEF ($<$ 35%)*	0.725	0.665	0.537	0.796
LVEF ($<$ 43%)	0.671	0.682	0.707	0.657

LVEF: left ventricular ejection fraction. (*): current clinical standard cut-off values of LVEF.

The edge-weighted GCN-Pooling model classified the 10 data for the interpretability evaluation as VA+. Table 3 shows the results of the best cut-off value for each method, according to the balanced accuracy ranking. The balanced accuracy ranking was used as the classification targets skewed toward the negative class. Due to the limited numbers of data, the proper validation of the cut-off values could not be done. The results, shown in the Table 3, were obtained from the 10 data used as training data. Overall, the interpretability methods performed considerably better than the WT baseline. The optimal cut-off for each interpretability methods were top-5, top-3 and 98th percentile for LIME quickshift, LIME grid and IG, respectively. The best performance method, based on the balanced accuracy, is LIME quickshift, with an average accuracy of 80% with 90% sensitivity and 78.9% specificity.

Table 3. Regional reentry channel classification performance ($mean$ ($\pm std$)) by LIME, IG and WT.

Method	Acc.	Bal. Acc.	Sens.	Spec.
QS: Top5	0.800 ± 0.075	0.845 ± 0.105	0.900 ± 0.200	0.789 ± 0.080
Grid: Top3	0.882 ± 0.037	0.829 ± 0.126	0.750 ± 0.261	0.908 ± 0.030
IG:98	0.853 ± 0.054	0.790 ± 0.135	0.700 ± 0.256	0.881 ± 0.040
WT:1	0.776 ± 0.073	0.693 ± 0.194	0.583 ± 0.352	0.802 ± 0.055

QS: LIME with quickshift segmentation; Grid: LIME with grid-base segmentation; IG: Integrated Gradients; WT: Wall thickness.

The visual comparison of the interpretability maps is shown in the Fig. 3. Overall, we observed a general agreement of the high coefficient regions among the three different interpretability maps. The visual assessment showed that the interpretability maps correctly highlighted the reentry regions in 7 out of the 10 cases.

5 Discussion

We showed that the addition of the neighbour distances in the adjacency matrix improved the classification accuracy. This highlights the importance of the geometry information in the model robustness. However, the current input did not contain any further anatomical details, such as basal, apical, anterior or posterior etc., which could provide additional information for the classification of VA.

Limited by the data availability, it is, furthermore, important to note that the scope of this study did not allow a conclusive evidence that the VA classification or the arrhythmogenic detection model will be generalisable on a larger scale clinical application. The current evaluation of the arrhythmogenic region localisation was also limited by the clinical understanding of the reentry channels. The "ground truth" used in this work was only based on the expert opinion according to the current understanding of this phenomenon. The AHA 17-segment model was selected to allow the regional comparison and for its anatomical standard in the clinical context. However, we acknowledge that the AHA segments are relatively large and that the grid-like division of the AHA model does not allow a flexible comparison of the reentry regions. Nonetheless, the correlation between the interpretability outputs and the reentry channels, as shown in Fig. 3, showed

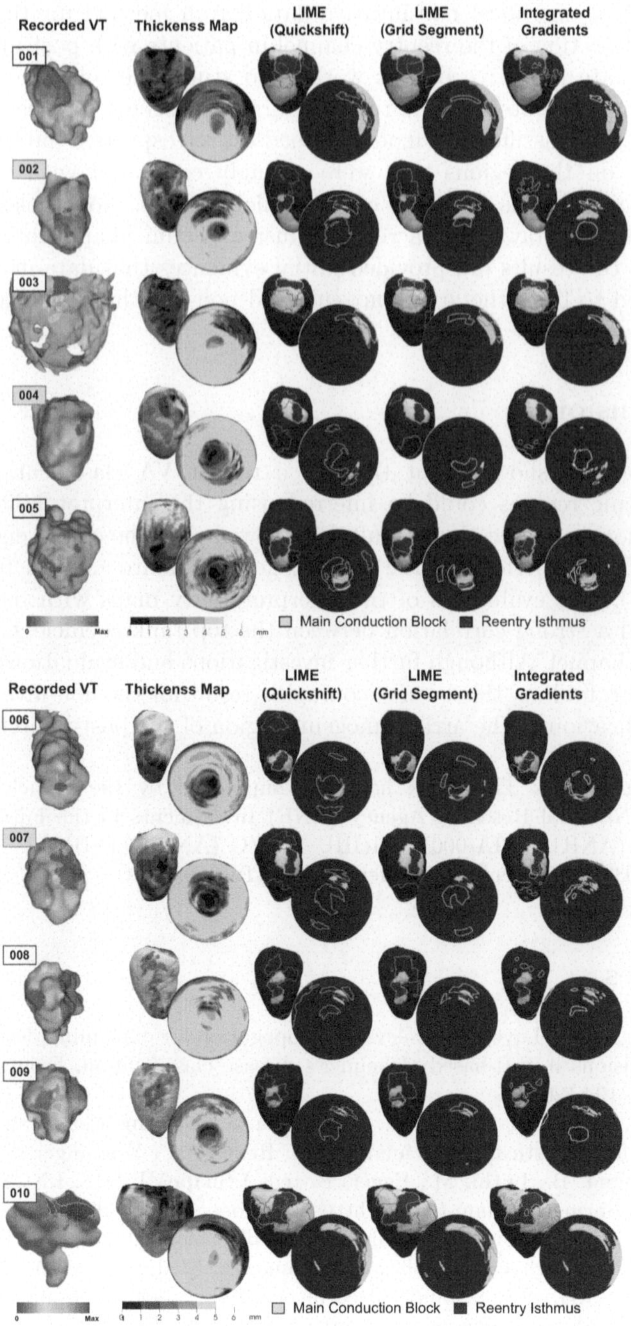

Fig. 3. Comparison between LAT map of the recorded VT and interpretability maps. *Red Box: fail detection of reentry regions based on the visual assessment.* (Color figure online)

the potential of our novel non-invasive data-driven method for the automatic and robust detection of the reentry channel in patients with post-infarct VA.

The ultimate objective of this work is to study the relation between the interpretability outputs and the recorded arrhythmogenic regions. The results showed that the classification model trained on retrospective dataset based its classification on the regions that were strongly correlated with the recorded arrhythmogenic regions in 7 out of 10 available cases. Nonetheless, the data was too limited to draw any concrete conclusion or clinical application proposal. Nonetheless, the results did provided initial evidences that data-driven method could be used to filter the arrythmogenic and non-arrythmogenic region of the myocardial scar on the LV.

6 Conclusion

In conclusion, we showed that through a robust VA classification network, arrhythmogenic regions could be inferred using the interpretability methods. The evaluation on the testing population showed that the edge-weighted GCN-Pooling model outperformed the current gold standard clinical marker (i.e., LVEF index). The evaluation of the interpretability maps with recorded LAT maps showed a strong correlation between the top rank coefficient regions and the reentry channel. Although further investigations and evaluations are needed to confirm this finding, this is an encouraging result for the non-invasive personalised identification of the arrhythmogenic region of the post-infarct VA.

Acknowledgements. This work has been supported by the French government through the National Research Agency (ANR) Investments in the Future with 3IA Côte d'Azur (ANR19-P3IA-000) and IHU LIRYC (ANR-10-IAHU-04) and through the France 2030 investment plan under the MediTwin project.

References

1. Böhle, M., et al.: Layer-wise relevance propagation for explaining deep neural network decisions in mri-based alzheimer's disease classification. Front. Aging Neurosci. **11**, 194 (2019)
2. Bône, A., Louis, M., Martin, B., Durrleman, S.: Deformetrica 4: an open-source software for statistical shape analysis. In: Reuter, M., Wachinger, C., Lombaert, H., Paniagua, B., Lüthi, M., Egger, B. (eds.) ShapeMI 2018. LNCS, vol. 11167, pp. 3–13. Springer, Cham (2018). https://doi.org/10.1007/978-3-030-04747-4_1
3. Cedilnik, N., et al.: Fast personalized electrophysiological models from CT images for ventricular tachycardia ablation planning. In: EP-Europace 20 (2018)
4. Cerqueira, M.D., et al.: Standardized myocardial segmentation and nomenclature for tomographic imaging of the heart. A statement for healthcare professionals from the Cardiac Imaging Committee of the Council on Clinical Cardiology of the American Heart Association. In: Circulation (2002)
5. Dong, Q., et al.: Laplacian2Mesh: laplacian-based mesh understanding. IEEE Trans. Visualizat. Comput. Graph. (2024)

6. Früh, M., et al.: Weakly supervised segmentation of tumor lesions in PET-CT hybrid imaging. J. Med. Imaging **8**, 054003 (2021)
7. Gilpin, L.H., et al.: Explaining explanations: an overview of interpretability of machine learning. In: 2018 IEEE 5th International Conference on Data Science and Advanced Analytics (DSAA) (2018)
8. Huang, Y., et al.: Diffusion convolutional recurrent neural network with rank influence learning for traffic forecasting. In: 2019 18th IEEE International Conference on Trust, Security and Privacy in Computing and Communications/13th IEEE International Conference on Big Data Science and Engineering (TrustCom/BigDataSE) (2019)
9. Al-Khatib, S.M., et al.: 2017 AHA/ACC/HRS guideline for management of patients with ventricular arrhythmias and the prevention of sudden cardiac death: a report of the American college of cardiology/American heart association task force on clinical practice guidelines and the heart rhythm society. J. Am. Coll. Cardiol. (2018)
10. Kipf, T.N., Welling, M.: Semi-supervised classification with graph convolutional networks (2016)
11. Klabunde, R.E." Cardiovascular Physiology Concepts. Second. Lippincott Williams & Wilkins (2011)
12. Komatsu, Y., et al.: Regional myocardial wall thinning at multidetector computed tomography correlates to arrhythmogenic substrate in postinfarction ventricular tachycardia: assessment of structural and electrical substrate. Circ. Arrhythm. Electrophysiol. (2013)
13. Luna, A., Coumel, P., Leclercq, J.: Ambulatory sudden cardiac death: mechanisms of production of fatal arrhythmia on the basis of data from 157 cases. Am. Heart J. (1989)
14. Ly, B., et al.: Interpretable prediction of post-infarct ventricular arrhythmia using graph convolutional network. In: STACOM 2022 - 13th Workshop on Statistical Atlases and Computational Modelling of the Heart, Singapore, Singapore (2022)
15. Nuñez-Garcia, M., Cedilnik, N., Jia, S., Sermesant, M., Cochet, H.: Automatic multiplanar CT reformatting from trans-axial into left ventricle short-axis view. In: Puyol Anton, E., et al. (eds.) STACOM 2020. LNCS, vol. 12592, pp. 14–22. Springer, Cham (2021). https://doi.org/10.1007/978-3-030-68107-4_2
16. Ribeiro, M.T., Singh, S., Guestrin, C.: “"Why Should I Trust You?": explaining the predictions of any classifier. In: Proceedings of the 22nd ACM SIGKDD International Conference on Knowledge Discovery and Data Mining, San Francisco, California, USA (2016)
17. Selvaraju, R.R., et al.: Grad-CAM: visual explanations from deep networks via gradient-based localization. In: Proceedings of the IEEE International Conference on Computer Vision (ICCV) (2017)
18. Simonovsky, M., Komodakis, N.: Dynamic edge-conditioned filters in convolutional neural networks on graphs. In: Proceedings of IEEE Computer Society Conference on Computer Vision and Pattern Recognition, pp. 29–38 (2017)
19. Sundararajan, M., Taly, A., Yan, Q.: Axiomatic attribution for deep networks. In: Proceedings of the 34th International Conference on Machine Learning, ICML'17, Sydney, Australia (2017)
20. Unal, I.: Defining an optimal cut-point value in ROC analysis: an alternative approach. Comput. Math. Methods Med. (2017)
21. Valette, S., Chassery, J.M.: Approximated centroidal voronoi diagrams for uniform polygonal mesh coarsening. Comput. Graph. Forum (2004)

22. Vedaldi, A., Soatto, S.: Quick shift and kernel methods for mode seeking. In: Forsyth, D., Torr, P., Zisserman, A. (eds.) ECCV 2008. LNCS, vol. 5305, pp. 705–718. Springer, Heidelberg (2008). https://doi.org/10.1007/978-3-540-88693-8_52
23. Veličković, P., et al.: Graph attention networks. In: International Conference on Learning Representations (2018)
24. Yezzi, A.J., Prince, J.L.: An Eulerian PDE approach for computing tissue thickness. IEEE Trans. Med. Imaging (2003)
25. Zeppenfeld, K., et al.: 2022 ESC Guidelines for the management of patients with ventricular arrhythmias and the prevention of sudden cardiac death. Eur. Heart J. (2022)

3D Left Ventricular Reconstruction from 2D Echocardiograms for Reliable Volume Estimation

Moïse Z. Jansen[1,2], Debbie Zhao[1(✉)], Edward Ferdian[1], Stephen A. Creamer[1], Joshua R. Dillon[1], Mathilde A. Verlyck[1], Gina M. Quill[1], Nicola C. Edwards[3,4], Malcolm E. Legget[4], Robert N. Doughty[3,4], Alistair A. Young[5], and Martyn P. Nash[1,6(✉)]

[1] Auckland Bioengineering Institute, University of Auckland, Auckland, New Zealand
{debbie.zhao,martyn.nash}@auckland.ac.nz

[2] Department of Biomedical Engineering, Eindhoven University of Technology, Eindhoven, The Netherlands
m.z.jansen@student.tue.nl

[3] Green Lane Cardiovascular Service, Auckland City Hospital, Auckland, New Zealand

[4] Department of Medicine, University of Auckland, Auckland, New Zealand

[5] School of Biomedical Engineering and Imaging Sciences, King's College London, London, UK

[6] Department of Engineering Science and Biomedical Engineering, University of Auckland, Auckland, New Zealand

Abstract. Volume quantification is an important step in the clinical assessment of left ventricular function using echocardiography. Current guidelines recommend the use of the biplane method, which is known to significantly underestimate cavity volumes. We explored the use of the Efficient Pix2Vox network to reconstruct 3D geometries from standard 2D echocardiographic views, from which volumes could be directly computed. Reconstruction performance was evaluated and compared to data derived from cardiac magnetic resonance (CMR) imaging. The DICE score was 0.856 ± 0.043 for end-diastolic geometries and 0.800 ± 0.052 for end-systolic geometries. Using CMR as the reference, the intraclass correlation coefficient for the proposed 3D reconstruction method compared to the conventional biplane method was greater for end-diastolic volume (0.860 [0.69, 0.94] versus 0.836 [0.21, 0.95], respectively), but not for end-systolic volume nor ejection fraction. Furthermore, reductions in absolute and proportional biases were observed for both end-diastolic and end-systolic volumes. This study demonstrates the potential of reconstructing 3D cardiac geometries from 2D echocardiograms for more reliable volumetric assessment.

Keywords: Machine learning · 2D echocardiography (2DE) · Left ventricular (LV) volume assessments · 3D reconstruction

O. Camara et al. (Eds.): STACOM 2024, LNCS 15448, pp. 433–442, 2025.
https://doi.org/10.1007/978-3-031-87756-8_42

1 Introduction

Accurate measurements of left ventricular (LV) end-diastolic (ED) and end-systolic (ES) volumes are important for the assessment of heart function. The gold standard for non-invasive LV volume measurement is cardiac magnetic resonance (CMR) imaging [14]. However, practical considerations, such as high costs, prolonged acquisition times, and restricted accessibility mean that this modality is generally not suitable for widespread clinical assessment. Meanwhile, two-dimensional echocardiography (2DE) is the frontline modality for cardiac assessment. Clinical guidelines recommend LV volume assessment using the 2DE Simpson's biplane method [6]. This method, however, results in significant underestimates of LV volumes compared to CMR [3,10], due to geometric assumptions.

Previous research has shown that 2D-to-3D reconstruction models can provide accurate representations of LV volumes. Rajan et al. [9] showed that the use of a curve-fitting-based reconstruction network using a combination of long-axis and short-axis input views resulted in accurate, patient-specific 3D models of the LV. More recently, Stojanovski et al. [11] showed that machine learning could be used on synthetically generated 2DE images to reconstruct complete 3D heart geometries. The present study investigates the use of the previously described Efficient Pix2Vox method [11,17] for 3D reconstruction when applied to 2DE data from human hearts, and compares the resultant end-diastolic volumes (EDV), end-systolic volumes (ESV), and ejection fractions (EF) to measurements derived from CMR.

2 Methods

Paired non-invasive 2DE and CMR images were prospectively acquired from 125 participants (72 healthy controls and 53 patients with cardiac disease) based on previously described protocols [2,18]. Ethical approval was granted by the Health and Disability Ethics Committee of New Zealand (17/CEN/226), and written informed consent was obtained from each participant. The following section describes the 3D LV reconstruction method, as summarised in Fig. 1.

2.1 Data Preparation

For 2DE, LV-focused images were selected for apical 2, 3, and 4 chamber (A2CH, A3CH, and A4CH) views. Where available ($n = 96$), a mid-ventricular parasternal short axis view was also selected. Cropping of the images to a square were applied, followed by zero padding to ensure the same physical dimensions across all images, allowing for comparison of LV size across the dataset. 3D masks of the LV cavity were generated semi-automatically using guide-point modelling from CMR [7,18]. Padding to a cube was similarly applied to all CMR labels such that the physical dimensions of the labels were equal for all participants. For both 2DE and CMR labels, ED and ES frames were extracted based on the

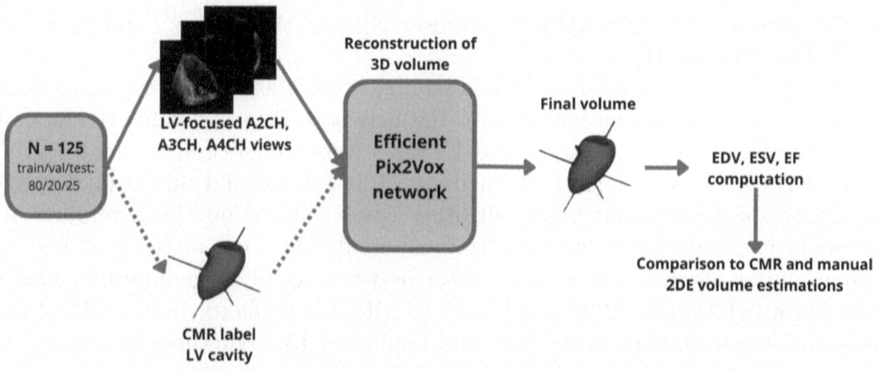

Fig. 1. Overview of the reconstruction of 3D geometries from 2DE images (apical 2, 3, and 4 chamber (A2CH, A3CH, and A4CH) views) using the Efficient Pix2Vox network [11,17]. 3D reconstruction was performed on ED and ES frames. Dotted arrows indicate steps used for training purposes only. End-diastolic volume (EDV), end-systolic volume (ESV), and ejection fraction (EF) comparisons were performed on testing cases and used as model evaluation.

LV cavity volumes (largest for ED, smallest for ES), which were used for model training and testing. Data were divided using an 80/20 ratio for training/testing, with the training data further divided into 80/20 for training/validation, and ED and ES frames from the same participant were grouped together.

2.2 Model Optimisation

To find the most optimal parameters for network training, several parameter settings were tested. The best parameter settings were found using a step-wise approach on ED frames only, comparing the results of each optimisation step to the previous best-performing model in terms of the DICE score, Intraclass Correlation Coefficient (ICC), and differences in volume estimates indexed to the body surface area (BSA) using the Mosteller equation [8], compared to CMR. Three factors influencing model performance were tested: i) the combination of different input views; ii) the application of different geometric augmentations; and iii) the use of different loss functions and their combinations.

2.3 Efficient Pix2Vox Network

3D LV reconstruction was performed using the Efficient Pix2Vox network [11, 17], where ED and ES frames were fed independently into the same network. This network consisted of several encoder branches to extract features from the 2DE images, decoder branches to generate coarse volumes and to re-align to the cardiac coordinate system based on those features, and context-aware fusion to generate a single volume. Finally, the volume was refined. The model was trained for 250 epochs, using the ADAM optimiser [4], and the Binary Cross Entropy (BCE) loss function [1].

Geometric augmentations including scaling, stretching, and compressing were applied before feeding the images into the network. To account for the lack of large LVs in the dataset, the original LV sizes were categorised as small (<100 ml for ED, and <30 ml for ES), medium (100–140 ml and 30–50 ml for ED and ES, respectively) and large (all other cases). Based on this grouping, an augmentation scale factor was determined; small LVs were augmented with a random uniformly distributed scale factor between 40–50%, medium LVs with a scale factor between 30–40%, and large LVs with a scale factor between 20–30%. Following augmentation, cases that had truncated LV structures in one of the images or the labels were excluded from the training dataset (n = 4).

2.4 Post-processing and Calculation of Clinical Indices

The output of the network was a down-sampled volume, which was up-sampled such that in the final volume, the size of each voxel was 1 mm^3. Gaussian blurring ($\sigma = 2$) followed by global thresholding (T = 0.4) was applied to produce smooth volume outputs. The model was evaluated on indexed EDV, indexed ESV, and EF, which were derived from the 3D geometries.

2.5 Model Evaluation

The predicted clinical indices were compared to the gold standard CMR-derived volumes and to volumes derived from conventional analysis of the 2DE images using the biplane method performed by a single expert using TOMTEC-ARENA 2.31 (TOMTEC Imaging Systems GmbH, Unterschleißheim, Germany).

DICE scores between the gold standard and reconstructed volumes were also computed to assess reconstruction accuracy. The agreement between the estimated clinical indices was quantified using a two-way mixed effects, absolute agreement, single measurement ICC [5] using CMR as the reference. Tukey Mean-Difference plots were used to illustrate these comparisons of volume estimates. Wilcoxon signed-rank tests were performed to classify statistically significant differences (p-values < 0.05) between volume estimates from the 3D reconstruction network and the biplane method with respect to CMR. Statistical tests were implemented using Python v3.10 [15], using Pingouin [13] and SciPy [16] packages.

3 Results

3.1 Model Optimisation

The results of the model optimisation procedure are presented in Tables 1, 2, and 3. The best model overall required three input views (A2CH, A3CH, and A4CH), including all geometric augmentations with selective augmentation factors, where larger LVs have a smaller augmentation factor compared to smaller LVs, and training using the BCE loss.

Table 1. Comparison of models with an increasing number of input views. Network training was performed using the BCE loss. Models based on 2-view and 3-view inputs (in bold) performed equally well, so both were selected and used for the next model optimisation step. Model performance is displayed as means and standard deviations for DICE and differences in indexed volumes (bias) compared to CMR. For the single measures intraclass correlation coefficient (ICC), the 95% confidence interval is shown in brackets.

Views (19 test cases)	DICE	ICC	Bias [ml/m^2]
A2CH, A4CH	**0.860±0.040**	**0.778 [0.47, 0.92]**	**-4.3±18.7**
A2CH, A3CH, A4CH	**0.862±0.042**	**0.720 [0.36, 0.89]**	**-2.2±19.7**
A2CH, A3CH, A4CH, mid SAX	0.846±0.044	0.580 [0.12, 0.83]	-0.9±23.7

Table 2. Results of the second step of the model optimisation procedure, applying different geometric augmentations (stretching, compressing, and uniform scaling). Network training was performed using the BCE loss. The best model (indicated in bold) was used for the next optimisation step. Model performance is displayed as means and standard deviations for DICE and differences in indexed volumes (bias) compared to CMR. For the single measures intraclass correlation coefficient (ICC), the 95% confidence interval is shown in brackets.

Views (25 test cases)	Augmentations	DICE	ICC	Bias [ml/m^2]
A2CH, A4CH	None	0.859±0.050	0.733 [0.44, 0.89]	-0.7±19.6
A2CH, A4CH	Stretching & Compressing	0.862±0.046	0.831 [0.63, 0.93]	-0.3±16.3
A2CH, A4CH	Stretching, Compressing & Uniform scaling	0.860±0.050	0.917 [0.80, 0.97]	-1.9±15.4
A2CH, A3CH, A4CH	None	0.858±0.045	0.691 [0.37, 0.87]	0.3±20.8
A2CH, A3CH, A4CH	Stretching & Compressing	0.858±0.050	0.885 [0.73, 0.95]	0.3±16.3
A2CH, A3CH, A4CH	**Stretching, Compressing & Uniform scaling**	**0.866±0.045**	**0.924 [0.81, 0.97]**	**2.7±13.4**

Table 3. Results of the final step of the model optimisation procedure, testing the differences between BCE loss, IOU loss, and a combination (equal weights) of IOU and BCE loss. The best model (indicated in bold) was used as the final model. Model performance is displayed as means and standard deviations for DICE and differences in indexed volumes (bias) compared to CMR. For the single measures intraclass correlation coefficient (ICC), the 95% confidence interval is shown in brackets.

Views (25 test cases)	Augmentations	Loss	DICE	ICC	Bias [ml/m^2]
A2CH, A3CH, A4CH	**Stretching, Compressing & Uniform scaling**	**BCE**	**0.866±0.045**	**0.924 [0.81, 0.97]**	**2.7±13.4**
A2CH, A3CH, A4CH	Stretching, Compressing & Uniform scaling	IOU	0.858±0.052	0.881 [0.63, 0.96]	2.1±13.9
A2CH, A3CH, A4CH	Stretching, Compressing & Uniform scaling	IOU× BCE	0.862±0.047	0.911 [0.79, 0.96]	-1.5±15.1

3.2 Model Performance

3D reconstructed labels were generated using a combined ED and ES model, resulting in average DICE scores of 0.856 ± 0.043 and 0.800 ± 0.052 for ED and

Table 4. Bias and its standard deviation of EDV, ESV and EF measurements compared to CMR after volume assessment using the reconstruction network and the expert's manual analysis of echo images, and single measurement intraclass correlation coefficients (ICC) and 95% confidence intervals displayed in brackets. Asterisks indicate statistically significant differences (p-value < 0.05) of volume estimates compared to CMR using a Wicoxon signed-rank test.

	Reconstruction model		Expert (manual analysis)	
20 test cases	Bias	ICC	Bias	ICC
EDV [ml/m^2]	–3.9±13.9	0.860 [0.69, 0.94]	*–13.2±12.0	0.836 [0.21, 0.95]
ESV [ml/m^2]	2.3±9.0	0.914 [0.80, 0.96]	*–5.6±6.8	0.927 [0.71, 0.98]
EF [%]	*–5.6±6.0	0.707 [0.14, 0.90]	1.0±5.8	0.841 [0.64, 0.93]

ES reconstructions, respectively. The evaluation metrics in terms of ICC and biases in indexed volumes and EF compared to CMR are given in Table 4. The reconstruction model outperformed manual analyses for EDV and ESV measurements. In terms of ICC classification, the reliability of EDV increased from poor-to-excellent to moderate-to-excellent. For ESV, there was a shift from moderate-to-excellent to good-to-excellent reliability with respect to the 95% confidence intervals. The biases were closer to zero using the 3D reconstruction network, although slightly larger standard deviations of the differences were observed. For EF, a statistically significant difference (p-value < 0.05) was found using the 3D reconstruction method. Additionally, ICC classification shifted from moderate-to-excellent to poor-to-excellent reliability. Tukey Mean-Difference plots (Fig. 2) show the distributions of differences in the indexed volumes and EF estimates. Proportional biases were apparent for both the model estimates and manual measurements of indexed volumes.

4 Discussion

By applying the Efficient Pix2Vox network, 2DE images were reconstructed to 3D geometries, and used for volumetric assessment, with the aim of obtaining more reliable volume estimates. Volume estimation using the reconstructed 3D volumes outperformed the conventional manual analysis of 2DE images in terms of a decreased bias, and better ICC classifications for EDV and ESV estimates. However, a higher bias with significant differences compared to CMR was found for EF estimates using reconstructed 3D volumes. The study performed by Stojanovski et al. [11] has shown that applying the Efficient Pix2Vox network to synthetically generated data resulted in IOU scores of 0.74 for the reconstruction of complete heart geometries. As our method used real 2DE images, results were expected to be worse due to the synthetic nature of the images used by Stojanovski et al. [11]. A previous deep-learning method has shown better performance for EF estimates compared the proposed method [12]. However, our

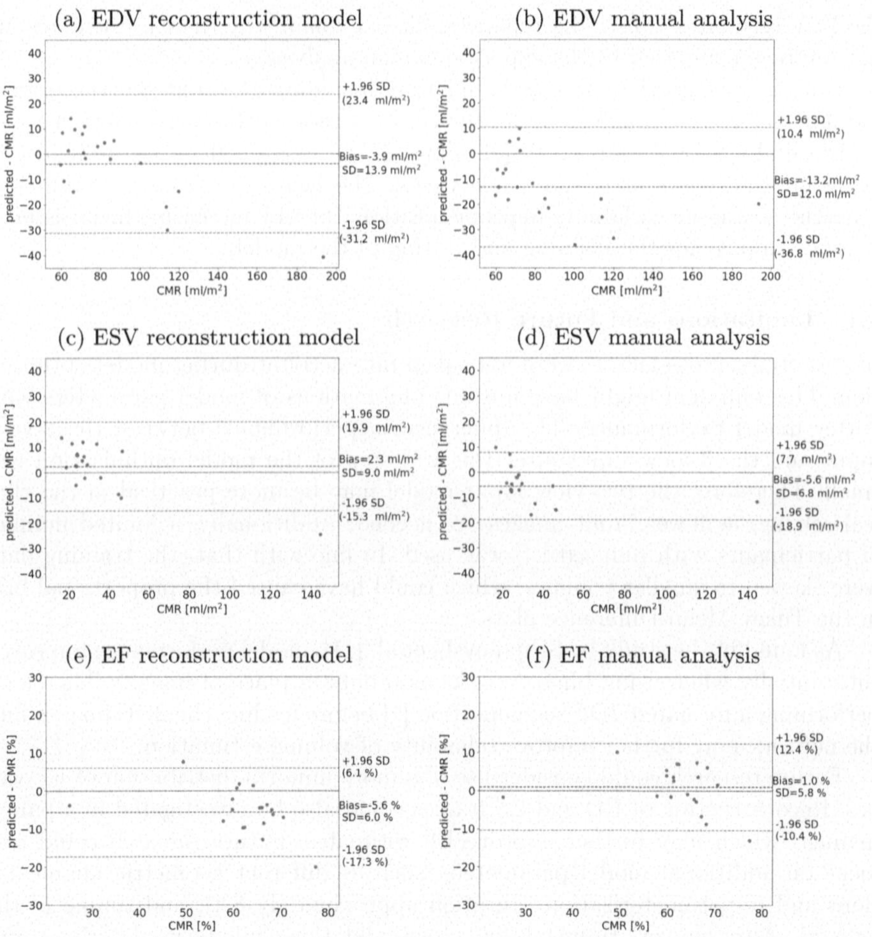

Fig. 2. Tukey Mean-Difference plots showing bias and limits of agreement of CMR volumes compared to volume estimations by the reconstruction model and manual analysis.

study converts the 2DE images to 3D volumes, thereby providing the advantage of enabling the use for advanced applications such as shape analysis or computational modelling.

When comparing ED and ES reconstruction using the DICE score, the reconstruction of ED geometries was on average more accurate than for ES, which is possibly due to the more complex geometry at ES, and because the model optimisation was performed on ED frames only. Additionally, the Tukey Mean-Difference plots (Fig. 2) show an apparent proportional bias for EDV assessment using the 3D reconstruction method. This was less obvious, but still present, for the ESV estimates. These discrepancies likely explain why the EF estimates did not improve compared to the expert's manual analysis.

During optimisation, it was found that including the mid short-axis view did not increase the model's performance, even though this was expected based on the findings by Rajan et al. [9]. This can be explained by the difficulty of acquiring mid-ventricular short-axis views. The lack of anatomical landmarks gives rise to a large variability in probe position, thereby increasing inconsistency in model inputs for the training and testing of the models.

4.1 Limitations and Future Research

In this study, cross-factors were not taken into account during model optimisation. This omission might have missed combinations of model parameters with better model performance. The difference in performance between the 2-view input and the 3-view input after the first step of the model optimisation was small. Therefore, the two-view input model may be more practical in the clinical setting, as fewer input images are needed. Additionally, a limited number of participants with reduced EF was used. In line with that, the training data were skewed to smaller volumes, which could have caused the proportional bias in the Tukey Mean-Difference plots.

As found in the study by Stojanovski et al. [11], model performance improved substantially when using binary masks as inputs in place of images. Therefore, performing automated 2DE segmentation [2] before feeding the 2DE images into the network may further improve reliability of volume estimation.

Future research could consider use of a more uniform distribution of EF values. Reconstruction of ED and ES frames could also be investigated in a paired manner, which may further improve EF estimates. Future research could also focus on additional model parameters, such as different geometric augmentations and re-parameterisation (duration approximately 8 h), and tuning of the network. After network tuning, k-fold cross-validation could be used during training of the final model.

5 Conclusion

In this study, 2DE-derived 3D volumes were created using the Efficient Pix2Vox network. Volumetric assessments improved using the 3D reconstruction method (compared to Simpson's biplane method) in terms of decreased bias and improved reliability with respect to gold standard CMR. However, EF estimation did not improve, suggesting that further work is needed for clinical adoption. Future research may focus on the application of segmentation algorithms, and acquiring a more uniform input dataset based on volumes.

Acknowledgments. This study was funded by the Health Research Council of New Zealand (grants 17/608 and 23/527). DZ is supported by a PÅńtahi Manawa | Healthy Hearts for Aotearoa New Zealand Centre of Research Excellence Research Fellowship (grant #: FLW23-001).

References

1. BCELoss - PyTorch 2.3 documentation (2023). https://pytorch.org/docs/stable/generated/torch.nn.BCELoss.html
2. Geven, B.W.M., et al.: Automated quality-controlled left heart segmentation from 2D echocardiography, pp. 98–107 (2024). https://doi.org/10.1007/978-3-031-52448-6_10
3. Greupner, J., et al.: Head-to-head comparison of left ventricular function assessment with 64-row computed tomography, biplane left cineventriculography, and both 2- and 3-dimensional transthoracic echocardiography: Comparison with magnetic resonance imaging as the reference standard. J. Am. Coll. Cardiol. **59**(21), 1897–1907 (2012). https://doi.org/10.1016/j.jacc.2012.01.046
4. Kingma, D.P., Ba, J.: Adam: a method for stochastic optimization (2014). http://arxiv.org/abs/1412.6980
5. Koo, T.K., Li, M.Y.: A guideline of selecting and reporting intraclass correlation coefficients for reliability research. J. Chiropr. Med. **15**(2), 155–163 (2016). https://doi.org/10.1016/j.jcm.2016.02.012
6. Lang, R.M., et al.: Recommendations for cardiac chamber quantification by echocardiography in adults: an update from the American Society of Echocardiography and the European Association of Cardiovascular Imaging. J. Am. Soc. Echocardiogr. **28**(1), 1–39 (2015). https://doi.org/10.1016/j.echo.2014.10.003
7. Li, B., Liu, Y., Occleshaw, C.J., Cowan, B.R., Young, A.A.: In-line automated tracking for ventricular function with magnetic resonance imaging. JACC Cardiovasc. Imaging **3**(8), 860–866 (2010). https://doi.org/10.1016/j.jcmg.2010.04.013
8. Mosteller, R.D.: Simplified calculation of body-surface area. N. Engl. J. Med. **317**(17), 1098 (1987). https://doi.org/10.1056/NEJM198710223171717
9. Rajan, N.K., Song, Z., Hoffmann, K.R., Belohlavek, M., McMahon, E.M., Borazjani, I.: Automated three-dimensional reconstruction of the left ventricle from multiple-axis echocardiography. J. Biomech. Eng. **138**(1) (2016). https://doi.org/10.1115/1.4031977
10. Rigolli, M., Anandabaskaran, S., Christiansen, J.P., Whalley, G.A.: Bias associated with left ventricular quantification by multimodality imaging: a systematic review and meta-analysis. Open Heart **3**(1), e000388 (2016). https://doi.org/10.1136/openhrt-2015-000388
11. Stojanovski, D., Hermida, U., Muffoletto, M., Lamata, P., Beqiri, A., Gomez, A.: Efficient Pix2Vox++ for 3D cardiac reconstruction from 2D echo views. In: Lecture Notes in Computer Science, vol. 13565 LNCS, pp. 86–95. Springer Science and Business Media Deutschland GmbH (2022). https://doi.org/10.1007/978-3-031-16902-1_9
12. Sveric, K.M., et al.: Improved assessment of left ventricular ejection fraction using artificial intelligence in echocardiography: a comparative analysis with cardiac magnetic resonance imaging. Int. J. Cardiol. **394**, 131383 (2024). https://doi.org/10.1016/j.ijcard.2023.131383
13. Vallat, R.: Pingouin: statistics in Python. J. Open Source Softw. **3**(31), 1026 (2018). https://doi.org/10.21105/joss.01026
14. Van Der Ven, J., et al.: Multicentre reference values for cardiac magnetic resonance imaging derived ventricular size and function for children aged 0–18 years. Eur. Heart J. Cardiovasc. Imaging **21**(1), 102–113 (2020). https://doi.org/10.1093/ehjci/jez164

15. Van Rossum, G., Drake, F.: Python reference manual, vol. 111. Centrum voor Wiskunde en Informatica Amsterdam (1995)
16. Virtanen, P., et al.: SciPy 1.0: fundamental algorithms for scientific computing in Python. Nat. Methods **17**(3), 261–272 (2020). https://doi.org/10.1038/s41592-019-0686-2
17. Xie, H., Yao, H., Sun, X., Zhou, S., Zhang, S.: Pix2vox: context-aware 3d reconstruction from single and multi-view images. In: ICCV (2019). https://doi.org/10.48550/arXiv.1901.11153
18. Zhao, D., et al.: MITEA: a dataset for machine learning segmentation of the left ventricle in 3D echocardiography using subject-specific labels from cardiac magnetic resonance imaging. Front. Cardiovasc. Med. **9** (2023). https://doi.org/10.3389/fcvm.2022.1016703

Comparing Left Atrial Spontaneous Echo Contrast Intensity with Gaussian Process Emulator Predictions

Paolo Melidoro[1](✉), Malak Sabry[1], Abdel Rahman Amr Sultan[2,3], Ahmed Qureshi[1], Gregory Y. H. Lip[4,5], Natalie Montarello[6], Nishant Lahoti[6], Ronak Rajani[1], Magdalena Klis[6], Steven E. Williams[1,7], Oleg Aslanidi[1], and Adelaide De Vecchi[1]

[1] School of Biomedical Engineering and Imaging Sciences, King's College London, London, UK
paolo.melidoro@kcl.ac.uk
[2] Department of Mechanical Engineering, The American University in Cairo, New Cairo 11835, Egypt
[3] Aswan Heart Research Centre, Magdi Yacoub Foundation, Aswan, Egypt
[4] Liverpool Centre for Cardiovascular Science at University of Liverpool, Liverpool John Moores University and Liverpool Heart & Chest Hospital, Liverpool, UK
[5] Danish Center for Health Services Research, Department of Clinical Medicine, Aalborg University, Aalborg, Denmark
[6] Cardiovascular Directorate, Guy's and St Thomas' NHS Foundation Trust, London SE1 7EH, UK
[7] Centre for Cardiovascular Science, The University of Edinburgh, Edinburgh, UK

Abstract. Atrial Fibrillation (AF) significantly increases the risk of ischemic stroke due to blood stasis and hypercoagulability in the left atrium (LA). Effective stroke risk stratification is crucial for identifying AF patients who require anticoagulation therapy. Spontaneous echo contrast (SEC), a phenomenon arising from blood stasis and fibrinogen-mediated red blood cell aggregation, serves as a strong predictor of stroke and can be observed via transesophageal echocardiograms (TEEs). This study employs Gaussian Process Emulators (GPEs) trained on non-Newtonian Computational Fluid Dynamics (CFD) simulations to predict the occurrence and intensity of LA SEC. Using haematocrit and fibrinogen concentration as inputs, the GPEs compute the average blood viscosity in the left atrial appendage (LAA). We also quantified geometric and functional parameters of the LAA to assess their relative impact on SEC intensity and compared these findings with those in patients without SEC. Our results indicate that both LAA motility and GPE-predicted LAA viscosity can distinguish between SEC and non-SEC groups ($p > 0.001$). In the SEC cohort, significant correlations were observed between grayscale intensity and LAA motility ($r = 0.8$), LAA orifice diameter ($r = 0.71$), and predicted LAA viscosity ($r = 0.57$). This study demonstrates the potential utility of GPEs for predicting LA SEC, thereby enhancing stroke risk stratification in AF patients. Additionally, we identify key geometric and functional features of the LAA that influence SEC formation.

Keywords: Atrial Fibrillation · Spontaneous Echo Contrast · Gaussian Process Emulators

O. Camara et al. (Eds.): STACOM 2024, LNCS 15448, pp. 443–452, 2025.
https://doi.org/10.1007/978-3-031-87756-8_43

1 Introduction

Atrial fibrillation (AF), the most common significant cardiac arrhythmia, is linked to high morbidity and mortality, to which stroke and thromboembolism greatly contribute [1, 2]. It is estimated that 89% of thrombi in AF patients originate in the left atrial appendage (LAA), a small pouch protruding from the left atrium (LA), which is subsequently a target of LAA occlusion [3, 4]. To identify AF patients at high risk of stroke, stroke risk stratification schemes such as CHA_2DS_2-VASc are routinely used clinically [5]. These schemes give each patient a score dependent on comorbidities and thus do not consider mechanistic factors known to influence stroke risk such as blood stasis in the LA [6].

Using Computational Fluid Dynamics (CFD) to develop mechanistic models of blood flow in the LA and LAA has been explored as a way of enhancing such risk scores [7, 8]. In addition to quantifying blood stasis, some models have also incorporated metrics designed to quantify the hypercoagulability of blood, an additional factor outlined by Virchow's triad that contributes to stroke risk [9, 10]. The non-Newtonian behavior of blood is an important consideration when modelling blood in areas of low shear rates, due to the shear thinning nature of blood meaning areas of high blood viscosity occur at low shear rates. Non-Newtonian models have subsequently been implemented into CFD simulations of the left atrium [11–13]. The primary mechanism contributing to the non-Newtonian behavior of blood at low shear rates is red blood cell (RBC) aggregation, which also responsible for Spontaneous Echo Contrast (SEC) formation [14, 15]. LA SEC can be detected via transesophageal echocardiograms (TEE), with the presence and intensity of SEC being highly associated with stroke risk [16]. Haematocrit and fibrinogen both influence SEC intensity and non-Newtonian blood behavior due to fibrinogen mediating RBC aggregation at low shear rates, however current CFD simulations of the LA have used non-Newtonian models dependent only on haematocrit [17–19].

Mechanistic models such as CFD are computationally expensive, limiting their use in healthcare. Gaussian Process Emulators (GPEs) offer a solution to this issue thanks to their ability to be trained on outputs of complex deterministic models to make accurate predictions at a fraction of the computation cost [20, 21]. GPEs have also shown potential in replicating outcomes of CFD models of the LAA to predict stroke risk [22].

In this study we compare predicted values for viscosity in the LAA of AF patients from GPEs trained on CFD simulations with a non-Newtonian blood model dependent on haematocrit and fibrinogen, with the SEC intensity measured from TEE data.

2 Methodology

2.1 Patient Characteristics

TEE data was acquired in 18 patients prior to catheter ablation for AF or LAA occlusion. Blood tests were performed within 6 months of the procedure. The patients' data was fully anonymized, and the echocardiograms and blood tests were performed at Guy's and St Thomas' NHS Trust. The patients were divided into two cohorts depending on the occurrence of SEC, with 9 patients in each group. Patient characteristics included: age (non-SEC group: 64.6 ± 14.4 years old; SEC group: 65.0 ± 10.0 years old, $p = 0.94$); haematocrit (non-SEC group: 0.40 ± 0.035; SEC group: 0.41 ± 0.05, $p = 0.63$); BMI (non-SEC group: 28.60 ± 5.24; SEC group: 32.96 ± 7.34, $p = 0.17$); sex (non-SEC group: 5/9 female; SEC group: 2/9 female); and type of AF, i.e. paroxysmal (non-SEC group: n = 4; SEC group: n = 2), persistent (non-SEC group: n = 4; SEC group: n = 6), and permanent (non-SEC group: n = 1; SEC group: n = 1).

2.2 TEE Measurements

Due to the importance of LAA geometric and functional characteristics on stroke risk in AF patients, the following anatomical features were measured: LAA orifice diameter, distance between orifice and first bend, and bend angle. Motility was also measured as a functional metric to assess LAA contractility. All these parameters have been associated with thrombus formation and stroke risk in AF patients [23, 24].

To measure the geometric and functional parameters of the LAA, RadiAnt DICOM Viewer (Medixant) was used. The orifice diameter, length from orifice to bend and bend angle were all measured when the LAA area was at its maximum value, while the motility was calculated by measuring the maximum and minimum area of the LAA and then using a formula previously used by Lacomis et al. [24]:

$$Motility = \frac{Max\,Area - Min\,Area}{Max\,Area} \tag{1}$$

Figure 1 shows the three anatomical measurements in a patient. The distance to first bend was measured perpendicularly to the LAA orifice and taken from the point at the center of the LAA to the point where the LAA starts to bend. The bend angle was calculated by taking three points: the center of the orifice, the tip of the LAA, and the point where the LAA starts to bend.

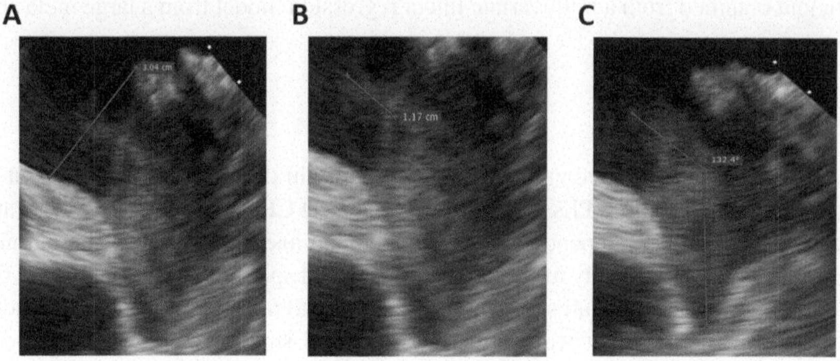

Fig. 1. Measurements of the LAA from TEE scans. A) Measuring the orifice diameter B) Measure the distance from the orifice to the first bend C) Measuring the bend angle

SEC was graded from 1 to 3, where grade 1 is mild SEC indicating minimal echogenicity, grade 2 is moderate where SEC is observable in the main LA cavity, and grade 3 indicates SEC where dense swirling patterns are observable in the main cavity [25]. SEC intensity was quantified using the ImageJ software package (National

Institutes of Health) to measure the average grayscale intensity throughout the LAA. This was computed by creating an outline of the LAA and then averaging the grayscale within the resulting region of interest, as shown in Fig. 2.

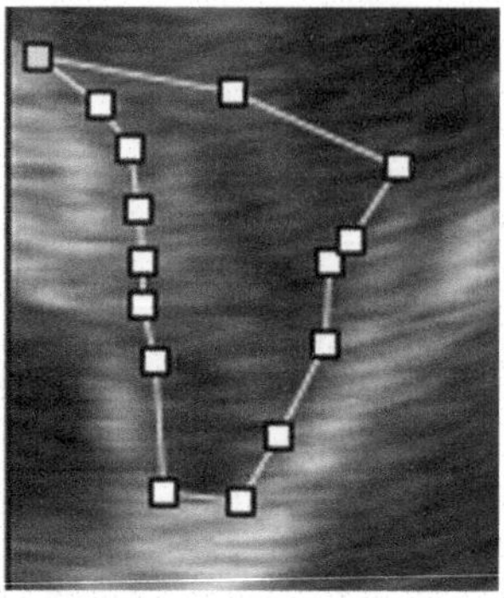

Fig. 2. Measuring intensity of SEC in the LAA using ImageJ

2.3 Fibrinogen Values

As fibrinogen was not available from the blood tests performed in this study, it was estimated using population data based on age, gender, BMI, and type of AF from previous studies [26, 27]. Patients with persistent and permanent AF were attributed the average fibrinogen concentration found experimentally in chronic patients (3.79 g/l) [26]. The values of fibrinogen were then adjusted based on age, gender, and BMI using a regression coefficient obtained from a multivariate linear regression model from a large meta study [27].

2.4 GPE Training

Two GPEs were used to predict the average viscosity in the LAA from fibrinogen and haematocrit values. Both GPEs were trained on 40 3D CFD simulations incorporating a non-Newtonian model dependent on fibrinogen and haematocrit [11]. The output of the GPEs was blood viscosity averaged temporally and spatially in the LAA. One GPE was trained using moving wall simulations attempting to mimic the contractile function in the LAA, while the other was trained on rigid wall simulations, comparable to the impaired contractile function that is observed in AF patients. The moving GPE and rigid wall GPE were trained on the most common LAA anatomy, i.e. the chicken wing type, which accounts for around 40% of the population [28] and achieved maximum R^2 of. 0.9866 and 0.9982, respectively. An example of the CFD simulations seen to train the GPE can be seen in Fig. 3.

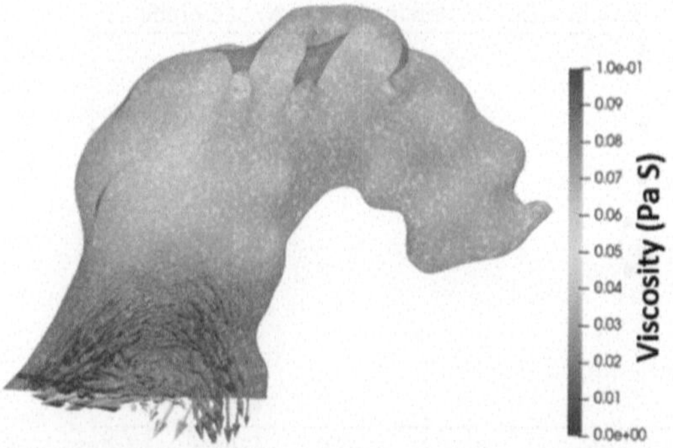

Fig. 3. Example of the CFD simulations of the left atrium used to train the GPE, illustrating how the viscosity varies throughout the LAA due to the non-Newtonian model

To determine which of the GPEs should be applied to each patient, the average motility across both cohorts was calculated as a threshold, where the moving wall GPE was applied to patients above average motility, while the rigid wall GPE was applied to patients below average motility. When calculating the correlation between the grayscale and the GPE viscosity prediction, the viscosity was divided by motility, due to the significant effect motility has on flow in the LAA.

2.5 Statistical Analysis

The GPE predictions for viscosity, and the functional and geometric parameters were compared between groups. For each parameter a t-test was used to determine statistical significance, except for the distance between the orifice and first bend, where a Mann-Whitney U test was used due to the Shapiro-Wilk test finding non-normality for this parameter in the non-SEC cohort. In the SEC group the correlation between grayscale and each of the geometric and functional parameters was quantified using Pearson's correlation coefficient.

3 Results

3.1 SEC Grade, Motility and GPE Predictions

The GPE prediction of average LAA viscosity for each patient was plotted against motility and then colored by the grade of SEC to compare how both motility and the viscosity prediction in the LAA distinguish between SEC and non-SEC and the qualitative grades of SEC (Fig. 4).

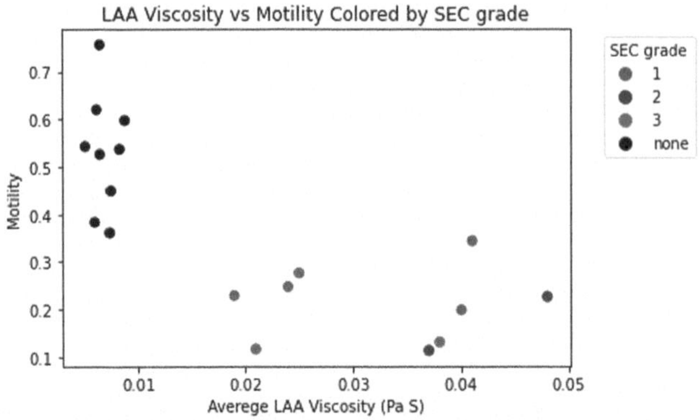

Fig. 4. Motility plotted against the average viscosity estimated by the GPEs. Each dot is colored based on the grade of SEC that was observed from the TEE scans

The average LAA viscosity in the non-SEC group was 0.0069 ± 0.0012 Pa S, significantly smaller than the value observed in the SEC group of 0.0324 ± 0.01 Pa S (p < 0.001). The mean motility observed in the non-SEC group (0.53 ± 0.07) also significantly exceeded the corresponding value in the SEC group (0.21 ± 0.12). Tables 1 illustrates the mean values of all the measured metrics in the SEC and non-SEC group.

Table 1. Statistical testing results comparing parameters observed in SEC and non-SEC groups

	non-SEC (mean ± SD)	SEC (mean ± SD)	p value
Motility	0.53 (0.07)	0.21 (0.12)	<0.001
Orifice diameter (mm)	23.02 (5.41)	21.12 (4.95)	0.45
Bend Angle (degrees)	119.88 (10.9)	113.92 (16.92)	0.39
LAA viscosity (Pa S)	0.0069 (0.0012)	0.0324 (0.01)	< 0.001
Distance from orifice to first bend	12.86 (6.48)	10.49 (6.07)	0.28

3.2 SEC Intensity Correlations

As illustrated in Figs. 5, the variable that demonstrated the strongest correlation with the grayscale intensity was the motility (r = 0.8), followed by the orifice diameter (r = 0.71) and by the LAA viscosity when normalized by motility (r = 0.57). No significant correlation was found between other anatomical measurements (i.e. distance to first bend and bend angle) and the SEC intensity. A strong correlation was also found between the LAA orifice diameter and the LAA motility (r = 0.96).

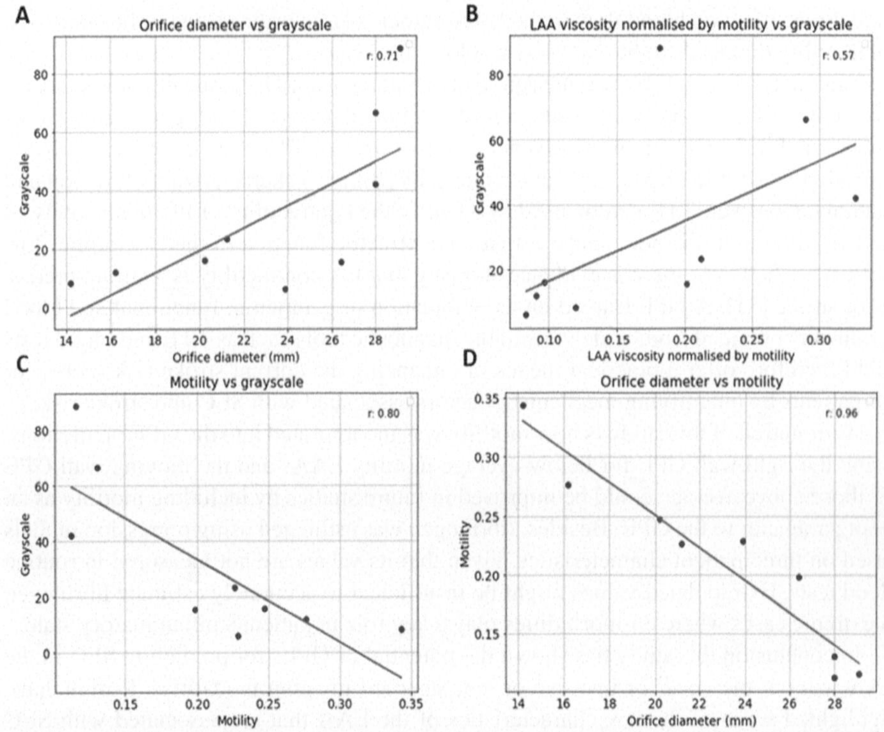

Fig. 5. Correlation between the grayscale intensity, used as a surrogate metric for SEC intensity and A) orifice diameter, B) LAA viscosity normalized by motility and C) motility. Correlation between orifice diameter and motility (D)

4 Discussion and Conclusion

This study has uniquely applied trained GPEs to predict non-Newtonian effects of blood viscosity due to red blood cell aggregation and fibrinogen, and used the resulting prediction of viscosity in the LAA to estimate the presence of SEC. The results show a statistically significant difference in the average LAA viscosity between the SEC and non-SEC groups. They also highlight the importance of the LAA motility effects on the presence and intensity of SEC. This can be explained considering that, although the effect of haematocrit and fibrinogen on SEC are well established, the shear rates present are also highly important for red blood cells aggregation [19]. The LAA contractility has been found to be the primary determinant of LAA flow characteristics, providing a rationale for the dominant role the motility had on SEC in this study [29].

Amongst the anatomical parameters measured, the orifice diameter was the one variable which showed a strong correlation with the SEC intensity. This is likely due to the orifice diameter being strongly correlated with motility in this study and more broadly with atrial mechanics. LAA orifice enlargement is associated with LA dilation and LAA remodeling in AF patients, which is known to have a deleterious effect on the contractile function of the LAA [30]. Therefore, LAA orifice size may not have a large direct effect on blood flow in the LAA; rather, a large orifice size implies impaired contractility that leads to blood stasis and low shear rates.

Although LAA motility was measured in this study using TEE, transthoracic echocardiography (TTE) is commonly used in routine clinical practice of AF patients due to its wide availability and lack of invasiveness. A drawback of TTE is its inability to provide a detailed view of the LAA, severely limiting its ability to characterize its function and anatomy. However, TTE can be used to quantify the contractility of the main cavity of the LA. Although it is not clear if impaired contractile function in the LA implies low motility in the LAA, there is evidence showing that LA contractility is a strong predictor of stroke [31]. A GPE trained on a combination of geometric, functional and blood parameters obtained from widely available diagnostic tools such as TTE and blood tests could therefore offer a practical means of enhancing the current stroke risk scores for AF patients by quantifying mechanistic factors associated with SEC and stroke risk.

A limitation of this study is how motility was incorporated into the GPE predictions. Using the rigid wall GPE for below-average motility LAAs and the moving wall GPE for those above average could be improved in future studies by including motility as an input parameter to the GPE. Besides, fibrinogen was estimated using regression models based on three patient characteristics, given that its values are not measured in routine blood tests. Extrapolated values might be insufficient to accurately estimate fibrinogen in extreme cases where co-morbidities play a key role in patient's inflammatory state.

In conclusion this study has shown the potential of GPEs for predicting SEC in the LAA without the need for invasive or non-standard imaging modalities. Results have highlighted some of the key characteristics of the LAA that are associated with SEC intensity, such as the motility of the LAA and orifice size. To advance the concept of a GPE to predict SEC for enhancing stroke risk stratification, the next steps will therefore be to train a GPE with LA geometric and functional data obtained from TTE.

Acknowledgements. Supported by King's Centre for Doctoral Studies through the CDT on Surgical and Interventional Engineering.

Funding. PM was funded by the EPSRC Research Council, part of the EPSRC CDT (grant Ref: EP/R513064/1). SW acknowledges support from the British Heart Foundation (FS/20/26/34952). This work was supported by the Wellcome/EPSRC Centre for Medical Engineering (WT203148/Z/16/Z) at King's College London. The funders had no role in the preparation, review, or approval of the manuscript.

References

1. O'Keefe, E.L., et al.: Prevention and treatment of atrial fibrillation via risk factor modification. Am. J. Cardiol. **160**, 46–52 (2021). https://doi.org/10.1016/j.amjcard.2021.08.042
2. Lip, G.Y.H., Lim, H.S.: Atrial fibrillation and stroke prevention. Lancet Neurol. **6**(11), 981–993 (2007). https://doi.org/10.1016/S1474-4422(07)70264-8
3. Lip, G.Y.H., et al.: Atrial fibrillation and stroke prevention: 25 years of research at EP Europace journal. Europace **25**(9), euad226 (2023). https://doi.org/10.1093/europace/euad226

4. Potpara, T., et al.: Practical guide on left atrial appendage closure for the non-implanting physician: an international consensus paper. Europace **26**(4), euae035 (2024). https://doi.org/10.1093/europace/euae035. Erratum in: Europace **26**(5), euae142 (2024). https://doi.org/10.1093/europace/euae142. PMID: 38291925; PMCID: PMC11009149
5. Lip, G.Y.H., Nieuwlaat, R., Pisters, R., Lane, D.A., Crijns, H.J.G.M.: Refining clinical risk stratification for predicting stroke and thromboembolism in atrial fibrillation using a novel risk factor-based approach: the euro heart survey on atrial fibrillation. Chest **137**(2), 263–272 (2010). https://doi.org/10.1378/chest.09-1584
6. Lee, D.C., et al.: Left atrial blood stasis by 4D flow MRI correlates with stroke risk estimation by CHA2DS2-VASc score. J. Cardiovasc. Magn. Reson. **16**(1), O51 (2014). https://doi.org/10.1186/1532-429X-16-S1-O51
7. Zingaro, A., et al.: A comprehensive stroke risk assessment by combining atrial computational fluid dynamics simulations and functional patient data. Sci. Rep. **14**, 9515 (2024). https://doi.org/10.1038/s41598-024-59997-2
8. Pons, M.I., et al.: Joint analysis of morphological parameters and in silico haemodynamics of the left atrial appendage for thrombogenic risk assessment. J. Intervent. Cardiol. **2022**, 9125224 (2022). https://doi.org/10.1155/2022/9125224
9. Qureshi, A., Lip, G.Y., Nordsletten, D.A., Williams, S.E., Aslanidi, O., De Vecchi, A.: Imaging and biophysical modelling of thrombogenic mechanisms in atrial fibrillation and stroke. Front. Cardiovasc. Med. **9** (2023). https://www.frontiersin.org/articles/10.3389/fcvm.2022.1074562. Accessed 05 Dec 2023
10. Ding, W.Y., Gupta, D., Lip, G.Y.H.: Atrial fibrillation and the prothrombotic state: revisiting Virchow's triad in 2020. Heart **106**(19), 1463–1468 (2020)
11. Melidoro, P., et al.: Enhancing stroke risk stratification in atrial fibrillation through non-Newtonian blood modelling and Gaussian process emulation. J. Physiol. https://doi.org/10.1113/JP287283
12. Gonzalo, A., et al.: Non-Newtonian blood rheology impacts left atrial stasis in patient-specific simulations. Int. J. Numer. Methods Biomed. Eng. **38**(6), e3597 (2022). https://doi.org/10.1002/cnm.3597
13. Zhang, Z., Zhu, J., Wu, M., Neidlin, M., Wu, W.-T., Wu, P.: Computational modeling of hemodynamics and risk of thrombosis in the left atrial appendage using patient-specific blood viscosity and boundary conditions at the mitral valve. Biomech. Model. Mechanobiol. **22**(4), 1447–1457 (2023). https://doi.org/10.1007/s10237-023-01731-4
14. Black, I.W.: Spontaneous echo contrast: where there's smoke there's fire. Echocardiography **17**(4), 373–382 (2000). https://doi.org/10.1111/j.1540-8175.2000.tb01153.x
15. Chien, S., Usami, S., Dellenback, R.J., Gregersen, M.I., Nanninga, L.B., Guest, M.M.: Blood viscosity: influence of erythrocyte aggregation. Science **157**(3790), 829–831 (1967). https://doi.org/10.1126/science.157.3790.829
16. Zhao, Y., et al.: Intensity of left atrial spontaneous echo contrast as a correlate for stroke risk stratification in patients with nonvalvular atrial fibrillation. Sci. Rep. **6**(1), Art. no. 1 (2016). https://doi.org/10.1038/srep27650
17. Fatkin, D., Loupas, T., Low, J., Feneley, M.: Inhibition of red cell aggregation prevents spontaneous echocardiographic contrast formation in human blood. Circulation **96**(3), 889–896 (1997). https://doi.org/10.1161/01.CIR.96.3.889
18. Rastegar, R., et al.: Spontaneous echo contrast videodensity isflow-related and is dependent on the relative concentrations of fibrinogen and red blood cells. J. Am. Coll. Cardiol. **41**(4), 603–610 (2003). https://doi.org/10.1016/S0735-1097(02)02898-X
19. Matsuda, T., Murakami, M.: Relationship between fibrinogen and blood viscosity. Thromb. Res. **8**, 25–33 (1976). https://doi.org/10.1016/0049-3848(76)90044-X

20. Longobardi, S., et al.: Predicting left ventricular contractile function via Gaussian process emulation in aortic-banded rats. Philos. Transact. A Math. Phys. Eng. Sci. **378**(2173), 20190334 (2020). https://doi.org/10.1098/rsta.2019.0334
21. Karabelas, E., et al.: Global sensitivity analysis of four chamber heart hemodynamics using surrogate models. IEEE Trans. Biomed. Eng. **69**(10), 3216–3223 (2022). https://doi.org/10.1109/TBME.2022.3163428
22. Smine, Z., et al.: Global sensitivity analysis of thrombus formation in the left atrial appendage of atrial fibrillation patients. In: Statistical Atlases and Computational Models of the Heart. Regular and CMRxRecon Challenge Papers, pp. 55–65 (2024). https://doi.org/10.1007/978-3-031-52448-6_6
23. Dudzińska-Szczerba, K., et al.: Assessment of the left atrial appendage morphology in patients after ischemic stroke — the ASSAM study. Int. J. Cardiol. **330**, 65–72 (2021). https://doi.org/10.1016/j.ijcard.2021.01.001
24. Lacomis, J.M., et al.: Dynamic multidimensional imaging of the human left atrial appendage. EP Eur. **9**(12), 1134–1140 (2007). https://doi.org/10.1093/europace/eum227
25. Watanabe, T., et al.: Dabigatran exhibits low intensity of left atrial spontaneous echo contrast in patients with nonvalvular atrial fibrillation as compared with warfarin. Heart Vessels **32**(3), 326–332 (2017). https://doi.org/10.1007/s00380-016-0871-5
26. Lip, G.Y., Lowe, G.D., Rumley, A., Dunn, F.G.: Fibrinogen and fibrin D-dimer levels in paroxysmal atrial fibrillation: evidence for intermediate elevated levels of intravascular thrombogenesis. Am. Heart J. **131**(4), 724–730 (1996). https://doi.org/10.1016/s0002-8703(96)90278-1
27. Fibrinogen Studies Collaboration et al.: Associations of plasma fibrinogen levels with established cardiovascular disease risk factors, inflammatory markers, and other characteristics: individual participant meta-analysis of 154,211 adults in 31 prospective studies: the fibrinogen studies collaboration. Am. J. Epidemiol. **166**(8), 867–879 (2007). https://doi.org/10.1093/aje/kwm191
28. Di Biase, L., et al.: Does the left atrial appendage morphology correlate with the risk of stroke in patients with atrial fibrillation? Results from a multicenter study. J. Am. Coll. Cardiol. **60**(6), 531–538 (2012). https://doi.org/10.1016/j.jacc.2012.04.032
29. Agmon, Y., Khandheria, B.K., Gentile, F., Seward, J.B.: Echocardiographic assessment of the left atrial appendage. J. Am. Coll. Cardiol. **34**(7), 1867–1877 (1999). https://doi.org/10.1016/S0735-1097(99)00472-6
30. Yosefy, C., et al.: Difference in left atrial appendage remodeling between diabetic and nondiabetic patients with atrial fibrillation. Clin. Cardiol. **43**(1), 71–77 (2019). https://doi.org/10.1002/clc.23292
31. Mannina, C., et al.: Association of left atrial strain with ischemic stroke risk in older adults. JAMA Cardiol. **8**(4), 317–325 (2023). https://doi.org/10.1001/jamacardio.2022.5449

UPCMR: A Universal Prompt-Guided Model for Random Sampling Cardiac MRI Reconstruction

Donghang Lyu[1(✉)], Chinmay Rao[1], Marius Staring[1], Matthias J. P. van Osch[1], Mariya Doneva[4], Hildo J. Lamb[1], and Nicola Pezzotti[2,3]

[1] Department of Radiology, Leiden University Medical Center, Leiden, The Netherlands
d.lyu@lumc.nl
[2] Cardiologs, Philips, Paris, France
[3] Faculty of Computer Science, Eindhoven University of Technology, Eindhoven, The Netherlands
[4] Philips Innovative Technologies, Hamburg, Germany

Abstract. Cardiac magnetic resonance imaging (CMR) is vital for diagnosing heart diseases, but long scan time remains a major drawback. To address this, accelerated imaging techniques have been introduced by undersampling k-space, which reduces the quality of the resulting images. Recent deep learning advancements aim to speed up scanning while preserving quality, but adapting to various sampling modes and undersampling factors remains challenging. Therefore, building a universal model is a promising direction. In this work, we introduce UPCMR, a universal unrolled model designed for CMR reconstruction. This model incorporates two kinds of learnable prompts, undersampling-specific prompt and spatial-specific prompt, and integrates them with a UNet structure in each block. Overall, by using the CMRxRecon2024 challenge dataset for training and validation, the UPCMR model highly enhances reconstructed image quality across all random sampling scenarios through an effective training strategy compared to some traditional methods, demonstrating strong adaptability potential for this task.

Keywords: CMR Reconstruction · Random Sampling · Prompts · UNet

1 Introduction

Cardiac Magnetic Resonance Imaging often requires multiple breath-holds for a comprehensive heart scan, leading to potential patient discomfort and slice misalignment. Moreover, the combination of data from multiple cardiac cycles can be problematic, especially in cases of arrhythmia, limiting real-time imaging to lower spatio-temporal resolutions. To address these challenges, undersampling techniques have been explored over the years to accelerate the whole scanning

O. Camara et al. (Eds.): STACOM 2024, LNCS 15448, pp. 453–463, 2025.
https://doi.org/10.1007/978-3-031-87756-8_44

process. In recent years, deep learning methods [12,15,16] have gained significant attention and are increasingly being used for MRI reconstruction, aiming to restore image details effectively, even with high acceleration factor. However, managing various undersampling scenarios with a single model remains challenging. The main difficulty lies in achieving and maintaining high adaptability across these varied conditions. Inspired by the rapid development of prompt learning [4,7,20], we introduce two kinds of learnable prompts, aiming to improve the adaptability across different contrasts, k-space trajectories, and acceleration factors.

Prompt learning adapts large pre-trained models, also known as foundation models, to new tasks by providing specific hints. It has been widely used in the natural language processing (NLP) [5,23] and computer vision (CV) [6,11,20] tasks. In the medical imaging field, prompts have been extensively utilized for segmentation tasks. Medical Segment Anything Model (MedSAM) [8] uses box prompt to provide spatial hints, enabling accurate segmentation of masks within the specified box. Uniseg [22] introduces a universal prompt, which generates corresponding task-specific prompts and improves the correlation among different tasks. For MRI reconstruction, PromptMR [20] employs learnable prompts to adapt itself to two different tasks: cine MRI reconstruction and T1-weighted (T1w) and T2-weighted (T2w) MRI reconstruction. Inspired by the general strategy of prompt-based guidance, we proposed an unrolled all-in-one model called UPCMR, which integrates learnable prompts into the UNet structure in each block so as to better adapt to different undersampling scenarios. In addition to model design, an effective training strategy is crucial for handling versatile random sampling. Inspired by the widespread use of curriculum learning [1] in similar tasks, we also explored its effectiveness with two formats and selected the most effective one as the final training strategy for the UPCMR method.

2 Preliminaries and Dataset

2.1 Preliminaries

Consider reconstructing a sequence of complex-valued MR images $x \in \mathbb{C}^{T \times H \times W}$ from multi-coil undersampled k-space measurements y, where H and W denote the height and width of each frame, respectively, T represents the sequence length. Therefore, the goal is to reconstruct x from y, formulated as

$$\mathrm{argmin}_x \|y - Ax\|_2^2 + \lambda \mathcal{R}(x), \tag{1}$$

where A is the linear forward operator composed of coil sensitivity encoding S, 2D Fourier transform $\mathcal{F}$, and undersampling mask M, $\mathcal{R}$ represents the regularisation terms with λ as a hyper-parameter controlling the regularization strength. For deep unrolled methods, $\mathcal{R}$ represents a trainable neural network block. Some previous deep unrolled models based on the ADMM optimization algorithm such as Deep-ADMM net [17] introduce an intermediate variable z, also computed via a learnable block. When constraining z to be equal to x, the above problem is reformulated as

$$\text{argmin}_{x,z} \|y - Ax\|_2^2 + \mu\|x - z\|_2^2 + \lambda\mathcal{R}(z). \tag{2}$$

The unrolled model thus learns a sequence of transition where x^i passes through a neural network block to produce z^i, and z^i is used to generate x^{i+1} through the data consistency (DC) layer. These operations are represented as

$$z^i = f_\theta^i(x^i), \tag{3}$$

$$x^{i+1} = DC(z^i, y, \lambda_0, \Omega) = A^\dagger \Lambda A z^i + \frac{\lambda_0}{1+\lambda_0} A^\dagger y \tag{4}$$

where $A^\dagger$ denotes the Hermitian operation of A, λ_0 is a regularization parameter and we make it approach infinity to ensure the preservation of sampled k-space information, Ω is an index set of the acquired k-space samples and Λ is a diagonal matrix, whose diagonal values are

$$\Lambda_{kk} = \begin{cases} 1 & if\ k \notin \Omega \\ \frac{1}{1+\lambda_0} & if\ k \in \Omega \end{cases} \tag{5}$$

2.2 CMRxRecon2024 Dataset

The CMRxRecon2024 dataset [19] includes data from 330 healthy volunteers scanned with 3 T magnets. Task 2 features multi-contrast k-space data (Cine, Aorta, Mapping, Tagging) with anatomical views such as long-axis, short-axis, LVOT, and aortic views (transversal and sagittal). Each contrast consists of 5–15 slices, segmented into 12–25 cardiac phases with a temporal resolution of 50 ms. Key geometrical parameters are a spatial resolution of $1.5 \times 1.5\,\mathbf{mm^2}$, slice thickness of 8.0 mm, and a slice gap of 4.0 mm. The dataset includes three k-space trajectories (uniform, Gaussian, pseudo radial) with temporal/parametric interleaving, and various acceleration factors (4×, 8×, 12×, 16×, 20×, 24×) without including autocalibration signal (ACS) for calculation. For the ACS area, it includes the central 16 lines for uniform and Gaussian types, and the central 16×16 regions for the pseudo radial type. Training data includes 200 subjects with fully-sampled k-space data across all combinations of k-space trajectories and acceleration factors, while 60 validation subjects are provided with random undersampling settings. The ground-truth of validation set is unavailable, and performance is assessed by submitting the reconstructed central cropped parts of the images to the platform.

3 Methodology

3.1 Model Structure

UPCMR. We follow an unrolled design used by some previous methods [12,20], as shown in Fig. 1. From a global perspective, in addition to common inputs like undersampling k-space, corresponding mask, coil-combined image sequence with 2 channels, and coil sensitivity map, an undersampling-specific prompt embedding is also provided as input. Given that both k-space trajectory and acceleration factor are prior information, we build two learnable prompt pools: one for

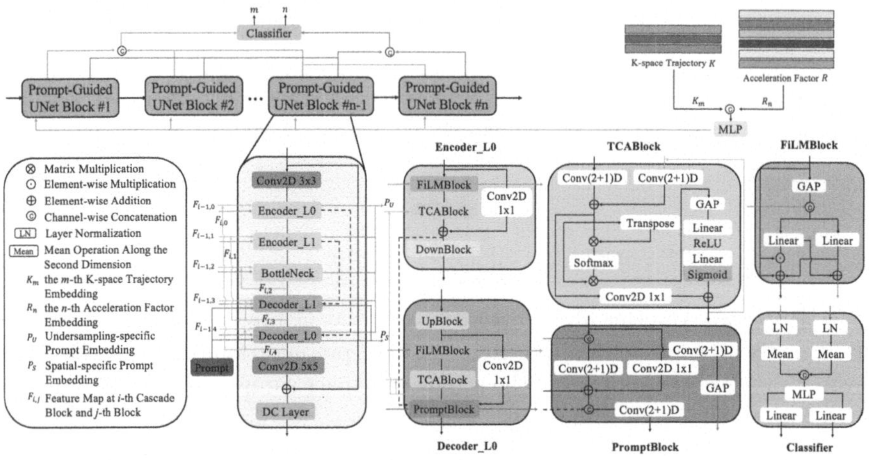

Fig. 1. Overview of the UPCMR model. In each cascade block, green lines represent feature maps transitioning from the previous one to the next block. Blue and red lines highlight the direction of undersampling-specific prompts and spatial-specific prompts, respectively. Black dashed lines refer to the skip connections between the corresponding encoder and decoder components. (Color figure online)

k-space trajectory $K \in \mathbb{R}^{3\times C}$ and another for the acceleration factor $R \in \mathbb{R}^{6\times C}$, where C is the embedding length, which matches the channel number of the UNet block. When selecting the mth k-space trajectory K_m and n^{th} acceleration factor R_n, these prompt embeddings are combined by a multi-layer perceptron (MLP) and fed into each cascade block to introduce undersampling-related information, enhancing model adaptability. UPCMR outputs include the reconstructed image series, as well as predictions for k-space trajectory and acceleration factor classes. Here, an additional classification block is used during training. The undersampling-specific prompt embedding P_U and the spatial-specific prompt embedding P_S from all cascade blocks are concatenated to make classification so as to further condition the model.

Prompt-Guided UNet Block. Each block of the cascading structure incorporates a 2-level UNet featuring some extra Conv2D layers and a bottleneck block that mirrors the encoder block's structure. The channel number is fixed at 64 across all layers, with downsampling and upsampling operations only occur at the first level. In the encoder part, for the input feature map at i^{th} UNet block and j^{th} encoder block, $F_{i,j} \in \mathbb{R}^{B\times T\times C\times H\times W}$, it first passes through a FiLMBlock [10], interacting with the undersampling-specific prompt $P_U \in \mathbb{R}^{B\times C}$, where B denotes batch size. $F_{i,j}$ is processed by global average pooling (GAP) operation to combine with P_U along the channel dimension. Then two separate linear layers generate weight embedding $W_P \in \mathbb{R}^{B\times T\times C}$ and bias embedding $B_P \in \mathbb{R}^{B\times T\times C}$. These embeddings are applied to $F_{i,j}$ through element-wise multiplication and addition to produce the output $F_{i,j+1}$ with prompt information.

Additionally, W_P and B_P add together and do average along the time dimension to form an updated undersampling-specific prompt embedding that integrates image sequence information. The updated undersampling-specific prompt embedding from each block are concatenated to form the final P_U of the current cascade block. After that, a temporal-channel attention block (TCABlock) is applied. Compared to traditional Conv3D or *Conv(2+1)D* [14] operations, the TCABlock incorporates a simple attention mechanism along temporal and spatial dimensions, aiming to better exploit the spatio-temporal correlation across the whole sequence. Furthermore, inspired by the CRNN-i operation from the CRNN-MRI [12] model, the feature maps from the previous cascade block are added to strengthen cascade connections and retain some important features. As shown in Fig. 1, two inputs, $F_{i,j}$ and $F_{i-1,j}$, pass through separate *Conv(2+1)D* layer and add together to obtain an intermediate feature map F_{tmp}. Then it multiplies with the transposed F_{tmp} to compute temporal correlations between frames and generate an attention map using the softmax function. This attention map is multiplied with F_{tmp} to produce the final output, incorporating temporal attention information. Channel attention is applied following the same design as the SE block [3]. After the TCABlock, the feature map is processed through a *Conv(2+1)D* layer, which either downsamples or maintains the current shape. Moreover, temporal circular padding [13] is used with all *Conv(2+1)D* layers to account for the circular nature of cardiac MRI.

The decoder blocks follow a design similar to the encoder blocks but include a PromptBlock with a learnable spatial-specific prompt $P_S \in \mathbb{R}^{B\times 1\times C\times H\times W}$, inspired by [11,20]. This addition helps the model adapt to different undersampling cases by enriching the spatial context. The PromptBlock's pipeline begins with concatenating $F_{i,j}$ with P_S along the channel dimension. This is followed by two branches: one generates the updated spatial-specific prompt embedding $P_S \in \mathbb{R}^{B\times C}$, while the other one uses convolutional layers to combine with the corresponding encoder output, generating the feature map for subsequent layers. Similar to the operation for undersampling-specific prompt embeddings, the updated spatial-specific prompt embedding from each block are concatenated to form the final P_S of the current cascade block.

3.2 Loss Function

Given the three outputs of UPCMR, the loss function comprises the classification loss $\mathcal{L}_{cls}$ and the reconstruction loss $\mathcal{L}_{rec}$. The $\mathcal{L}_{cls}$ is the sum of the cross-entropy losses for the k-space trajectory class and the acceleration factor class. The $\mathcal{L}_{rec}$ is the weighted sum of an L1 loss term and an SSIM loss term, defined as follows:

$$\mathcal{L}_{rec} = \lambda_{l1}\|I_{rec} - I_{gnd}\|_1 + \lambda_{ssim}(1 - SSIM(|I_{rec}|, |I_{gnd}|)), \tag{6}$$

where I_{rec} denotes the reconstructed CMR image sequence and I_{gnd} represents the original ground-truth image sequence, both of which are double-channeled. For the SSIM loss calculation, we use the absolute value to compute the loss. Based on the empirical study from [2], we set λ_{l1}=0.16 and λ_{ssim}=0.84. Finally, the overall loss function is as follows:

$$\mathcal{L} = \lambda_{cls}\mathcal{L}_{cls} + \mathcal{L}_{rec} \tag{7}$$

Considering classification as a minor task relative to the main reconstruction task, we set a relatively small weight of $\lambda_{cls} = 0.025$.

3.3 Training Strategy

For random undersampling, an efficient training strategy is essential. Given thousands of slices, each with $3 \times 6 = 18$ different undersampling scenarios, iterating through all in each epoch is time-consuming and impractical under the situation of limited GPU resource. Therefore, our policy is to randomly pick a slice with a random k-space trajectory and acceleration factor for each training sample. This approach highly reduces training time and ensures the model can still handle diverse scenarios.

Another challenge is the varying difficulty of reconstruction across different undersampling modes, where, for example, an acceleration factor of 24 is much more difficult than one of 4. Therefore, equally random sampling might not be an optimal strategy for the model training. Inspired by the concept of curriculum learning [1] and its extensive applications [7,21], we explore incorporating this method into our model training process. Firstly, we trained the model for 90 epochs, divided into four stages:

1. Train the model for 10 epochs using {*uniform*} k-space sampling and an acceleration factor of {4}.
2. Train the model for 20 epochs using {*uniform, gaussian*} k-space with the sampling probabilities {0.2, 0.8} and acceleration factors {4, 8, 12} with the sampling probabilities {0.04, 0.48, 0.48}.
3. Train the model for 30 epochs using {*uniform, gaussian, pseudo_radial*} k-space with the sampling probabilities {0.1, 0.1, 0.8} and acceleration factors {4, 8, 12, 16, 20, 24} with the sampling probabilities {0.02, 0.02, 0.03, 0.31, 0.31, 0.31}.
4. Train the model for 30 epochs using all the k-space and all the acceleration factors with the equal sampling probability.

In the above 4 stages, we start with easier cases to help the model grasp basic reconstruction principles and adapt to less noisy data. Gradually, we introduce more complex k-space trajectories and higher acceleration factors to increase difficulty. The final stage involves combined training, which further boosts the model's adaptability.

Although the above training strategy aligns with curriculum learning principles, using the same full model structure and expecting it to handle multiple acceleration factors may not be optimal. This approach increases the training difficulty of the UPCMR model, making it harder to learn and differentiate between these internal variations. Moreover, assigning the optimal sampling probabilities at each stage can be challenging, which may hinder the model's overall performance. Therefore, we propose an alternative training strategy applying curriculum learning in a sequential way, divided into 7 stages. In each stage, all three k-space trajectories are equally likely to be included, with a total of 50 training epochs. The key differences across stages lie in the number of cascade blocks in the UPCMR model and the introduction of acceleration factors, as detailed below:

1. Initialize the model with 3 cascade blocks and train for 50 epochs with an acceleration factor of $\{4\}$.
2. Add a new block to the model (total 4 blocks), train for 40 epochs with an acceleration factor of $\{8\}$, followed by 10 epochs using acceleration factors $\{4, 8\}$, with the sampling probabilities of $\{1/2, 1/2\}$.
3. Add a new block to the model (total 5 blocks), train for 40 epochs with an acceleration factor of $\{12\}$, followed by 10 epochs using acceleration factors $\{4, 8, 12\}$, with the sampling probabilities of $\{1/3, 1/3, 1/3\}$.
4. Add a new block to the model (total 6 blocks), train for 40 epochs with an acceleration factor of $\{16\}$, followed by 10 epochs using acceleration factors $\{4, 8, 12, 16\}$, with the sampling probabilities of $\{1/4, 1/4, 1/4, 1/4\}$.
5. Add a new block to the model (total 7 blocks), train for 40 epochs with an acceleration factor of $\{20\}$, followed by 10 epochs using acceleration factors $\{4, 8, 12, 16, 20\}$, with the sampling probabilities of $\{1/5, 1/5, 1/5, 1/5, 1/5\}$.
6. Add a new block to the model (total 8 blocks), train for 40 epochs with an acceleration factor of $\{24\}$, followed by 10 epochs using acceleration factors $\{4, 8, 12, 16, 20, 24\}$, with the sampling probabilities of $\{1/6, 1/6, 1/6, 1/6, 1/6, 1/6\}$.
7. Train the complete model for 50 epochs using all the k-space trajectories and all the acceleration factors with the equal sampling probability.

In the first 6 stages, the model learns one acceleration factor at a time in the first 40 epochs, followed by 10 epochs of fine-tuning with all learned acceleration factors for comprehensive adjustment. Within each stage, a new cascade block is added to align with the increasing difficulty of the acceleration factor. Additionally, initializing the previous blocks with the trained weights from the last stage, we ensure smoother progression and better model adaptation. In the final stage, the model is trained with all possible scenarios to further refine its performance and ensure robust adaptability. In general, this training strategy follows a sequential approach with respect to model structure and acceleration factors, gradually increasing the model's complexity and training samples' difficulty to align with the principles of curriculum learning, helping the model solidify its understanding of internal differences between various undersampling scenarios.

4 Experiments

4.1 Implementation Details

During training, we generated the coil sensitivity map (CSM) from the time-averaged autocalibration signals (ACS) using the ESPIRIT algorithm [18]. The training data was normalized by first transforming the k-space data to the image domain with the Inverse Fast Fourier Transform (IFFT), normalizing the multi-coil images by their maximum absolute value, and then transforming the data back to k-space domain with the Fast Fourier Transform (FFT). Additionally, to further stabilize the model training, we incorporated z-score normalization and un-normalization at the start and end of each block in the cascade.

The models were implemented in Pytorch and trained on Nvidia A100 GPU with 80 GB of GPU memory. To optimize GPU memory usage, we utilized mixed precision training [9], enabling us to use a UPCMR model with up to 8 blocks. In the first training strategy, we used AdamW as the optimizer with an initial learning rate of 2e-4 and a weight decay of 1e-3. The learning rate was reduced by a factor of 0.8 every 10 epochs. The second training strategy also used AdamW with an initial learning rate of 2e-4 and a weight decay of 1e-2 but introduced a cosine scheduler with a 6-epoch warm-up, and the minimal learning rate was set to 2e-5. In the final stage, the initial learning rate was 1e-4, reduced by a factor of 0.8 every 5 epochs, with a minimum of 8e-6, dropping to 1e-6 in the last epoch. The batch size was set to 1. During training, we split 1380 samples for training and 16 samples for validation. Each validation sample was subjected to 18 different undersampling scenarios and each slice to increase the number of validation samples and better measure the model's overall performance. For evaluation, peak signal-to-noise ratio (PSNR), structural similarity (SSIM) and normalized mean squared error (NMSE) were used as image quality metrics, focusing on the cropped central region of each validation case.

4.2 Results

Table 1. Results of the methods on the central cropped area of validation set from the leaderboard. CL_1 refers to the first training strategy, while CL_2 is the second one. The best result is highlighted in bold.

Method	PSNR ↑	SSIM ↑	NMSE ↓
ZF	19.4182	0.4610	0.2853
SENSE	24.4780	0.6200	0.1734
GRAPPA	24.9678	0.6449	0.1554
UPCMR w/o CL	25.3694	0.6854	0.1422
UPCMR-CL_1 w/o FiLMBlock	25.2132	0.6755	0.1484
UPCMR-CL_1 w/o TCABlock	25.0871	0.6718	0.1487
UPCMR-CL_1 w/o PromptBlock	25.4463	0.6858	0.1409
UPCMR-CL_1	25.6070	0.6904	0.1391
UPCMR-CL_2	**26.6361**	**0.7338**	**0.1036**

We compare the proposed model with two traditional methods, SENSE and GRAPPA, on the given validation set. Some ablation studies evaluate (1) the effectiveness of curriculum learning versus a combined learning strategy, where

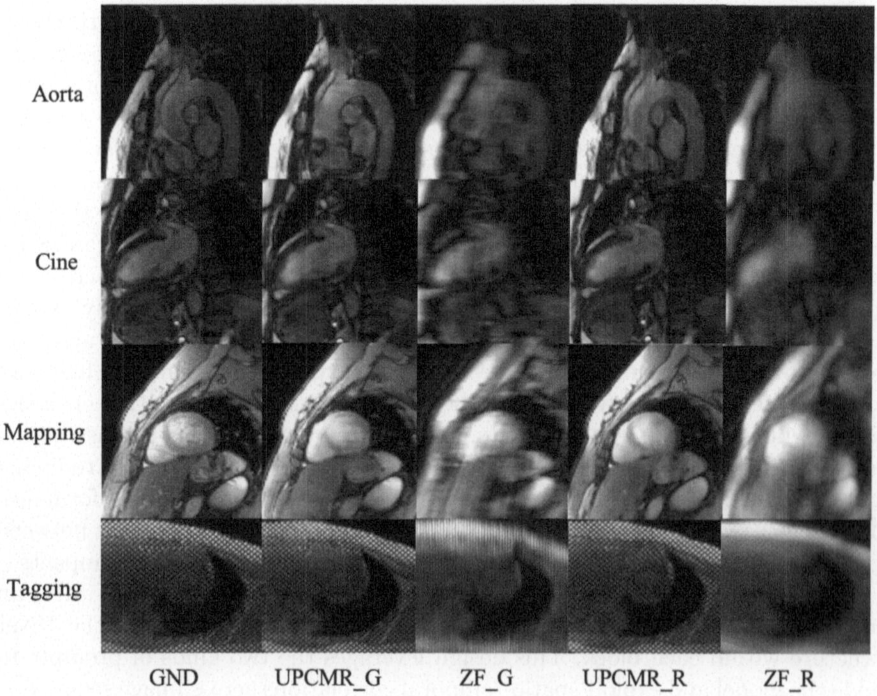

Fig. 2. Visualization of UPCMR reconstruction results for four contrasts under two k-space trajectories with an acceleration factor of 20. 'GND' indicates images from fully-sampled k-space, 'ZF' denotes images from zero-filled undersampled k-space, 'G' refers to Gaussian k-space, and 'R' stands for pseudo radial k-space. Images are cropped to focus on the central cardiac region. Rows from top to bottom show Aorta (sagittal), Cine (LAX), Mapping (T1), and Tagging.

equal sampling probabilities are assigned to each combination of k-space trajectories and acceleration factors in the whole training procedure, (2) the impact of key components, FiLMBlock, TCABlock, and PromptBlock, on UPCMR's performance under the first curriculum learning strategy, TCABlock is replaced with a Conv3D layer to assess its effect, and (3) the relative performance of two different curriculum learning strategies. Results are summarized in Table 1. Then Fig. 2 visualizes the reconstruction performance of UPCMR using the second curriculum learning strategy under some challenging undersampling scenarios from the splitting validation set.

5 Discussion and Conclusion

Table 1 shows that each key component block enhances performance, and the curriculum learning strategy outperforms the combined learning approach. Moreover, the second curriculum training strategy has been proven more effective, emphasizing the importance of learning each acceleration factor gradually and making the model size align with it. However, as shown in Fig. 2, UPCMR struggles with high acceleration factors, resulting in detail loss and blurriness in the cardiac region, indicating room for improvement. For the second curriculum learning strategy, due to its time-consuming nature, we only set 50 training epochs for each stage. Consequently, the model did not fully converge at each stage, limiting the upper bound of final reconstruction performance. Training the UPCMR model with more epochs at each stage could further improve the final performance. Additionally, the UPCMR model may have some design drawbacks. While TCABlock is intended to capture spatio-temporal correlations across the entire sequence, its simple structure and operation might not effectively exploit these features, potentially limiting performance, especially at higher acceleration factors. To further enhance the model, incorporating techniques like convolutional recurrent operations could be considered. Additionally, there is a significant difference between contrasts, which can complicate the reconstruction process. The current UPCMR model doesn't consider about it, and introducing contrast-specific prompts might further enhance the reconstruction performance for each contrast. Finally, considering the imbalance in training samples between contrasts, additional data processing is required to mitigate negative impacts.

To summarize the contribution of this work, we introduce UPCMR, an unrolled model that integrates two kinds of learnable prompts into the UNet structure within each block. This design leverages the two kinds of prompts to guide the model and exploit spatio-temporal correlations across image sequences, thereby enhancing reconstruction performance and improving adaptability to various random sampling scenarios. Furthermore, we explored two variants of the curriculum learning training strategy and selected the more effective one for versatile random sampling scenarios. Nonetheless, further refinements to the training process and model design are still needed to realize the full potential of UPCMR method.

Acknowledgement. This work is part of the project ROBUST: Trustworthy AI-based Systems for Sustainable Growth with project number KICH3.LTP.20.006, which is (partly) financed by the Dutch Research Council (NWO), Philips Research, and the Dutch Ministry of Economic Affairs and Climate Policy (EZK) under the program LTP KIC 2020-2023.

References

1. Bengio, Y., Louradour, J., Collobert, R., et al.: Curriculum learning. In: Proceedings of the 26th Annual International Conference on Machine Learning, pp. 41–48 (2009)
2. Zhao, H., Gallo, O., Frosio, I., Kautz, J.: Loss functions for image restoration with neural networks. IEEE Trans. Comput. Imag. **3**(1), 47–57 (2017)
3. Hu, J., Shen, L., Sun, G.: Squeeze-and-excitation networks. In: CVPR, pp. 7132–7141 (2018)
4. Jia, M., et al.: Visual prompt tuning. In: European Conference on Computer Vision, pp. 709–727. Springer (2022)

5. Wei, J., et al.: Chain-of-thought prompting elicits reasoning in large language models. In: Advances in Neural Information Processing Systems, vol. 35, pp. 24824–24837 (2022)
6. Kirillov, A., Mintun, E., Ravi, N., et al.: Segment anything. In: Proceedings of the IEEE/CVF International Conference on Computer Vision, pp. 4015–4026 (2023)
7. Kong, X., Dong, C., Zhang, L.: Towards Effective Multiple-in-One Image Restoration: A Sequential and Prompt Learning Strategy. arXiv preprint arXiv:2401.03379 (2024)
8. Ma, J., He, Y., Li, F., et al.: Segment anything in medical images. Nat. Commun. **15**(1), 654 (2024)
9. Micikevicius, P., Narang, S., Alben, J., et al.: Mixed precision training. In: ICLR (2018)
10. Perez, E., Strub, F., De Vries, H., et al.: Film: visual reasoning with a general conditioning layer. In: Proceedings of the AAAI Conference on Artificial Intelligence, vol. 32, no. 1 (2018)
11. Potlapalli, V., et al.: Promptir: prompting for all-in-one image restoration. In: Advances in Neural Information Processing Systems, vol. 36 (2024)
12. Qin, C., et al.: Convolutional recurrent neural networks for dynamic MR image reconstruction. IEEE TMI **38**(1), 280–290 (2018)
13. Sandino, C.M., et al.: Compressed sensing: From research to clinical practice with deep neural networks: shortening scan times for magnetic resonance imaging. IEEE Signal Process. Mag. **37**(1), 117–127 (2020)
14. Sandino, C.M., et al.: Accelerating cardiac cine MRI using a deep learning-based ESPIRiT reconstruction. Magn. Reson. Med. **85**(1), 152–167 (2021)
15. Schlemper, J., et al.: A deep cascade of convolutional neural networks for dynamic MR image reconstruction. IEEE TMI **37**(2), 491–503 (2017)
16. Sriram, A., et al.: End-to-end variational networks for accelerated MRI reconstruction. In: Martel, A.L., et al. (eds.) MICCAI 2020. LNCS, vol. 12262, pp. 64–73. Springer, Cham (2020). https://doi.org/10.1007/978-3-030-59713-9_7
17. Sun, J., Li, H., Xu, Z.: Deep ADMM-Net for compressive sensing MRI. In: Advances in Neural Information Processing Systems, vol. 29 (2016)
18. Uecker, M., Lai, P., Murphy, M.J., et al.: ESPIRiT-an eigenvalue approach to autocalibrating parallel MRI: where SENSE meets GRAPPA. Magn. Reson. Med. **71**(3), 990–1001 (2014)
19. Wang, Z., Wang, F., Qin, C., et al.: CMRxRecon2024: A Multi-Modality, Multi-View K-Space Dataset Boosting Universal Machine Learning for Accelerated Cardiac MRI. arXiv preprint arXiv:2406.19043 (2024)
20. Xin, B., et al.: Fill the k-space and refine the image: prompting for dynamic and multi-contrast MRI reconstruction. In: International Workshop on Statistical Atlases and Computational Models of the Heart, pp. 261–273. Springer, Cham (2023). https://doi.org/10.1007/978-3-031-52448-6_25
21. Xu, B., Zhang, L., Mao, Z., et al.: Curriculum learning for natural language understanding. In: Proceedings of the 58th Annual Meeting of the Association for Computational Linguistics, pp. 6095–6104 (2020)
22. Ye, Y., Xie, Y., Zhang, J., Chen, Z., Xia, Y.: UniSeg: a prompt-driven universal segmentation model as well as a strong representation learner. In: Greenspan, H., et al. (eds.) Medical Image Computing and Computer Assisted Intervention - MICCAI 2023. MICCAI 2023. Lecture Notes in Computer Science, vol. 14222. Springer, Cham (2023). https://doi.org/10.1007/978-3-031-43898-1_49
23. Zhang, Z., Zhang, A., Li, M., et al.: Automatic chain of thought prompting in large language models. In: The 11th International Conference on Learning Representations (ICLR) (2023)

An All-in-One Approach for Accelerated Cardiac MRI Reconstruction

Kian Anvari Hamedani, Narges Razizadeh, Shahabedin Nabavi(✉), and Mohsen Ebrahimi Moghaddam

Faculty of Computer Science and Engineering, Shahid Beheshti University, Tehran, Iran
{k.anvarihamedani,n.razizade}@mail.sbu.ac.ir,
{s_nabavi,m_moghadam}@sbu.ac.ir
https://github.com/kiananvari/CMRRecon

Abstract. Cardiovascular magnetic resonance (CMR) imaging is the gold standard for diagnosing several heart diseases due to its non-invasive nature and proper contrast. MR imaging is time-consuming because of signal acquisition and image formation issues. Prolonging the imaging process can result in the appearance of artefacts in the final image, which can affect the diagnosis. It is possible to speed up CMR imaging using image reconstruction based on deep learning. For this purpose, the high-quality clinical interpretable images can be reconstructed by acquiring highly undersampled k-space data, that is only partially filled, and using a deep learning model. In this study, we proposed a stepwise reconstruction approach based on the Patch-GAN structure for highly undersampled k-space data compatible with the multi-contrast nature, various anatomical views and trajectories of CMR imaging. The proposed approach was validated using the CMRxRecon2024 challenge dataset and outperformed previous studies. The structural similarity index measure (SSIM) values for the first and second tasks of the challenge are 99.07 and 97.99, respectively. This approach can accelerate CMR imaging to obtain high-quality images, more accurate diagnosis and a pleasant patient experience.

Keywords: Fast MRI · Cardiac MRI · MR Reconstruction

1 Introduction

Cardiovascular magnetic resonance (CMR) imaging is a powerful and versatile tool in cardiology, offering detailed and comprehensive information on cardiac anatomy, function, and tissue characteristics [1]. Its non-invasive nature, lack of ionizing radiation, and ability to provide quantitative and reproducible data make it an invaluable resource in diagnosing, treating, and managing a wide range of cardiac conditions [22]. Despite its many advantages, CMR imaging faces challenges related to patient factors and technical limitations [18]. Patients are often required to hold their breath during image acquisition to minimize

O. Camara et al. (Eds.): STACOM 2024, LNCS 15448, pp. 464–475, 2025.
https://doi.org/10.1007/978-3-031-87756-8_45

motion artefacts [13,14]. This can be challenging for patients with respiratory issues or those who cannot hold their breath for prolonged periods. The bounded space of the MR scanner can induce claustrophobia in some patients, making it difficult for them to remain still and calm during the scan time [6]. Thus, reducing acquisition time is one of the main objectives of researchers in this field.

Several reconstruction techniques, such as parallel imaging, compressed sensing, real-time imaging, optimized k-space sampling, AI-enhanced methods, and optimized imaging protocols, can significantly reduce scan times in CMR imaging. Traditional CMR image reconstruction relies on the inverse Fourier transform, which requires full k-space data for high-quality images. Parallel Imaging techniques like sensitivity encoding (SENSE) [16] and generalized auto-calibrating partially parallel acquisitions (GRAPPA) [7] reduced scan time by under-sampling k-space and reconstructing images using coil sensitivity profiles. Compressed sensing (CS) [11] proposed a reconstruction method from under-sampled k-space data utilizing the sparsity of MR images in specified domains. Non-Cartesian k-space sampling [3] and partial Fourier imaging [4] can also be mentioned among the fast MRI reconstruction methods. In recent years, deep learning algorithms have been widely exploited for high-quality image reconstruction from sparsely sampled k-spaces, reducing extensive data acquisition [12]. Convolutional neural networks (CNNs) [20,26,28], generative adversarial networks (GANs) [27], recurrent neural networks (RNNs) [17] and hybrid approaches [8] have demonstrated superior performance than other methods.

The current study proposes a universal approach to multi-contrast CMR reconstruction with various acceleration factors and k-space trajectories in the form of an integrated model. The contributions of the study include the following:

- An all-in-one approach is proposed based on the Patch-GAN architecture that reconstructs the subsampled k-space in a stepwise manner.
- The curriculum learning approach is used in the model training process to achieve better performance.
- The combined loss function was defined by considering the physical properties of the k-space and features of the image domain.
- The stepwise loss calculation was used to overcome the vanishing gradient problem.

2 Materials and Methods

The details of the datasets, the proposed approach, and the training and evaluation procedure are described in this section.

2.1 Datasets

The datasets used for training the proposed method include CMRxRecon2023 [23] and CMRxRecon2024 [25]. Both were acquired using a 3T scanner (MAGNETOM Vida, Siemens Healthineers, Germany) with a dedicated 32-channel cardiac coil, which was compressed into ten virtual coils. Participants were scanned in a prone position, with the cardiac cycle segmented into 12–25 phases, achieving a temporal resolution of approximately 50 ms, depending on the heart rate. Both datasets feature a spatial resolution of $1.5 \times 1.5\,\text{mm}^2$ and a slice thickness of 8 mm. The CMRxRecon2023 includes cine imaging using the TrueFISP sequence and T1/T2 mapping with MOLLI-FLASH and T2prep-FLASH sequences, respectively, across various anatomical views such as short-axis and long-axis (2, 3, and 4 chambers). This dataset consists of a training set with 120 fully sampled subjects and uses 4x, 8x, and 10x uniform acceleration masks for under-sampling the complete k-spaces. The CMRxRecon2024 expands upon this with additional contrasts and views, including tagging, aorta, flow2D, and BlackBlood contrasts, incorporating diverse anatomical perspectives such as the left ventricular outflow tract (LVOT) for the cine contrast and aortic views (transversal and sagittal). Flow2D and BlackBlood contrast images are used exclusively for validation as unseen data. This dataset contains a larger training set with 200 fully sampled subjects. It provides a variety of k-space under-sampling trajectories (uniform, Gaussian, pseudo-radial) with multiple acceleration factors (4x, 8x, 12x, 16x, 20x, 24x) than the CMRxRecon2023 under-sampling trajectories.

2.2 The Proposed Approach

The optimization problem of Eq. (1) must be solved based on the compressed sensing theory [5] to obtain the high-quality clinical interpretable image x from the highly subsampled multi-coil k-space y.

$$\min_{x} \frac{1}{2}\|y - Ax\|_2^2 + \lambda R(x) \tag{1}$$

where A and $R(x)$ are measurement matrix and sparsity regularization term of x, respectively. λ is used as a weighting coefficient for the $R(x)$.

This optimization problem has been solved iteratively by [21] in the form of Eq. (2).

$$k^{(t+1)} \leftarrow k^{(t)} - \eta^{(t)} * \mathcal{M} * (k^{(t)} - k^{(0)}) + G_k \tag{2}$$

where in an iterative quality improvement process, $k^{(t)}$ and $k^{(t+1)}$ are the k-spaces of t and $t+1$ stages. So, $k^{(0)}$ is the initial adjacent multi-coil subsampled k-spaces $[..., k^{(t)}_{central-1}, k^{(t)}_{central}, k^{(t)}_{central+1}, ...]$ at $t = 0$. These adjacent k-spaces are similar slices in adjacent cardiac phases/weightings of $k^{(t)}_{central}$. $\eta^{(t)}$ and $\mathcal{M}$ are also learnable step sizes and subsampling masks, respectively. G_k, the gradient of the sparsity regularization term $R(x)$, is obtained in a sequence of processes to complete the above equation (See Algorithm 1).

Figure 1 shows the overview of the proposed approach. This approach is proposed based on the Patch-GAN architecture [9] which ensures that regions of the reconstructed image are indistinguishable from the corresponding ones in

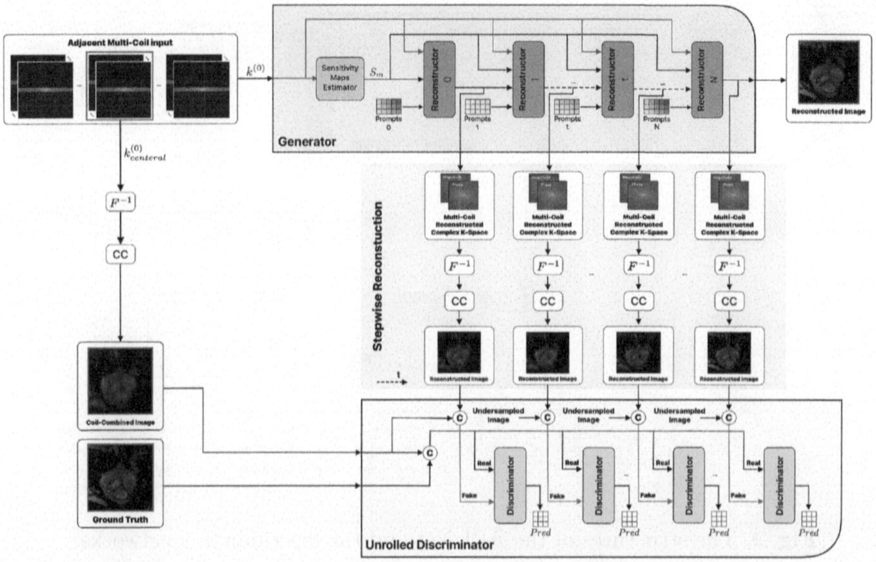

Fig. 1. The overview of the proposed approach.

the ground-truth image. $k^{(0)}$ is received from the input, and in a stepwise reconstruction process, a high-quality image is obtained. In the stepwise reconstruction process, $k^{(t+1)}$ is generated through Eq. (2) as an output of reconstructor t. In Fig. 1, each reconstructor module performs a sequence of operations to reconstruct $k^{(t+1)}$, in part of which an attention-based prompt UNet (APUNet) including prompt blocks inspired by [15, 26] is used. A fixed prompt is a vector of fixed length and random values. Each APUNet receives fixed prompts as many as its levels. At each level of APUNet, a fixed prompt and features obtained from a lower-level block are entered into the discriminative prompt block module. This module generates an input-type adaptive prompt that helps APUNet adjust the reconstruction process to the input type at each level. Upon receiving $k^{(0)}$, $k^{(t)}$, conjugate symmetric sensitivity map $S_m^{'}$, and prompt $P^{(t)}$, each reconstructor module tries to reconstruct $k^{(t+1)}$. From the output of each reconstructor module, a physical loss $\mathscr{L}_{Phys}^{(t)}$ is calculated in the frequency space, including the sum of the mean squared errors (MSE) of the magnitude (Mag) and phase (Φ) of $k_{centeral}^{(t+1)}$ and the ground-truth k-space $k^{(G)}$. $k_{centeral}^{(t+1)}$ represents the central k-space, which the reconstruction process is performed to refine, without considering its adjacent k-spaces. Besides, the loss function $\mathscr{L}_{SSIM}^{(t)}$ is computed using the structural similarity index measure (SSIM) [24] by transforming $k_{centeral}^{(t+1)}$ and $k^{(G)}$ to the image space using the inverse fast Fourier transform (FFT) and then coil combination (CC), which is added to the physical loss $\mathscr{L}_{Phys}^{(t)}$ to form stepwise loss $\mathscr{L}_{step}^{(t)}$.

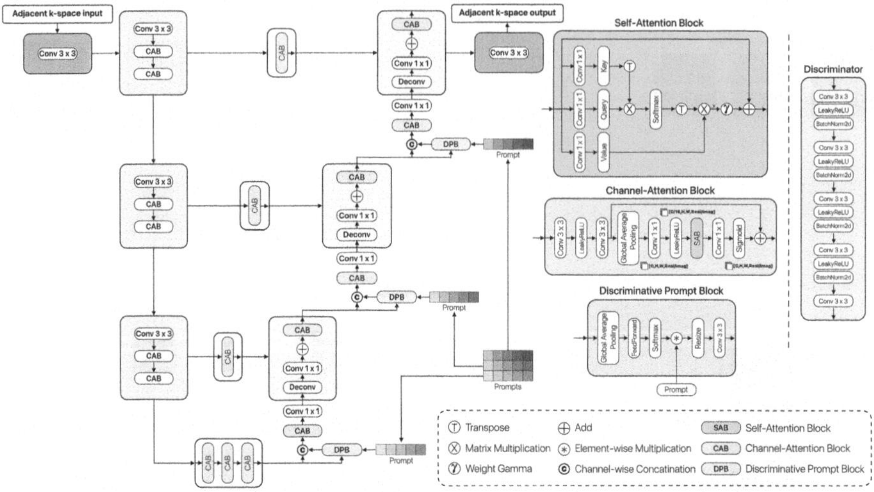

Fig. 2. The structures of the APUNet and the discriminator networks.

Zero-filled image concatenation with the output of each reconstructor and the ground-truth image to create $Input_{real}$ and $Input_{fake}$ causes the discriminator to learn the distortions resulting from subsampling. This issue can help the generator to achieve more efficient performance in the face of unseen data. Given $Input_{real}$ and $Input_{fake}$, the discriminator network predicts $Pred_{real}$ and $Pred_{fake}$, then uses binary cross-entropy (BCE) [19] loss to approximate the discriminator loss $\mathscr{L}_{Disc}$, which is used to learn the discriminator network. This process is repeated t times until the final $k^{(t+1)}$ is achieved. Using the output of each reconstructor module for the discriminator network training provides faster convergence. In the last stage, the generator loss $\mathscr{L}_{GAN-Gen}$ should be calculated in the image space and added to the sum of the step losses $\mathscr{L}_{step}^{(t)}$. The generator's loss function is estimated by Eq. (3). This loss function is leveraged to learn the generator network, learnable step size $\eta^{(t)}$, and the sensitivity map estimator network, which is an APUNet. The task of the sensitivity map estimator is to receive the concatenated coil data to predict the sensitivity maps. This concatenation approach allows for flexibility in receiving data from the different number of coils. The APUNet and discriminator network structures are shown in Fig. 2. The reason for using channel-attention blocks is the presence of adjacent k-spaces as input so that the features extracted from the adjacent k-spaces can be considered for the reconstruction of the central k-space. The training process of the proposed approach is summarized in Algorithm 1.

$$\mathscr{L}_{Gen} = \lambda * \sum_{i=1}^{N} \mathscr{L}_{step}^{(t)} + \mathscr{L}_{GAN-Gen} \tag{3}$$

where λ indicates the stepwise loss adjustment coefficient.

Algorithm 1. The Proposed All-in-one Approach for CMRI Reconstruction

Require:

▷ $k^{(0)}$: Adjacent multi-coil subsampled k-spaces
▷ $k^{(G)}$: Ground-truth multi-coil k-space
▷ $P^{(t)}$: Discriminative prompt
▷ $\mathscr{M}$: Subsampling mask
▷ $\eta^{(t)}$: Learnable step size
▷ $Label_{real}$: Zero-filled matrix
▷ $Label_{fake}$: One-filled matrix
▷ λ : Stepwise loss adjustment coefficient

Ensure:

▷ θ_{Gen} : Weights of the generator network

$k^{(ACS)} \leftarrow$ Auto_calibration_signal($k^{(0)}$)
$I^{(ACS)} \leftarrow \mathrm{FFT}^{-1}(k^{(ACS)})$
$S_m \leftarrow \mathrm{APUNet}^{(\mathrm{SME})}(I^{(ACS)})$
$S_m^{'} \leftarrow$ Conjugate_Symmetry(S_m)
for *iteration* $= 1, 2, \ldots$ **do**
  for $t = 0, 1, 2, \ldots$ **do**
    $I_{MC} \leftarrow \mathrm{FFT}^{-1}(k^{(t)})$
    $I_{SC} \leftarrow \mathrm{Multiplication}(S_m^{'}, I_{MC})$
    $I_{RF} \leftarrow \mathrm{APUNet}^{(t)}(I_{SC}, P^{(t)})$
    $I_{RF}^{'} \leftarrow$ Repeat_interleaved(I_{RF})
    $I_{SS} \leftarrow \mathrm{Multiplication}(I_{RF}^{'}, S_m)$
    $G_k \leftarrow \mathrm{FFT}(I_{SS})$
    $k^{(t+1)} \leftarrow k^{(t)} - \eta^{(t)} * \mathscr{M} * (k^{(t)} - k^{(0)}) + G_k$
    $\mathscr{L}_{Phys}^{(t)} \leftarrow \mathrm{MSE}(Mag(k_{centeral}^{(t+1)}), Mag(k^{(G)})) + \mathrm{MSE}(\Phi(k_{centeral}^{(t+1)}), \Phi(k^{(G)}))$
    $\mathscr{L}_{SSIM}^{(t)} \leftarrow$ SSIM_Loss($CC(FFT^{-1}(k_{centeral}^{(t+1)})), CC(FFT^{-1}(k^{(G)}))$)
    $\mathscr{L}_{step}^{(t)} \leftarrow \mathscr{L}_{Phys}^{(t)} + \mathscr{L}_{SSIM}^{(t)}$
    $Input_{real} \leftarrow \mathrm{Concat}(CC(FFT^{-1}(k^{(G)})), CC(FFT^{-1}(k_{centeral}^{(0)})))$
    $Input_{fake} \leftarrow \mathrm{Concat}(CC(FFT^{-1}(k_{centeral}^{(t+1)})), CC(FFT^{-1}(k_{centeral}^{(0)})))$
    $Pred_{real}, Pred_{fake} \leftarrow \mathrm{Discriminator}(Input_{real}), \mathrm{Discriminator}(Input_{fake})$
    $\mathscr{L}_{real} \leftarrow$ BCE_Loss($Label_{real}, Pred_{real}$)
    $\mathscr{L}_{fake} \leftarrow$ BCE_Loss($Label_{fake}, Pred_{fake}$)
    $\mathscr{L}_{Disc} \leftarrow \frac{1}{2}(\mathscr{L}_{real} + \mathscr{L}_{fake})$
    $\theta_{Disc} \leftarrow$ Adamw_Optimizer ($\mathscr{L}_{Disc}, \theta_{Disc}$)
  end for
  $Input_{fake} \leftarrow \mathrm{Concat}(CC(FFT^{-1}(k_{centeral}^{(t+1)})), CC(FFT^{-1}(k_{centeral}^{(0)})))$
  $Pred_{fake} \leftarrow \mathrm{Discriminator}(Input_{fake})$
  $\mathscr{L}_{GAN-Gen} \leftarrow$ BCE_Loss($Label_{real}, Pred_{fake}$)
  $\mathscr{L}_{Gen} \leftarrow \lambda * \sum_{i=1}^{N} \mathscr{L}_{step}^{(t)} + \mathscr{L}_{GAN-Gen}$
  $\theta_{Gen}, \theta_{SME} \leftarrow$ Adamw_Optimizer ($[\mathscr{L}_{Gen}], \theta_{Gen}, \theta_{SME}$)
end for
return θ_{Gen} ▷ Weights of the generator network

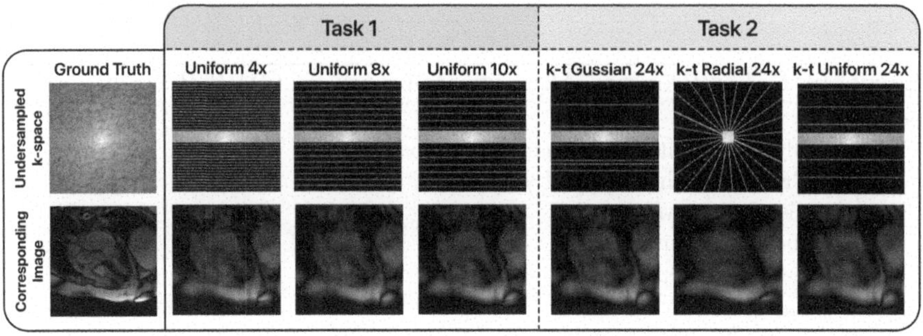

Fig. 3. An example of several k-space subsampling masks and corresponding images.

In our approach to reconstructing undersampled k-spaces using a stepwise Patch-GAN architecture, specific hyperparameters were employed in both the generator and discriminator networks. We deployed the AdamW optimizer [10] with a learning rate of 0.002, weight decay of 0.1, and a gradient clipping value of 0.1. The learning rate was scheduled to be adjusted by a scheduler with a step size of 11 and a gamma of 0.1. The generator network consists of 12 reconstructor modules, and we used 16 auto-calibration lines for the undersampled k-spaces with an adjacent k-space length of 5. Curriculum learning [2] starts with lower acceleration factors, allowing the model to grasp the fundamentals of reconstruction in an easy-to-hard manner. As the model progressively learned to handle these tasks, we used transfer learning to fine-tune it with higher acceleration factors. This method enabled the model to build on its initial understanding and adapt to more complex scenarios, gradually increasing the complexity of the task and improving overall performance and efficiency. The training was done in 12 epochs with a batch size of 1. Besides, the performance of our proposed method is evaluated using the SSIM, peak signal-to-noise ratio (PSNR), and normalized MSE (NMSE) metrics [24]. The model has about 100 million parameters, so to support the computational requirements of the proposed model, 2x NVIDIA GPU H100 were used for training and evaluations.

3 Experimental Results and Discussion

In this section, the evaluation results of the proposed approach are presented with extensive experiments, and then the role of each component in the approach is analyzed by ablation studies. An example of several k-space subsampling patterns of the CMRxRecon2024 challenge is shown in Fig. 3. In this figure, the result of k-space masking using the masks introduced is depicted in the image space after the inverse Fourier transform.

The experimental results related to evaluating the proposed approach in the two tasks introduced by the CMRxRecon2024 challenge are shown in Table 1. In Task 1, three different models for three acceleration factors of 4, 8, and 10x of uniform subsampling have been trained using the curriculum learning method.

The average results of the three trained models are presented separately for different contrasts based on NMSE, PSNR and SSIM metrics. In task 2, a model has been trained using k-space data with various trajectories and subsampling acceleration factors of 4 to 24x using the curriculum learning method. Besides, the results of the proposed approach have been compared with two image reconstruction methods and the state-of-the-art PromptMR [26] method, which is the first rank of the CMRxRecon2023 challenge. An example of a qualitative comparison of images reconstructed from subsampled k-space, reconstructed by the proposed approach and ground truth images is shown in Fig. 4.

Table 1. The results of the proposed approach versus various MRI reconstruction methods and the state-of-the-art one for the first and second tasks of the CMRxRecon2024 challenge. The results are reported based on NMSE/PSNR/SSIM metrics, and the best values are shown in boldface.

Task	Method	Cine			Mapping		Aorta		Tagging
		SAX	LAX	LVOT	T1W	T2W	TRA	SAG	SAX
1	PromptIR [15]	2.6/40.05/0.9643	2.8/38.42/0.9567	2.9/38.12/0.9513	2.4/38.87/0.9612	1.6/42.38/0.9778	2.0/41.21/0.9689	2.2/40.61/0.9743	1.9/41.41/0.9786
	E2E-Varnet [21]	1.7/41.95/0.9723	2.3/39.23/0.9632	2.1/39.63/0.9648	1.7/42.27/0.9733	1.6/42.32/0.9763	1.9/41.62/0.9754	2.0/41.59/0.9739	1.5/42.56/0.9791
	PromptMR [26]	1.1/45.34/0.9821	1.2/43.27/0.9814	1.2/43.32/0.9829	1.1/45.23/0.9806	0.8/46.15/0.9878	0.9/45.67/0.9873	1.1/45.12/0.9813	0.7/46.34/0.9912
	Proposed Approach	**0.8/47.13/0.9903**	**0.9/46.12/0.9893**	**0.9/46.40/0.9901**	**0.6/48.34/0.9897**	**0.6/48.50/0.9919**	**0.7/47.54/0.9914**	**0.8/46.81/0.9897**	**0.5/49.80/0.9931**
2	PromptIR [15]	3.2/37.95/0.9468	3.3/37.12/0.9351	3.4/37.00/0.9300	3.0/36.87/0.9505	2.5/40.32/0.9578	2.8/39.62/0.9589	3.0/38.42/0.9545	2.5/40.41/0.9686
	E2E-Varnet [21]	2.7/38.42/0.9531	3.0/38.29/0.9472	2.9/38.63/0.9450	2.6/39.27/0.9533	2.2/41.38/0.9653	2.6/40.21/0.9654	2.8/39.61/0.9643	2.5/41.56/0.9691
	PromptMR [26]	2.1/40.13/0.9649	2.2/41.25/0.9654	2.2/41.32/0.9629	2.1/42.23/0.9606	1.5/44.15/0.9778	1.6/44.21/0.9773	1.8/43.12/0.9713	1.4/45.34/0.9812
	Proposed Approach	**1.4/42.52/0.9774**	**1.4/42.43/0.9782**	**1.3/43.40/0.9781**	**1.0/45.34/0.9717**	**1.0/46.50/0.9829**	**1.1/45.54/0.9824**	**1.2/44.81/0.9797**	**0.9/48.80/0.9891**

Comparing our approach with the E2E-Varnet and PromptMR methods, a notable distinction lies in the strategies employed for image reconstruction. E2E-Varnet integrates the learning of sensitivity maps during model training and k-space intermediate quantities utilization for reconstruction. This approach has shown superior performance in estimating sensitivity maps when presented with limited sample lines for auto-calibration signals. PromptMR extends the E2E-Varnet method using the discriminative prompt block module for handling various input types. It further refines reconstructed images utilizing adjacent inputs and employing a video restoration model to enhance image quality, albeit at the cost of added complexity and a departure from a singular model architecture. In contrast, our approach leverages a Patch-GAN architecture for CMR image reconstruction, emphasizing parity between reconstructed and ground truth images through GAN architecture. By adopting a stepwise reconstruction method and incorporating two distinct physical and spatial loss functions at each reconstruction step, we address the issue of the vanishing gradient. Compared to the PromptMR, our method does not require a quality improvement step through a video restoration model and performs the reconstruction process in an integrated manner.

Several experiments were conducted to evaluate the role of different components in the proposed approach. These ablation studies determined the role of self-attention blocks, physical loss function, stepwise loss calculations, curriculum learning and discriminator network. These experiments were performed on cine CMR images of the first task of the CMRxRecon2024 challenge. The results of these ablation studies are given in Table 2.

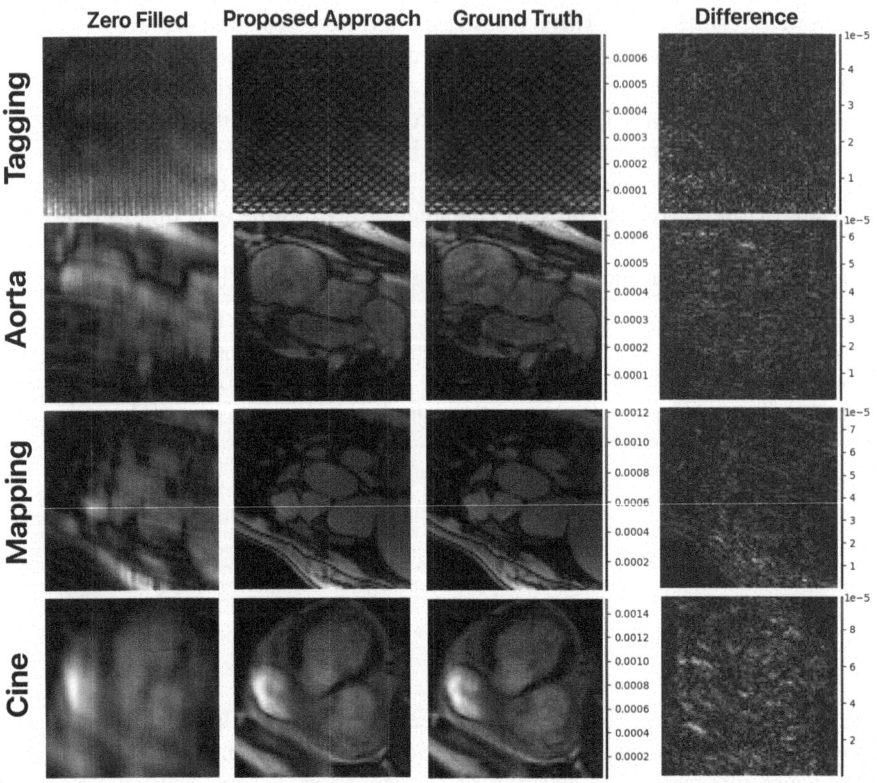

Fig. 4. An example of qualitative comparison of images reconstructed from subsampled k-space using a uniform sampling pattern with a 10x acceleration factor, reconstructed by the proposed approach and ground truth images.

By simultaneously considering the various acceleration factors, the multi-contrast characteristics of CMR imaging, the diverse anatomical views, and different k-space sampling trajectories, our all-in-one approach represents a holistic solution for accelerating CMR imaging. This approach facilitates the generation of high-quality images, ensures more accurate diagnoses, and enhances the overall patient experience.

Table 2. The results of ablation studies to investigate the role of different components of the proposed approach. The results are reported based on NMSE/PSNR/SSIM metrics, and the best values are shown in boldface.

	Cine		
	SAX	LAX	LVOT
Excluding Self-Attention Blocks	1.1/45.69/0.9795	1.2/45.21/0.9803	1.2/45.01/0.9843
Without Physical Loss	0.9/46.54/0.9892	1.0/45.82/0.9848	0.9/46.06/0.9861
Without Stepwise Loss	0.9/46.47/0.9878	1.0/45.71/0.9832	1.0/45.99/0.9848
Without Curriculum Learning	0.9/46.92/0.9887	1.0/45.98/0.9845	1.0/46.11/0.9862
Generator Only	1.0/46.52/0.9885	1.1/45.89/0.9834	1.1/46.05/0.9862
The Proposed Approach	**0.8/47.13/0.9903**	**0.9/46.12/0.9893**	**0.9/46.40/0.9901**

4 Conclusion

A universal approach for multi-contrast CMR image reconstruction was proposed in this study. Based on the Patch-GAN architecture, the proposed all-in-one approach receives subsampled multi-contrast k-space with different acceleration factors and trajectories and generates a high-quality clinical interpretable image. In this image reconstruction process, the advantages of k-space and image domain features are simultaneously used through physical and SSIM loss functions. The image refinement is based on a stepwise process, and the stepwise loss function is defined to avoid the vanishing gradient problem, which is finally added as a term to the generator loss function. Besides, the curriculum learning was leveraged in the proposed approach training, so the learning process from simple to hard tasks improved the overall performance. Evaluation of the proposed approach with extensive experiments and ablation studies showed that this approach outperforms previous methods. The future plan will be to improve the model's generality in the domain shift conditions between input datasets by relying on domain adaptation methods.

References

1. Arnold, J.R., McCann, G.P.: Cardiovascular magnetic resonance: applications and practical considerations for the general cardiologist. Heart **106**(3), 174–181 (2020)
2. Bengio, Y., Louradour, J., Collobert, R., Weston, J.: Curriculum learning. In: Proceedings of the 26th Annual International Conference on Machine Learning, pp. 41–48 (2009)
3. Block, K.T., Uecker, M., Frahm, J.: Undersampled radial MRI with multiple coils. Iterative image reconstruction using a total variation constraint. Magn. Reson. Med. Off. J. Int. Soc. Magn. Reson. Med. **57**(6), 1086–1098 (2007)
4. Bydder, M., Robson, M.D.: Partial fourier partially parallel imaging. Magn. Reson. Med. Off. J. Int. Soc. Magn. Reson. Med. **53**(6), 1393–1401 (2005)
5. Donoho, D.L.: Compressed sensing. IEEE Trans. Inf. Theory **52**(4), 1289–1306 (2006)

6. Enders, J., et al.: Reduction of claustrophobia during magnetic resonance imaging: methods and design of the "claustro" randomized controlled trial. BMC Med. Imaging **11**, 1–15 (2011)
7. Griswold, M.A., et al.: Generalized autocalibrating partially parallel acquisitions (GRAPPA). Magn. Reson. Med. Off. J. Int. Soc. Magn. Reson. Med. **47**(6), 1202–1210 (2002)
8. Hammernik, K., et al.: Learning a variational network for reconstruction of accelerated MRI data. Magn. Reson. Med. **79**(6), 3055–3071 (2018)
9. Isola, P., Zhu, J.Y., Zhou, T., Efros, A.A.: Image-to-image translation with conditional adversarial networks. In: Proceedings of the IEEE Conference on Computer Vision and Pattern Recognition, pp. 1125–1134 (2017)
10. Loshchilov, I., Hutter, F.: Decoupled weight decay regularization. arXiv preprint arXiv:1711.05101 (2017)
11. Lustig, M., Donoho, D., Pauly, J.M.: Sparse MRI: the application of compressed sensing for rapid MR imaging. Magn. Reson. Med. Off. J. Int. Soc. Magn. Reson. Med. **58**(6), 1182–1195 (2007)
12. Lyu, J., et al.: The state-of-the-art in cardiac MRI reconstruction: results of the cmrxrecon challenge in MICCAI 2023. arXiv preprint arXiv:2404.01082 (2024)
13. Nabavi, S., Hamedani, K.A., Moghaddam, M.E., Abin, A.A., Frangi, A.F.: Statistical distance-guided unsupervised domain adaptation for automated multi-class cardiovascular magnetic resonance image quality assessment. arXiv preprint arXiv:2409.00375 (2024)
14. Nabavi, S., Simchi, H., Moghaddam, M.E., Abin, A.A., Frangi, A.F.: A generalised deep meta-learning model for automated quality control of cardiovascular magnetic resonance images. Comput. Methods Programs Biomed. **242**, 107770 (2023)
15. Potlapalli, V., Zamir, S.W., Khan, S.H., Shahbaz Khan, F.: Promptir: prompting for all-in-one image restoration. In: Advances in Neural Information Processing Systems, vol. 36 (2024)
16. Pruessmann, K.P., Weiger, M., Scheidegger, M.B., Boesiger, P.: Sense: sensitivity encoding for fast MRI. Magn. Reson. Med. Off. J. Int. Soc. Magn. Reson. Med. **42**(5), 952–962 (1999)
17. Qin, C., Schlemper, J., Caballero, J., Price, A.N., Hajnal, J.V., Rueckert, D.: Convolutional recurrent neural networks for dynamic MR image reconstruction. IEEE Trans. Med. Imaging **38**(1), 280–290 (2018)
18. Raman, S.V., et al.: 30-minute CMR for common clinical indications: Â aâ society for cardiovascular magnetic resonance white paper. J. Cardiovasc. Magn. Reson. **24**(1), 13 (2022)
19. Reid, M.D., Williamson, R.C.: Composite binary losses. J. Mach. Learn. Res. **11**, 2387–2422 (2010)
20. Schlemper, J., Caballero, J., Hajnal, J.V., Price, A.N., Rueckert, D.: A deep cascade of convolutional neural networks for dynamic MR image reconstruction. IEEE Trans. Med. Imaging **37**(2), 491–503 (2017)
21. Sriram, A., et al.: End-to-end variational networks for accelerated MRI reconstruction. In: Martel, A.L., et al. (eds.) MICCAI 2020. LNCS, vol. 12262, pp. 64–73. Springer, Cham (2020). https://doi.org/10.1007/978-3-030-59713-9_7
22. Vasquez, M., Nagel, E.: Clinical indications for cardiovascular magnetic resonance. Heart **105**(22), 1755–1762 (2019)
23. Wang, C., et al.: Cmrxrecon: a publicly available k-space dataset and benchmark to advance deep learning for cardiac MRI. Sci. Data **11**(1), 687 (2024)

24. Wang, Z., Bovik, A.C., Sheikh, H.R., Simoncelli, E.P.: Image quality assessment: from error visibility to structural similarity. IEEE Trans. Image Process. **13**(4), 600–612 (2004)
25. Wang, Z., et al.: Cmrxrecon2024: a multi-modality, multi-view k-space dataset boosting universal machine learning for accelerated cardiac MRI. arXiv preprint arXiv:2406.19043 (2024)
26. Xin, B., Ye, M., Axel, L., Metaxas, D.N.: Fill the k-space and refine the image: prompting for dynamic and multi-contrast MRI reconstruction. In: International Workshop on Statistical Atlases and Computational Models of the Heart, pp. 261–273. Springer (2023)
27. Yang, G., et al.: Dagan: deep de-aliasing generative adversarial networks for fast compressed sensing MRI reconstruction. IEEE Trans. Med. Imaging **37**(6), 1310–1321 (2017)
28. Yiasemis, G., Moriakov, N., Sonke, J.J., Teuwen, J.: Deep cardiac MRI reconstruction with ADMM. In: International Workshop on Statistical Atlases and Computational Models of the Heart, pp. 479–490. Springer (2023)

Improving the Scan-Rescan Precision of AI-Based CMR Biomarker Estimation

Dewmini Hasara Wickremasinghe[1(✉)], Yiyang Xu[1], Esther Puyol-Antón[1], Paul Aljabar[2], Reza Razavi[1], and Andrew P. King[1]

[1] School of Biomedical Engineering and Imaging Sciences, King's College London, London, UK
dewmini.wickremasinghe@kcl.ac.uk
[2] Perspectum Ltd., Oxford, UK

Abstract. Quantification of cardiac biomarkers from cine cardiovascular magnetic resonance (CMR) data using deep learning (DL) methods offers many advantages, such as increased accuracy and faster analysis. However, only a few studies have focused on the scan-rescan precision of the biomarker estimates, which is important for reproducibility and longitudinal analysis. Here, we propose a cardiac biomarker estimation pipeline that not only focuses on achieving high segmentation accuracy but also on improving the scan-rescan precision of the computed biomarkers, namely left and right ventricular ejection fraction, and left ventricular myocardial mass. We evaluate two approaches to improve the apical-basal resolution of the segmentations used for estimating the biomarkers: one based on image interpolation and one based on segmentation interpolation. Using a database comprising scan-rescan cine CMR data acquired from 92 subjects, we compare the performance of these two methods against ground truth (GT) segmentations and DL segmentations obtained before interpolation (baseline). The results demonstrate that both the image-based and segmentation-based interpolation methods were able to narrow Bland-Altman scan-rescan confidence intervals for all biomarkers compared to the GT and baseline performances. Our findings highlight the importance of focusing not only on segmentation accuracy but also on the consistency of biomarkers across repeated scans, which is crucial for longitudinal analysis of cardiac function.

Keywords: Scan-rescan · Cardiac biomarkers · Precision

1 Introduction

Cardiovascular magnetic resonance (CMR) is considered the gold-standard for the assessment of the function and morphology of the heart [8]. The accurate delineation of the left ventricular (LV) and right ventricular (RV) blood pools (BP), and the LV myocardium from cine CMR data allows for the computation of cardiac biomarkers such as LV and RV ejection fraction (LVEF and RVEF)

O. Camara et al. (Eds.): STACOM 2024, LNCS 15448, pp. 476–485, 2025.
https://doi.org/10.1007/978-3-031-87756-8_46

and LV myocardial mass (LVM), which are crucial parameters for the assessment of cardiac health. CMR has also shown better interstudy reproducibility of cardiac biomarkers (based on manual contouring) compared to echocardiography [6]. However, manual segmentation of cardiac structures heavily depends on the training and experience of the clinicians, introducing both intra- and inter-observer variability [13,21].

With the growing availability of CMR imaging data and the technological advancements in deep learning (DL) methods, many studies have investigated the advantages of using DL models for the analysis of CMR data [1,10,16]. The integration of DL methods into clinical workflows holds the potential to significantly reduce the time required for the analysis of the images, achieve a more accurate delineation of the segmented structures, and eliminate intra- and inter- observer variability [5,18]. As demonstrated in [22], commercially available DL tools are now available for automated segmentation of cine CMR data, which have shown accuracy comparable to that of clinicians.

However, although the use of DL for the analysis of CMR imaging data presents many benefits and is able to achieve a high level of consistency with clinicians in quantifying cardiac function, many studies overlook the importance of the repeatability (i.e., precision) of the estimated biomarkers. The precision of biomarker estimates is an important measure of the reliability and robustness of the developed methods and can be measured from scan-rescan data acquired from the same subjects. A small number of papers have reported investigations into scan-rescan repeatability of DL-based CMR segmentation and biomarker estimation. Bhuva et al. [2] performed a scan-rescan analysis on LV biomarkers from cine CMR and demonstrated that their fully automated convolutional neural network (CNN) was able to obtain precision comparable to that of trained junior annotators, but slightly worse than that of experts. Davies et al. [4] also assessed scan-rescan variability in CNN-based LV biomarker estimation from cine CMR compared to human analysis. They reported that the precision of a DL model surpassed that of human annotators, obtaining slightly lower coefficients of variation.

These studies show that state-of-the-art DL-based segmentation and biomarker estimation methods have precision similar to that of human annotators. However, in many CMR imaging examinations clinicians are interested in assessing the *change* in biomarkers (compared to a previous scan), rather than just the absolute value. In such situations, because the changes can be very small, very high precision is needed and arguably even the precision of a human expert annotation may be insufficient. For instance, during expert segmentation, there can be significant variability in the choice of which short axis slice to stop segmenting the myocardium at the basal/apical ends of the LV [9,11]. This motivates the case for automated tools to have *better* precision than experts to enable such longitudinal analyses [2].

In this paper we (i) assess the accuracy and precision of DL-based biomarker estimation on an external validation set of scan-rescan cine CMR data, and (ii) propose methods to improve the precision of biomarker estimation. This work, to the best of our knowledge, is the first to specifically investigate methods for minimising scan-rescan variability in DL-based cardiac biomarker estimation. By addressing this gap in the literature, our study aims to enable a more robust DL-based analysis of longitudinal CMR data.

2 Materials and Methods

2.1 Scan-Rescan Dataset

The data used in the experiments consisted of a dataset of 184 scan-rescan cine CMR acquisitions obtained from 92 healthy volunteers provided by Perspectum. Each subject underwent the CMR acquisition twice within a short period of time, usually within the same day and using the same scanner. Repositioning of the patient and replanning were carried out for each of the two scans. These CMR scans were acquired using three scanners from Siemens Healthcare (1.5T MAGNETOM Aera, 3T MAGNETOM Vida and MAGNETOM Prisma). For each subject, short-axis (SAX), two-chamber and four-chamber long-axis (LAX) views were acquired. Ground truth (GT) segmentations of the LVBP, RVBP and LV myocardium for the SAX data were manually obtained for the end-diastole (ED) and end-systole (ES) frames. The other frames were not used in the experiments.

2.2 Overview

An overview of the two methods evaluated for making estimates of cardiac biomarkers with improved precision is shown in Fig. 1. We describe the components of the two approaches below. The image-based interpolation approach (box (c)) is a novel pipeline that we introduce here. The segmentation-based interpolation approach (box (d)) is based on the work described in [12] but is evaluated here for the first time on scan-rescan data.

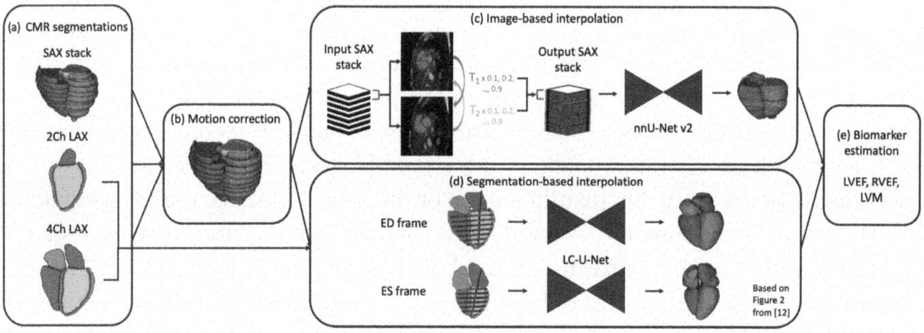

Fig. 1. Overview of the motion correction and interpolation pipelines.

2.3 Segmentation Models

The segmentations of the LVBP, RVBP, and LV myocardium for the SAX data were obtained using a 2D nnU-Net v2 model [7]. The model was trained on 4872 SAX cine CMR scans and manual annotations from the UK Biobank [20]. Two more 2D nnU-Net v2 models trained on a subset of 200 cases from the UK Biobank data were used to obtain segmentations of the LVBP, left atrium (LA), and LV myocardium from the two-chamber long-axis view and the LVBP, RVBP, LA, right atrium (RA), and LV myocardium from the four-chamber long-axis view, respectively. The SAX segmentations were used for biomarker estimation for all methods whereas both the SAX and LAX segmentations were used for motion correction (box (a) in Fig. 1) as well as in the segmentation interpolation method (box (d)).

2.4 Motion Correction

A motion correction algorithm [24] was applied to correct the misalignments between slices in the SAX stack due to differing breath-hold positions. This algorithm aimed to find the in-plane translations that maximised the intersection between the DL segmentations of the SAX stack data with those of the two-chamber and four-chamber long-axis data. The translations obtained from this registration process were then applied to the SAX images and DL-produced segmentations, and these were used as input data for the interpolation methods described in the following section.

2.5 Interpolation

Our approach to improving the precision of biomarker estimates is based on increasing the resolution of SAX images or segmentations in the through-plane direction. Two different methods were investigated: the first was an image-based interpolation of the SAX images, and the second was an interpolation of the SAX segmentations.

Image-Based Interpolation: In this approach, we first estimated the non-linear deformations between each pair of adjacent slices in the SAX stack. These estimates were made in two directions, base-to-apex and apex-to-base. The employed registration algorithm was a multi-level 2D B-spline free-form deformation (MFFD) algorithm [14,15,17] implemented in MIRTK. A region of interest was used for the registration, based on masks of the heart formed by the union of the DL segmentations of the LVBP, RVBP, and LV myocardium over all slices. The resulting B-spline control point displacements for each transformation were then scaled by factors in the range $[0.1, 0.2, \ldots, 0.9]$. The scaled transforms were applied to the source images and resampled using linear interpolation. As a result, interpolated slices were generated at distances proportional to the scaling coefficients. This process was repeated for both registration directions and the slices from each were averaged to form each interpolated image.

These interpolated images were then fed to the DL SAX segmentation model to obtain segmentations of the LVBP, RVBP, and LV myocardium. LV and RV BP volumes at ED and ES (LVEDV and LVESV, RVEDV and RVESV), and LVM were derived from the segmentations, and LVEF and RVEF values were computed from the cardiac volumes.

Segmentation-Based Interpolation: This method, described in [12], used a 3D U-Net to predict dense segmentation label maps from sparse 2D SAX and LAX segmentations. The input data for this model was a sparse representation of the heart obtained from the intersection of the SAX segmentations with the two- and four-chamber LAX segmentations. The dense segmentation estimates produced by the model were then used to compute the same cardiac biomarkers described above.

2.6 Evaluation

First, to assess accuracy, the mean error (ME) and mean absolute error (MAE) between the ground truth (GT) and predicted biomarkers were compared for the three methods: baseline (pre-interpolation), image-based interpolation, and segmentation-based interpolation.

Second, to assess precision, the scan-rescan variability of estimates of LVEF, RVEF, and LVM was assessed using Bland-Altman analysis. This was performed for the (human) GT, the DL segmentations before interpolation (baseline) and the segmentations produced using both of the interpolation techniques. Coefficient of variation between scan and rescan was computed for each cardiac biomarker. Finally, a qualitative analysis of the results was carried out by generating visual representations of the DL produced segmentations.

3 Experiments and Results

3.1 Quantitative Analysis

Table 1 shows the ME and MAE obtained from the comparison of LVEDV, LVESV, RVEDV, RVESV, and LVM computed from the GT and DL-produced values. These values were generated to assess the accuracy of the obtained cardiac biomarkers, and verify that the interpolated biomarkers were within acceptable ranges. The errors computed after the two interpolation methods are similar to the baseline ones, except for RVEDV and RVESV computed from the segmentations obtained from the segmentation-based interpolation, which are slightly higher. It is also noticeable how, for most biomarkers, the image-based interpolation method obtained the lowest ME and MAE.

To quantify precision, Bland-Altman plots for each biomarker are shown in Fig. 2 for the GT and the DL-based methods, and Table 2 shows the Bland-Altman confidence intervals (expressed as mean $\pm$ 1.96 $\times$ standard deviation). For LVEF and LVM, both interpolation methods showed the best agreement

Table 1. Mean error (ME) and mean absolute error (MAE) for LVEDV, LVESV, RVEDV, RVESV, and LVM computed between the ground truth (GT) and the baseline (pre-interpolation), image-based interpolation, and segmentation-based interpolation segmentations. Lowest values for each biomarker shown in bold.

Accuracy						
	Baseline		Image interpolation		Segmentation interpolation	
	ME	MAE	ME	MAE	ME	MAE
LVEDV (ml)	4.39	7.48	5.69	8.20	**2.46**	**6.96**
LVESV (ml)	0.99	**3.94**	**0.09**	4.04	−2.65	4.84
RVEDV (ml)	−8.00	9.88	**−3.49**	**8.30**	−12.19	13.21
RVESV (ml)	−10.91	12.18	**−8.67**	**10.29**	−17.34	17.79
LVM (g)	−0.66	**5.48**	**−0.51**	5.91	−2.44	5.93

between scan and rescan, with narrow confidence intervals and fewer outliers compared to the GT and the baseline (pre-interpolation) approach. The confidence interval of RVEF computed from the image-based interpolation method was slightly wider than the GT one, which is likely due to the two outliers that can be seen in the plot. It is also noticeable that the baseline confidence intervals are wider than the ones for the GT for LVEF and RVEF.

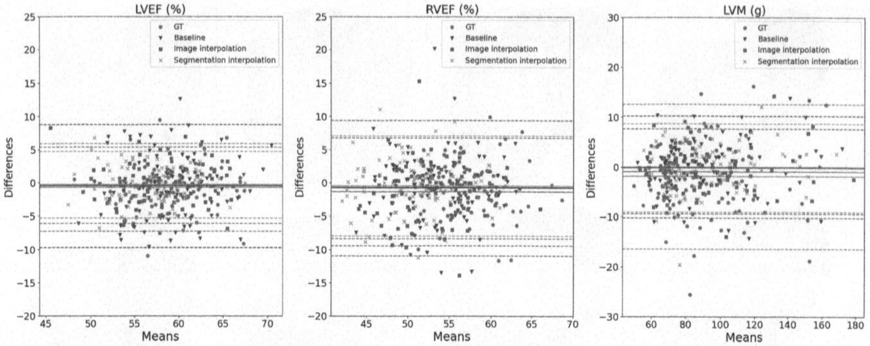

Fig. 2. Bland-Altman plots for LVEF, RVEF, and LVM between scan and rescan for human ground truth (GT, red), baseline pre-interpolation (blue), image-based interpolation (green) and segmentation-based interpolation (orange). (Color figure online)

Table 2 also shows the coefficient of variation between scan and rescan for LVEF, RVEF, and LVM computed from GT, baseline (pre-interpolation), and the two interpolation methods. Both interpolation methods obtained a lower coefficient of variation for LVEF and LVM. For RVEF the best coefficient of variation was obtained for the GT segmentations, although the interpolation methods obtained similar values to the GT and improved on the performance of the baseline approach.

Table 2. Bland-Altman confidence intervals and coefficient of variation values between scan and rescan for the LVEF, RVEF, and LVM computed from the ground truth (GT), baseline (pre-interpolation), image-based interpolation, and segmentation-based interpolation. Lowest values for each biomarker shown in bold.

Precision				
Bland-Altman Confidence Intervals				
	GT	Baseline	Image interpolation	Segmentation interpolation
LVEF (%)	−0.71 ± 6.57	−0.49 ± 9.23	−0.38 ± 5.71	**−0.31 ± 5.03**
RVEF (%)	−0.66 ± 7.73	−0.79 ± 10.17	−1.35 ± 8.12	**−0.48 ± 7.50**
LVM (g)	−2.00 ± 14.48	−0.06 ± 10.26	**−0.94 ± 8.54**	−0.27 ± 8.90
Coefficient of variation				
	GT	Baseline	Image interpolation	Segmentation interpolation
LVEF (%)	2.10	3.25	2.01	**1.88**
RVEF (%)	**2.66**	3.73	3.09	2.86
LVM (%)	3.38	2.22	**1.91**	2.01

3.2 Qualitative Analysis

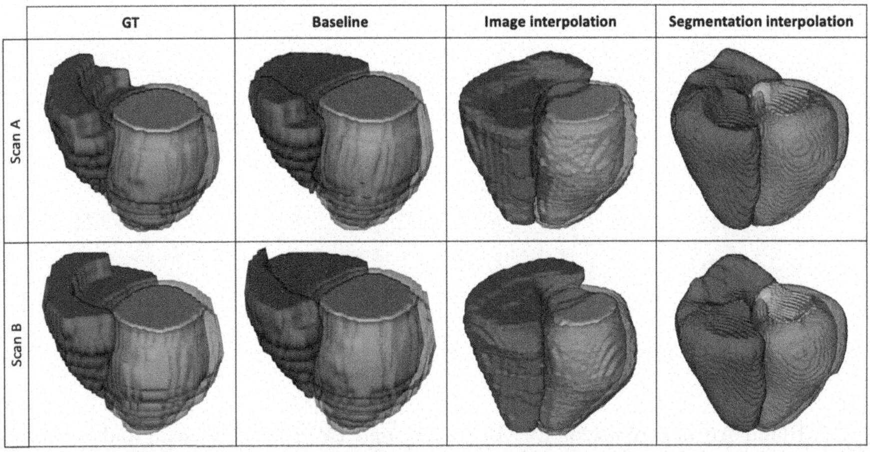

Fig. 3. Visual representations of the ground truth (GT), baseline (pre-interpolation), image-based interpolation, and segmentation-based interpolation segmentations.

Figure 3 shows visual examples of the GT and DL-derived segmentations of the LVBP, RVBP, and LV myocardium for sample scan and rescan images. Both interpolation approaches produce smoother and more realistic looking segmentations.

4 Discussion and Conclusion

This study has focused on using interpolation methods to estimate cardiac biomarkers from cine CMR data, with the aim to improve scan-rescan precision. Two different methods were assessed to evaluate the effect of through-plane interpolation of the SAX data on the repeatability of biomarker estimation. The first one was an image-based interpolation approach, and the second was a segmentation-based interpolation method. All experiments were performed on a database of scan-rescan cine CMR data from 92 subjects.

First, to assess the accuracy of the segmentations obtained after applying the two interpolation methods, the ME and MAE were computed between the GT cardiac volumes at ED and ES, and the DL-derived ones. Due to the different granularity (i.e. through-plane resolution) of the original SAX stack (and hence the GT segmentations) and the interpolated data, direct comparisons of error values between lower resolution and higher resolution data should be treated with caution. However, such a comparison does serve as a general indicator of the accuracy of the computed biomarkers. As seen in Table 1, both post-interpolation segmentations obtained error values similar to the ones obtained from the baseline approach, with the image-based interpolation method showing accuracy higher than the baseline for most of the cardiac biomarkers.

Scan-rescan precision was evaluated through Bland-Altman confidence limits and coefficient of variation. As summarised in Fig. 2 and Table 2, both interpolation methods showed good agreement between scan and rescan for LVEF, RVEF, and LVM. For LVEF and RVEF, the GT scan-rescan agreement was higher than the baseline, whereas the use of interpolation, whether image- or segmentation-based, always improved the agreement of scan-rescan biomarkers. This outcome not only shows the potential of the presented methods to enhance the precision of biomarker estimation, but also highlights the need for methods that focus on improving agreement between scans of the same subject, rather than solely prioritising accuracy against GT segmentations.

In line with the results shown from the Bland-Altman analysis, the coefficient of variation values in Table 2 indicate that both interpolation approaches achieved better precision between scan and rescan biomarkers compared to the baseline approach and, in most cases, were also able to improve on the GT precision.

Other methods that focus on increasing the through-plane resolution of the SAX stack have been proposed [3,19,23] but these have not yet been evaluated for their impact on scan-rescan precision. Future work will focus on improving the proposed image-based interpolation method by including LAX data to enable better identification of the basal plane of the heart, which can be used as cut off when segmenting the SAX stack, and on evaluating other learning-based image interpolation approaches, such as the ones mentioned above. However, we believe that the results presented here serve as a valuable proof of concept, demonstrating that interpolation can enhance the agreement in scan-rescan biomarkers, bringing us closer to reliable longitudinal analysis of changes in cardiac function.

Acknowledgments. We would like to acknowledge funding from the EPSRC Centre for Doctoral Training in Medical Imaging (EP/S022104/1) and the Wellcome/EPSRC Centre for Medical Engineering at King's College London (WT 203148/Z/16/Z). We also acknowledge funding from Perspectum Ltd., Oxford, UK.

Disclosure of Interests. The authors have no competing interests to declare that are relevant to the content of this article.

References

1. Bai, W., et al.: Automated cardiovascular magnetic resonance image analysis with fully convolutional networks. J. Cardiovasc. Magn. Reson. **20**(1), 65 (2018)
2. Bhuva, A.N., et al.: A multicenter, scan-rescan, human and machine learning CMR study to test generalizability and precision in imaging biomarker analysis. Circ. Cardiovasc. Imaging **12**(10), e009214 (2019)
3. Chen, Z., Ren, H., Li, Q., Li, X.: Motion correction and super-resolution for multi-slice cardiac magnetic resonance imaging via an end-to-end deep learning approach. Comput. Med. Imaging Graph. **115**, 102389 (2024)
4. Davies, R.H., et al.: Precision measurement of cardiac structure and function in cardiovascular magnetic resonance using machine learning. J. Cardiovasc. Magn. Reson. **24**(1), 16 (2022)
5. Galati, F., Ourselin, S., Zuluaga, M.A.: From accuracy to reliability and robustness in cardiac magnetic resonance image segmentation: a review. Appl. Sci. **12**(8), 3936 (2022)
6. Grothues, F., et al.: Comparison of interstudy reproducibility of cardiovascular magnetic resonance with two-dimensional echocardiography in normal subjects and in patients with heart failure or left ventricular hypertrophy. Am. J. Cardiol. **90**(1), 29–34 (2002)
7. Isensee, F., Jaeger, P.F., Kohl, S.A., Petersen, J., Maier-Hein, K.H.: nnU-Net: a self-configuring method for deep learning-based biomedical image segmentation. Nat. Methods **18**(2), 203–211 (2021)
8. von Knobelsdorff-Brenkenhoff, F., Pilz, G., Schulz-Menger, J.: Representation of cardiovascular magnetic resonance in the AHA/ACC guidelines. J. Cardiovasc. Magn. Reson. **19**(1), 70 (2016)
9. Marcus, J., et al.: The influence of through-plane motion on left ventricular volumes measured by magnetic resonance imaging: implications for image acquisition and analysis. J. Cardiovasc. Magn. Reson. **1**(1), 1–6 (1999)
10. Mariscal-Harana, J., et al.: An artificial intelligence tool for automated analysis of large-scale unstructured clinical cine cardiac magnetic resonance databases. Eur. Heart J. Digit. Health **4**(5), 370–383 (2023)
11. Marwick, T.H.: Ejection fraction pros and cons: JACC state-of-the-art review. J. Am. Coll. Cardiol. **72**(19), 2360–2379 (2018)
12. Muffoletto, M., et al.: Evaluation of deep learning estimation of whole heart anatomy from automated cardiovascular magnetic resonance short-and long-axis analyses in UK Biobank. Eur. Heart J. Cardiovasc. Imaging jeae123 (2024)
13. Petitjean, C., Dacher, J.N.: A review of segmentation methods in short axis cardiac MR images. Med. Image Anal. **15**(2), 169–184 (2011)
14. Rueckert, D., et al.: Comparison and evaluation of rigid and nonrigid registration of breast MR images. In: Medical Imaging 1999: Image Processing, vol. 3661, pp. 78–88. SPIE (1999)

15. Rueckert, D., Sonoda, L.I., Hayes, C., Hill, D.L., Leach, M.O., Hawkes, D.J.: Non-rigid registration using free-form deformations: application to breast MR images. IEEE Trans. Med. Imaging **18**(8), 712–721 (1999)
16. Ruijsink, B., et al.: Fully automated, quality-controlled cardiac analysis from CMR: validation and large-scale application to characterize cardiac function. Cardiovasc. Imaging **13**(3), 684–695 (2020)
17. Schnabel, J.A., et al.: A generic framework for non-rigid registration based on non-uniform multi-level free-form deformations. In: Niessen, W.J., Viergever, M.A. (eds.) MICCAI 2001. LNCS, vol. 2208, pp. 573–581. Springer, Heidelberg (2001). https://doi.org/10.1007/3-540-45468-3_69
18. Seetharam, K., Brito, D., Farjo, P.D., Sengupta, P.P.: The role of artificial intelligence in cardiovascular imaging: state of the art review. Front. Cardiovasc. Med. **7**, 618849 (2020)
19. Stolt-Ansó, N., McGinnis, J., Pan, J., Hammernik, K., Rueckert, D.: NISF: neural implicit segmentation functions. In: International Conference on Medical Image Computing and Computer-Assisted Intervention, pp. 734–744. Springer (2023)
20. Sudlow, C., et al.: UK Biobank: an open access resource for identifying the causes of a wide range of complex diseases of middle and old age. PLoS Med. **12**(3), e1001779 (2015)
21. Suinesiaputra, A., et al.: Quantification of LV function and mass by cardiovascular magnetic resonance: multi-center variability and consensus contours. J. Cardiovasc. Magn. Reson. **17**, 1–8 (2015)
22. Wang, S., et al.: AI based CMR assessment of biventricular function: clinical significance of intervendor variability and measurement errors. Cardiovasc. Imaging **15**(3), 413–427 (2022)
23. Xia, Y., Ravikumar, N., Greenwood, J.P., Neubauer, S., Petersen, S.E., Frangi, A.F.: Super-resolution of cardiac MR cine imaging using conditional GANs and unsupervised transfer learning. Med. Image Anal. **71**, 102037 (2021)
24. Xu, Y., et al.: Improved 3D whole heart geometry from sparse CMR slices. In: Statistical Atlases and Computational Models of the Heart (STACOM) Workshop (2024)

Author Index

A
Aguilar-Minguez, Adria 236
Alberola-López, Carlos 53
Alblas, Dieuwertje 269
Aljabar, Paul 476
Alkan, Cagan 226
Almar-Munoz, Enrique 110, 236
Anvari Hamedani, Kian 464
Anwaar, Waqas 280
Aslanidi, Oleg 313, 443
Atwany, Mohammad 1
Audigier, Chloé 323
Axel, Leon 289

B
Balle Sánchez, Marcos 87
Bandaru, Raja Sekhar 87
Banerjee, Abhirup 1, 214
Beveridge, Lucas 193
Bharath, A. A. 245
Boman, Magnus 301
Bos, Daniel 122
Bron, Esther E. 122
Bruchfeld, Samuel 301
Brune, Christoph 269
Bueno-Orovio, Alfonso 214

C
Cademartiri, Filippo 378
Camara, Oscar 75
Carlhäll, Carl-Johan 368
Casademunt, Paula 75
Castañeda, Eduardo 323
Cechnicka, Sarah 98
Celi, Simona 378
Cepas-Guillén, Pedro 75
Chen, Zhurong 280
Cheng, Yuan 390
Chiesa, Marco 301
Chiew, Mark 334
Chiribiri, Amedeo 43
Chiu, C. 200
Chiu, Ching-En 98
Choudhury, Robin P. 1
Christensen, Kim 98
Cochet, Hubert 12, 421
Cole, G. 245
Creamer, Stephen A. 433

D
De Backer, Ole 33
De Vecchi, Adelaide 443
Deprez, Maria 313
Diez, Paula López 22, 33
Dillon, Joshua R. 433
Doltra, Ada 75
Doneva, Mariya 453
Doughty, Robert N. 433
Dummer, Sven 378

E
Ebrahimi Moghaddam, Mohsen 464
Edwards, Nicola C. 433
Ennis, Daniel B. 226

F
Feng, Fan 368
Ferdian, Edward 433
Finsterbach, Sonny 421
Flores-Umanzor, Eduardo 75
Freixa, Xavier 75
Fuchs, Andreas 22

G
Galazis, C. 200
Galazis, Christoforos 98
Garzia, Simone 378
Gomez, Juan F. 64
Gómez-Herrero, Javier 33
Grau, Vicente 1
Gu, Yun 357, 368

O. Camara et al. (Eds.): STACOM 2024, LNCS 15448, pp. 487–490, 2025.
https://doi.org/10.1007/978-3-031-87756-8

Gunawardhana, Malitha 357
Guo, Fumin 411
Guo, Xinrui 411
Gupta, Ashvin 98

H
Haltmeier, Markus 110, 236
Hansen, Bjørn S. 172
Hasaballa, Abdallah 368
Hennersperger, Christoph 87
Hickler, Julia 87
Holler, Martin 257
Howard, J. 245
Hu, Jiongtong 280
Huang, Yanxin 164

J
Jacob, Athira 323
Jami, Jitin 323
Jansen, Moïse Z. 433
Johnson, Kiane 209

K
Kadota, Brenden T. 334
Kapper, Kathrin Lisa 257
Karabelas, Elias 257
Kavousi, Maryam 122
Khan, Abbas 209
Kholmovski, Eugene G. 345
King, Andrew P. 476
Klis, Magdalena 443
Koch, Sabine 301
Kofoed, Klaus F. 22, 33
Kofoed, Klaus Fuglsang 172
Kostic, Dejan 301
Kremser, Christian 110, 236

L
Lahoti, Nishant 443
Lamb, Hildo J. 453
Lashgari, Mojtaba 1
Leening, Maarten J. G. 122
Lefebvre, Arthur L. 345
Legget, Malcolm E. 433
Leon Contreras, Nohemi Sofia 236
Li, Liwen 411
Lindeskog, Mathias B. 22
Linton, N. 245
Lip, Gregory Y. H. 443
Lluch, Èric 323
Loecher, Michael 226
Lowes, Mathias Micheelsen 172
Lulic, Davorka 33
Ly, Buntheng 12, 64, 421
Lyu, Donghang 453

M
Maguire Jr., Gerald Q. 301
Malyala, Anurag 140
Mandic, Danilo 98
Margeta, Jan 33
Martín-Fernández, Marcos 53
Mayr, Agnes 110, 236
Mazher, Moona 132
Meister, Felix 323
Melidoro, Paolo 443
Merino-Caviedes, Susana 53
Metaxas, Dimitris N. 289
Mihalef, Viorel 323
Mill, Jordi 75
Mo, Hyunho 122
Montarello, Natalie 443
Morales, Xabier 75
Moriakov, Nikita 183
Moya-Sáez, Elisa 53

N
Nabavi, Shahabedin 464
Nash, Martyn P. 433
Neary-Zajiczek, Lydia 87
Ni, Dong 280
Niederer, Steven A. 43, 132
Njomgue Fotso, Evariste 12
Nnadozie, Emmanuel Chibuikem 53
Nuñez-Garcia, Marta 421

O
Obada, George 313
Ogbomo-Harmitt, Shaheim 313
Oksuz, Ilkay 152
On, Y. 200, 245
Özer, Caner 152

P
Pamminger, Mathias 110
Passerini, Tiziano 323
Patel, Jaykumar H. 334
Paulsen, Rasmus R. 22, 33, 172

Pedersen, Jonas J. 22
Pedersen, Jonas Jalili 172
Perez-Ramirez, Daniel F. 301
Pezzotti, Nicola 453
Pham, Michael H. C. 22
Pinto, Arieh Levy 98
Pop, Mihaela 64
Puyol-Antón, Esther 43, 476

Q
Qayyum, Abdul 132
Qi, Jin 164
Qi, Yuan 390
Qin, Chen 140
Quill, Gina M. 433
Qureshi, Ahmed 443

R
Radutoiu, Ana-Teodora 22
Rajani, Ronak 443
Rao, Chinmay 453
Razavi, Reza 476
Razizadeh, Narges 464
Rodríguez Padilla, Jesus Jairo 64
Roney, Caroline 209
Roy, Aditi 98
Rygiel, Patryk 269, 378

S
Sabry, Malak 443
Saiz-Vivó, Marta 75
Sakata, Kensuke 345
Sanchis, Laura 75
Sermesant, Maxime 12, 64, 172, 421
Sheagren, Calder D. 334
Shun-Shin, M. 245
Sinclair, Matthew 43
Slabaugh, Gregory 209
Sonke, Jan-Jakob 183
Sørensen, Kristine A. 22
Staring, Marius 453
Sultan, Abdel Rahman Amr 443

T
Teuwen, Jonas 183
Thaler, Franz 257
Thesing, Claas 214
Tice, Brock 345
Tiefenthaler, Markus 236
Trayanova, Natalia A. 345
Trenor, Beatriz 64
Tziritas, Georgios 398

U
Urschler, Martin 257

V
Van Den Abeelen, Robin 313
van Osch, Matthias J. P. 453
Van, Manh The 280
Vandersickel, Nele 313
Varela, M. 200, 245
Varela, Marta 98
Vasanawala, Shreyas S. 226
Vatsaraj, Ishan 345
Verardo, Giacomo 301
Verlyck, Mathilde A. 433
Villar-Valero, Javier 64
Vimalesvaran, K. 245
Viti, Bruno 257
Vizitiu, Anamaria 323

W
Wang, Chengyan 140
Weiss, Maximilian E. R. 323
Wen, Grace 368
Wickremasinghe, Dewmini Hasara 476
Willemen, Yannick 33
Williams, Michelle C. 43
Williams, Steven E. 43, 443
Wolterink, Jelmer M. 269, 378
Wörz, Stefan 87
Wright, Graham A. 334

X
Xin, Bingyu 289
Xu, Fangqiang 357, 368

Xu, Hao 43
Xu, Ruru 152
Xu, Yiyang 43, 476
Xu, Zhixin 164
Xue, Wufeng 280

Y
Yamamoto, Carolyna A. P. 345
Yang, Jiayuan 368
Ye, Meng 289
Yeung, Kak Khee 269
Yiasemis, George 183
Young, Alistair A. 43, 433
Youssef, Yasmin W. E. 22
Yurt, Mahmut 226

Z
Zaman, S. 245
Zhang, Chi 226
Zhang, Le 193
Zhang, Yajing 164
Zhang, Yichi 390
Zhang, Zhenlin 140
Zhao, Debbie 433
Zhao, Jichao 357, 368
Zolotarev, Alexander 209

The manufacturer's authorised representative in the EU is Springer Nature Customer Service Centre GmbH, Europaplatz 3, 69115 Heidelberg, Germany. If you have any concerns regarding our products, please contact ProductSafety@springernature.com

Printed and bound by CPI Group (UK) Ltd, Croydon, CR0 4YY
01/10/2026
02225952-0017